Power Plant Evaluation
and
Design Reference Guide

The McGraw-Hill Engineering Reference Guide Series

This series makes available to professionals and students a wide variety of engineering information and data available in McGraw-Hill's library of highly acclaimed books and publications. The books in the Series are drawn directly from this vast resource of titles. Each one is either a condensation of a single title or a collection of sections culled from several titles. The Project Editors responsible for the books in the Series are highly respected professionals in the engineering areas covered. Each Editor selected only the most relevant and current information available in the McGraw-Hill library, adding further details and commentary where necessary.

Hicks • PLUMBING DESIGN AND INSTALLATION REFERENCE GUIDE

Covers the fundamentals of plumbing design and installation for a wide variety of industrial buildings and structures. Culled by Tyler G. Hicks from several McGraw-Hill books.

Hicks • POWER PLANT EVALUATION AND DESIGN REFERENCE GUIDE

Provides concise evaluation and design information for power plants serving many different needs—utility, industrial, and commercial. Culled by Tyler G. Hicks from several McGraw-Hill books and magazine articles.

Johnson & Jasik • ANTENNA APPLICATIONS REFERENCE GUIDE

Includes practical information and guidelines to antenna applications in all areas of communication. Comprised of one full section of Johnson and Jasik's Antenna Engineering Handbook, *Second Edition. Prepared by Richard C. Johnson.*

Markus and Weston • CLASSIC CIRCUITS REFERENCE GUIDE

Collects in one source hundreds of electronic circuits immediately useful in a wide variety of applications. Culled by Charles D. Weston from Markus's Sourcebook of Electronic Circuits, Electronics Circuits Manual, *and* Guidebook of Electronics Circuits.

Merritt • CIVIL ENGINEERING REFERENCE GUIDE

Offers quick reference to major civil engineering fields: structural design, surveying; geotechnical, environmental, and water engineering. A condensation by Max Kurtz of Merritt's Standard Handbook for Civil Engineers, *Third Edition.*

Woodson • HUMAN FACTORS REFERENCE GUIDE FOR ELECTRONICS AND COMPUTER PROFESSIONALS

Presents all essential data on human factors (ergonomics) relevant to the electronics and computer fields. Compiled by Wesley E. Woodson from his Human Factors Design Handbook.

Woodson • HUMAN FACTORS REFERENCE GUIDE FOR PROCESS PLANTS

Makes available to engineers and specialists all essential data on human factors (ergonomics) relevant to the process industries. Compiled by Nicholas P. Chopey from Woodson's Human Factors Design Handbook.

Power Plant Evaluation
and
Design Reference Guide

Edited by
Tyler G. Hicks, P. E.
International Engineering Associates

McGraw-Hill Book Company

New York   St. Louis   San Francisco   Auckland   Bogotá   Hamburg
Johannesburg   London   Madrid   Mexico   Milan   Montreal   New Delhi
Panama   Paris   São Paulo   Singapore   Sydney   Tokyo   Toronto

Library of Congress Cataloging in Publication Data

Power plant evaluation and design reference guide.

  (McGraw-Hill engineering reference guide series)
  Includes index.
  1. Power-plants—Design and construction—Handbooks, manuals, etc. 2. Electric power-plants—Design and construction—Handbooks, manuals, etc. I. Hicks, Tyler Gregory, 1921- . II. Series.
TJ164.P64   1986   621.31   86-7412
ISBN 0-07-028794-5

Copyright © 1986 by McGraw-Hill, Inc. All rights reserved. Printed in the United States of America. Except as permitted under the United States Copyright Act of 1976, no part of this publication may be reproduced or distributed in any form or by any means, or stored in a data base or retrieval system, without the prior written permission of the publisher.

1234567890 Hal/Hal 8932109876

ISBN 0-07-028794-5

Produced by Publishing Synthesis, Ltd. for McGraw-Hill, Inc.

Printed and bound by Arcata Graphics/Halliday.

Contents

	Foreword	vii
1	Industrial Power Plants and Steam Systems	1-1
2	Steam Power Plants for Utility Stations	2-1
3	Power Plant Costs—Their Estimation and Use	3-1
4	Cogeneration Engineering and Design	4-1
5	Hydroelectric Power Systems	5-1
6	Geothermal, Solar, Wind, Wave, and Ocean Thermal Difference Energy Systems	6-1
7	Energy from Municipal Wastes	7-1
8	Energy Storage Systems	8-1
9	Environmental Aspects of Power Generation	9-1
10	Economic Operation of Power Systems	10-1
11	Power System Reliability Factors	11-1
	Bibliography	11-16
	Index	I-1

Foreword

Power plant design is both a demanding and a rewarding task that attracts mechanical, electrical, civil, chemical, petroleum, and nuclear engineers. These engineers are helped by a score of other specialists—drafters, stress analysts, metallurgical engineers, welding contractors, economists, financial experts, etc. The results of their joint efforts are seen throughout the world in a variety of power plants—large, small, and in-between.

Despite the huge need for power plants in both developed and developing nations, there are few books to which the designer can refer for assistance and ideas. This book is aimed at meeting the need for an up-to-date work covering the entire field of power-plant design.

But instead of being written by one person—or a few—this book has a different genesis, drawing from a number of recently published works and magazine articles, blending their content into a unified treatment of modern power plant design. Using this approach, the editor was able to use the best materials from a number of works to produce a valuable design reference guide for the working power plant designer—no matter what engineering or technical background the designer might bring to the task. Thus, the book will be valuable to mechanical, electrical, chemical, petroleum, civil, textile, nuclear, and consulting engineers throughout the world, whether they design industrial or utility plants, or any combination thereof.

To meet the need for published material in the field, the content of this reference guide has been carefully chosen to embrace the entire field of power plant design. Thus, the user of this guide will find detailed coverage of steam, hydro, nuclear, and internal-combustion engine plants. Additional coverage serves the needs of designers working in solar, wind, tidal, and a variety of other alternative power-source fields. So the guide serves to meet a long-felt need for a modern reference for power plant designers throughout the world.

The guide starts with a comprehensive analysis of industrial power plants and steam systems. Such plants and systems probably demand more design creativity and skill than almost any other type, since there is little standardization amongst industrial power plants. Some are nearly utilities in their design and operation. Others are simply "reducing valves" which reduce the pressure of steam between a boiler plant and an industrial steam system, generating electricity as a byproduct of the process. Such plants were truly the cogeneration leaders. They squeezed the last Btu of energy out of a pound of steam long before the word cogeneration was coined. And with today's greater emphasis on energy conservation, the industrial power plant makes greater demands than ever on the designer's skill and vision. This section provides valuable tips to any designer who does work for industrial firms of all types.

Next, the guide covers steam power plants for central-station utility installations. Since the steam plant is the most widely used central-station facility, the coverage is in great depth. A variety of plants are described and their design approach analyzed. Data in this chapter are useful to the industrial plant designer as well.

Power plant costs—their estimation and use—is the third major topic discussed in this guide. Since "it always comes down to money" in every design situation, power plant costs are of primary concern to every designer. Both specific costs and trends of these costs are considered in this section.

Cogeneration—a new buzzword in power-plant design—is the subject of Section 4. As noted earlier, the concept of cogeneration has been with power-plant designers for many years. But the new emphasis on byproduct power has produced other views of this important subject. They are covered in detail in this section of the guide.

Hydroelectric power systems, next in popularity to steam systems, are covered in Section 5. Since the rise in uncertainty of supply and price of fossil fuels, hydro has made a comeback. Long neglected low-head sites once thought uneconomic now sport brand new generating facilities. The design and cost aspects of this new trend are covered in great detail in this section of the guide. Much useful design information is provided.

A variety of alternative energy sources—geothermal, solar, wind, wave, and ocean thermal difference—are covered in the sixth section of this guide. Specific design information which can be used in the engineering office is presented for each type of energy source considered.

Municipal wastes from residential and industrial sources continue to grow every year. Coupled with the need to deal with shrinking landfill availability and the relatively high energy content of each pound of waste, there is a persistent desire to utilize such waste to conserve natural resources. Section 7 gives the latest thinking on how waste can be used to generate power while reducing the problem of where to discard waste.

Energy storage is becoming more important as the cost of both plants and fuel rises. To evaluate any proposed storage method, a designer must know what methods are available, their relative efficiency, and the cost of each method. These topics are covered in detail in Section 8 of the guide.

Growing concern for the environment, and the effects of power generation on the surrounding area, are of critical importance to designers today. Acid rain, snow, and fog get enormous attention from various environmental groups. Every power-plant designer must be equipped with a knowledge of how to reduce pollution of the environment. Section 9 covers the environmental aspects of power generation in great detail.

Economic operation of power systems is discussed in the next section of the guide. Since fuel costs are such a critical factor in every plant today, the economic operation of every plant must be considered by the designer when planning a new installation. Typical design factors critical to economic operation are discussed in this section.

Finally, Section 11 covers power system reliability factors. With so much attention being paid to the elimination of power failures of all kinds, this section provides every designer with key ideas useful during the design process.

The works used in compiling this guide are acknowledged in each section. The editor thanks the various authors, manufacturers, and agencies for allowing access to their material. It is truly hoped that the guide proves useful to every power plant designer during every stage of the planning, design, and operation of modern power plants of all types.

Tyler G. Hicks

Power Plant Evaluation
and
Design Reference Guide

Section 1

Industrial Power Plants and Steam Systems

Steam power plants comprise the major generating and process steam sources throughout the world today. Internal-combustion engine and hydro plants generate less electricity and steam than steam power plants. For this reason we will give our initial attention in this book to steam power plants and their design and application.

In the steam power field two major types of plants serve the energy needs of customers—industrial plants for factories and other production facilities—and central-station utility plants for residential, commercial, and industrial demands. Of these two types of plants, the industrial power plant probably has more design variations than the utility plant. The reason for this is that the demands of industry tend to be more varied than the demands of the typical utility customer.

To assist the power-plant designer in understanding better the many variations in plant design, industrial power plants are considered first in this book. And to provide the widest design variables, a power plant serving several process operations and all utilities is considered.

In the usual industrial plant, a steam generation and distribution system must be capable of responding to a wide range of operating conditions, and often must be more reliable than the plant's electrical system. The system design is often the last to be settled but the first needed for equipment procurement and plant startup. Because of these complications the power plant design evolves slowly, changing over the life of a project.

From "Steam-System Design: How it Evolves," by John F. Peterson and William L. Mann, *Chemical Engineering,* vol. 92, no. 21, October 14, 1985, Copyright © 1985. Used by permission of McGraw-Hill, Inc. All rights reserved.

A steam generation and distribution system must be capable of responding to a wide range of operating conditions, and often must also be more reliable than the plant's electrical system. Its design is the last to be settled but the first needed for procurement and startup. Because of these complications, the design evolves slowly, changing constantly over the life of a project.

The plant for which a steam system is being designed is assumed here to consist of several process operations, a unit that supplies all the utilities (including boiler feedwater and steam), and offsite facilities, such as a wharf, a relief system, a tankcar and truck loading station, pipe-racks, and administration buildings. Fig. 1-1 shows the site plan, which is assumed to cover an area of 3,000 feet by 3,000 feet.

A typical steam and condensate system diagram for a major process plant is illustrated in Fig. 1-2. Its elements, which are normally found in a large chemical complex or oil refinery, include: (1) three pressure levels of steam—high, medium and low—which are distributed throughout the facility, (2) a boiler plant, (3) process consumers of steam, (4) steam turbines, (5) waste-heat recovery systems, (6) pressure-letdown stations, (7) boiler feedwater deaerator, and (8) condensate-recovery system. The integration of all these into a reliably functioning system is the objective of the steam-system designer.

Process steam loads

Steam is a source of power and heating, and may be involved in process reactions. Its applications include serving as a stripping, fluidizing, agitating, atomizing, ejector-motive and direct-heating stream. Its quantities, pressure levels and degrees of superheat are set by such process needs.

As reaction steam, it becomes a part of the process kinetics, as in H_2, ammonia and coal-gasification plants. Although such plants may generate all the steam needed, steam from another source must be provided for startup and backup.

The second major process consumption of steam is for indirect heating, such as in distillation-tower reboilers, amine-system reboilers, process heaters, piping tracing and building heating. Because the fluids in these applications generally do not need to be above 350°F, steam is a convenient heat source.

Again, the quantities of steam required for these services are set by the process design of the facility. There are many options available to the process designer in supplying some of these low-level heat requirements, including heat-exchange systems, and circulating heat-transfer-fluid systems, as well as steam and electricity. The selection of an option is made early in the design stage and is based predominantly on economic trade-off studies.

Generating steam from process heat affords a means of increasing the overall thermal efficiency of a plant. After providing for the recovery of all the heat possible via exchanges, the process designer may be able to reduce cooling requirements by making provisions for the generation of low-pressure (50-150 psig) steam. Although generation at this level may be feasible from a process-design standpoint, the impact of this on the overall steam balance must be considered, because low-pressure steam is excessive in most steam balances, and the generation of additional quantities may worsen the design. Decisions of this type call for close coordination between the process and utility engineers.

Steam is often generated in the convection section of fired process heaters in order to improve a plant's thermal efficiency. High-pressure steam can be generated in the furnace convection section of process heaters, which have radiant heat duty only (e.g., H_2-plant reformers, catalytic reformer/heaters).

Adding a selective-catalytic-reduction unit for the purpose of lowering NO_x emissions may require the generation of waste-heat steam to maintain the correct operating temperatures to the catalytic-reduction unit.

Heat from the incineration of waste gases represents still another source of process steam. Waste-heat flues from the CO boilers of fluid-catalytic crackers and from fluid-coking units, for example, are hot enough to provide the highest pressure level in a steam system.

Selecting pressure and temperature levels

The selection of pressure and temperature levels for a process steam system is based on: (1) moisture content in condensing-steam turbines, (2) metallurgy of the system, (3) turbine water rates, (4) process requirements, (5) water treatment costs, and (6) type of distribution system.

Moisture content in condensing-steam turbines — The selection of pressure and temperature levels normally starts with the premise that somewhere in the system there will be a condensing turbine. Consequently, the pressure and temperature of the steam must be selected so that its moisture content in the last row of turbine blades will be less than 10-13%. In high-speed turbines (greater than 9,000 rpm), a moisture content of 10% or less is desirable. This restriction is imposed in order to minimize erosion of the blades by water particles. This, in turn, means that there will be a minimum superheat temperature for a given pressure level, turbine efficiency and condenser pressure for which the system can be designed.

To quantify the selection, let it be assumed that a superheat temperature is to be determined on the basis of a turbine condensing at 0.7 psia (1½ in. Hg) and having an efficiency of 80% (reasonable for large turbines). The minimum superheat temperature can be calculated from a Mollier diagram. The results for various pressure levels are listed in Table 1-1. Based on these data, the following minimum supply-side superheat temperatures for the indicated pressures are required in order to attain no less than 11.5% moisture in the exhaust: 1,500 psig—930°F; 900 psig—825°F; 600 psig—750°F; and 150 psig—490°F.

System metallurgy — A second pressure-temperature concern in selecting the appropriate steam levels is the limitation imposed by metallurgy. Carbon steel flanges, for example, are limited to a maximum temperature of 750°F because of the threat of graphite (carbides) precipitating at grain boundaries. Hence, at 600 psig and less, carbon-steel piping is acceptable in steam distribution systems. Above 600 psig, alloy piping is required. In a 900- to 1,500-psig steam system, the piping must be either a ½ carbon-½ molybdenum or a ½ chromium-½ molybdenum alloy.

Turbine water rates — Steam requirements for a turbine are expressed as water rate, i.e., lb of steam/bhp, or lb of

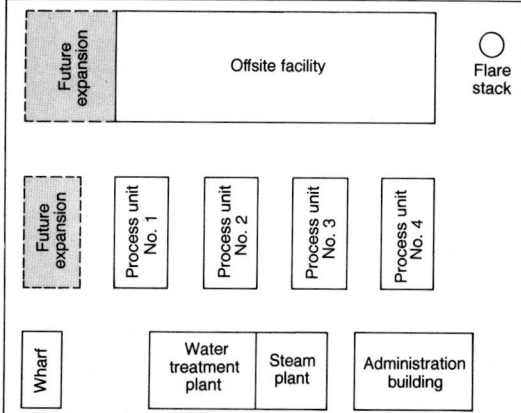

Figure 1-1—Site plan of the hypothetical plant for which steam system is being designed

steam/kWh. Actual water rate is a function of two factors: theoretical water rate and turbine efficiency.

The first is directly related to the energy difference between the inlet and outlet of a turbine, based on the isentropic expansion of the steam. It is, therefore, a function of the turbine inlet and outlet pressures and temperatures.

The second is a function of size of the turbine and the steam pressure at the inlet, and of turbine operation (i.e., whether the turbine condenses steam, or exhausts some of it to an intermediate pressure level). From an energy standpoint, the higher the pressure and temperature, the higher the overall cycle efficiency.

Process requirements — When steam levels are being established, consideration must be given to process requirements other than for turbine drivers. For example, steam for process heating will have to be at a high-enough pressure to prevent process fluids from leaking into the steam. Steam for pipe tracing must be at a certain minimum pressure so that low-pressure condensate can be recovered.

Water treatment costs — The higher the steam pressure, the costlier the boiler feedwater treatment. Above 600 psig, the feedwater almost always must be demineralized; below 600 psig, softening may be adequate. It may have to be of high quality if the steam is used in the process, such as in reactions over a catalyst bed (e.g., in hydrogen production).

Type of distribution system — There are two types of systems: local, as exemplified by powerhouse distribution; and complex, by which steam is distributed to many units in a process plant. For a small local system, it is not impractical from a cost standpoint for steam pressures to be in the 600-1,500-psig range. For a large system, maintaining pressures within the 150-600-psig range is desirable because of the cost of meeting the alloy requirements for higher-pressure steam distribution system.

Because of all these foregoing factors, the steam system in a chemical process complex or oil refinery frequently ends up as a three-level arrangement. The highest level, 600 psig, serves primarily as a source of power. The intermediate level, 150 psig, is ideally suitable for small emergency turbines, tracing off the plot, and process heating. The low level, normally 50 psig, can be used for heating services, tracing within the plot, and process requirements. A higher fourth level is normally not justified, except in special cases as when a large amount of electric power must be generated.

Whether or not an extraction turbine will be included in the process will have a bearing on the intermediate-pressure level selected, because the extraction pressure should be less than 50% of the high-pressure level, to take into account the pressure drop through the throttle valve and the nozzles of the high-pressure section of the turbine.

Drivers for pumps and compressors

The choice between a steam and an electric driver for a particular pump or compressor depends on a number of things, including the operational philosophy. In the event of a power failure, it must be possible to shut down a plant orderly and safely if normal operation cannot be continued. For an orderly and safe shutdown, certain services must be available during a power failure: (1) instrument air, (2) cooling water, (3) relief and blowdown pumpout systems, (4) boiler feedwater pumps, (5) boiler fans, (6) emergency power generators, and (7) fire water pumps.

These services are normally supplied by steam or diesel drivers because a plant's steam or diesel emergency system is considered more reliable than an electrical tie-line.

The procedure for shutting down process units must be analyzed for each type of process plant and specific design. In general, the following represent the minimum services for which spare pumps driven by steam must be provided: column reflux, bottoms and purge-oil circulation, and heater charging. Most important is to maintain cooling; next, to be able to safely pump the plant's inventory into tanks.

Driver selection cannot be generalized; a plan and procedure must be developed for each process unit.

The control required for a process is at times another consideration in the selection of a driver. For example, a compressor may be controlled via flow or suction pressure. The ability to vary driver speed, easily obtained with a steam turbine, may be basis for selecting a steam driver instead of a constant-speed induction electric motor. This is especially important when the molecular weight of the gas being compressed may vary, as in catalytic-cracking and catalytic-reforming processes.

In certain types of plants, gas flow must be maintained to prevent uncontrollable high-temperature excursions during shutdown. For example, hydrocrackers are purged of heavy hydrocarbon with recycle gas to prevent the exothermic reactions from producing high bed temperatures. Steam-driven compressors can do this during a power failure.

Each process operation must be analyzed from such a safety viewpoint when selecting drivers for critical equipment. The size of a relief and blowdown system can be reduced by installing steam drivers. In most cases, the size of such a system is based on a total power failure. If heat-removal is powered by steam drivers, the relief system can be smaller. For example, a steam driver will maintain flow in the pump-around circuit for removing heat from a column during a power failure, reducing the relief load imposed on the flare system.

Table 1-1—Superheat-temperature minimums and moisture-content maximums for an 80%-efficient turbine condensing at 1½-in. Hg

Inlet-steam pressure, psig	Inlet-steam temperature, °F	Moisture content, %
1,500	1,000	9.5
	900	12.5
	800	16.5
900	900	9.3
	800	12.1
	700	15.1
600	800	10.1
	700	13.0
	600	16.5
150	600	8.1
	500	11.2
	400	14.8

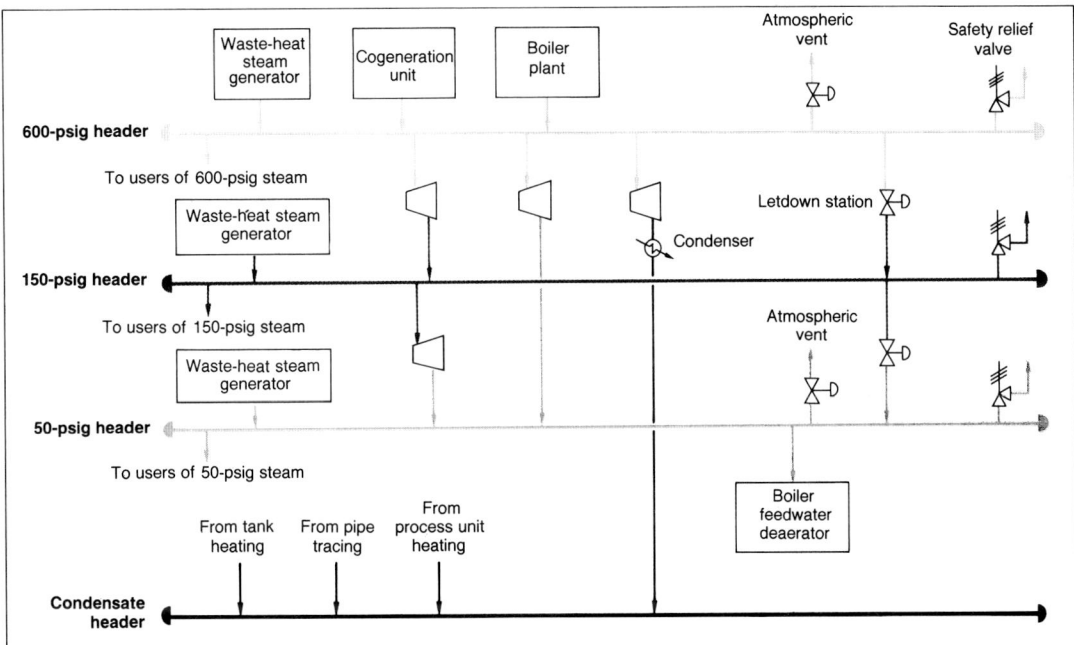

Figure 1-2—This steam and condensate system is typical of that found in a large chemical complex or oil refinery

Equipment support services (such as lubrication and seal-oil systems for compressors) that could be damaged during a loss of power should also be powered by steam drivers.

Driver size can also be a factor. An induction electric motor requires large starting currents — typically six times the normal load. The drop in voltage caused by the startup of such a motor imposes a heavy transient demand on the electrical distribution system. For this reason, drivers larger than 10,000 hp are normally steam turbines, although synchronous motors as large as 25,000 hp are used.

The reliability of life-support facilities — e.g., building heat, potable water, pipe tracing, emergency lighting — during power failures is of particular concern in cold climates. In such a case, at least one boiler should be equipped with steam-driven auxiliaries to provide these services.

Lastly, steam drivers are also selected for the purpose of balancing steam systems and avoiding large amounts of letdown between steam levels. Such decisions regarding drivers are made after the steam balances have been refined and the distribution system has been fully defined. There must be sufficient flexibility to allow balancing the steam system under all operating conditions.

Selecting steam drivers

After the number of steam drivers and their services have been established, the utility or process engineer will estimate the steam consumption for making the steam balance. The standard method of doing this is to use the isentropic

Table 1-2—Typical theoretical steam rates, in lb/kWh, for determining turbine actual steam consumption

Exhaust pressure	Steam inlet conditions, psig/°F				
	1,500/925	900/825	600/750	150/500	50/400
600 psig	29.99	68.2			
150 psig	13.97	18.18	23.83		
50 psig	10.7	12.90	15.36	39.9	
1 atm	8.09	9.25	10.40	17.51	29.10
4.0 in. Hg abs.	6.368	7.03	7.65	10.78	14.00
1.5 in. Hg abs.	5.845	6.388	6.888	9.30	11.52

efficiency for backpressure and condensing turbines are shown in Fig. 1-3 [9].

When exhaust steam can be used for process heating, the highest thermodynamic efficiency can be achieved by means of backpressure turbines. Large drivers, which are of high efficiency and require low theoretical steam rates, are normally supplied by the high-pressure header, thus minimizing steam consumption.

Small turbines that operate only in emergencies can be allowed to exhaust to atmosphere. Although their water rates are poor, the water lost in short-duration operations

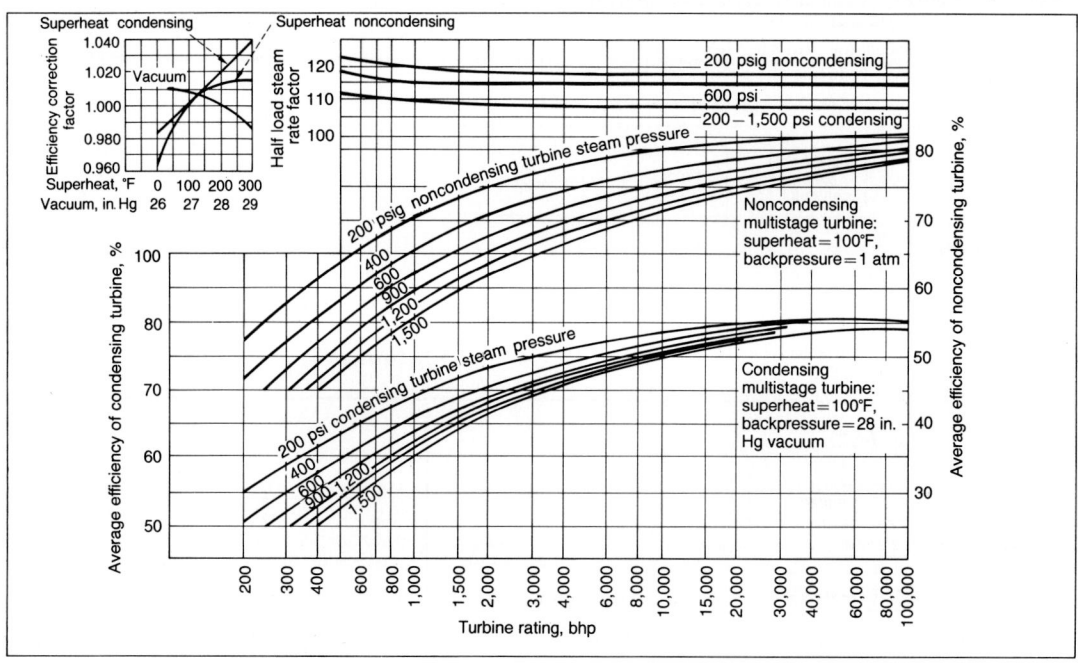

Figure 1-3—Average efficiencies of multistage backpressure and condensing turbines

expansion of steam corrected for turbine efficiency. The isentropic expansion rates, called theoretical steam rates, are tabulated in the Keenan and Keyes "Theoretical Steam Rate Tables" [1].

Actual steam consumption by a turbine is determined via:

$$SR = (TSR)(bhp)/E$$

Here, SR = actual steam rate, lb/h; TSR = theoretical steam rate, lb/hr/bhp; bhp = turbine brake horsepower; and E = turbine efficiency.

Typical theoretical steam rates are presented in Table 1-2. To convert the usually tabulated lb/kWh value to lb/hr/bhp, multiply it by 0.746 (the factor for converting bhp to kW).

Turbine efficiency foremostly depends on size: that of small turbines (10-50 bhp) range from 30% to 40%, and that of large turbines (10,000-50,000 bhp) from 70% to 80%. Efficiency is also a function of steam superheat, turbine speed and (in the case of noncondensing turbines) pressure ratio. These corrections are normally minor. Examples of

may not represent a significant cost. Such turbines obviously play a small role in steam-balance planning.

Constructing steam balances

After the process and steam-turbine demands have been established, the next step is to construct a steam balance for the chemical complex or oil refinery. A sample balance is shown in Fig. 1-4. It shows steam production and consumption, the header systems, letdown stations, and boiler plant. It illustrates a normal (winter) case.

It should be emphasized that there is not one balance but a series, representing a variety of operating modes. The object of the balances is to determine the design basis for establishing boiler size, letdown station and deaerator capacities, boiler feedwater requirements, and steam flows in various parts of the system.

The steam balance should cover the following operating modes: normal, all units operating; winter and summer conditions; shutdown of major units; startup of major units;

loss of largest condensate source; power failure with flare in service; loss of large process steam generators; and variations in consumption by large steam users.

From 50 to 100 steam balances could be required to adequately cover all the major impacts on the steam system of a large complex.

At this point, the general basis of the steam system design should have been developed by the completion of the following work:

1. All significant loads have been examined, with particular attention focused on those for which there is relatively little design freedom — i.e., reboilers, sparging steam for process units, large turbines required because of electric power limitation and for shutdown safety.

2. Loads have been listed for which the designer has some liberty in selecting drivers. These selections are based on analyses of cost competitiveness.

3. Steam pressure and temperature levels have been established.

4. The site plan has been reviewed to ascertain where it is not feasible to deliver steam or recover condensate, because piping costs would be excessive.

5. Data on the process units are collected according to the pressure level and use of steam — i.e., for the process, condensing drivers and backpressure drivers. A summary of such information is shown in Fig. 1-5.

6. After Step 5, the system is balanced by trial-and-error calculations or computerized techniques to determine boiler, letdown, deaerator and boiler feedwater requirements.

7. Because the possibility of an electric power failure normally imposes one of the major steam requirements, normal operation and the eventuality of such a failure must both be investigated, as a minimum.

Checking the design basis

After the foregoing steps have been completed, the following should be checked:

Boiler capacity — Installed boiler capacity would be the maximum calculated (with an allowance of 10-20% for uncertainties in the balance), corrected for the number of boilers operating (and on standby).

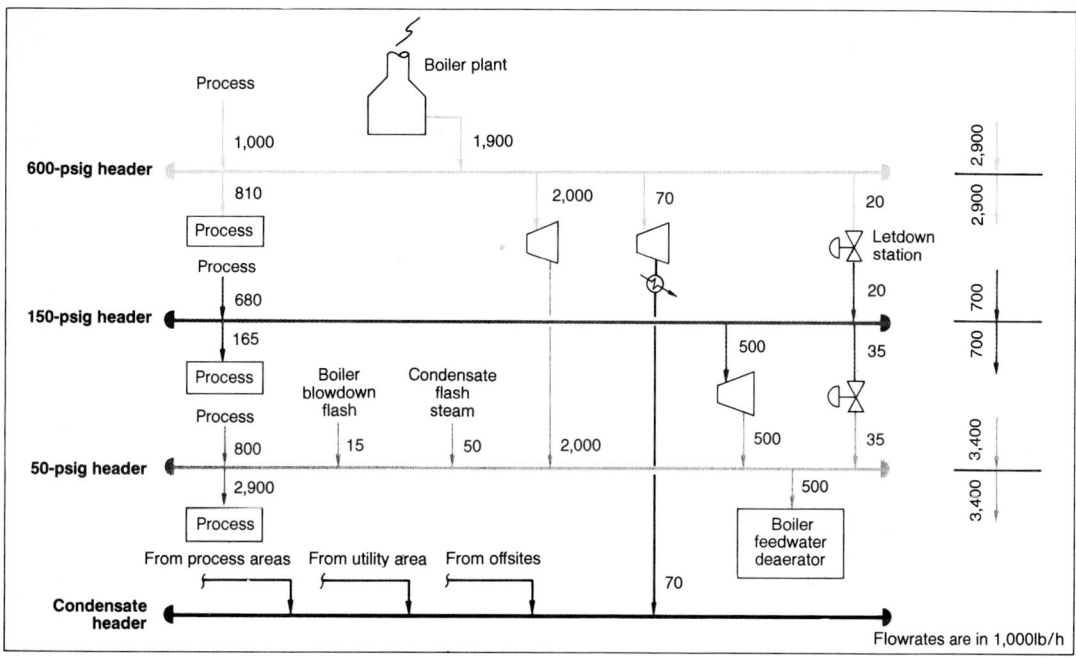

Figure 1-4 — A normal (winter) case steam balance is constructed after process and turbine demands have been established

The balance plays a major role in establishing normal-case boiler specifications, both number and size. Maximum firing typically is based on the emergency case. Normal firing typically establishes the number of boilers required, because each boiler will have to be shut down once a year for the code-required drum inspection. Full-firing levels of the remaining boilers will be set by the normal steam demand. The number of units required (e.g., three 50% units, four 33% units, etc.) in establishing installed boiler capacity is determined from cost studies. It is generally considered double-jeopardy design to assume that a boiler will be out of service during a power failure.

Minimum boiler turndown — Most fuel-fired boilers can be operated down to approximately 20% of the maximum continuous rate. The minimum load should not be expected to be below this level.

Differences between normal and maximum loads — If the maximum load results from an emergency (such as power failure), consideration should be given to shedding process steam loads under this condition in order to minimize installed boiler capacity. However, the consequences of shedding should be investigated by the process designer and the operating engineers to ensure the safe operation of the entire process.

Low-level steam consumption — The key to any steam balance is the disposition of low-level steam. Surplus low-level steam can be reduced only by including more condensing steam turbines in the system, or devising more process applications for it, such as absorption refrigeration for cooling process streams and Rankine-cycle systems for generating power. In general, balancing the supply and consumption of low-level steam is a critical factor in the design of the steam system.

Quantity of steam at pressure-reducing stations — Because useful work is not recovered from the steam passing through a pressure-reducing station, such flow should be kept at a minimum. In the Fig. 1-5 150/50-psig station, a flow of only 35,000 lb/h was established as normal for this steam-balance case (normal, winter). The loss of steam users on the 50-psig systems should be considered, particularly of the large users, because a shutdown of one may demand that the 150/50-psig station close off beyond its controllable limit. If this happened, the 50-psig header would be out of control, and an immediate-pressure buildup in the header would begin, setting off the safety relief valves.

The station's full-open capacity should also be checked to ensure that it can make up any 50-psig steam that may be lost through the shutdown of a single large 50-psig source (a turbine sparing a large electric motor, for example). It would be undesirable for the station to be sized so that it opens more than 80%. In some cases, rangeability requirements may dictate two valves (one small and one large).

Intermediate pressure level — If large steam users or suppliers may come onstream or go offstream, the normal (day-to-day) operation should be checked. No such change in normal operation should result in a significant upset (e.g., relief valves set off, or the system pressure control lost).

If a large load is lost, the steam supply should be reduced by the letdown-station. If the load suddenly increases, the 600/150-psig station must be capable of supplying the additional steam. If steam generated via the process disappears, the station must be capable of making up the load. If 150-psig steam is generated unexpectedly, the 600/150-psig station must be able to handle the cutback.

The important point here is that where the steam flow could rise to 700,000 lb/h, this flow should be reduced by a cutback at the 600/150-psig station, not by an increase in the flow to the lower-pressure level, because this steam would have nowhere to go. The normal (600/150-psig) letdown station must be capable of handling some of the negative load swings, even though, overall, this letdown needs to be kept to a minimum.

On the other hand, shortages of steam at the 150-psig level can be made up relatively easily via the 600/150-psig station. Such shortages are routinely small in quantity or duration, or both·(startup, purging, electric drive maintenance, process unit shutdown, etc.)

High-pressure level — Checking the high-pressure level is generally more straightforward because rate control takes

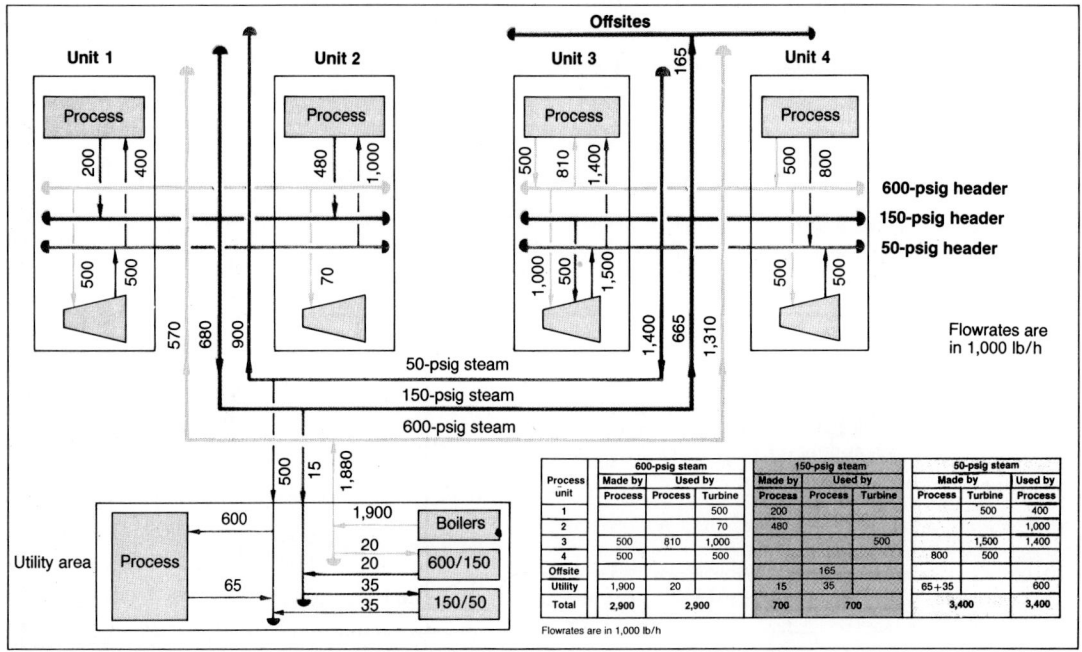

Figure 1-5 — Data on the steam generated and consumed by each process can be summarized in a table or diagram

place directly at the boilers. Firing can be increased or lowered to accommodate a shortage or surplus.

Typical steam-balance cases

The Fig. 1-4 steam balance represents steady-state condition, winter operation, all process units operating, and no significant unusual demands for steam.

An analysis similar to the foregoing might also be required for the normal summertime case, in which a single upset must not jeopardize control but the load may be less (no tank heating, pipe tracing, etc.).

The balance representing an emergency (e.g., loss of electric power) is significant. In this case, the pertinent test point is the system's ability to simply weather the upset, not to maintain normal, stable operation. The maximum relief pressure that would develop in any of the headers represents the basis for sizing relief valves. The loss of boiler feed water or condensate return, or both, could result in a major upset, or even a shutdown.

Header pressure control during upsets

At the steady-state conditions associated with the multiplicity of balances, boiler capacity can be adjusted to meet user demands. However, boiler load cannot be changed quickly to accommodate a sharp upset. Response rate is typically limited to 20% of capacity per minute. Therefore, other elements must be relied on to control header pressures during transient conditions.

The roles of several such elements in controlling pressures in the three main headers during transient conditions are listed in Table 1-3. A control system having these elements will result in a steam system capable of dealing with the transient conditions experienced in moving from one balance point to another.

Tracking steam balances

Because of schedule constraints, steam balances and boiler size are normally established early in the design stage. These determinations are based on assumptions regarding turbine efficiencies, process steam generated in waste-heat furnaces, and other quantities of steam that depend on purchased equipment. Therefore, a sufficient number of steam balances should be tracked through the design period to ensure that the equipment purchased will satisfy the original design concept of the steam system.

This tracking represents an excellent application for a utility data-base system and a system linear programming model. During the course of the mechanical design of a large "grass roots" complex, 40 steam balances were continuously updated for changes in steam loads via such an application.

Cost tradeoffs

To design an efficient but least-expensive system, the designer ideally develops a total minimum-cost curve — which incorporates all the pertinent costs related to capital expenditures, installation, fuel, utilities, operations and maintenance — and performs a cost study of the final system. However, because the designer is under the constraint of keeping to a project schedule, major, highly expensive equipment must be ordered early in the project, when many key parts of the design puzzle are not available (e.g., a complete load summary, turbine water rates, equipment efficiencies and utility costs).

A practical alternative is to rely on comparative-cost estimates, as are conventionally used in assisting with engineering decision points. This approach is particularly useful in making early equipment selections when fine-tuning is not likely to alter decisions, such as regarding the number of boilers required, whether boilers should be shop-fabricated or field-erected, and the practicality of generating steam from waste heat or via cogeneration.

Table 1-3 — Roles played by the various control elements in regulating header pressures

	Headers					
	High pressure		Medium pressure		Low pressure	
Control element	+	−	+	−	+	−
600/150-psig pressure-reducing station	↓①	↓	↓①		↓①	
150/50-psig pressure-reducing station			↓②	↓②	↓①	↑
600-psig atmospheric vent valve	↑②					
50-psig atmospheric vent valve					↑②	
600-psig safety relief valve	↑③					
50-psig safety relief valve					↑③	

Legend: + = pressure is rising
− = pressure is falling
○ = step sequence 1, 2, 3, etc.
↓ = control element opens
↑ = control element closes

The significant elements of a steam-system cost-comparative study are costs for: equipment and installation; ancillaries (i.e., miscellaneous items required to support the equipment, such as additional stacks, upgraded combustion control, more extensive blowdown facilities, etc.); operation (annual); maintenance (annual); and utilities.

The first two costs may be obtained from in-house data or from vendors. Operational and maintenance costs can be factored from the capital cost for equipment based on an assessment of the reliability of the purchased equipment.

Utility costs are generally the most difficult to establish at an early stage because sources frequently depend on the site of the plant. Some examples of such costs are: purchased fuel gas — $5.35/million Btu, raw water — $0.60/1,000 gal, electricity — $0.07/kWh, and demineralized boiler feedwater — $1.50/1,000 gal. The value of steam at the various pressure levels can be developed [5].

The comparative-cost estimate technique is highly effective in determining the power-boiler design basis. A normal power-boiler demand of 1,900,000 lb/h is indicated in Fig. 1-2. The choice of boiler number and size will depend on their ability to provide 1,900,000 lb/h of 600-psig steam when one boiler is shutdown for inspection or maintenance.

Let it be further assumed that the emergency balance requires 2,200,000 lb/h of steam (all boilers available). Listed in Table 1-4 are some combinations of boiler installations that meet the design conditions previously stipulated.

Table 1-4 indicates that any of the several combinations of power-boiler number and size could meet both normal and emergency demand. Therefore, a comparative-cost analysis would be made to assist in making an early decision regarding the number and size of the power boilers.

(Table 1-4 is based on field-erected, industrial-type boilers. Conventional sizing of this type of boiler might range from 100,000 lb/h through 2,000,000 lb/h for each.)

An alternative would be the packaged-boiler option (although it does not seem practical at this load level). Because it is shop-fabricated, this type of boiler affords a significant saving in terms of field installation cost. Such boilers are available up to a nominal capacity of 100,000 lb/h, with some versions up to 250,000 lb/h.

Selecting turbine water rate (i.e., efficiency) represents another major cost concern. Beyond the recognized payout period (e.g., 3 years), the cost of drive steam can be significant in comparison with the equipment capital cost. The typical 30% efficiency of the medium-pressure backpressure turbine can be boosted significantly.

Driver selections are frequently made with the help of cost-tradeoff studies, unless overriding considerations preclude a drive medium. Electric pump drives are typically recommended on the basis of such studies. Turbine spares may be specified to help operations get through an upset. Offsite pump drives are usually electric, because of the high cost of installing steam and condensate-recovery piping.

Steam tracing has long been the standard way of winterizing piping, not only because of its history of successful performance but also because it is an efficient way to use low-pressure steam.

Design considerations

As the steam system evolves, the designer identifies steam loads and pressure levels, locates steam loads, checks safety aspects, and prepares cost-tradeoff studies, in order to provide low-cost energy safely, always remaining aware of the physical entity that will arise from the design.

How are design concepts translated into a design document? And what basic guidelines will ensure that the physical plant will represent what was intended conceptually?

Basic to achieving these ends is the *piping and instrument diagram* (familiar as the P&ID). Although it is drawn up primarily for the piping designer's benefit, it also plays a major role in communicating to the instrumentation designer the process-control strategy, as well as in conveying specialty information to electrical, civil, structural, mechanical and architectural engineers. It is the most important document for representing the specification of the steam system.

On it are shown all the major equipment items, identified by number. Also included in it are such significant mechanical features as pump design capacities and pressure outputs, motor horsepowers, exchanger duties, vessel diameters and tangent-to-tangent dimensions, and insulation thicknesses.

Instruments are numbered, located, and identified as to function. Flows, pressures, temperatures and abnormal operating conditions are taken from line designation tables (described later), which form an important adjunct to the piping and instrument diagram.

Piping is identified by line number, the line size and pipe specification identification reference being included. This information, together with the data in the line designation table, allows the piping designer, the pipe-support designer, and the stress analyst to complete the detail design of all the steam and condensate piping systems.

The piping-specification document represents a source of American National Standards Institute (ANSI) code references and standards, and procurement descriptions and specifications, as well as of piping and valve material, temperature range, maximum pressure range, corrosion allowance, gasket material and pipefitting connections for each service (class).

The line designation table carries design information that cannot be easily portrayed in the piping and instrument diagram. Every numbered process, utility and blowdown line is tabulated in it, along with pertinent normal and upset characteristics. Mass or volumetric flows, or both, are entered in it for line sizing and instrument selection purposes. The identification of each line's source and destination make it possible to determine absolute line loss and commodity flow velocity. Upset pressures and temperatures form the bases for sizing relief valves, and provide data for making stress analyses, for choosing the type of pipe support (guides, slide-shoes, anchor points, hangers and spring supports), and for designing and locating expansion loops. Insulation requirements and thickness, whether for saving energy or protecting personnel, also are indicated.

Table 1-4 — Boiler sizing basis for some combinations of installations

To meet normal demand (1,900,000 lb/h)		To meet upset demand (2,200,000 lb/h)	
No. of boilers normally operating	Rated capacity of each boiler, 1,000 lb/h	No. of boilers installed	Installed capacity, 1,000 lb/h
3	650	4	2,600
4	500	5	2,500
5	400	6	2,400
6	350	7	2,450

Table 1-5 — Typical pressure setpoints in a three-header system

	Steam-system setpoints, psig		
	50-psig system	150-psig system	600-psig system
Turbine-casing pressure safety valve	100	185	None
Header pressure safety valve	80	175	715
Atmospheric vent valve	60	None	660
Pressure-reducing station	53	155	615
Line-loss allowance	5	8	30
Distant subheader operating point	48	147	585

The applicable piping code for chemical plants and petroleum refineries is ANSI B31.3.

Establishing steam pressure profiles

In setting up the program for implementing the design of the overall steam system, early consideration should be given to the multiplicity of pressure settings that must be established. Such things as safety-valve settings for each header, setpoints for pressure-reducing stations, atmospheric-vent and turbine-casing safety valves, and actual operating pressures are all important in promulgating the procurement of the proper equipment and instrumentation, which inevitably must be completed early in the development of the design.

A typical pattern of pressure points for each system is presented in Table 1-5. Although many of the settings are arbitrarily designated, some general guidelines can be applied in their selection.

Profile of the 50-psig header

The 53-psig operating point at the reducing station provides a minimum operating point of 48 psig at distant subheader takeoffs. Thus, a maximum of 5 psig has been allowed for sizing the header line (using the normal, winter steam balance as the design basis; of course, other balance cases may override this basis for line sizing).

The atmospheric-vent setpoint at 60 psig was selected with these thoughts in mind: (1) the pressure-reducing station will accommodate normal hour-by-hour pressure swings in the 50-psig header, including any fluctuations caused by a significant change in the supply or consumption of 50-psig steam; (2) the system furnishes deaeration steam to the water-treatment facility, and a backup supply to the pressure-reducing station will prevent pressure surges in the deaeration steam; and (3) the atmospheric-vent system is not to be the first line of control.

The header safety relief valve is set at 80 psig because this is well within the pressure-temperature profile for 150-psig carbon-steel flanges (even with a superheat temperature as high as 750°F), and it affords a reasonable range between the setpoints of the turbine safety valve and the atmospheric vent (consideration being given to the fact that the pressure-reducing station's accumulation pressure is +10%, i.e., 88 psig, and its reseating pressure is –7%, i.e., 74 psig). Note that, because of the quantity of steam passing through the 50-psig header, it may be desirable to install more than one pressure-relief valve, with settings slightly staggered to smooth out the popping and reseating of relief valves.

By code, one pressure-relief valve for each turbine is required to protect the casing at (or below) the maximum allowable working-pressure rating, in case the discharge valve were closed. This rating, which varies with manufacturer, is related to the casing material (cast iron, carbon steel, alloy and stainless steel, etc.). It is mandatory that the turbine safety valve be set to relieve at a pressure higher than that of the 50-psig steam system relief, to avoid having the turbine's relief valve act as the system's relief valve.

Profile of the 150-psig header

The same general considerations apply to the pressure profile of the 150-psig header. However, in the interest of economy, this system has not been provided with an atmospheric vent. A pressure buildup in the 150-psig header is, instead, relieved by reducing the letdown at the 600/150 psig station, which will initiate a boiler cutback, or by increasing the letdown through the station. With either action, atmospheric venting will probably occur until boiler steam generation has been cut back to a manageable level.

Noted that, when the 150-psig flange pressure/temperature curve is addressed at a 490°F superheat, the continuous pressure rating of the 150-psig header flange network is approximately 175 psig. Although the code does permit

Table 1-6 — Steam generation for three levels of electric-power generation

Pressure/temperature, psig/°F	Steam generation, 1,000 lb/h		
	9,000 kW	25,000 kW	80,000 kW
Unfired boilers:			
160/371	69.9	155.8	393.0
630/755	53.5	107.6	301.0
895/830	50.4	100.0	283.0
1,525/955	—	—	—
Supplementary fired to 1,400°F exhaust gas:			
630/755	97.1	226	539
895/830	94.0	219	522
1,525/955	88.9	207	494
Excess air and 300°F stack temperature, fully fired to 10%:			
630/775	327	755	1,739
895/830	318	736	1,694
1,525/955	306	706	1,626

temporary excursions above the curve, the 175-psig rating was selected because it still provides ample margin above the header setpoint of 155 psig.

The turbine safety valve's setpoint, which again is based on the manufacturer's recommendation, must protect against the turbine being blocked-in, and cannot be set so low as to relieve before the header safety valve (175 psig).

Profile of the 600-psig header

Establishing the pressure profile of the 600-psig header follows the same pattern. An atmospheric vent has been included to relieve modest pressure rises until boiler output has been adjusted. A setpoint of 615 psig has been set for the letdown station, with a line pressure loss of 5% (30 psig) allowed for distant points of steam consumption.

It is likely that the boiler safety relief valves will provide header relief. Design up to the boiler's first block valve is governed by the ASME Pressure Vessel Code, rather than by ANSI 16.3, the first typically being more conservative.

Overview of steam-system control

Because an upset at one point in the steam system has an impact throughout the entire system, the controls designer must establish the sequence of control remedies that will restore normalcy. This will ensure that the highest priority areas will be protected first, and that two control-system elements will not oppose each other.

A simplified overview is illustrated in Table 1-3. The transient upset conditions is indicated at the top of Table 1-3. and the control responses, in order of priority, are noted for each of the major pressure-reducing stations, atmospheric relief valves and header safety relief valves.

Sizing piping

The balancing of steam requirements and its safe handling having been considered in some detail, there remains the important matter of delivering the steam to the points of consumption, and of returning depressed steam or condensate to the utility area efficiently and inexpensively.

Many plant problems arise because the steam distribution has been inadequately designed. Steam quality may be below standard, pressures too low, or isolation difficult because of physical shortcomings.

The first step in providing an acceptable service is to size the header properly. Although an oversized header may make operations easier, its capital cost will be excessive. Furthermore, a domino effect on other costs follows from an oversided header, boosting costs of insulation, supports, hangers and expansion loops. Also, the additional heat loss will increase condensate recovery, tending to complicate this system's design, and a host of problems related to installing large elements in already crowded spaces will arise.

It is more likely, however, that shortcomings will come about because of lines being undersized. Although it is acceptable to arrive at general conclusions early, basing pipe sizes on the 100-ft/s velocity guideline, all major piping should be checked for line loss by means of conventional fluid-flow calculations.

If steam supply lines are not suitably arranged, condensate will collect at low points. This tends to reduce a pipe's cross-sectional area, increasing the pressure-drop loss and, more significantly, local velocity. Eventually, condensate splash will culminate in a mass of condensation being propelled through the line, creating a water hammer at the first turning point. Water hammer will, at the least, cause noise and flow irregularity and, at the worst, flange leaks or anchor strains.

In laying out the header network, the designer is concerned with low points, where condensate trapping, collecting and returning can be more effectively handled. Steam lines should be inclined on a descending gradient, so that condensate will flow with the steam. If the line must climb a pipeway waterfall or it is to be isolated by means of a valve, trapping must be incorporated. If it must rise (e.g., because the offsite terrain is hilly), reverse condensate flow must be accommodated by oversizing the line.

Traps should not be tied directly to the line (because condensate is generally swept along by the steam) but to a pocket or separator that forms a collection sump. A boot (a downward leg of pipe) performs this function particularly well on the supply side of steam-turbine piping. Because the steam supply line typically runs downward, a boot trap can be easily installed at the line's transition to the horizontal section immediately adjacent to the turbine. Branch connections should be taken from the top of steam headers to ensure that the steam will be dry.

Condensate collection

Condensate recovery is important because of the high cost of treating boiler feed water, particularly when higher steam pressures require that makeup water be demineralized. Because condensate is returned to the boiler plant to recover not only treated water but also the sensible heat, recovery should be evaluated even when it must be transported long distances.

The following guidelines apply to the design of the condensate collection system:
• All the condensate that can be returned should be recovered, unless there is a genuine risk that it may contaminate the boiler feedwater.
• Backpressures on steam traps should be minimized by sizing return lines adequately.
• Condensate flashing should be avoided because it adds pressure drop in the return line.
• Condensate lines should be located lower than steam traps to provide gravity drainage.

Alternate steam sources

The generation of steam in the stacks of process heaters has become a common way of recovering heat that would other-

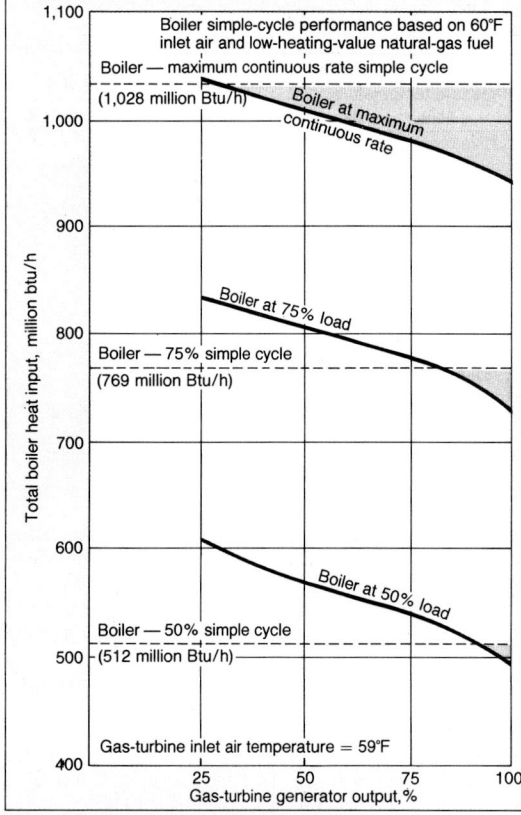

Figure 1-6—Simple-cycle boiler performance is based on 60°F inlet air and low-heating-value natural-gas fuel

wise be lost to the atmosphere. Steam can be generated via this method at any of the three pressure levels (600, 150 or 50 psig), depending on the stack-gas temperature. An estimate of how steam will be generated can be based on a 150°F difference between the stack-gas and steam temperatures at the exit of the convection coil. Costs of pumps, steam drum, and supplying boiler feedwater have to be balanced against the usefulness of the steam so produced. These costs usually limit this method of steam generation to large heaters having high stack-gas temperatures.

In these cogeneration systems, a generator driven by a production from a simple waste-heat-recovery unit is limited to the amount of heat available from the turbine exhaust.

If the fully-fired boilers are sized on the basis of a single boiler being out of service, the boiler load in normal operation will be considerably lower than maximum continuous rate. The combined efficiency of a gas turbine and boiler can be less than that of a simple-cycle boiler because of the large quantities of excess air from the gas turbine. As shown in Fig. 1-6 for a 900-psig, 925°F, 750,000-lb/h maximum-continuous-rate boiler with a 25-megawatt gas turbine operating in a combined cycle, there is a relationship between gas-tur-

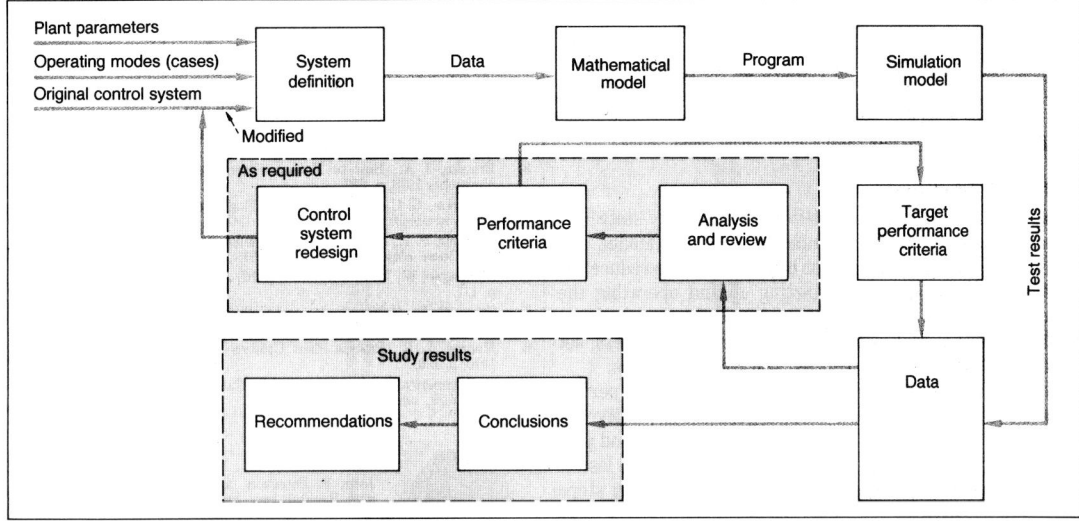

Figure 1-7—Sequence of steps involved in conducting a dynamic-simulation study

gas turbine, either the aircraft-derivative or industrial type, produces electricity, and the exhaust from the gas turbine, which is 800-1,000°F and 15% oxygen, is a source of hot combustion air or waste heat for generating steam. In the latter case, the waste-heat boiler can be fired additionally to produce larger amounts of steam than could be generated from only the sensible heat of the turbine exhaust.

Typical quantities of steam can be generated by three types of combined-cycle boilers: an unfired waste-heat-recovery boiler, a boiler supplementary-fired to 1,400°F, and a fully-fired boiler with 10% excess air (see Table 1-6).

After the decision has been made to resort to cogeneration, the next step is to select the boiler and gas-turbine size. A gas turbine is a constant-flow machine, the air flow and speed being fixed, the latter being based on the electrical generator. Depending on the configuration: (1) in the case of a heat-recovery boiler, the amount of steam produced can be decreased by going to a smaller size; (2) if the boiler is supplementary-fired, fuel firing can be increased to maintain steam production, and (3) if the boiler is fully fired, fuel firing can be increased to meet steam requirements.

At constant power demand, a varying steam demand is possible only with a supplementary-fired or fully-fired boiler. Without the turbine exhaust being bypassed, steam bine output and boiler load for which a boiler operating in simple cycle is more efficient — i.e., less boiler fuel is consumed. This is caused by the large quantity of excess air and the lower gas-turbine discharge temperature at reduced boiler and gas-turbine load.

If the normal boiler and gas-turbine loads are 60% of design loads, operating the boiler in a simple cycle, instead of in the combined cycle, is more efficient.

Which type of boiler to install is a second concern. There can be two types of combined-cycle-boiler design: a pressurized-firebox boiler without an induced-draft fan, and a balanced-draft boiler with such a fan. The first type imposes a backpressure on the gas turbine that reduces power output. The large fan of the latter type poses control problems and implosion and explosion hazards. Operating the latter requires considerable transient analysis to account for fan or damper failure. In case such a failure should occur, the means must be provided for diverting the gas-turbine exhaust quickly so as to prevent overpressuring the boiler. Because high-temperature dampers are difficult to operate and maintain, these design considerations are critical to safe and reliable operation of the system.

A third concern is operation during power failure. In such an emergency, the generators of the gas turbines are discon-

Table 1-7—Test-run matrix facilitates analysis of simulation-model results

Upset number	Upset description	Operating cases tested								
		A	B	C	D	E	F	G	H	I
1	Backpressure-turbine trip	x	x	x	x	x	x	x	x	x
2	Condensing-turbine trip	x	x	x	x	x	x	x	x	x
3	Boiler trip	x	x	x	x	x	x	x	x	x
4	Total electric-power failure	x		x	x		x		x	
5	Coker air-blower trip	x	x	x	x	x	x	x	x	x
6	Coke-superheater trip	x	x	x	x	x	x	x	x	x
7	Boiler trip from maximum load	x	x	x						
8	600/150 valve fails open								x	x
9	150/50 valve fails open			x						

nected, but the boilers still require air to maintain steam production. This calls for a gas turbine that will operate at the same air capacity at full or zero load — i.e., a single-shaft gas turbine rather than a dual-shaft one.

Dynamic simulation

In a dynamic computer simulation, both steady-state and transient behavior are represented by means of a model. The knowledge gained via simulation can be applied to reduce the cost of designing, building, starting up and operating the system studied. A model can be used to design and evaluate control systems, analyze system behavior during upsets, set up and check startup procedures, and train operators.

Simulation can be an effective tool for verifying the performance of a control scheme as steam generators (utility and waste-heat boilers) and consumers (steam turbines and process units) respond to possible disturbances (i.e., the loss of boilers, turbines or process units). Fig. 1-7 shows the steps involved in analyzing or designing a control system. The first, and most crucial step, is defining the system, which involves not only identifying its physical boundaries but also establishing the variables and types of behavior that the mathematical model must represent.

The more rigorous the model, the larger (in terms of number of equations) it is, and the more expensive it is to program and run on a computer. Because cost and complexity are related, a model should be as simple as possible, yet detailed enough to meet the objectives. So, it is important to define what information is expected from the simulation before starting the modeling and programming.

After the boundaries of the system and the performance criteria have been established, the mathematical model is written and programmed on a computer. After the programming errors have been eliminated, a period of model evaluation follows. This involves making a series of computer runs to observe the dynamic behavior of the model and to determine if the behavior is realistic. If the model is not yet adequate, it is revised and retested until it accurately represents the system. Table 1-7 shows a typical test-run matrix.

Test results are analyzed and checked against the target performance criteria. If these criteria are not met, then the control system must undergo revision. If the steam-system design does not exhibit good controllability during transient conditions, the plant may be liable to unnecessary shutdowns, which can have severe economic implications.

References

1. Keenan, J. H., Keyes, F. G., Hill, P. G., and Moore, J. G., "Steam Tables," John Wiley & Sons, New York, 1969.
2. Keenan, J. H., and Keyes, F. G., "Theoretical Steam Rate Tables," American Soc. of Mechanical Engineers.
3. Brinsko, J. A., How to Make a Steam Balance," *Hydrocarbon Proc.*, November 1978, p. 227.
4. Patterson, G. C., Fundamentals of Engineering Offsite and Utilities for HPI, *Petrochem. Engineer.*, August 1967, p. 39.
5. Gambhir, S. P., Heil, T. J., and Schuelke, T. F., Steam Use and Distribution, *Chem. Eng.*, Dec. 18, 1978, p. 91.
6. Campagne, W., What is Steam Worth?, *Hydrocarbon Proc.*, August 1981, p. 117.
7. Stacy, G. D., Gaines, L. D., and Collis, S., Optimize Steam Systems by Computer, *Hydrocarbon Proc.*, August 1981, p. 75.
8. Monroe, L. R., Process Plant Utliities—Steam, *Chem. Eng.*, Dec. 14, 1970, p. 130.
9. "Transamerica Delaval Engineering Handbook," 4th ed., McGraw-Hill, New York, 1983, p. 5-11.

Section 2

Steam Power Plants for Utility Stations

Steam power plants for utility service are principally fossil-fuel fired (coal, oil, gas), though nuclear plants are making their appearance, despite a variety of problems with both the plant and environmentalists. And with both fossil-fuel and nuclear plants, a number of design approaches exist and are used.

This section covers several types of steam plants serving the utility field. Major design features are tabulated and discussed.

Principal design parameters for utility steam power plants are also considered. These parameters involve the compromises which must be made in plant design to balance improvements in thermal efficiency against increases in capital costs. Further, the compromises vary, depending on the type of service the plant is designed to render. A variety of these compromises are discussed in this section, as are the environmental aspects of various fuels used.

From *Engineering Evaluation of Energy Systems,* by Arthur P. Fraas. Copyright © 1982. Used by permission of McGraw-Hill, Inc. All rights reserved.

In view of the fact that there has been no increase in the temperature and pressure employed in steam power plants since about 1950 and no improvement in the thermal efficiency, one might ask why a chapter on steam plants is included in a text on advanced power plants. This has been done because steam plants are truly highly developed systems, they provide the bulk of our power, and they provide a standard against which any advanced concept must be compared to determine whether it is sufficiently attractive to justify its selection.

HISTORICAL DEVELOPMENT

It is instructive to trace the development of steam power from its inception with the Newcomen engine nearly 300 years ago. Interestingly enough, that engine was so inefficient (less than 1 percent) that it was competitive with horses in only one field: pumping water from mines. Even there it gained only limited acceptance until James Watt in 1765 made the first major improvement: separation of the steam condenser from the cylinder to cut heat losses and increase the thermal efficiency by a factor of 4—to a few percent! People were so skeptical that this big improvement could be effected that for the first 20 years the Watt engines were not sold; they were erected at the site, and Watt's company was paid royalties on the basis of the coal savings that were effected through the use of the more efficient engine that Watt had patented.[1] This gave an opportunity for cheating on the part of the engine operator, who of course got more power from a given size of engine and hence more useful work with the former fuel consumption—a situation that led Watt to develop the Prony brake for measuring horsepower and defining that quantity.

Reciprocating-Steam-Engine Development

Refinements in the engine mechanisms and improvements in machine tools to give better fitting parts followed rapidly in the next 100 years. Thus the reciprocating steam engine was developed to essentially its natural limits by the time Edison invented the electric light bulb. The reciprocating steam engine is very different from gasoline or diesel engines in that the condensing-heat-transfer coefficient for steam is of the order of 5.7 W/(m$^2 \cdot$ °C [10,000 Btu/(h $\cdot$ ft$^2 \cdot$ °F)], about 100 times as high as the corresponding value in a gasoline or diesel engine cylinder. As a consequence, the upper cylinder wall in a steam engine is heated by steam condensation to very nearly the saturation temperature of the steam supplied to the cylinder. The best lubricants available will keep wear rates low only if the rubbing surfaces are below about 204°C(400°F). With care in piston design, it has proved possible to keep the temperature at the piston rings about 40°C below that of the inlet steam and thus to operate with a steam saturation temperature and pressure of 242°C and 34 bars (467°F and 500 psia). Significantly, no improvements in this temperature limit have been achieved in the past 100 years; repeated efforts to exceed it in automotive steam engines have led to high wear rates and short engine life.

The development of steam plants for utility service in the United States began with the construction of the Brush Electric Light Company plant in Philadelphia in 1881.[2] Construction of other plants followed rapidly throughout the country to satisfy the need for electric power for lighting and for electric street railways. Initially, single-expansion reciprocating steam engines were employed. By 1890, double-expansion engines and shortly thereafter triple-expansion engines became common. Note that because of their low cost and simplicity of construction, triple-expansion steam engines were used in the huge fleet of Liberty ships built during World War II.

Steam-Turbine Development

By 1900, the advantages of a steam turbine in permitting higher temperatures and pressures with greater expansion ratios became apparent, and in 1904 the first steam turbine in U.S. utility service went into operation, giving a higher thermal efficiency than any reciprocating engine in service.[3-8] Although pressures up to 41 bars (600 psia) were used in steam engines for mobile power plants in order to reduce the engine size and weight while sacrificing engine life, overall expansion ratios of

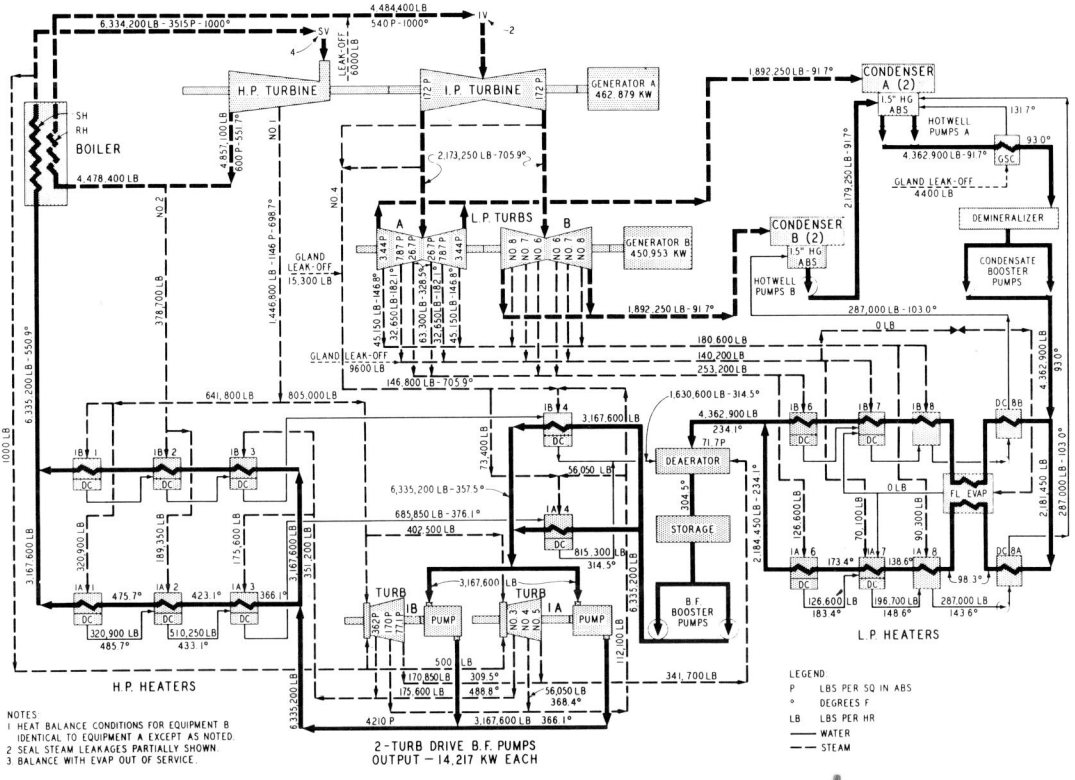

Figure 2-1 Flowsheet and heat balance diagram for the TVA Bull Run coal-fired steam plant operating at 914,402 kW. (*Courtesy TVA.*)

more than 100 were not practicable even in triple-expansion reciprocating steam engines, and hence there was no incentive from the standpoint of thermal efficiency to go to steam pressures in excess of about 10 bars (145 psia). A steam turbine could employ much higher expansion ratios to good advantage, thus leading to fairly rapid increases in both temperature and superheat.[5,6,7]

Increasing the steam pressure and the resulting expansion ratio in steam turbines increased the number of stages and rotor length required to the point that bearing and shaft dynamics problems became limiting. (The critical speed of a shaft is inversely proportional to the square of its length.) This limitation led to the use of first two and finally as many as four turbines in tandem. In a supercritical-pressure steam plant, for example, there may be a superhigh-pressure, a high-pressure, an intermediate-pressure, and a low-pressure turbine. Other factors favoring the use of a multiplicity of turbines in series are the large differences in casing thickness required to withstand the internal pressure and the large differences in rotor diameter stemming from the enormous change in steam specific volume associated with the steam expansion (a factor of over 2000). In fact, the flow-passage area required for the volume of steam leaving the lower-pressure turbines becomes so great in units of large capacity that double-flow turbines are employed with the steam entering in the center and splitting into two streams that flow in opposite directions toward the outboard ends of the rotor. This has the additional advantage that the axial forces on the two sets of rotor blades balance each other, thus avoiding a difficult thrust-bearing problem. For units of more than 300 MWe, the size of the low-pressure turbine becomes so great that two double-flow turbines are used in parallel, and so there are actually a total of four stages of rotor blades operating in parallel. Note that the last-stage turbine rotor size required is greatly reduced by bleeding off steam for regenerative feed-water heating. (Examination of the flowsheet of Fig. 2-1 shows that in this case the steam flow to the condenser is only 60 percent the flow into the high-pressure turbine.)

The high-, intermediate-, and low-pressure turbines may be coupled in series on a single shaft having a total length (including the generator) of as much as 77 m (250 ft) for a 1300-MW unit. An alternative arrangement is to reduce the size of the higher-pressure units by designing them to operate at 3600 rpm and mounting them on one shaft and mounting the inherently large low-pressure units on a separate shaft that operates at 1800 rpm. The latter arrangement is referred to as a cross-compound turbine-generator unit. The steam pressure available from PWR or BWR nuclear plants is enough lower than in fossil fuel plants—i.e., 65 bars (~950 psia) as opposed to 240 bars (~3500 psia)—that an 1800-rpm single-shaft machine is employed with a number of turbines in tandem.

Steam Generator-Furnace Units

The rapidly increasing size of early steam power plants led to the development of traveling grate and other types of mechanical stoker, and these rapidly displaced hand-fired boilers.[3] Improvements in metallurgy made possible higher-strength steels with greater resistance to corrosion, thus leading to further increases in pressure and temperature throughout the 1920s together with concurrent increases in unit size.[4] At the same time, refinements in the thermodynamic cycle were made to improve the thermal efficiency by reheating,[9] regenerative feed heating, and air preheating. Improved methods of feed-water treatment eased the corrosion problems on the steam side and reduced maintenance.[2]

By 1930, developments in both electric welding techniques and x-ray inspection of welds enabled the advancement from riveted header drums with rolled tube-to-header joints and flanged joints in the piping to virtually all-welded construction of the header drums, tube-to-header joints, and plant piping.[2] This, in turn, permitted marked further increases in the system temperature and pressure to around 170 bars (2450 psi) and 565°C (1050°F), so that by 1950 the new central station steam plants had virtually reached the operating conditions current at the time of writing.[10,11] Concurrently, the reliability of boilers and turbines increased sufficiently that the initial practice of manifolding a multiplicity of boilers to a multiplicity of turbines was dropped in favor of independent units with each boiler coupled to a particular turbine. This simplified the piping (particularly provisions for thermal expansion) and eliminated many large shutoff valves which gave a great deal of trouble when used at high temperatures and pressures. In the 1950s, the once-through boiler came into use and was almost immediately applied to produce supercritical-pressure steam, commonly 240 bars (3500 psi) and 565°C (1050°F).[12-16] A simplified flowsheet showing the major components in a typical modern plant is presented in Fig. 2-24 (p. 2-42), and a more detailed flowsheet for a supercritical-pressure steam system is shown in Fig. 2-1.

A section through a pulverized-coal-fired-furnace-steam generator unit is shown in Fig. 2-2. Combustion takes place in the large open chamber at the lower right, which is surrounded by vertical boiler tubes joined together with short thick fins as shown in Fig. 2-3 to form a continuous "waterwall" that ab-

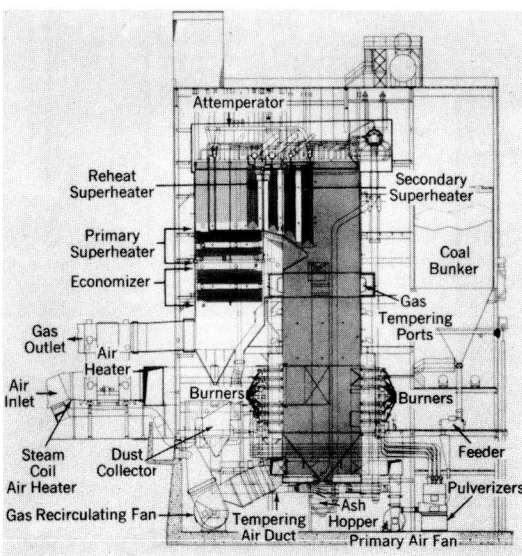

Figure 2-2 Radiant boiler for pulverized-coal firing. Design pressure 2875 psi; primary and reheat steam temperatures 1000°F; maximum continuous steam output 1,750,000 lb/h.[2] (Courtesy Babcock & Wilcox Co.)

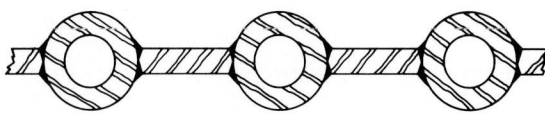

Figure 2-3 Section through the boiler tubes in a waterwall for a boiler furnace such as that of Fig. 2-2.

TABLE 2-1 Summary of Principal Developments in Steam Power Plants for Utility Service

Period	Typical parameters for new plants				Principal developments
	Peak temp., °C (°F)	Peak press., bars (psia)	Unit output, MW	Full-load thermal efficiency, %	
1880–1890	173 (344)	8.6 (125)	0.3	5	Generators belt-driven from reciprocating steam engines with hand-fired boilers.
1890–1900	173 (344)	9.0 (130)	0.7	7	Double- and triple-expansion engines with Corliss poppet valve gear introduced.
1900–1910	181 (358)	10 (145)	1	10	Steam turbine with steam superheat came into common use with mechanical coal stokers.
1910–1920	293 (560)	19 (275)	10	15	Steam pressures and temperatures increased somewhat and sizes and efficiency of turbines much increased.
1920–1930	385 (725)	38 (550)	25	25	Large pulverized-coal-fired boilers introduced with new furnace designs employing water-cooled walls with steam reheat, regenerative feed-water heating, and combustion air preheat. Turbine sizes increased to as much as 100 MW. Feed water treated.
1930–1940	482 (900)	62 (900)	50	32	Arc-welded construction and new alloys permitted marked increase in pressures and temperatures: large steam generators and turbines combined into single units (to eliminate troublesome valves) replaced former multiplicity of small boilers (as many as 60 in one powerhouse) manifolded to a multiplicity of turbines. Control of feed-water chemistry improved.
1940–1950	538 (1000)	83 (1200)	150	36	Higher pressures and temperatures came into use with further improvements in alloys and control of feed-water chemistry. Further increases in unit size.
1950–1960	566 (1050)	165 (3500)	300	38	Once-through and supercritical steam pressure systems came into use. Turbine designs improved to tolerate and remove much more moisture, thus permitting use of saturated steam from nuclear plants. Unit sizes increased further, and still better control of feed-water chemistry.
1960–1970	566 (1050)	240 (3500)	500	39	Unit size further increased. Nuclear reactors come into use. Cost of electricity continues to drop so that by 1970 the inflation-adjusted cost was $\frac{1}{5}$ that in 1930.
1970–1980	538 (1000)	240 (3500)	1000	36	EPA requirements lead to widespread introduction of cooling towers, stack gas scrubbers, and improved particulate removal equipment, all causing losses in plant efficiency and increased costs. Further increases in unit size served to reduce costs, but increasing shortages of gas and petroleum lead to large increases in the overall cost of electricity.

sorbs the intense thermal radiation heat flux from the burning fuel. The heat flux is high: ~ 63 W/cm² [200,000 Btu/(h · ft²)]. Thus a large fraction of the heat of combustion is removed in a few seconds as the hot gases pass upward through the furnace into the second-stage superheater, then across through the reheater and downward through the first-stage superheater and economizer, and finally laterally through the air preheater to the stack.

Chronological Development

The highlights[18] of this long developmental history are summarized in Table 2-1 and Fig. 2-4. Note the steady increases in pressure and temperature up to 1957, with a concomitant increase in thermal efficiency, but with no further significant increases in any of these three parameters in the subsequent 20 years. Efforts in the 1960s and early 1970s were directed

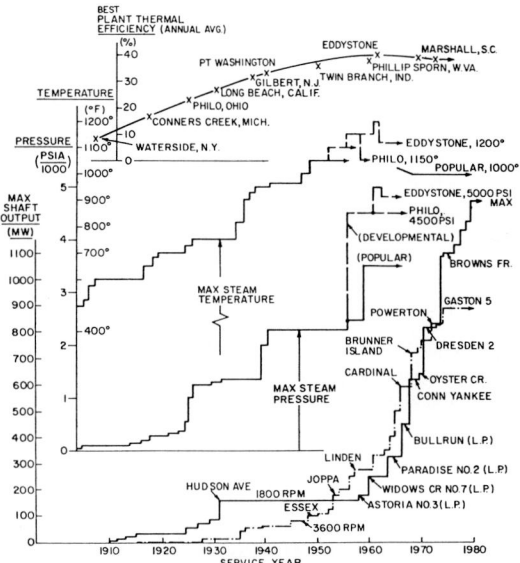

Figure 2-4 Historical development of advanced turbine generators.[18] (*Courtesy General Electric Company.*)

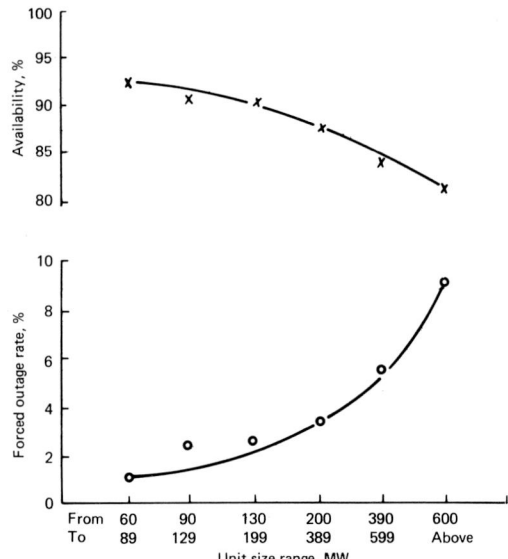

Figure 2-5 Averages for 1965–1974 EEI availability data for fossil-fuel steam-turbine units as a function of unit size.[19]

primarily toward increases in both reliability and in unit size; the latter effect is also shown in Fig. 2-4.[18] However, the reliability and availability[19] of units seem to have declined with increasing size, as indicated in Fig. 2-5. This stems in part from the increase in the number of boiler tubes; for a given probability of a tube failure the probability of a forced outage from a boiler tube failure increases with unit size, so that boiler tube failures become a large fraction of the total outages. Partly for this reason the trend toward even larger units seems to have tapered off in recent years (Fig. 2-6).[19]

Emissions

Since passage of the Environmental Protection Act of 1968, the prime emphasis in fossil-fuel steam power plant development has been on the reduction of stack gas emissions and minimizing the environmental effects of waste-heat rejection; both efforts have led to reductions in thermal efficiency. The environmental effects involved and something of the cost/benefit aspects of the EPA regulations are well-known. This section is concerned with the equipment involved, the problems it poses, and the prospects for future development.

Work on the reduction of stack gas emissions and their effects has included the use of stacks as much as 366 m (1200 ft) tall to reduce concentrations at ground level, marked improvements in electrostatic precipitators to remove particulates, bag houses for even better removal of particulates with fabric filters, stack gas scrubbers to remove SO_2, and improvements in furnace and burner design to reduce the emission of oxides of nitrogen, e.g., via staged combustion.[20,21,22] Most of this work has been carried out by superimposing new equipment on existing types of power plant and selection of low-sulfur fuels. In addition, work has been under way to remove sulfur from the fuel before it is fed to the burners, either by solvent extraction of sulfur from coal or by a coal gasification or liquefaction process that includes removal of the sulfur.[23] Work has also been under way to reduce the sulfur content in fairly low sulfur coals by coal-washing operations that appear to be effective in removing perhaps half of the ash and ~15 percent of the sulfur.[24]

Several efforts to develop fluidized-bed coal combustion systems have been under way since the middle 1950s.[25,26,27] This process is the most promising development in sight for the economical and efficient use of high-sulfur fuels in coal-fired steam plants at the time of writing. A series of both generalized and specific design studies of fluidized-bed coal combustion systems for utility, as well as industrial and institutional applications, have been carried out to show the potential of fluidized-bed furnaces coupled to steam boilers.[16,28-36]

The characteristics of the principal means for reducing the amount and effects of stack gas emissions are summarized in

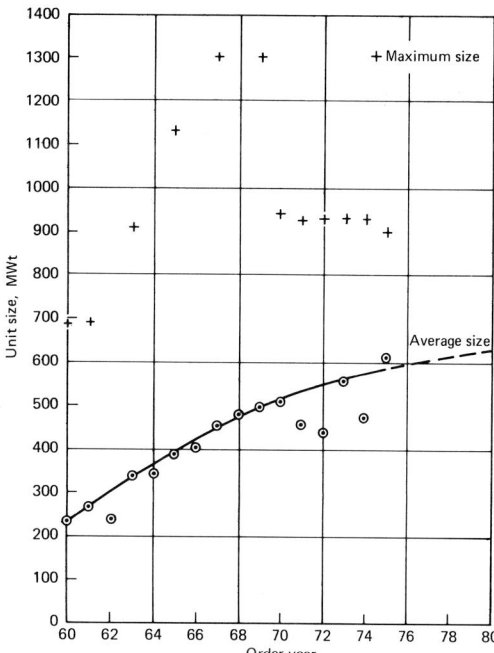

Figure 2-6 Average and maximum size of fossil-fuel steam-turbine units as a function of the year in which they were ordered.[19]

TABLE 2-2 Summary of Characteristics of the Principal Approaches to Reducing the Amount and Effects of Sulfur Emissions from Steam Power Plants

System	Relative capital cost	Efficiency of coal utilization	Relative plant availability	Provisions for retrofit	Estimated date for commercial operations	R&D uncertainties
Low-sulfur coal	1.0	0.99	0.80	No problem	Current	None
Fluidized-bed comb.						
AFB	1.2–1.5	0.9–0.98	0.8	Furnace vol. larger than for pulv. coal	1984–1986	Corrosion in bed, coal metering and feed, reliability
Supercharged bed	0.8–1.1	0.99	0.8	Furnace small	1986–1988	Ditto above, turbine erosion and deposits
PFB	?	0.99	?	Furnace small	?	Ditto above, hot gas cleanup system
Wet limestone stack gas scrubbers	1.2–1.3	0.93	0.7?	Site-dependent	Current	Reliability and life of components, high capital and operating costs
Solvent refined coal	1.6–2.5	0.6–0.8	0.8	No problem	1985?	Reliability and life of plant; conversion efficiency
Coal liquefaction	1.6–2.5	0.6–0.7	0.8	No problem	?	Ditto above
High-Btu gas coal gasification	1.6–2.5	0.5–0.6	0.8	No problem	?	Ditto above
Low-Btu gas coal gasification	1.4–2.0	0.6–0.8	0.6	New plant required for good integration	1980–1982	Reliability and life of plant, conversion efficiency
Tall stack	1.01–1.02	0.99	0.80	No problem	Current	None

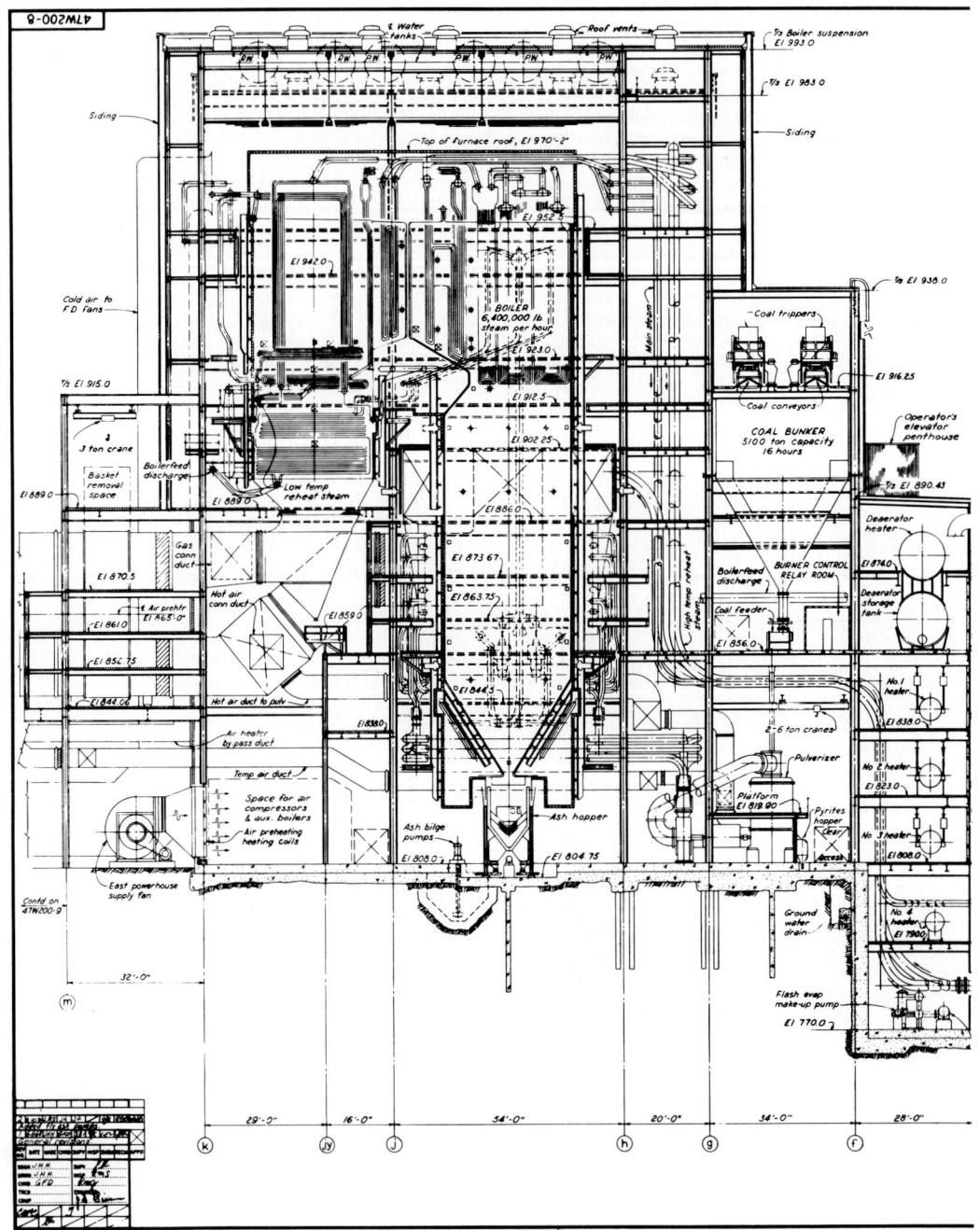

Figure 2-7 Elevation showing the arrangement of the principal components in the 950-MWe Bull Run power plant unit of TVA near Knoxville, Tenn. (*Courtesy TVA.*)

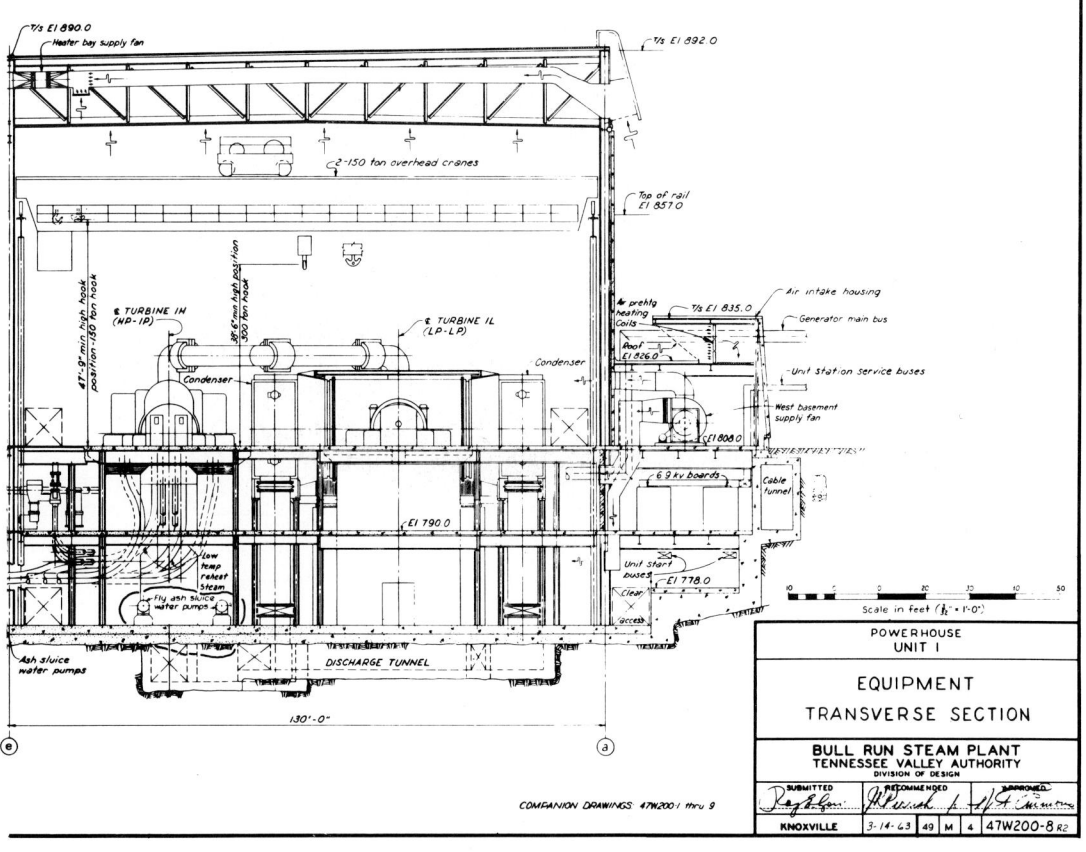

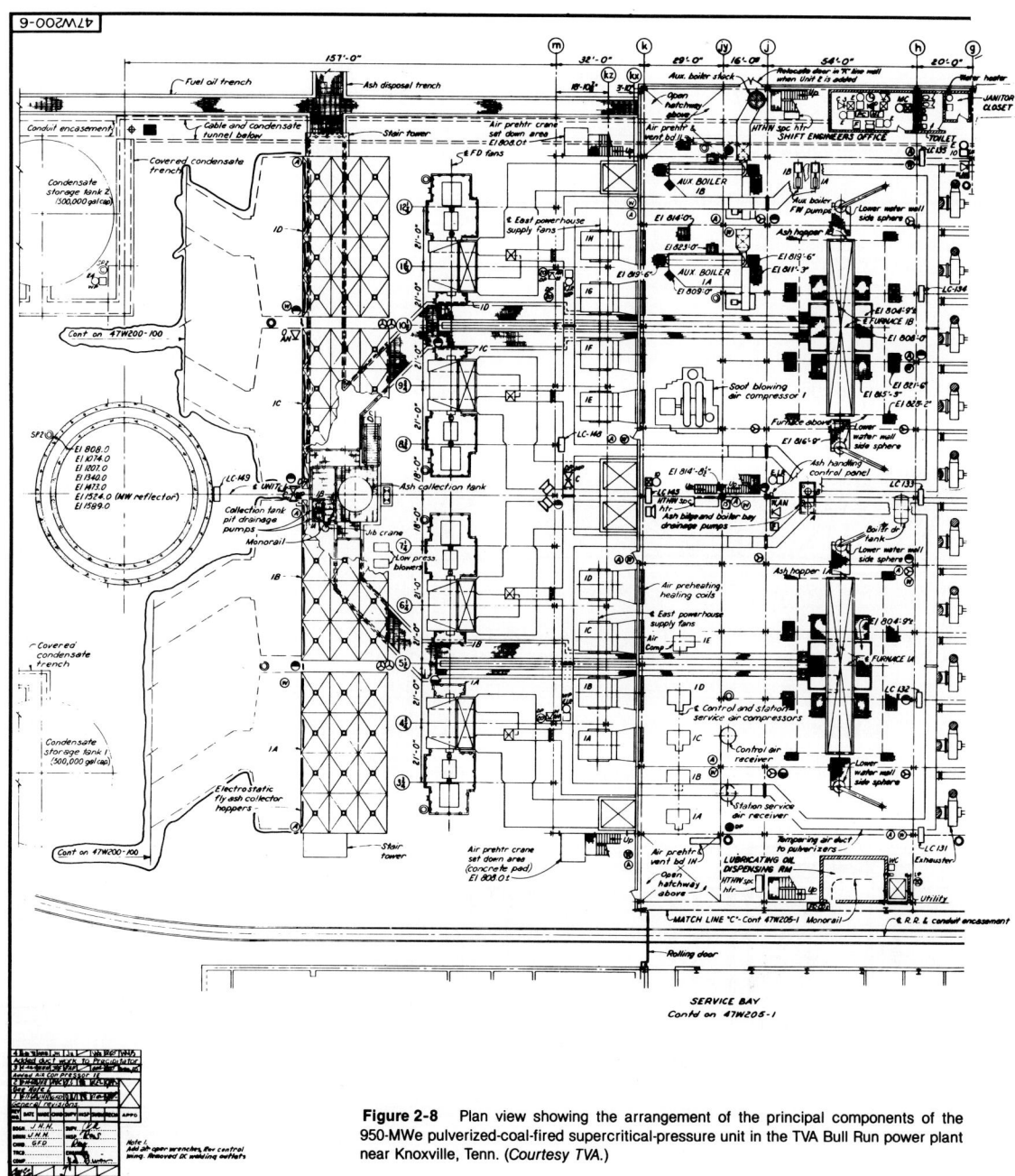

Figure 2-8 Plan view showing the arrangement of the principal components of the 950-MWe pulverized-coal-fired supercritical-pressure unit in the TVA Bull Run power plant near Knoxville, Tenn. (*Courtesy TVA.*)

2–10

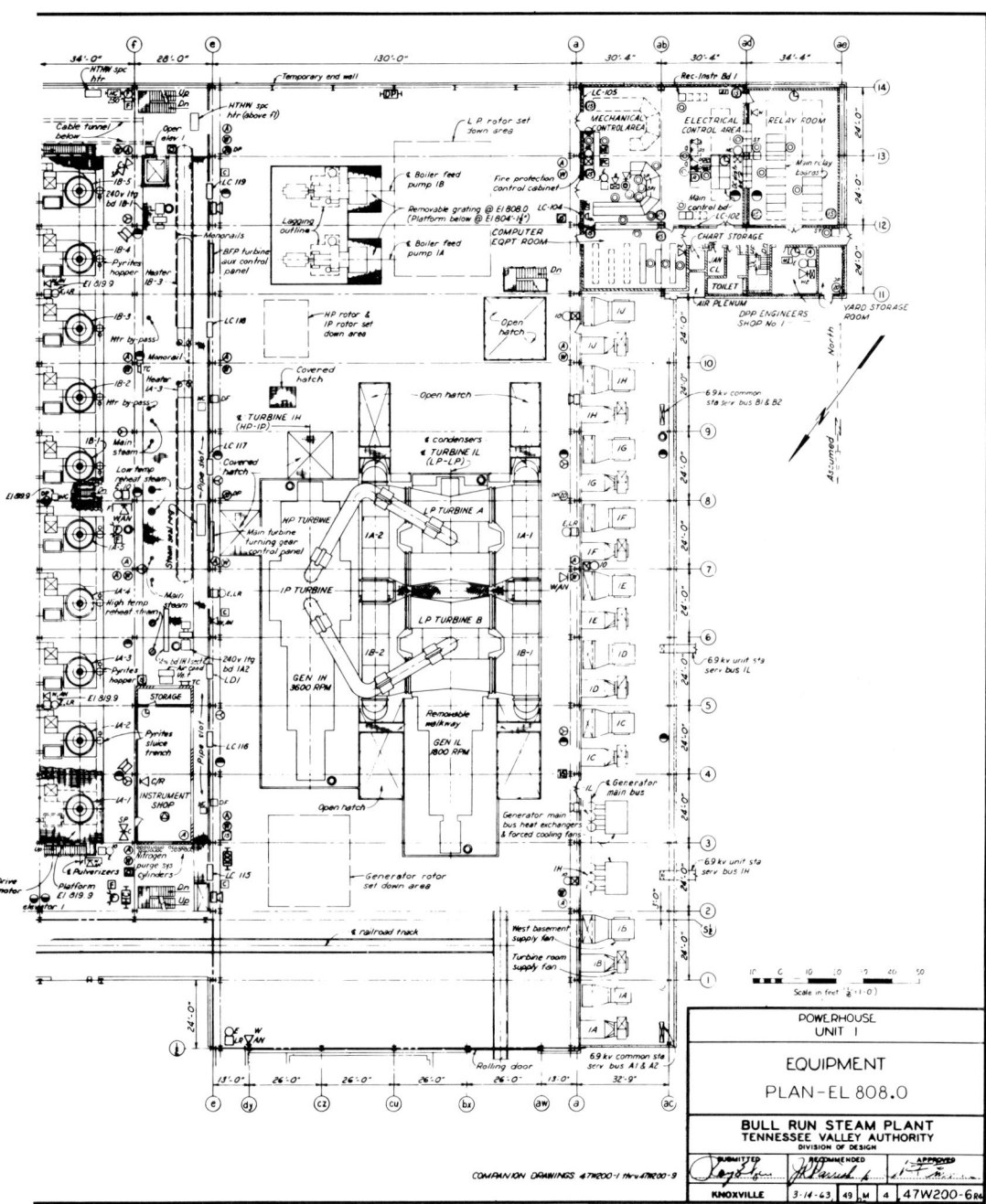

Table 2-2. The special problems of stack gas scrubbers and fluidized-bed furnace-steam generators are treated in later sections of this chapter.

TYPICAL PLANTS

The size, complexity, and cost of modern steam power plants can be appreciated only by a fairly detailed look at the principal components in the system coupled with visits to plants during both construction and operation. In 1980 the cost of a 1000-MWe power plant unit was approaching a billion dollars, and the size and variety of the equipment required are even more impressive than the cost. Further, in developing and appraising conceptual designs for advanced power plants, it is often helpful to make comparisons between the equipment required and that of existing power plants; hence fairly detailed descriptions of some typical examples are presented in this section. The first, the TVA Bull Run plant, is one of the 10 most efficient units in the world and is widely used as an example in engineering studies and comparisons. The second example is the TVA Sequoyah Unit No. 1 PWR nuclear plant. The third is a high-temperature gas-cooled reactor (GCR), and the fourth is a liquid-metal-cooled fast-breeder reactor (LMFBR).

Pulverized-Coal-Fired Supercritical-Pressure Plant

At the time it began commercial operation in 1967, the 950-MWe Bull Run steam plant was the largest steam-boiler-turbine-generator unit in the world with many new features in the detail design of its components. Thus it was not surprising that the solution of numerous mechanical problems required several years of shakedown operation. These problems were solved, and at the time of writing the plant is considered to be one of the most reliable and efficient coal-fired plants in the country.

A flowsheet for the plant is presented in Fig. 2-1, and both a vertical section through the plant and a plan view are shown in Figs. 2-7 and 2-8. Diagrams outlining the coal yard and fuel

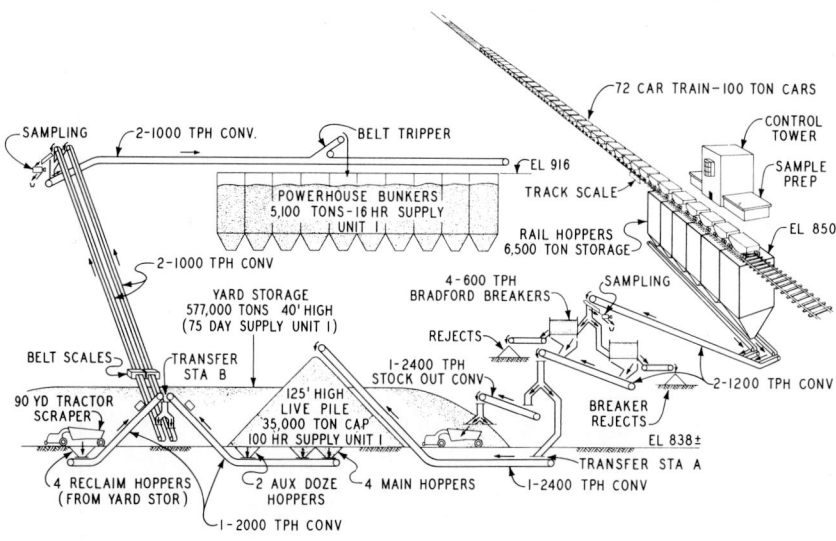

Figure 2-9 Schematic diagram for the entire coal-handling operation of the TVA Bull Run plant. The coal arrives from the coal field in 72-car unit trains at the right, is handled and processed, and is delivered to the bunkers beside the furnaces shown near the top of the diagram.[17] (*Courtesy TVA.*)

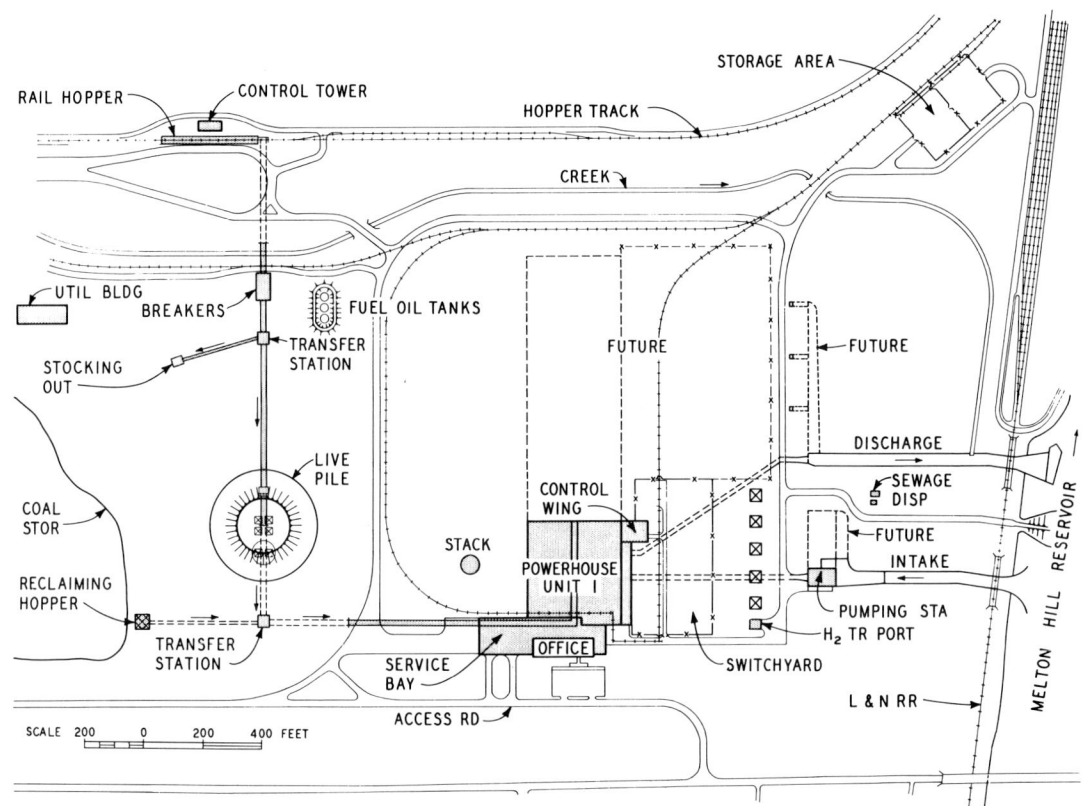

Figure 2-10 Site layout for the TVA Bull Run Power Plant. (*Courtesy TVA.*)

supply system are given in Figs. 2-9 and 2-10. Table 2-3 summarizes the principal design parameters. (This table represents 7 of 35 pages of data in Ref. 17.) The author has found that this table and these figures have provided invaluable data for making equipment comparisons in design studies and analysis work on new types of power plants.

A detailed account relating the functions of the various items in Fig. 2-1, Figs. 2-7 to 2-9, and Table 2-3 is probably not needed by most readers, would require many pages, and does not appear necessary. Readers not familiar with steam plants will probably find it more fruitful to take the time to trace through the drawings and table themselves and thus become familiar with the principal components of the plant and their size, costs, and interrelationships.

In making comparisons with other fossil-fuel-fired steam plants, it should be remembered that oil- and gas-fired units do not require coal- and ash-handling equipment that consume a substantial amount of power and increase both capital and operating costs. On the other hand, the stack loss chargeable to the heat of vaporization of the H_2O in the combustion products increases with the H_2 content of the fuel so that, while coal-fired furnaces have efficiencies of ~ 90 percent, the values for oil- and

TABLE 2-3 Design and Construction Data for the 950-MWe Pulverized-Coal-Fired Bull Run Power Plant Designed, Built, and Operated by TVA near Knoxville, Tenn.[17]

TURBOGENERATOR

Number of units installed (HP and LP)—1
Manufacturer—General Electric Company
Maximum generator nameplate rating—950,000 kW
Turbine, type—horizontal, impulse reaction, cross compound, 4-flow exhaust extraction
  3600-rpm shaft—HP and IP turbines
  1800-rpm shaft—LP turbines with 4-flow exhaust and 43-in last-stage buckets, 22 stages total
Generator, type and maximum nameplate rating—2 direct connected, hydrogen-cooled rotor, water-cooled stator, 527,778 kVA, 0.9 pf, 45 psig H_2, 3 ph, 60 cps, 24,000 V, 12,696 A, 0.640 minimum test scr, Y-connected
  Temperature rise—stator 45°C; rotor 74°C
Exciter, type and capacity—2 shaft driven, HP 1880 kW, 500 V, 603 rpm, and LP 970 kW, 500 V, 892 rpm
Spacing of shafts (HP and LP)—50 ft 0 in
Overall dimensions above turbine room floor
  Length—HP 106 ft 7¼ in; LP 132 ft 3⅛ in
  Width—HP 28 ft 0 in; LP 27 ft 10 in
  Height (HP and LP)—22 ft (approximate)

Low-Pressure Turbine Weights, lb

Diaphragms (upper and lower combined)

Low pressure A—210,160
Low pressure B—210,160

Oil tank

Without oil—52,240
Weight of oil—48,375

Rotors (with buckets and wheels)

Low pressure A—281,950
Low pressure B—292,910

Rotors (with 7th and 8th stage buckets removed)

Low pressure A— 258,935
Low pressure B— 269,895
  Subtotal   1,624,625

Inner shells, A or B

	Unassembled	Diaphragms Stages	Diaphragms Total	Assembled
Turbine end, inner casing, upper	8,415	—	—	8,415
Turbine end, inner casing, lower	16,110	—	—	16,110
Generator end, inner casing, upper	8,415	—	—	8,415
Generator end, inner casing, lower	16,110	—	—	16,110
Turbine inner shell, upper	5,525	2-4	20,700	26,225
Turbine inner shell, lower	5,580	—	—	5,580
Generator inner shell, upper	5,525	2-4	20,700	26,225
Generator inner shell, lower	5,580	—	—	5,580
Turbine packing casing cone extension, upper	4,465	—	—	4,465
Turbine packing casing cone extension, lower	4,465	—	—	4,465
Gen packing casing cone extension, upper	4,465	—	—	4,465
Gen packing casing cone extension, lower	4,465	—	—	4,465
Subtotal				130,520

Total weight

Estimated total weight of all LP turbine parts—3,275,000

Exhaust hoods A and B

	Unassembled	Inner casing	Diaphragms	Steam guides	Assembled
Upper turbine end	86,250	—	—	—	86,250
Generator end	86,250	—	—	—	86,250
Middle	27,330	—	—	—	27,330
(Middle) inner	88,565	16,830	60,670	14,210	180,275
Lower turbine end (A)	136,365	—	—	—	136,365
Lower turbine end (B)	137,890	—	—	—	137,890
Generator end (A)	141,500	—	—	—	141,500
Generator end (B)	135,115	—	—	—	135,115
Middle	29,140	—	—	—	29,140
(Middle) inner	101,035	32,220	—	14,210	147,465
Subtotal					1,107,580

TURBOGENERATOR (Continued)

High-Pressure Turbine Weights, lb

Outer shell

	Unassembled	Inner shells	Misc.	Assembled
HP, upper	99,895	—	19,275	119,170
HP, lower	100,675	46,085	—	146,760
Reheat, upper	97,565	—	17,775	115,340
Reheat, lower	100,095	23,880	—	123,975
Subtotal				505,245

Inner shell

	Unassembled	Misc.	Assembled
HP, upper	30,685	870*	31,555
HP, lower	34,765	6180*	40,945†
No. 2, upper	5,040	—	5,040
No. 2, lower	5,140	—	5,140
No. 1 reheat turbine, upper	5,605	—	5,605
No. 1 reheat turbine, lower	5,660	—	5,660†
No. 2 reheat turbine, upper	6,240	—	6,240
No. 2 reheat turbine, lower	6,280	—	6,280†
No. 1 reheat generator, upper	5,605	—	5,605
No. 1 reheat generator, lower	5,660	—	5,660†
No.2 reheat generator, upper	6,240	—	6,240
No. 2 reheat generator, lower	6,280	—	6,280†
Subtotal			130,190

*Control, intercept, etc., when mounted on shell
†No diaphragms included

Diaphragms

High pressure—15,270
Reheat pressure—27,800

Oil tank

Without oil—48,160
Weight of oil—40,875
360-degree nozzle box, estimated—3740

Steam chest

A—89,790
B—90,100
Combined reheat stop valves—108,400

Rotors (with buckets and wheels)

HP—38,105
IP, double flow—51,445
Hydraulic enclosure (servomotors)— 24,000
Subtotal   538,785

Total weight

Estimated total weight of all HP turbine parts—1,380,000

TURBINE

Exhaust size (inside dimensions)—Eight at 20 ft 9½ in by 11 ft 3½ in
Number of steam chests—2
Number of steam control valves—8 (4 in each chest)
Number of stop valves—4
Number of combined reheat valves (stop and intercept)—2
Type—horizontal
Throttle pressure—3500 psig
Throttle temperature—1000°F
Reheat temperature—1000°F
Design back pressure—1½ in Hg abs
Extraction stages, five points—No. 1, 5th HP; No. 2, 8th HP exhaust; No. 4, 14th IP (exhaust); No. 6, 4th LP; No. 7, 6th LP; No. 8, 7th LP

Speed—No-load steam, 3500 psig, 800°F, at throttle and 4 in Hg abs exhaust conditions, full excitation, pph (approximate)
   speed-matching valve open—175,000
   speed-matching valve closed—95,000
Heat rate—Guaranteed performance based on extraction for feed-water heating, including all losses in the unit, also exciter and rheostat losses, rated throttle steam conditions, and 1½ in of Hg abs exhaust pressure, with zero makeup
   kW—914,402
   Btu per kWh—7142

TABLE 2-3 Design and Construction Data for the 950-MWe Pulverized-Coal-Fired Bull Run Power Plant Designed, Built, and Operated by TVA near Knoxville, Tenn.[17] (*Continued*)

TURBINE (*Continued*)

Turning gears

Number—2 (1 each HP and LP element)
Location—adjacent to each turbine-generator coupling
Type—electric motor-driven, transmitted through a link belt and a reducing gear train to the turbine shaft
Speed—2-3 rpm
Drive—15 hp for HP shaft, 20 hp for LP shaft, 3 ph, 60 cps, 120 rpm, 440 V ac

Main turbine oil tanks

Number—2 (1 each HP and LP elements)
Lubricating oil requirements, gal:

	HP	LP
Oil tank capacity	5450	6100
Piping runback	2100	1800
Seal oil system	300	300
Trap drain	350	350

Total lubricating oil—16,750 gal

Vapor extractors

Number—2 (1 each HP and LP tanks)
Location—one mounted externally on roof of each turbine oil tank
Type—electric motor driven, horizontal, gear-type blower
Drive—3 hp, 3 ph, 60 cps, 900 rpm, 440 V ac

Lubricating oil coolers

Number—4 (2 each for HP and LP tanks)
Manufacturer—General Electric Company
Location—supported at roof of main turbine oil tank with shell inside tank and main head outside
Type—vertical tank, straight tubes (through-type), water-cooled, 2-pass
Design data—surface 2080 ft^2, 1272 tubes, length 9 ft 11$\frac{1}{4}$ in; $\frac{5}{8}$-in OD, No. 18 Bwg inhibited admiralty, aluminum bronze tube sheets; cast-iron channel; steel cover; steel shell; 40-psig operating pressure, oil side; 125-psig operating pressure, water side
Weight of tube bundle (estimated), lb—5800
Cooling water required, raw—HP 1700 gpm; LP 1980 gpm at 90°F inlet water temperature

Shaft seals (steam)

Number of packings—8 (4 each HP and LP element)
Sealing steam required at startup, normal (including boiler feed-water pump turbine requirements)— 14,000 pph
Sealing steam leakoff, normal (including boiler feed-water pump turbine)— 13,600 pph
Air leakage to steam packing exhauster (including boiler feed-water pump turbine)— 1900 pph

Pumps, turbine accessory

Service	No.	Location	Type	Drive hp	ph	cps	rpm	V ac
Main oil HP element	1	(1)	(6)	........................(13)........................				
Main oil LP element	1	(2)	(7)	........................(14)........................				
Auxiliary oil	2[11]	(3)	(8)	250	3	60	3600	440
Booster oil	2[11]	(4)	(9)	........................(15)........................				
Turning gear oil	2[11]	(3)	(9)	30	3	60	1800	400
Emergency bearing oil	2[11]	(3)	(9)	30	—	—	1700	240 dc
Lift oil	6[12]	(5)	(10)	5	3	60	1200	440

Location:
1. Mounted on HP turbine shaft.
2. Mounted on LP turbine shaft.
3. One pump in each turbine oil tank with electric motor drive mounted externally on tank roof.
4. One in each turbine oil tank.
5. Two packages of three pumps and motors each; assembly A for bearing Nos. 1, 2, and 3 mounted on platform El. 780.5 adjacent to LP oil tank; assembly B for bearing Nos. 4, 5, and 6 mounted on floor El. 790 adjacent to No. 5 bearing of LP element.

Type:
6. Horizontal, double suction, single stage, centrifugal.
7. Horizontal, single suction, 2 stage, centrifugal.
8. Vertical centrifugal, single stage.
9. Vertical, centrifugal.
10. Horizontal, axial piston, constant volume.

Data:
11. One each HP and LP tanks.
12. LP element only.   14. LP turbine shaft.
13. HP turbine shaft.   15. Oil turbine.

TURBINE (*Continued*)

Generators

	HP generator	LP generator		HP generator	LP generator
Speed	3600 rpm	1800 rpm	Main exciters		
Total losses, kW @ 37 psig H_2 (at			Drive	gear driven	gear driven
500,000 kVA, 0.9 pf)	5732	4714	Rating—kW	1880	970
Reactances (based on 500,000 kVA)			—V	500	500
Direct axis synchronous	160%	164%	—rpm	603	892
Direct axis subtransient	24%	30.5%	Pilot exciters		
Negative sequence	16%	16.5%	Type	amplidyne	amplidyne
Zero sequence	13%	12.0%	Rating—kW	10	3
Resistance at 75°C			—V	250	250
Stator per phase	0.00295 Ω	0.00311 Ω	Motor—hp	25	10
Field	0.11600 Ω	0.22100 Ω	Neutral grounding transformer		
Negative sequence	2.10%	2.50%	kVA	75	75
Field currents			V	24,000/220	24,000/220
Full load—1.0 pf	2630 A	1400 A	Neutral resistor		
—0.9 pf	3400 A	1760 A	Amp (1 min)	345	440
Short-circuit ratio test values	0.640	0.648	Ohm	0.4	0.29
Hydrogen treatment	vacuum detraining	vacuum detraining			

Voltage regulators—automatic high-speed, continuous-acting, dynamic type
Surge protection—lightning arresters only

	HP generator	LP generator
Hydrogen cooling		
Number of coolers	4	8
Cooling water (gpm at 90°F)	1500	1400
Gas space in generator casing	3600 cu ft	2500 cu ft
Length of generator	30 ft 0½ in	32 ft 11 in
Weight of principal parts		
Stator	269.79 tons	258.66 tons
Rotor	58.4 tons	145.9 tons
Resistance temperature detectors (number installed)		
Stator windings	48	60
Stator gas ducts	8	16
Thermocouples (number installed)		
Stator cooling water	48	60
Stator iron	28	28
Stator gas ducts	8	16

STEAM GENERATOR

General data

Manufacturer—Combustion Engineering, Incorporated
Type—supercritical, combined circulation, twin divided furnaces, reheat
Rated capacity, pph—6,400,000
Steam pressure (at superheater outlet), psig—3650
Steam temperature (at superheater outlet), °F—1003
Efficiency (guaranteed at rated load), %—90.08

Furnaces

Type—pressurized, twin, waterwall, dry bottom
Principal dimensions—60 ft 9¼ in wide by 29 ft 9 1/16 in deep by 92 ft high (each furnace)
Total heating surface, ft^2—74,590
Total volume, ft^3—382,000

TABLE 2-3 Design and Construction Data for the 950-MWe Pulverized-Coal-Fired Bull Run Power Plant Designed, Built, and Operated by TVA near Knoxville, Tenn.[17] (*Continued*)

STEAM GENERATOR (*Continued*)

Superheaters
Type—4-stage, horizontal, partition panel, platen, and pendant
Tube size—2 × $1\frac{3}{4}$ in OD
Heating surface, ft^2—255,200
Design pressure, psi—3840
Design temperature, °F—1003

Boiler recirculating pumps (4 total, 2 spares)
Type—canned motor
Capacity, gpm—6500
Manufacturer—Westinghouse Electric Corporation
Motors—375 kW

Economizers
Type—continuous-loop, finned-tube
Tube size—2-in OD
Design pressure, psig—4340
Total heating surface, ft^2—270,000

Reheaters
Type—three-stage, horizontal, inlet and outlet pendant
Tube size—$2\frac{1}{2}$- and $2\frac{1}{8}$-in OD
Rated capacity, pph—4,500,000
Design pressure, psig—725
Design temperature, °F—1003
Operating inlet pressure, psig—575
Operating outlet pressure, psig—545
Operating inlet temperature, °F—552
Operating outlet temperature, °F—1003
Heating surface, ft^2—153,000

Air preheaters
Number—4
Manufacturer—Air Preheater Corporation
Type and size—Ljungstrom, regenerative, counterflow, $29\frac{1}{2}$ HX
Heating surface, each, ft^2—302,800
Element length—$75\frac{1}{2}$ in (hot end 36 in; intermediate $27\frac{1}{2}$ in; and cold end 12 in)
Design temperature, gases
  Entering, °F—656
  Leaving, °F—290
Design temperature, air
  Entering, °F—110
  Leaving, °F—591

Firing equipment
Burners—80, tilting tangential
Pulverizers—10, Raymond No. 843RPS
Feeders—10, volumetric
Pilot oil torches—48
Burner oil guns—48, retractable

Boiler safety valves
Number and size
  Intermediate superheater—three 3 in; three $2\frac{1}{2}$ in
  Superheater outlet—four $2\frac{1}{2}$ in
  Low temperature reheat—twelve 4 in
  High temperature reheat—three 4 in
Manufacturer—Consolidated
Type—maxiflow

Power control valves
Number and size
  On superheater outlet—two $2\frac{1}{2}$ in
Manufacturer—Consolidated
Type—electromatic

Startup system safety valves
Number and size
  On separator spillover line—two 4 in
Manufacturer—Consolidated
Type—maxiflow

Soot blowers
Number installed
  Long retractable—20
  Waterwall deslaggers—72
Blowing medium—air, with steam standby
Drive—electric motor
Manufacturer—Copes-Vulcan
Control—automatic sequential or manual
Blowing pressure—125-200 psig

COAL FEEDERS

Number—10
Type—volumetric
Drive—variable speed motor
Manufacturer—Stock Equipment Company

COAL VALVES

Number—10
Size—24 in
Type—slide gate, rack and pinion drive, motor operated
Manufacturer—Fairfield Engineering Company

FORCED DRAFT FANS

Number—4
Manufacturer—Buffalo Forge Company
Type—double width, double inlet
Capacity, each (at test block)—545,000 cfm
Static pressure (at test block)—32 in water
Temperature (at test block)—120°F
Control—variable speed (fluid drive) and inlet dampers
Motor—3500 hp, General Electric

FLY ASH COLLECTORS

Number—4
Manufacturer—American-Standard Industrial Division
Type—electrostatic
Capacity, each—594,000 cfm
Efficiency—99 %
Pressure drop—0.5 in water

CONDENSING EQUIPMENT

Condenser

Number—1 (2 twin shells)
Manufacturer—Foster Wheeler Corporation
Type—horizontal, twin shell, modified double flow, single pass, side inlet
Location—axial, along each side of two, double flow, LP turbines
Surface, sq ft—320,000
Tube data—41,412 (3552 90-10 cupronickel; 37,860 inhibited admiralty) tubes, 33-ft $8\frac{1}{4}$-in effective length, approximate weight of tubes 683,000 lb, $\frac{7}{8}$-in OD, No. 18 Bwg
Shell-and-tube support plates—36 (9 per section), $\frac{3}{4}$-in copper bearing plates
Tube sheets—8 (2 per section), $1\frac{1}{2}$-in rolled steel
Waterboxes—nondivided, design pressure of 40 psi, steel, 2 inlet (6 by 11 ft inside) and 4 outlet (78-in diam) bottom connections
Hotwell data—full rectangular pattern, deaerating-type, storage capacity hotwell B sections 5500 gal each, A sections 11,000 gal each

Total weight (approximate), lb		
Empty—2,640,000		
Operating—3,520,000		
Condenser performance		
Tube cleanliness, %	85	100
Steam load, pph	3,787,000	3,787,000
Btu rejected per lb steam	940.3	940.3
Absolute pressure, in Hg	1.30	1.18
Circulating water temperature, °F	55	55
Circulating water flow, gpm	397,500	397,500
Water velocity, fps	6.5	6.5
Friction loss through water passages, ft	10.95	10.95
Maximum oxygen content of effluent, cc per liter	0.01	0.01

TABLE 2-3 Design and Construction Data for the 950-MWe Pulverized-Coal-Fired Bull Run Power Plant Designed, Built, and Operated by TVA near Knoxville, Tenn.[17] (Continued)

CONDENSING EQUIPMENT (Continued)

Vacuum pumps

Number—4
Manufacturer—Nash Engineering Company
Type—Model H-8 Nash rotary vacuum pump with atmospheric air ejector

Operating performance
  Suction pressure, in Hg abs—1.0
  Suction temperature, °F—74.0
  Rated capacity, each—7.5 scfm
  Motor hp—50
  Motor manufacturer—General Electric Company

CONDENSATE AND FEED-WATER SYSTEMS

Deaerating heater and storage tank

Number—1
Manufacturer—Graver Water Conditioning Company
Type—nonmetering, cylindrical, spray tray, nonstorage-type, horizontal heater, and horizontal storage tank
Location—storage tank supported at El. 856.0 floor in heater bay; heater supported on storage tank but accessible from El. 874.0 floor
Performance—under conditions of service outlined below, the heater will deliver 6,335,150 pph, heated to a temperature equal to that of saturated steam in the heater and containing not more than 0.005 cc per liter of dissolved oxygen; when operated at overload conditions and delivering up to a maximum of 6,650,000 pph for short periods of peaking operation, the oxygen content of the fluid will not exceed 0.03 cc per liter of dissolved oxygen
Conditions of service—performing under the conditions of service given above, the proportions of water and steam quantities are as follows
  Condensate, pph (rated)—4,368,490
  HP heater drains, pph (rated)—1,624,460
  Extraction steam to heater, pph (rated)—342,200
Construction details

	Heater	Storage tank
Shell diameter	11-ft 0-in OD	12-ft 0-in OD
Length (overall)	37 ft 10 in	94 ft 5 in
Material	Copper bearing steel plate	
Thickness of heads	$\frac{11}{16}$ in	$\frac{11}{16}$ in spherical
Thickness of shell	$\frac{9}{16}$ in	$\frac{1}{2}$ in
Design pressure shells, psi	85	85
Design temperature, °F	650	650

Capacity of deaerator storage tank (minimum)—48,000 lb at 304.4°F
Weight, lb
  Entire unit, dry—192,500
  Unit at maximum operating level—785,000
  Complete unit, flooded—990,000

Flash evaporator

Number—1
Manufacturer—Westinghouse Electric Corporation
Type—horizontal, single stage, flash type
Location—turbine bay, on El. 790 floor
Approximate weight, lb
  Empty—60,000
  Operating—78,000
  Flooded—129,500
Design data
  Heater section (integral with evaporator condenser and flash chamber)
    Total surface, ft²—4710

	Shell	Tubes
Design pressure, psig	15.0*	50
Design temperature, °F	300	300
Test pressure, psig	22.5	75
Quantity fluid, pph	60,000 (steam)	2,625,000 (brine)
Tubes, OD, in		$\frac{3}{4}$-18 Bwg
Number		630
Average effective length		$\frac{3}{4}$-18 Bwg
Material		Admiralty
Pitch, in		1 Tri
Passes		1

*At 30 in Hg vac

CONDENSATE AND FEED-WATER SYSTEMS (Continued)

Condenser section (integral with evaporator heater and flash chamber)
Total surface, ft² — 3200

	Shell	Tubes
Design pressure, psig	15.0	300
Design temperature, °F	300	300
Test pressure, psig	22.5	450
Quantity fluid, pph	60,000	4,368,000
Tubes, OD, in		$1\frac{1}{2}$-22 Bwg
Number		214
Average effective length		38 ft 0 in
Material		304 SS
Pitch, in		$1\text{-}\frac{3}{4}$ Tri
Passes		1

Flash evaporator chamber (knitted wire mesh mat moisture separators between flash chamber and evaporator condenser)

Injection water coolers

Number — 2
Manufacturer — The Whitlock Manufacturing Company
Type — closed, horizontal, straight tube, 2 pass
Location — turbine bay floor El. 770.0
Design performance data

	Shell	Tubes
Total effective surface, sq ft		299
Medium flowing	injection water	cooling water
Quantity, gpm	90	255
Design temperature, °F	350	350
Design pressure, psig	450	100
Test pressure, psig	675	150
Shell ID, in	10.136	
Tubes OD, in		$\frac{5}{8}$-22 Bwg
Material	steel	304 SS
Pitch, in		0.833 Tri
Number		80
Length, in		276

Shell and bundle weight, dry, lb — 2800
Shell and bundle weight, flooded, lb — 3650

Auxiliary deaerator

Number — 1
Manufacturer — Allis-Chalmers Manufacturing Company
Type — nonmetering contact, spray tray type, vertical heater with integral storage capacity
Location — floor El. 856.0 in heater bay
Performance — The heater will deliver 180,000 pph, heated to a temperature equal to that of saturated steam at operating pressures from 1 to 10 psig with inlet distilled water at temperatures varying from 40° to 212°F, and inlet steam enthalpy of 1140 Btu per lb and containing not more than 0.005 cc per liter of dissolved oxygen
Construction details
  Shell diameter — 8 ft 0 in OD
  Length (overall) — 12 ft $8\frac{7}{8}$ in
  Material — steel plate
  Thickness of heads, in — $\frac{5}{8}$
  Thickness of shell, in — $\frac{7}{16}$
  Design pressure, psi — 85
  Storage capacity, gal — 1800
  Weight, lb
    Entire unit, dry — 11,900
    Unit at operating level — 28,000
    Unit flooded — 47,000

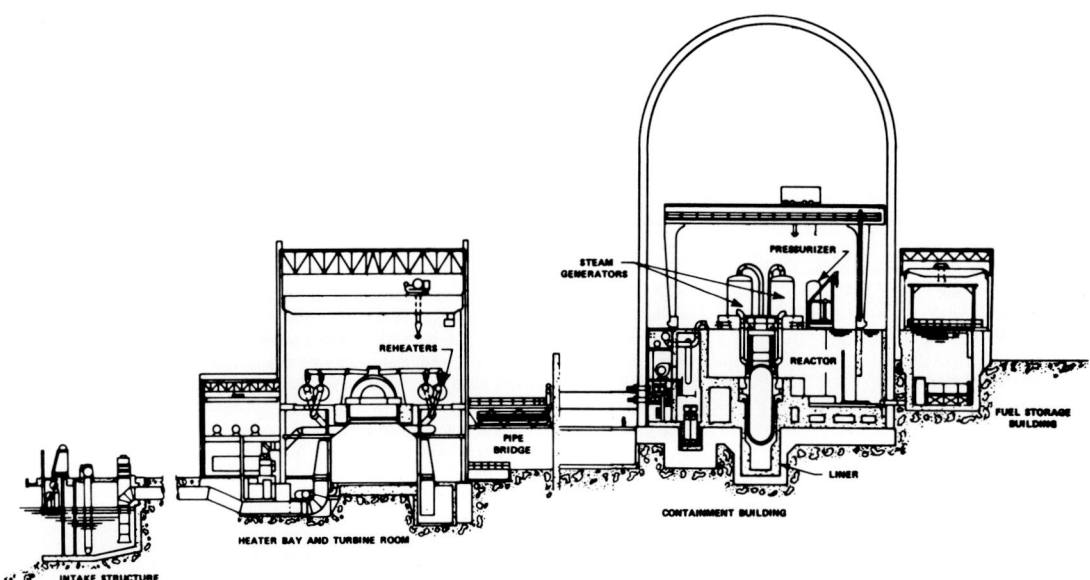

Figure 2-11 Longitudinal section through a typical PWR power plant. (*From "Pressurized Water Reactor Fundamentals Manual," Inspection and Enforcement Training Center, NRC.*)

gas-fired units are ~87 and ~85 percent, respectively, when calculated on the basis of the lower heating value of the fuel.

Light-Water Reactor Plants

TVA, the largest electric utility system in the United States, began the design and construction of nuclear plants later than many large utilities, but by 1990 about half of its power output will be nuclear. The first plant, that at Brown's Ferry, consists of three boiling-water reactor (BWR) units of 1097 MWe each. That plant is in operation at the time of writing and, in spite of high capital costs stemming in part from the facts that these units represented a new type of plant for TVA and were larger than any previous BWR units, this plant produces power at a substantially lower cost than any of the TVA coal-fired plants, and it is not vulnerable to coal shortages caused by strikes or other problems. A second, and, from the potential-accident-containment standpoint, more advanced TVA nuclear installation is the Sequoyah pressurized-water reactor (PWR) plant that consists of two 1125-MWe units, the first of which began operation in 1980. The design of a nuclear plant is dominated by reactor safety considerations and provisions for the containment of radioactivity that might otherwise be released in the event of an accident. Because of this, and because of their interesting reactor safety characteristics, one of these units was chosen as an example for inclusion in this section.

The Sequoyah PWR units are the first water reactors to make use of an ice condenser system in a free-standing steel containment vessel as a means for absorbing the energy that might conceivably be released in the form of steam by a rupture in the primary reactor cooling system.[37] There is so much energy in the sensible heat in the inventory of high-temperature water in the reactor cooling system that, if the pressure were relieved by a rupture in the system, the steam formed would produce a high pressure in the containment building. It is for this reason that reactor containment structures are commonly made in the form of

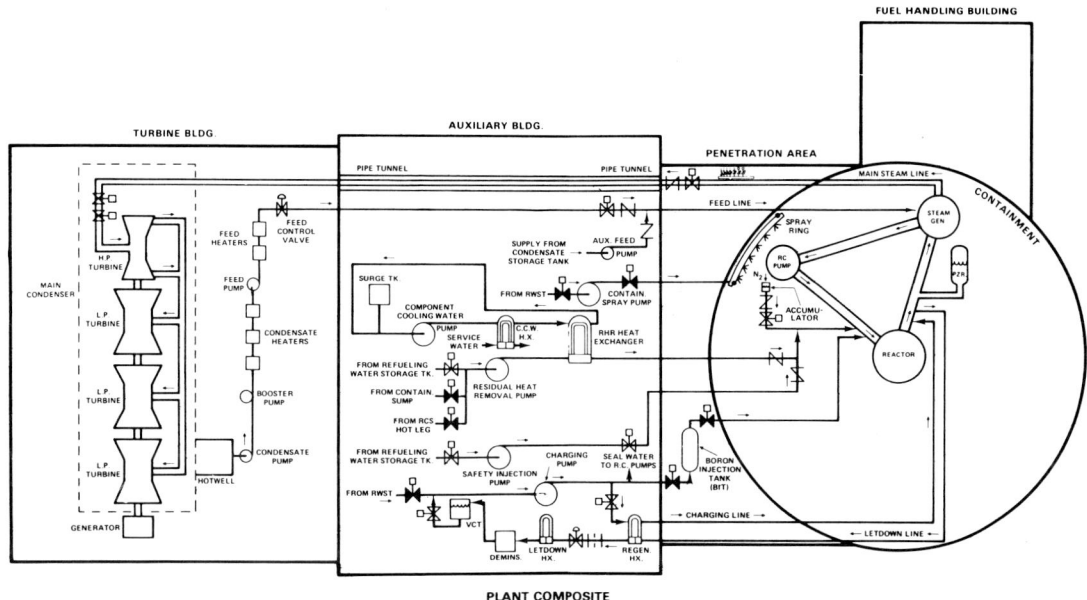

Figure 2-12 Schematic diagram of a pressurized-water reactor plant showing the way in which the principal components are coupled. (*From "Pressurized Water Reactor Fundamentals Manual," Inspection and Enforcement Training Center, NRC.*)

either spherical or vertical cylindrical shells with hemispherical or ellipsoidal heads. (The first such structure was a 200-ft-diameter sphere built by the General Electric Company near Schenectady in the latter 1940s for testing the prototype sodium-cooled reactor subsequently used in the Sea Wolf submarine.) The pressure that might conceivably build up in the containment vessel in the event of the maximum credible accident can be reduced drastically by using a chilled water spray or by a heat-transfer matrix made of ice. In the Sequoyah plant the ice is in the form of an array of vertical columns similar to a bank of tubes in a shell-and-tube heat exchanger. These units are arranged in the containment shell along with the other reactor components. The size and overall cost of the containment system can be reduced substantially through the use of this ice condenser system.

The longitudinal section of Fig. 2-11 and the schematic diagram of Fig. 2-12 show the way in which the principal components of a PWR plant are coupled. As is the case for coal-fired plants, numerous detailed sheets are required to show the host of lesser components. The steam system is similar to that for a coal-fired plant except for the steam generator. Table 2-4 gives the principal design data for the plant.

A familiarity with and an appreciation for the relations between the principal components, their characteristics, proportions, and general construction are probably best obtained by a detailed examination of Figs. 2-11 and 2-12 and Table 2-4. This, coupled with a visit to a nuclear plant (preferably in the latter stages of construction when the key components can be seen before they become radioactive), the reader will probably find more effective than 10,000 words on the subject—which the space available in this text does not permit.

Gas-Cooled Reactor Plants

The evidence of the marketplace is that light-water reactors give lower costs but a poorer neutron economy, whereas

TABLE 2-4 Summary of Design Data for the 1125-MWe PWR Unit No. 1 of the TVA Sequoyah Power Plant near Chattanooga, Tenn.[37]

Power	
Net electric output	1125 MWe
Gross electric output	1171 MWe
Gross thermal output	3411 MWt
Reactor core	
Core diameter (equivalent)	3.40 m (133.7 in)
Core height (active)	3.66 m (144 in)
Number of fuel assemblies	193
Fuel pin lattice pitch	14.3 mm (0.563 in)
Average thermal output	589,200 kcal/(m²·h)
	217,200 Btu/(ft²·h)
Maximum thermal output	1,573,200 kcal/(m²·h)
	579,600 Btu/(ft²·h)
Weight of fuel as UO_2	97.6 te (215,400 lb)
Fuel assemblies	
Fuel material	UO_2
Pellet diameter	9.29 mm (0.366 in)
Clad material	Zr-4
Clad thickness	0.61 mm (0.024 in)
Pin diameter	10.7 mm (0.422 in)
Number of pins per assembly	204
Maximum fuel central temperature	2282°C (4140°F)
Maximum clad surface temperature	247°C (657°F)
Feed enrichment (equilibrium)	3.2%
Fuel discharge burnup (equilibrium)	31,000 MWd/t
Control rods	
Neutron absorber	Ag In Cd
Cladding material	S.S. type 304
Number, full length	53
part length	8
Shape	Rod cluster
Length of poison section	3.62 m (142.7 in)

Primary coolant system	
Type	Forced circulation
Operating pressure	158 kg/cm² (2250 psia)
Reactor inlet temperature	285°C (545°F)
Reactor outlet temperature	321°C (610°F)
Coolant pumps	4
Total reactor flow	61×10^6 kg/h (134×10^6 lb/h)
Reactor pressure vessel	
Inside diameter	4.39 m (173 in)
Inside height	12.6 m (495 in)
Wall thickness (core region)	219 mm (8.625 in)
Material	ASTM A-508 Class II
Design pressure	176 kg/cm² (2500 psia)
Design temperature	343°C (650°F)
Containment building	
Type	Double (steel vessel, concrete shield)
Pressure suppression	Ice condenser
Design pressure	0.76 kg/cm² (10.8 psi)
Inside diameter (steel vessel)	34.4 m (115 ft)
Inside height (steel vessel)	47.5 m (156 ft)
Turbogenerator	
Rating	1220 MWe
Speed	1800 rpm
TSV pressure	55 kg/cm² (782 psi)
TSV temperature	268°C (514°F)

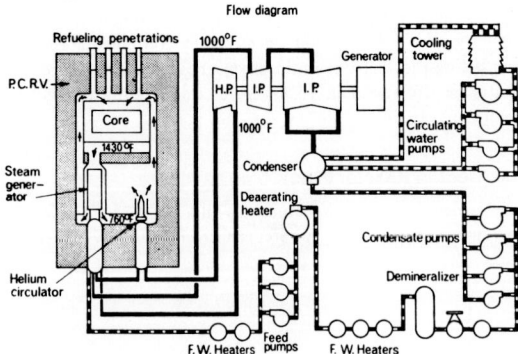

Figure 2-13 Flowsheet for the Fort St. Vrain GCR power plant. *(Courtesy General Atomic Corp.)*

sodium-cooled reactors show greater promise for breeding but give higher capital costs. As a consequence, although many gas-cooled reactors have been built in Great Britain, only a few have been built elsewhere. Only one large commercial gas-cooled reactor is in operation in the United States at the time of writing, and no others are planned. However, because of their excellent fuel economy potential, they may prove more attractive in the future. Hence it seems best to include one example, and thus the 330-MWe Fort Saint Vrain reactor is included here. The system is similar to that of a PWR. The principal differences are that the reactor core must be larger for a given power output, while the reactor system pressure is lower, and the reactor coolant outlet temperature is higher, permitting the steam system temperature to be as high as in a fossil fuel plant. A flowsheet is shown in Fig. 2-13, and the principal design data are given in Table 2-5. For a detailed description of this plant the reader is referred to Ref. 38.

TABLE 2-5 Summary of Design Data for the 330-MWe GCR of the Fort Saint Vrain Power Plant of the Public Service Co. of Colorado near Plattville, Colo.[38]

Capacity	
Net electric output	330 MWe
Gross generation	342 MWe
Overall station net efficiency	39.2%
Design life of plant	30 years

Reactor core	
Reactor output	841.7 MWt
Core diameter	5.9 m (19.5 ft)
Active core height	4.7 m (15.6 ft)
Number of fuel elements	1482
Element lattice pitch	360 mm (14.2 in)

Fuel	
Fuel material	Th/U^{235} (93% enriched)
Total quantity of thorium	19,458 kg
Total quantity of uranium	882 kg
Fuel form	Coated particles in cylindrical beds
Number of fuel elements per refuel region	42
Element (hexagonal across flats)	355 mm (14 in)
Element (length)	787 mm (31 in)
Fuel bed diameter	12.4 mm (0.491 in)
Coolant channel diameter	15.8 mm (0.625 in)
Burnup	100,000 MWd/t

Control	
Control rods	37 pairs
Active length	4.7 m (186 in)
Absorber material	B$_4$C/Graphite
Canning material	Incoloy
Shape	Hollow cylindrical
Drive, normal	Electric motor
Scram	Gravity

Thermal data	
Primary steam flow	2,305,300 lb/h
Primary steam pressure	168 kg/cm^2 (2400 psig)
Feed-water temperature	206°C (403°F)
Primary coolant flow	3.39 × 10^6 lb/h
Primary coolant pressure	49 kg/cm^2 (700 psia)
Coolant temp. reactor inlet	406°C (762°F)
Coolant temp. reactor outlet	785°C (1444°F)
Avg. heat flux	142 kW/m^2 (45,000 Btu/h · ft^2)
Max. heat flux	441 kW/m^2 (140,000 Btu/h · ft^2)
Max. fuel temperature	1260°C (2300°F)
No. of steam generator modules	12
No. of steam generators	2

Reactor vessel	
Type	Prestressed concrete
Internal clearance dimension	9.4 × 23 m (31 ft ID × 75 ft IH)
Max external dimension	(21 m × 32 m) (68 ft across corner × 106 ft high)
Normal working pressure	48 kg/cm^2 (688 psig)

Circulators	
Type	Axial-flow compressor with integral driver
Drive	Single-stage steam turbine
Flow control	Variable speed
Number of circulators	4 (2 per loop)
Rated steam flow	70 kg/s (155 lb/s/circulator)
Speed	9550 rpm
Compressor press. rise (helium)	1.0 kg/cm^2 (14 psi)
Compressor inlet temperature	395°C (742°F)
Power	5200 hp

Steam generators (per module)	
Total heat transfer	6 × 10^{11} Cal/h (2.4 × 10^8 Btu/h)
Bulk gas inlet temperature	776°C (1427°F)
Gas mass flow	128,000 kg/h (284,170 lb/h)
Superheater steam flow	87,000 kg/h (192,110 lb/h)
Superheater outlet pressure	2512 psia
Superheater outlet temp.	540°C (1005°F)
Reheater steam flow	84,000 kg/h (187,150 lb/h)
Reheater inlet pressure	178 kg/cm^2 (650 psi)
Reheater inlet temperature	356°C (673°F)
Reheater outlet pressure	42 kg/cm^2 (600 psi)
Reheater outlet temperature	539°C (1002°F)
Number of steam generator modules	12

Turbine generator	
Type	Cross-compound
Gross output	342 MWe
TSV pressure	168 kg/cm^2 (2400 psig)
TSV temp.	538°C (1000°F)
IP cylinder TSV pressure	399 kg/cm^2 (567.5 psia)
IP cylinder TSV temperature	538°C (1000°F)
Vacuum	63 mm Hg (2.5 in Hg)

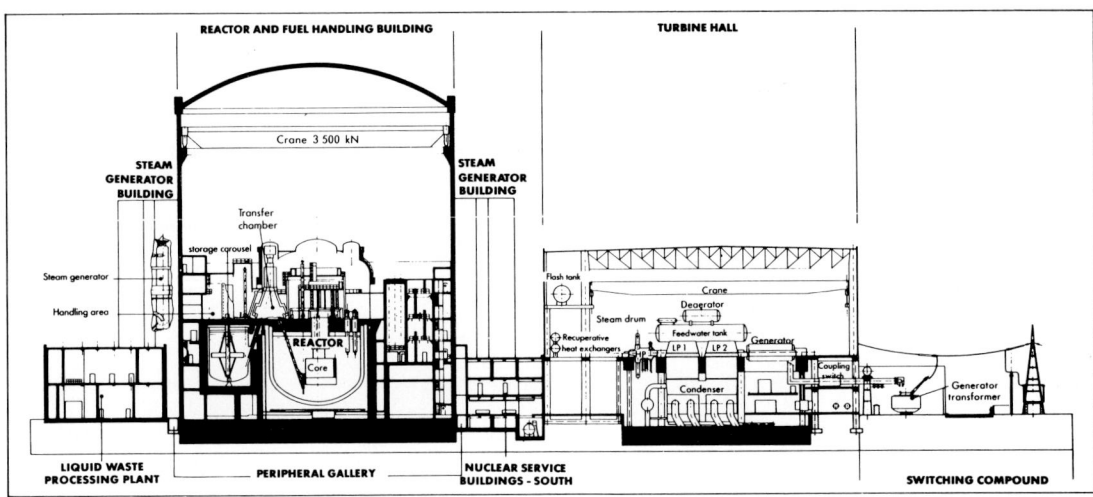

Figure 2-14 Longitudinal cross section of the Creys-Malville power station.[39] (*Courtesy Nuclear Engineering International.*)

Liquid-Metal-Cooled Fast-Breeder Reactor (LMFBR) Plant

The world's first full-scale commercial LMFBR plant is the 1200-MWe unit at Creys-Malville in southwestern France, scheduled for commissioning in 1983. The design of this plant was based on that of its prototype, the 250-MWe Phoenix (so named because, as a breeder reactor, it is designed to generate more fuel than it burns), which was started in 1973. Figures 2-14 to 2-17 show, respectively, vertical and horizontal sections through the plant, a flowsheet for the main power circuits, and a section through the steam generator. Table 2-6 gives the principal design data.[39] The materials limitations on the reactor operating temperature turn out to be such that the plant produces steam at almost as high a temperature as a fossil fuel plant.

The plant design is typical of most of the designs for LMFBRs in that both the reactor core and the intermediate heat exchangers (IHXs) are mounted within a single pressure vessel along with the sodium pumps for the primary cooling circuit. There are eight intermediate sodium circuits that transport the heat out to four steam generators, thus providing a high degree of redundancy to assure cooling of the reactor under all conditions. The reactor is surrounded by ~4 m of concrete shielding, and all the circuits containing radioactivity are housed within the reactor building, which is 64 m in diameter, 84 m high, and has reinforced-concrete walls 1 m thick. The concrete walls of the vaults for equipment containing sodium are lined with sheet metal so that liquid sodium could not contact concrete and react with moisture in the concrete. A nitrogen buffer gas region lies between the main vessel containing the primary sodium system and a safety tank that surrounds it, in effect providing a double-walled tank.

As was the case for the two previous plant designs presented in this section, it seems best to leave it to the reader to trace through the flowsheet, table, and the drawings to become familiar with the relationship of the various components and their characteristics. For further details refer to Ref. 39.

PRINCIPAL DESIGN PARAMETERS

Numerous compromises must be made in power plant design to balance improvements in thermal efficiency against increases in capital costs. These compromises differ, depending on the type of service expected, i.e., base load, intermediate load, or peaking power. Fossil-fuel base-load plants typically employ high

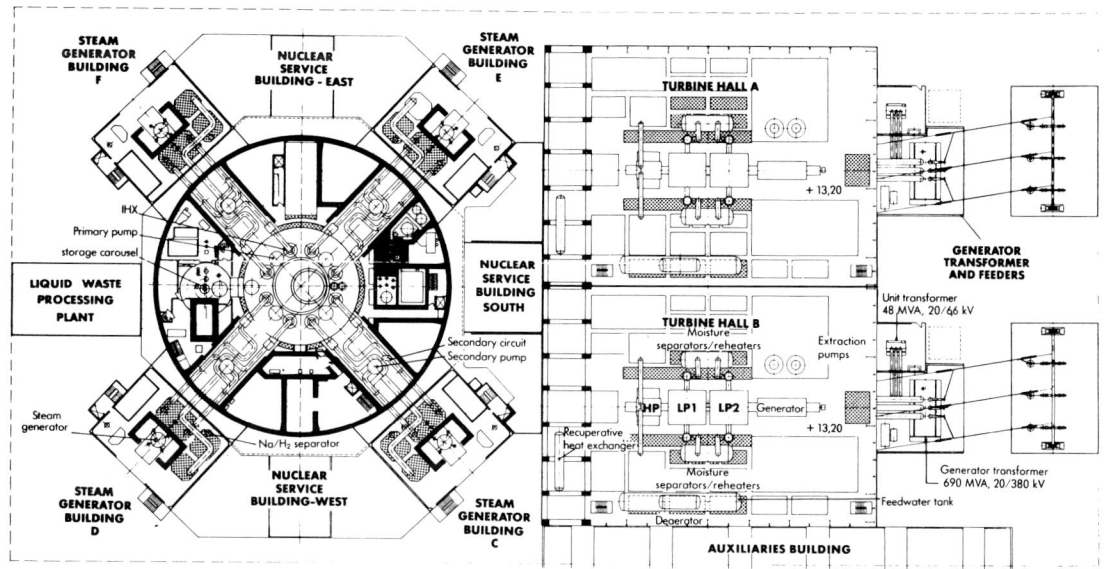

Figure 2-15 Plan view of the 1200-MWe LMFBR power plant at Creys-Malville on the Rhône River in France.[39] (*Courtesy Nuclear Engineering International.*)

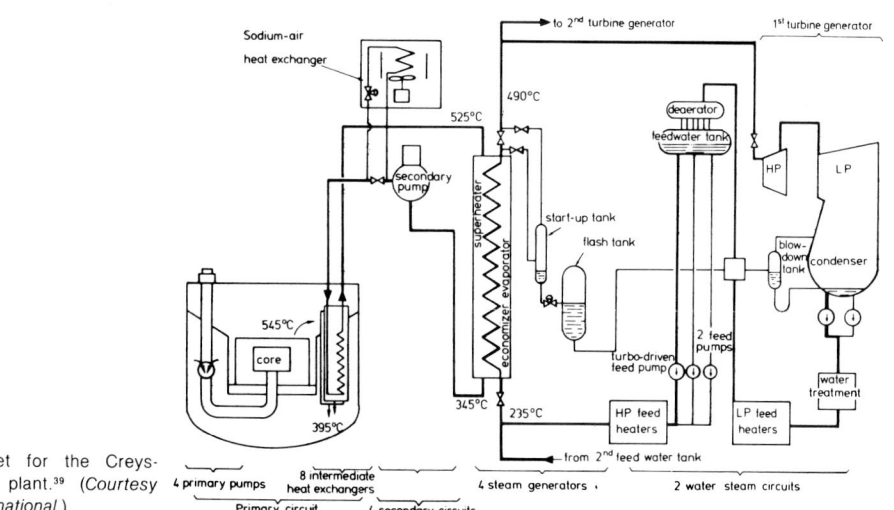

Figure 2-16 Flowsheet for the Creys-Malville LMFBR power plant.[39] (*Courtesy Nuclear Engineering International.*)

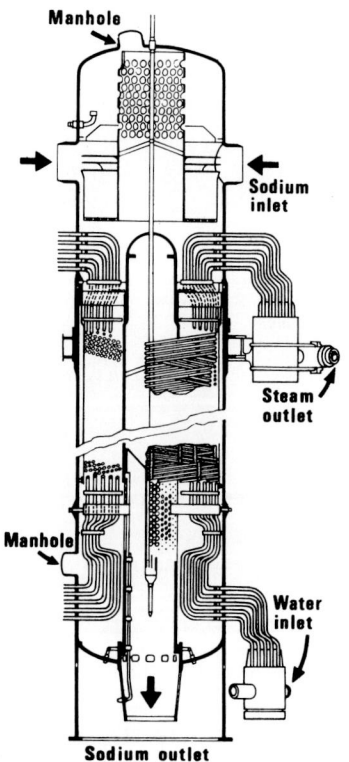
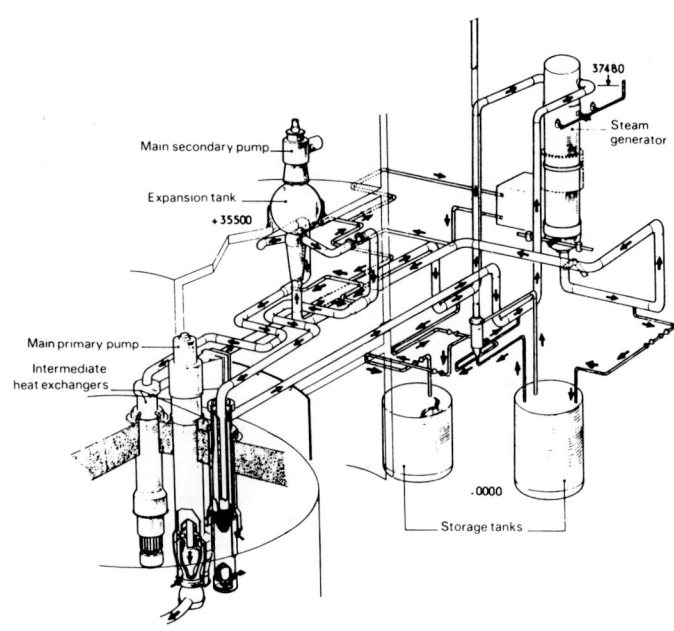

Figure 2-17 Isometric view of the secondary sodium circuit of the Creys-Malville power plant, together with a detail showing a section through the steam generator.[39] (*Courtesy Nuclear Engineering International.*)

TABLE 2-6 Summary of Design Data for the 1200-MWe LMFBR of the Centrale Nucléaire Européenne à Neutrons Rapides SA Creys-Malville Plant on the Rhône River near Creys-Malville, France[39]

Reactor type (Type de reacteur)	
Sodium cooled, pool, fast breeder (à neutrons rapides, du type intégrè, á refroidissement par sodium)	

Fuel (Combustible)	
Composition	sintered UO_2-PuO_2 mixed oxide (oxide mixte fritté UO_2-PuO_2)
Average Pu (Enrichissement moyen)	16%
Mass of Pu (Masse du Pu)	5600 kg
Breeding gain in core (Gain de régénération dans le coeur)	0.24
Maximum burnup (guaranteed) [Taux de combustion maximal (garanti)]	70,000 MWd/te
Target (objectif visé)	100,000 MWd/te

Fuel assemblies (Assemblages combustibles)	
Assemblies in core (Assemblages dans le coeur)	364
Pins per assembly (Aiguilles par assemblage)	271
Pin length (Longueur de l'aiguille)	2700 mm
Assembly length (Longueur de l'assemblage)	5400 mm
Cladding (Matériau de gainage)	stainless steel (acier inoxydable)
Nominal maximum cladding temperature (Température maximale nominale de gaine)	620°C

Blanket assemblies (Assemblages fertiles)	
Assemblies in core (Assemblages dans le coeur)	233
Pins per assembly (Aiguilles par assemblage)	91
Pin length (Longueur totale de l'aiguille)	1950 mm
Assembly length (Longueur totale de l'assemblage)	5400 mm
Cladding material (Matériau de gainage)	stainless steel (acier inoxydable)

Control rod assemblies (Assemblages de commande)	
Main shutdown system (Système d'arrêt principal)	
Assemblies in core (Assemblages dans le coeur)	21
Absorber pins per assembly (Aiguilles absorbantes par assemblage)	31
Pin length (Longueur de l'aiguille)	1300 mm
Cladding material (Matériau de gainage)	stainless steel (acier inoxydable)
Backup shutdown system (Système d'arrêt complémentaire)	

Control rod assemblies (Assemblages de commande) (Cont.)	
Assemblies in core (Assemblages dans le coeur)	3
Elements for assembly (Nombre d'éléments par assemblage)	3
Cladding material (Matériau de gainage)	stainless steel (acier inoxydable)

Main reactor tank (Cuve de réacteur)	
Inside diameter (Diamètre intérieur)	21,000 mm
Height (Hauteur)	19,500 mm
Material (Matériau)	stainless steel (acier inoxydable)

Primary circuits (Circuits primaires)	
Total mass of sodium in primary circuits (Masse totale de sodium dans les circuits primaires)	3.500 te
Nominal flow rate (Débit nominal)	4×4.1 te/s
IHX outlet temperature (Température de sortie des échangeurs intermédiares)	392°C
Core inlet temperature (Température d'entrée dans le coeur)	395°C
Core outlet temperature (Temperature de sortie du coeur)	545°C
IHX inlet temperature (Température d'entrée aux échangeurs intermédiaires)	542°C

Secondary circuits (Circuits secondaires)	
Total mass of sodium in secondary circuits (Masse totale de sodium dans les circuits secondaires)	1.500 te
Nominal flow rate (Débit nominal)	4×3.3 te/s
Steam generator outlet temperature (Température de sortie des générateurs de vapeur) IHX inlet temperature (Température d'entrée aux échangeurs intermédiaires)	345°C
IHX outlet temperature (Température de sortie des échangeurs intermédiaires) Steam generator inlet temperature (Température d'entrée aux générateurs de vapeur)	525°C

Water-steam circuits (Circuits eau-vapeur)	
Water temperature/pressure at steam generator inlet (Température/pression de l'eau à l'entrée des générateurs de vapeur)	235°C/210 bars
Steam temperature/pressure at turbine stop valves (Température/pression de la vapeur à l'admission des turbines)	487°C/177 bars
Nominal flow rate (Débit nominal)	4×340 kg/s

steam temperatures and pressures, i.e., 538 to 565°C (1000 to 1050°F) and 100 to 270 bars (1500 to 4000 psi), whereas intermediate- or peaking-load steam plants commonly employ peak steam temperatures that are lower by 50 to 100°C and pressures of 70 to 120 bars. The steam temperatures and pressures used in nuclear plants are determined by the characteristics of the reactor (see Tables 2-4 to 2-6).

Effects of Unit Size on Efficiency and Cost

Increasing the size of a unit leads to reductions in capital costs and small increases in the efficiency stemming largely from the higher Reynolds numbers inherent in the larger sizes. An important savings in operating costs stems from the reduction in the amount of manpower required per unit of output (Table 3-5); an increase in unit size by a factor of 10 commonly leads to an increase in manpower by only a factor of ~6. Other savings result from the fact that some compounds, such as instrumentation and control equipment, increase relatively little in cost and complexity as the unit size is increased. As shown in Fig. 2-18, these effects are especially pronounced for nuclear plants.[40] Other data on these effects are given in Section 3, e.g., Figs. 3-4 and 3-5.

Plant Thermal Efficiency

The overall thermal efficiency of a steam power plant is normally taken as the ratio of the useful electric output delivered to the grid divided by the chemical energy available in the fuel. The principal component efficiencies commonly run around 90 percent. This is the usual value for a coal-fired steam generator in which most of the loss is in the form of heat in the stack gases. Aerodynamic and moisture-churning losses in the turbine and pressure losses in the piping between turbines reduce the turbine efficiency to 80 to 85 percent. The heat losses inherent in the thermodynamic cycle run about 45 percent of the heat in the steam delivered to the turbine. In nuclear reactors the pumping power for the reactor cooling circuit commonly runs ~2 percent of the plant gross output.[41] Other smaller losses include electrical losses in the generator (about 1 percent) and energy to the steam-turbine-driven feed pumps (~1.5 percent of the useful output). The electric power required to drive induced- and forced-draft fans, coal feed and preparation equipment, condensate pumps, condenser cooling-water circulating pumps, miscellaneous minor equipment, and general station service and lighting totals about 3 percent of the plant net power output.[41]

Effects of Peak Steam Temperature on Thermal Efficiency

Any effort to increase plant thermal efficiency by increasing the steam pressure and/or temperature entails many compromises that must be made in an effort to avoid excessive increases in

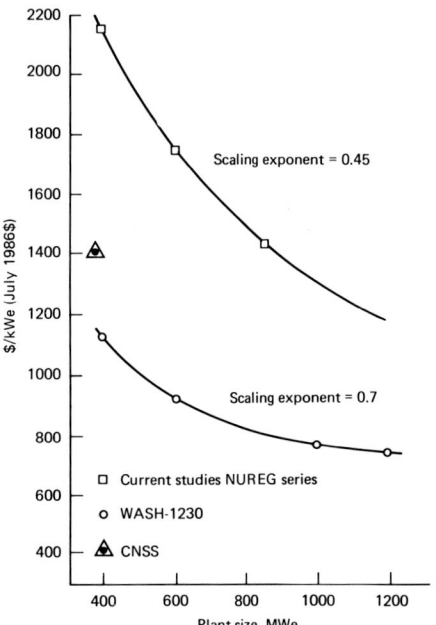

Figure 2-18 Effects of unit size on the estimated capital cost of nuclear power plants as obtained in two different studies.[40]

TABLE 2-7 Auxiliary Power Requirements for the TVA 950-MW Bull Run Coal-fired Plant[17]

	Fraction of net output, %
Forced-draft fans	1
Pulverizers	0.6
Condenser cooling water	0.25
Miscellaneous pumps	0.5
Station lighting and service	0.65
Total auxiliary power	3.0

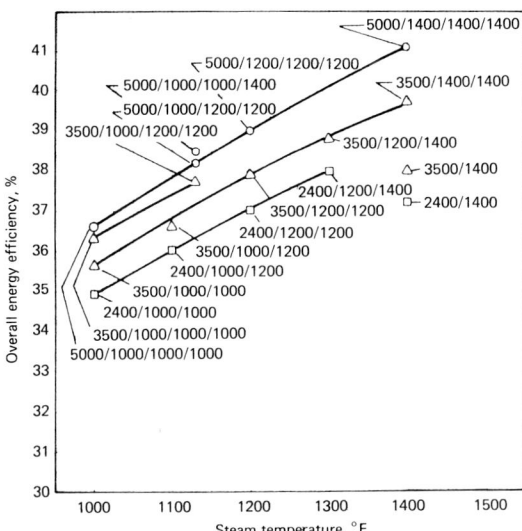

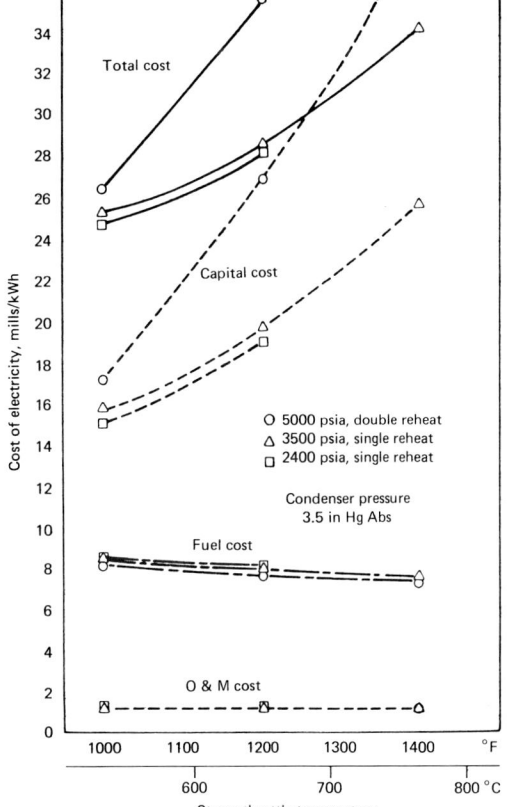

Figure 2-19 *(Above)* Effect of steam throttle conditions on overall efficiency for a 500-MWe steam plant with an atmospheric furnace. The steam conditions given on the curves are: pressure, psi/superheat, °F/first reheat, °F/second reheat, °F.[16]

Figure 2-20 *(Right)* Effect of steam-turbine throttle conditions on the cost of electricity from a 500-MWe steam plant with an atmospheric pressure furnace on the basis of 1975 dollars.[16]

capital cost. A nice set of curves showing these effects for a large fossil fuel plant is given in Figs. 2-19 and 2-20, taken from a Westinghouse study.[16] This shows the relatively small improvements in the thermal efficiency obtainable through the use of increased steam temperatures and pressures beyond 550°C/240 bars together with the relatively large increases in capital costs, so that the minimum cost of electricity is obtained with a peak steam temperature of about 538°C (1000°F). This result is consistent with numerous other studies that have been carried out in the past 25 years.

The curves of Fig. 2-20 suggest that the minimum capital cost of a plant would be found at steam pressures and temperatures lower than those considered in preparing Fig. 2-19. A review of the whole development of steam power plants shows that increasing the steam pressure and temperature from the low levels of the 1920s has led to reduction in the size and unit cost per kilowatt of electric output for the combustion air and stack gas equipment. On the other hand, increasing the steam temperature beyond 482°C (900°F) leads to a fairly rapid loss in strength of the relatively inexpensive low-alloy steels. The best resolution of the various compromises that must be made in selecting a steam temperature and pressure for minimum costs is indicated by the actual sales of steam plants for intermediate- and peaking-power purposes;[38] these plants are often designed for 60 to 120 bars (900 to 1800 psi) at 482 to 510° C (900 to 950° F), with many having steam conditions of 160 bars (2400 psi) and 540/540° C (1000/1000° F).

Effects of Superheating and Reheating

Superheating leads to an increase in the peak temperature in the cycle for a given peak steam pressure, and this increases the thermal efficiency. The increase is not as great as one might expect from simple peak temperature considerations because, when the amount of superheat is increased, the temperature-entropy diagram deviates farther and farther from the ideal rectangular Carnot cycle. Reheating has the effect of giving a somewhat closer approach to the ideal Carnot cycle.

One of the most important reasons for employing superheat and reheat relatively early in the development of steam power plants was that steam expansion in the turbine from a saturated-steam condition leads to moisture formation, which in turn may lead to turbine blade erosion, coupled with losses in turbine efficiency stemming from moisture-churning energy losses. Refined design of the turbine blading to prevent the accumulation of liquid slugs in stagnation regions near the tips of stator blades, coupled with bleed-off slots in the turbine casing and the use of bleed-off steam for regenerative feed heating, has greatly reduced the moisture-churning losses and tendencies toward blade erosion. In addition, the use of erosion-resistant materials, such as stellite and titanium, in the lower-stage turbine blades has made it possible to increase the turbine tip speed to ~500 m/s (1640 ft/s) in the last stage without getting into difficulty with serious moisture-churning losses or turbine bucket erosion even when operating with saturated steam at the turbine inlet.[42]

Air Preheat

The principal loss from the furnace–steam generator units in fossil fuel plants is heat in the stack gas. By using a recuperative heat exchanger, a pronounced energy savings can be effected by transferring heat from the stack gas to the incoming combustion air. This is commonly accomplished with a large cylindrical matrix of alternate corrugated and flat plates placed so that one portion is in the hot gas stream and the other in the incoming air. The cylinder is rotated at low speed so that heat is picked up and stored in the metal plates on the hot side, moved over and transmitted to the incoming air on the other side, giving a heat exchanger that has a low cost per unit of surface area. Preheating the air also serves to improve combustion conditions and increases the temperature difference in the furnace and thus the heat flux. This is particularly important because the SO_3 in the stack gas combines with moisture to give corrosive sulfuric acid if the stack gas temperature is reduced below the dew point. Depending on the sulfur content of the fuel, this ordinarily places a lower limit on the stack gas temperature of about 150° C (300° F).

FLUE GAS DESULFURIZATION

Most of the developmental activity on steam power plants at the time of writing is concerned with problems stemming from environmental legislation and efforts to use lower-grade coals, particularly those high in sulfur and ash.[43-47] Chief among the steps being taken is the development of stack gas scrubbers. The basic concept is simple—i.e., spray an alkaline solution or slurry into the flue gas to react with the SO_3 and SO_2 to form a sulfate or sulfite—but in practice such systems present serious corrosion, scaling, plugging, and waste-disposal problems.

Background of Experience

The reduction of sulfur emissions by stack gas scrubbing began in England in 1925 at the Battersea station with lime used to neutralize the acidity of the scrub water, which was discharged to the Thames River in a once-through process.[43] In the late 1930s a closed-loop lime-limestone scrubber system was installed at the nearby Fulham station that burned 1 to 1.5% S coal. Scaling and corrosion proved to be severe problems, and the FGD system was removed from service in 1942. Further work in England on both stack gas scrubbing and environmental effects led to the choice of tall stacks rather than scrubbers, and that course is being followed in Great Britain at the time of writing.[43]

Serious work on flue gas desulfurization began in the United States in the early 1940s, and small-scale efforts were continued on both lime-limestone scrubbing and processes for recovery of the products. In the 1960s, experiments were conducted with pulverized limestone injected into the furnace to calcine it and obtain very reactive lime particles in the flue gas, but the lime formed low-melting-point carbonate glasses that caused severe slagging of the boiler tubes, making this approach clearly unsatisfactory. Subsequent work has been directed mainly to various stack-gas scrubbing processes together with processes for disposing of the sludge produced. Extensive work has also been under way in Japan, largely for use with high-sulfur fuel oils.

Parameters and Criteria

The problem of sulfur emissions has been highly politicized and the public bombarded with a host of contradictory statements. Moreover, the courts have been flooded with suits brought by government agencies against utilities for failing to comply with regulations, by utilities against government agencies claiming that regulations are unrealistic, by environmentalist groups against government agencies charging laxity in enforcement, and by consumers against utilities to protest high rates stemming from expensive measures to reduce emissions. Much of both the acrimony and the litigation stems from preoccupation with one or a few of the parameters and criteria involved without recognition of their dependence on other parameters. The situation is enormously complex with no clear-cut solution in sight, but it is instructive to review the key technical facts, parameters, and criteria, their relationships, and the experience with operating systems. The scope of these problems is indicated in Table 2-2.

Sulfur Content of Fuel

The sulfur content of the fuel is a major parameter. The higher the sulfur content, the greater the droplet density required in the spray tower, the more rapid the sulfur pickup, the greater the difficulties with scale formation and mist flow up the stack, the more difficult the removal of a high percentage of sulfur, and the greater the amount of waste material.

Ash Content of Fuel

Both the percentage of ash and its chemistry affect the desulfurization process. Eastern coals are generally acidic, whereas western coals are usually basic so that their alkali content contributes to the conversion of sulfurous and sulfuric acid to sulfates. In addition, the chloride content of the coal is an important factor affecting corrosion, particularly in closed-loop systems where the chloride concentration builds up.

In some systems electrostatic precipitators are installed ahead of the scrubbers to reduce the adverse effects of the ash on scaling and corrosion in the air preheaters and scrubbers.

Type of Sorbent

Lime is both more soluble in water and more reactive than limestone, but it is more expensive. Further, in either case the presence of other minerals may have adverse or beneficial effects; e.g., magnesium that may be present in locally available limestone acts to improve its effectiveness. In fact, MgO is used as the sorbent in some processes. Lime obtained as a by-product from acetylene plants has been used at some sites and has been found to be much less prone to cause scaling because of the presence of a small amount of sodium thiosulfate.

Sodium carbonate is available at an acceptable cost at some sites in the western United States and has been found to give much less trouble with scaling and plugging. It is normally regenerated by reacting with lime or limestone and recirculated. A disadvantage is that the resulting calcium sulfite and sulfate waste has a substantial content of soluble sodium compounds that present a leaching problem, as well as requiring a makeup stream of sodium carbonate. In the arid west, where evaporation from a disposal pond exceeds the rainfall, the leaching problem can be handled; but in the east, contamination of streams by effluent from settling ponds is a serious problem.

Fraction of Sulfur Removed

The costs and problems of FGD processes increase with the fraction of the sulfur removed. Not only is it necessary to employ progressively greater amounts of sorbent to drive the reaction, but the fraction of the unreacted sorbent in the waste increases, thus increasing both the cost for the sorbent and the disposal problem more rapidly than linearly.

Water Recycle

There is a step change in difficulties with scaling, plugging, and corrosion if one shifts from an open cycle with waste-water discharge to a closed cycle in which impurities such as chloride build up. Significantly, no full-scale system handling high-sulfur coal has been operated on a closed cycle up to the time of writing.[43]

Energy Requirements

Scrubbing the flue gas reduces its temperature below the dew point, thus introducing major corrosion problems in damper mechanisms, induced-draft fans, stacks, piping, and in heat exchangers employed to raise the stack gas emission temperature to avoid moisture precipitation and high ground concentrations of stack gases immediately downwind. Reheating the stack gas after scrubbing is sometimes done with gas or oil burners, sometimes with steam, and sometimes with regenerative heat exchangers. All such measures entail energy losses equivalent to several percent of the energy consumption of the plant. In addition, extra power is required for the induced- and forced-draft fans, pumps, etc., so losses total 3 to 7 percent of the plant energy input. Additional energy is required if the waste is converted to a salable product such as H_2S or elemental sulfur. In the latter case the energy consumption of the FGD system has run as much as 25 percent of the energy input to the plant.[43]

Product

In principle, the sulfur removed from the stack gas can be made into a salable product, and efforts to do this have been under way in the United States and other countries since the 1930s. The only really successful commercial operations have been in Japan, where there is a good market for gypsum. Other potential products for which the waste might be used include portland cement, building and road base materials, H_2S, and elemental sulfur. However, the total market for these products is far too small to utilize the output from the U.S. utility industry if all its coal-fired plants were fitted with FGD systems, and even for the limited number of FGD installations at the time of writing (a total of ~ 15,000 MWe of capacity), it has been found that the lowest-cost method is to use either a settling pond or a landfill after dewatering the sludge and treating it chemically to give a stable material. The calcium sulfite is gelatinous and tends to flow under load. Oxidation to calcium sulfate gives a stable product. Depending on the chemistry of the ash, mixing with fly ash may also give a structurally stable material.

Capital Cost

The capital cost of an FGD system depends on all the above parameters, generally running from $60 to $130/kWe (giving annualized capital costs of 3 to 6 mills per kilowatthour) for new plants in 1975 dollars according to an EPA study.[46] Cost estimates by industrial organizations are generally higher.[47] Operating costs vary even more widely, depending on maintenance problems, running as much as 26 mills per kilowatthour in some plants. Another indication of costs is given by a 1979 compromise agreement on new FGD installations between TVA and EPA, which by one government estimate will increase TVA's cost of electricity generation by 30 percent in coal-fired plants.

Availability

The continual maintenance required to cope with scaling, plugging, corrosion, etc., is expensive not only in man-hours, replacement parts, and supplies but also in reduced availability of the plant. In some cases the normal load on the plant is sufficiently low at night so that dampers can be operated to close off perhaps half of the units to permit cleaning and servicing, and in some plants this is done every night.[43] For a base-load plant it is necessary to install extra FGD capacity to reduce losses in the availability of the full capacity.

Life of Equipment

Operating any type of equipment in a wet flue gas atmosphere is bound to give trouble because the gas is inherently corrosive and the erosive effects of slurries aggravate the corrosion problems. Many different methods of coping with these problems have been tried, including lining pipes with rubber, plastic, or special coatings, but up to 1980 none has proved satisfactory when employed over a period of several years.[43] At the time of writing, expensive high-nickel alloy parts are being tried.

Operating Experience

The Clean Air Act was amended for the third time in August 1977 to require that all new U.S. power plants employ the "best available technology" to reduce emissions, "taking into consideration the cost of achieving such emission reduction, any nonair quality health and environmental impacts and energy requirements." There have been widely different interpretations of the Congressional language. Those demanding tight controls on emissions can point to satisfactory service of flue gas scrubbers in a plant that has operated a short time on relatively low-sulfur coal with a low-cost supply of a sorbent having good characteristics and a plant load pattern that permits operation at low nighttime loads with maintenance of the FGD equipment every night. Others can point to the high costs that everyone has experienced, the short life of the equipment, the severe loss in availability, and the inability of any plant to operate on high-sulfur coal with a closed system. A detailed and comprehensive review of all the U.S. and Japanese experience up to 1979 is given in Ref. 43 and clearly indicates that the technology is not well in hand. In fact, none of the systems installed by 1976 was still being offered for sale by May 1978—another indication that the technology had not yet been developed.

Efforts to apply cost-benefit criteria have foundered on the refusal of some activists to accept such studies as long as any medical authorities express concern over possible health effects. At the time of writing it is not clear what systems will prove viable, how great costs will prove to be, or what standards will emerge, though the trend to date has been a continual tightening of EPA requirements irrespective of the costs. At the time of writing (1981) it appears that this trend is changing, in part because of rapidly escalating costs and in part because of recent studies by such organizations as the National Academy of Sciences indicating that earlier inferences of severe health effects were exaggerated or in error.[48] Significantly, at a symposium conducted by the New York Academy of Medicine on the health effects of sulfur oxides and related particulates in

December 1978, 72 percent of the participants felt that then-current standards were too stringent.[48]

FLUIDIZED-BED COAL COMBUSTION FOR STEAM GENERATION

The most promising approach to future improvements in steam plants at the time of writing appears to be through the development of fluidized-bed combustors (FBCs) that will give lower costs and higher efficiencies for coal-fired operation than conventional pulverized-coal-fired furnaces (PCFs) fitted with FGD equipment.[49] Experiments with fluidized-bed combustion systems have been under way since the middle 1950s, yet none is in regular commercial service at the time of writing. Some insight into the problems involved is given by Table 2-8, which summarizes the principal design data for three TVA plants having conventional PCFs and eight FBC plants.[49-52] The first fairly large-scale FBC pilot plant to be built, the 30-MWe experimental unit at Rivesville, West Virginia, is the first of the four atmospheric fluidized-bed combustors (AFBCs) in the table, the other three being design studies prepared for TVA. Data from design studies for one supercharged fluidized-bed combuster (SFBC) and three pressurized fluidized-bed combustors (PFBCs) are also included, the latter three being combined-cycle plants in which part of the electric output would be delivered by steam turbines and part by gas turbines. Note that the use of an FBC affects only the furnace and stack gas clean-up equipment; the rest of the steam system is unaffected. In fact, the steam conditions chosen for the FBC plants of Table 2-8 range from 6MPa/440°C (800 psi/825°F) to 26 MPa/540°C (3800 psi/1005°F), a substantially wider range than for the three PCF plants shown.

A good notion of the furnace geometry envisioned for AFBCs is given by Fig. 2-21, while Fig. 2-22 shows the principal components required in the system. Table 2-9 gives design data for this furnace.

Capital Cost Parameters

For FBCs to be attractive, their capital costs ought to be less than for conventional plants. Costs for new systems are always difficult to estimate, but several parameters in Table 2-8 provide a basis for appraising some of the major cost factors.

Heat-Transfer Surface

One of the most important factors affecting the capital cost is the heat-transfer surface area required per unit of output. Note that the heat-transfer surface area parameter is about the same for the AFBC as for the PCFs, but that increasing the furnace pressure reduces the surface area requirements for the PFBCs. The reason is that, while the gas-side heat-transfer coefficient in fluidized beds is high and acts to reduce the surface area requirements, the average temperature difference available is less than for a conventional furnace, so that the heat-transfer surface area requirements are about the same for PCF and AFBC furnaces. For PFBCs, increasing the furnace pressure improves the heat-transfer coefficient in the heat-transfer matrices above the bed and thus reduces their size, weight, and cost—a major advantage of furnace pressurization.

Floor Space

The amount of floor space required for the furnace is another important parameter from the cost standpoint, particularly if an AFBC is being retrofitted to an existing PCF plant. Table 2-8 indicates that this parameter is twice as great for full-scale AFBCs as for the PCF plants, but that supercharging the FBC to 3 atm gives about the same floor area requirement, and pressurizing the furnace to 10 atm cuts the floor space required to roughly half that for a conventional furnace. (Note that the value for the Rivesville furnace is not consistent with the others because it does not include ancillary equipment.) The crux of the problem is that the heat release per unit of fluidized-bed cross-sectional area is directly proportional to the airflow rate, and the superficial gas velocity leaving the bed is usually limited to ~ 1.8 m/s (6 ft/s) to avoid excessive elutriation of fine particles. The shallow depth (~ 1 m) and large cross-sectional area of an AFBC gives an excessive floor space requirement unless a series of beds is stacked one above the other in tiers, as in the AFBC furnace of Fig. 2-21, but even this appears to give about double the floor space requirement of a conventional furnace. In addition, the floor space required for auxiliaries, particularly the coal feed equipment, seems higher than for conventional plants.

Number of Fuel Feed Points

The cost, complexity, and maintenance problems of a furnace increase with the number of fuel feed points, a parameter treated in the bottom line of Table 2-8. Pressurizing a fluidized-bed furnace reduces the bed area, and—other considerations being equal—should make the number of coal feed points inversely proportional to the pressure. The wide range of design values given in Table 2-8 indicates that the information available to the designers was insufficient to form a firm basis for the designs. There is no question but that, to minimize both the cost and complexity of the coal feed system for a fluidized-bed fur-

TABLE 2-8 Major Design Parameters for Both Typical Pulverized Coal-Fired Power Plants and Conceptual Designs for Fluidized-Bed Combustion Systems [49]

Plant	Conventional pulverized coal			Atmospheric fluidized bed			
	Bull Run	Colbert #5	Kingston #9	Rivesville	Foster-Wheeler	Combustion engineering	Babcock & Wilcox
Design steam pressure and temperature, MPa/°C/°C (psig/°F/°F)	24/538/538 (3500/1000/1000)	17/566/538 (2400/1050/1000)	12/566/566 (1800/1050/1050)	9/496/538 (1350/925/1000)	26/540/540 (3800/1005/1005)	18/540/540 (2600/1005/1005)	18/539/539 (2581/1003/1003)
Design output, MWe	850	550	200	30	150	200	200
Overall plant thermal efficiency, %	40	38.6	26.7				
Furnace arrangement	Conventional	Conventional	Conventional	Single bed 4 compartments	Stacked 4 main 1 CBC†	Ranch 7 main 1 CBC	Stacked 4 main 2 CBC
Furnace pressure, MPa (atm)	0.10 (1)	0.10 (1)	0.10 (1)	0.10 (1)	0.10 (1)	0.10 (1)	0.10 (1)
Surface area, m² (ft²)	70,000 (753,000)		20,500 (221,000)			22,000 (237,000)	
Surface area, m²/MWe (ft²/MWe)	82 (886)		103 (1105)			110 (1185)	
Furnace plan area, m²/MWe (ft²/MWe)	0.051 (5.5)		0.052 (5.6)	0.14* (1.52)	0.97 (10.4)	1.24 (13.4)	1.80 (19.4)
Heat-transfer surface weight, kg (lb)		2,270,000 (5,000,000)	1,000,000 (2,200,000)				
Total furnace weight, kg (lb)	10,900,000 (24,000,000)		3,800,000 (8,400,000)				
Heat-transfer surface weight, kg/MWe (lb/MWe)		4140 (9100)	5000 (11,000)				
Total furnace weight, kg/MWe (lb/MWe)	12,800 (28,230)		19,000 (42,000)				
Number of burners or fuel feed points	176		56	44	16	118	640
Number of burners or fuel feed points/100 MWe	21		28	147	11	59	320

*Does not include space for coal feed and other equipment.
†Carbon burnup cell

Supercharged fluidized bed	Pressurized fluidized bed		
ORNL	GE—ECAS	Westinghouse ECAS	Curtiss-Wright
12/566/566 (1800/1050/1050)	24/538 (3515/1000)	24/538 (3515/1000)	6/440 (800/825)
200	904	679	500
	39.2	39.0	38.8
Single bed vertical tubes	Stacked 7 bed	Stacked 4 bed	
0.30 (3)	1.0 (10)	1.0 (10)	0.7 (7)
8000 (86,700)	8300 (89,400)	5900 (62,980)	
40 (434)	9 (99)	9 (93)	
0.632 (6.8)	0.055 (0.59)	0.17 (1.85)	
404,500 (890,000)			
	1,825,000 (4,016,000)		
2000 (4450)			
	2000 (4400)	1872 (4120)	
144	335	64	
72	37	9.4	

TABLE 2-9 Fluidized-Bed Design Parameters for Fig. 2-21

Superficial velocity	
Main beds	12 ft/s
Carbon burnup cells	9 ft/s
Excess air	
Main beds	20%
Carbon burnup cells	25%
Temperature	
Main beds	1550 F
Carbon burnup cells	2000 F
Bed depth	
Static	24 in
Full load	48 in
Grid plate pressure drop	16 in H_2O
Bed pressure drop	30 in H_2O
Coal/limestone feed spacing	18 ft^2
Freeboard	6 ft
Limestone feed rate	
Design coal—0.9% S	2.3 Ca/S mol ratio
3.25% S coal	4.0 Ca/S mol ratio
Heat-transfer coefficient	
Vertical tubes	50 Btu/(h · ft^2 · °F)
Horizontal tubes	45 Btu/(h · ft^2 · °F)
Combustion efficiency	
Main beds	90%
Carbon burnup cells	90%
Elutriation	
Carbon	10% of heat input weight equivalent
Coal ash	100% of input coal ash weight
Limestone	40% of solid weight after calcination

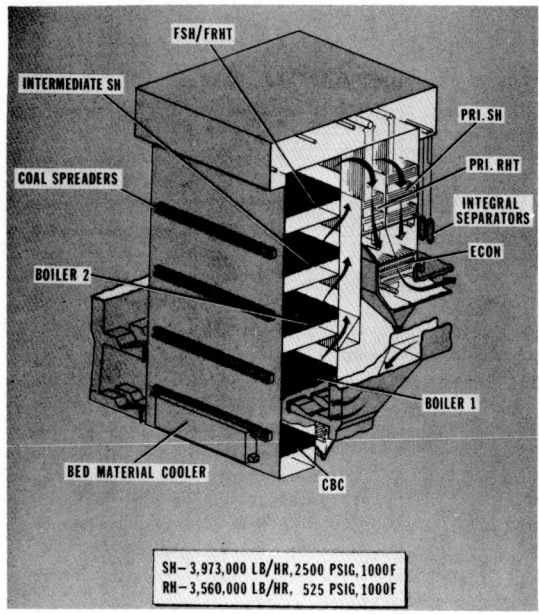

nace, one should minimize the number of coal feed ports. However, a number of other considerations, such as tube corrosion, improved sulfur retention, and carbon and lime utilization, favor close spacing of the coal feed ports. Of these, corrosion of metal tubes in the bed is probably controlling; rapid corrosion of metal in the fluidized bed will occur if there are local regions in which there are vacillations between oxidizing and reducing conditions—e.g., in fuel-rich zones above coal feed ports. Tendencies toward such conditions can be reduced by increasing the nominal amount of excess air, reducing the coal feed port spacing, increasing the depth of the mixing region between the coal feed ports and the heat-transfer matrix in the bed, reducing the amount of volatile matter in the coal, increasing the amount of air preheat, and modifying the geometries of both the air tuyeres and the coal feed ports to increase the rate of lateral mixing of the coal injected into the bed.

Major Problem Areas

Hundreds of fluidized-bed combustion systems are in use commercially to burn low-grade solid waste, such as sawdust and

Figure 2-21 Schematic isometric view of an atmospheric fluidized-bed steam generator with five beds arranged in vertical tiers for a 570-MWe power plant.[50] (*Courtesy Foster Wheeler Energy Corp.*)

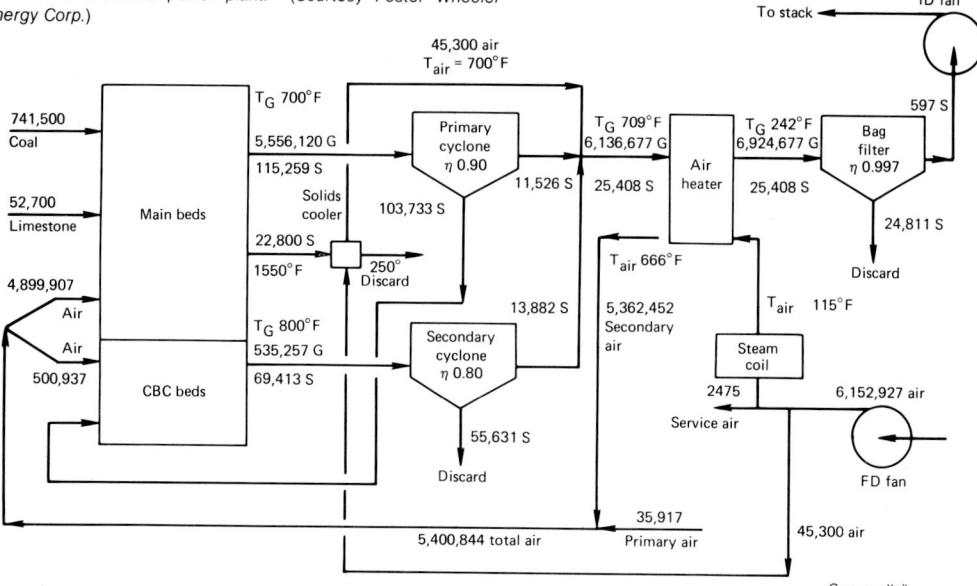

Figure 2-22 Flowsheet for the fluidized-bed furnace system of Fig. 11.21.[50] (*Courtesy Foster Wheeler Energy Corp.*)

sludge from sewage plants, while other hundreds of units are in commercial operation roasting pyrite ores. In view of this, it is surprising that there are no fluidized-bed coal combustion systems in commercial operation with steam boilers, although quite a number of experimental units have been operated. The principal reasons for this situation are treated in this section.

Boiler Stability and Control

There have been serious problems in controlling the combination of a fluidized-bed coal combustion system and a steam boiler with its tubes immersed in the bed. For good sulfur removal the bed must be operated in the 816 to 927 °C (1500 to 1700 °F) range, and to maintain this temperature about two-thirds of the heat released by combustion of the coal must be removed from the fluidized bed by a tube matrix immersed in the bed. Herein lies the crux of the problem. The heat-transfer coefficient between the bed and the tube wall is high—around 300 J/(s · m² · °C) [50 to 60 Btu/(h · ft² · °F)]. At the same time the heat-transfer coefficient from the tube to the boiling water under nucleate boiling conditions is very high—of the order of 30,000 J/(s · m² · °C) [5000 Btu/(h · ft² · °F)]. Thus, the tube wall tends to run close to the water temperature. Both the fluid-bed-side and water-side coefficients are essentially independent of the combustion gas and water flow rates, respectively. Inasmuch as $Q/A = u \Delta T$ and the overall heat-transfer coefficient u is constant, if one attempts to reduce the rate of heat release in the bed, the bed temperature will drop until it reaches about 540 °C (1000 °F), at which point the bed will quench; i.e., "the fire will go out." If the coal feed is continued, coke will form and clog the bed. If the water flow to the boiler is reduced in an effort to keep the bed temperature constant at reduced outputs at 870 °C (1600 °F), the water-side heat-transfer mechanism shifts from a nucleate boiling condition to a vapor-film blanket condition. This effect is illustrated by Fig. 2-23, which shows a typical curve for the heat flux to boiling water as a function of the temperature difference between the boiling point of the water and the metal wall from which the heat is being conducted.[33] If the fluid-bed temperature is held at 870 °C (1600 °F) and the metal tube wall is held at 316 °C (600 °F), the heat flux through the tube wall under normal nucleate boiling conditions would run about 300,000 J/(s · m²) [50,000 Btu/(h · ft²)]. This point has been plotted in Fig. 2-23 to indicate the normal operating point to be expected. Note that operation in this region is stable, because any small increase in the tube-wall temperature will result in a greatly increased heat flux from the tube wall to the boiling water. If one attempts to reduce the heat flux from the bed to the tube wall by allowing the tube-wall temperature to increase to, say, 540 °C (1000 °F), the

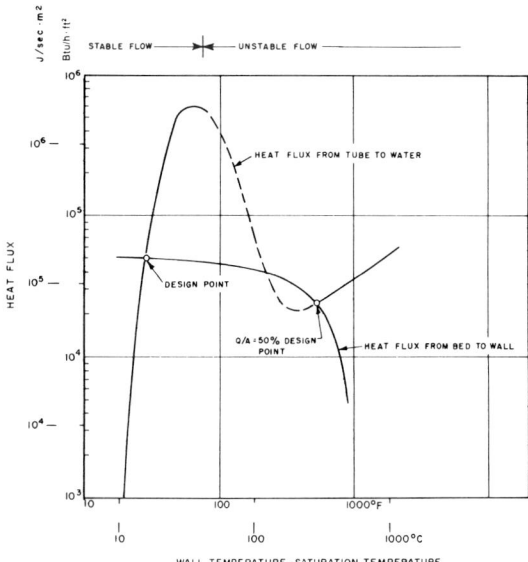

Figure 2-23 Curve for the calculated heat flux from the bed to the tube in a fluidized-bed steam boiler with a bed temperature 555 °C (1000 °F) above the saturation temperature of the steam is shown superimposed on a typical curve for the heat flux from the tube wall to boiling water. Both heat fluxes are plotted against the difference between the temperature of the tube wall and the saturation temperature of the boiling water.[33]

heat flux will be cut roughly in half, but the heat-transfer mode will be via vapor-blanket boiling, which will be very unstable and will lead to wide fluctuations in the local metal temperature. Even worse, the water flow rate through the tube will become highly unstable, and this in turn will lead to even more severe temperature fluctuations and thermal stresses. Thermal-strain cycling failure of the tubes will result if the tubes do not overheat so much that their loss in strength leads to failure from pressure stresses. Note that this unstable operating region is indicated in Fig. 2-23. Note, too, that intermediate conditions between these two regions would be unstable because any small increase in the tube-wall temperature would lead to a reduction in the heat flux from the tube wall to the boiling water.

For good reliability, operation must be constrained to the nucleate boiling regime with a fixed bed temperature of around 870 °C (1600 °F). Basic heat-transfer considerations indicate that the only good way in which the heat input to the boiler can be reduced is to reduce the amount of surface area of the tube matrix immersed in the bed. This, in turn, implies that either the bed should be compartmentalized so that the output can be changed in quantum jumps or the level of the fluid bed should be varied. The former approach presents difficulties in the relative phasing of the flow of red-hot "sand" and feed water to a compartment being activated or deactivated. Unless this phasing is carried out in just the right fashion, one runs the danger either of chilling the bed in that region and building up heavy coke deposits or of getting into a vapor-blanket boiling region, generating hot spots in the tubes, and of failure of the tubes after a series of perhaps 10 to 100 thermal cycles.

Varying the bed depth presents a problem in that it is difficult to tell just where the surface of the bed is during operation, because in the regime of interest the fluid bed will be bubbling violently. Experience at the BCURA Laboratory in England indicates that the bed depth can be determined and controlled within 7 to 10 cm (3 or 4 in) by measuring the pressure drop across the bed. For an atmospheric fluidized bed for which the maximum bed depth with an acceptable pumping-power loss would be around 1 m (40 in), the uncertainty in bed depth would be around 10 percent of the bed depth at full power and progressively greater if the bed depth were reduced at part power. If, on the other hand, the bed depth at full power is made around 3 m (10 ft), the uncertainty in bed depth becomes only a few percent and provides a good basis for control of the power output. That is, the power output will not be too sensitive to modest additions or removal of hot "sand" to or from the bed. Thus, just from the control standpoint, there is a strong incentive to supercharge the bed so that the bed depth can be made great enough to permit good control and yet not be so great as to entail excessive pumping-power requirements for the combustion air.

Coal Feed and Metering System

Pulverized-coal and fluidized-bed coal feed and metering systems differ in two important respects; namely, the requirements on particle size and requirements on the uniformity in fuel distribution in the furnace. Pulverized-coal firing requires not only crushing and screening but also pulverizing to a size less than 2 μm. The pulverizers are expensive, require considerable maintenance, and consume a substantial amount of electric power—typically 0.5 to 1.0 percent of the gross plant output. On the other hand, the open furnace can tolerate substantial variations and fluctuations in the fuel flow to individual burners, whereas in the fluidized bed such variations and fluctuations affect the local fuel-air ratio and may lead to local fluctuations between oxidizing and reducing conditions in the vicinity of the coal feed ports. (This would be much less of a problem if char were the fuel rather than raw coal with its volatile content of ~35 percent.) Getting a high degree of uniformity in coal feed rate and distribution to a large number of feed points in a fluidized bed is an extremely formidable problem.[53]

Forcing the coal into the furnace against the furnace pressure becomes progressively more difficult as the furnace pressure is increased from atmospheric to the 10 atm projected in some designs. The pressure barrier can be handled with lock hoppers of the sort employed for solids handling in many chemical processes (and in blast furnaces), but to get good sealing in the valves of lock hopper systems has generally proved troublesome. A variety of helical screw feeders have been developed, and these have proved satisfactory for feeding against small pressure differentials. Gas backflow through the material in the screw has been a problem in some types during operation against pressures of more than 1.5 atm, though at least one reportedly has functioned well in bench tests against 4 atm. Rotary feeders with vanes or pockets have been used to pump the crushed coal into a pressurized furnace in several experimental FBCs, but these machines seem to be subject to rapid wear unless the solids feed rate is kept well below their capacity. This experience indicates that they are suitable for metering and might be used as a pressure lock, but not for both functions.

Dividing the coal stream so that it is uniformly distributed to dozens or hundreds of coal feed points is difficult for any given flow rate. It is even more difficult to get the same set of devices to perform well over a wide range of flow rates and do so consistently on both a short-time and long-time basis, but such performance is required in order to prevent the local fluctuations that might cause serious fire-side corrosion. Experience in bench tests shows that the only way that it can be achieved is to dry the coal to less than 1 percent superficial moisture so that the granular solid flows freely.[53]

After feeding the coal through a pressure lock, it can be metered and then divided before it is entrained in an airstream for pneumatic transport to the furnace, or the solid particles can be entrained in an airstream and the stream divided. In either case, the airstream velocity required depends on the particle size, generally running about 6 m/s for −3 mm (−$\frac{1}{8}$ in) particles. At this velocity erosion in straight pipes has not been a problem, but severe erosion has been experienced in bends and elbows. Interestingly, use of a tee with a blocked leg facing the inflowing stream has proved to be an entirely satisfactory replacement for elbows. Solid particles deposited in the dead leg

TABLE 2-10 Comparison of Relative Advantages and Disadvantages of Atmospheric and Pressurized Fluidized-Bed Coal Combustion Systems

Furnace pressure	Atmospheric 0.10 MPa (1 atm)	Pressurized (turbosupercharged) 0.3-0.5 MPa (3-5 atm)	Pressurized (gas turbine-generator) 1.0 MPa (10 atm)
Advantages	Fewest feasibility problems Greatest amount of test experience Coal feed against only 0.014 MPa (~2 psi) Solid waste disposal much less difficult than for wet scrubbers Plant efficiency higher than for wet scrubbers by ~1 point Capital cost modestly improved over conventional plant with scrubbers	Capital cost of furnace and steam generator appears lower than for conventional pulverized-coal unit Coal feed points reduced by a factor of 3-5 over AFB Large bed depth improves combustion efficiency and SO_2 removal, and permits vertical boiler tubes with simpler furnace structure Plant thermal efficiency about 1 point higher than for AFB Control characteristics appear to be excellent	Potential cycle thermal efficiency is about 3.5 points higher than AFB Capital cost of furnace and steam generator appears lower than for conventional pulverized-coal unit Coal feed points reduced by about a factor of 10 relative to AFB Large bed depth gives better coal utilization and SO_2 removal as well as permitting vertical tube array and simpler furnace structure
Disadvantages	Large number of coal feed points gives complex coal feed system Shallow beds require horizontal tubes with support structure at bed temperature Startup and control problems appear to be complex	Feasibility of acceptable turbine erosion and deposits with 538°C (1000°F) turbine inlet temperature seems likely but has not yet been proven Coal feed against 0.3-0.5 MPa (3-5 atm) poses some reliability problems Test experience much less than for AFB	Feasibility of acceptable turbine erosion and deposit rates is highly doubtful Coal feed against 1.0 MPa (10 atm) poses difficult reliability problems Test experience much less than for AFB Large heat capacity of granular bed filters for hot gas cleanup gives very slow response times Particulate removal equipment may entail excessive capital cost

of the tee act as a cushion to absorb the impact of particles in the airstream so that they bounce around the bend without eroding the metal wall.

Advantages and Disadvantages of Pressurized Furnaces

As indicated by Table 2-8, pressurizing the furnace for a fluidized bed reduces the furnace size and capital costs, and it permits an increase in bed depth which improves both sulfur retention and combustion efficiency. However, pressurization requires such a large power input to the compressor that it can be accomplished economically only if a gas turbine is employed to recover power from the hot pressurized gas leaving the furnace, and this, in turn, presents problems with gas-turbine blade erosion and deposits caused by ash particles entrained from the bed. These problems are treated in the next chapter. Combined gas turbine-steam power plants, including those employing FBCs, are treated in Section 4. However, a brief summary of the principal advantages and disadvantages of the three basic types of plants is presented in Table 2-10.

COOLING TOWERS

Heat rejection from a steam power plant to rivers, lakes, or estuaries may have adverse effects on the marine biota. To avoid possible ecological damage, natural-draft cooling towers with an intermediate coolant circuit have been required by the EPA on new plants as the least expensive alternative. The incremental capital cost of wet cooling towers has proved to be ~$40/kWe in 1978 dollars. The higher condenser temperature reduces the thermal efficiency ~3 percent, and these factors coupled with added maintenance and troubles such as freezing and solids buildup from evaporation increase the cost of electricity by ~7 percent. Efforts to investigate the ecological justification for these costs have shown that in many cases there is no significant ecological advantage to the use of cooling towers, and, in fact, mandating their use may be counterproductive.[54]

Cooling towers function by evaporating water and hence have a substantial water consumption, which is a problem in arid areas. They also introduce a fog problem under some conditions at certain sites. Both of these problems can be avoided through the use of "dry" cooling towers, which employ heat-transfer matrices in the air inlet region. In principle, these cooling towers could act as direct condensers for the steam, but the steam volume flow is so great that, to reduce the size and cost of piping, an intermediate fluid is ordinarily employed—usually either water or a water-glycol solution to avoid difficulties with freezing. A few such installations have been built, one of them a 300-MWe plant in the Dakotas that uses forced-draft air cooling. The dry cooling towers have proved very expensive, increasing the plant capital cost by ~30 percent. Further, corrosion has caused trouble. In an experimental dry cooling tower installation near Manchester, England, the aluminum heat exchangers corroded and leaked so badly that the system was removed from service after only 2 years.

An effort to design a special finned tube tailored to suit the special requirements of dry cooling towers indicates that costs might be reduced somewhat. Depending on the amount of pumping power and the temperature differences between the air and cooling water considered acceptable, a design study has yielded an estimate of $100 to $136/kWe for the proposed type of extruded-machined aluminum finned-tube system in 1978 dollars.[55]

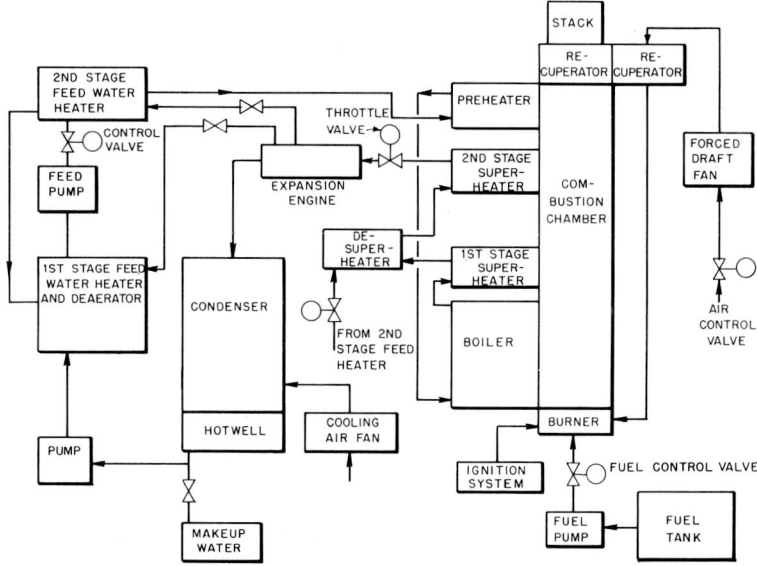

Figure 2-24 Schematic diagram of a steam power plant with a once-through boiler and the principal components usually required. In this case an air-cooled condenser is employed.

REFERENCES

1. Orrok, G.A.: "James Watt, 1736–1819," *Mechanical Engineering,* vol. 58, 1936, pp. 75–80.

2. *Steam: Its Generation and Use,* 38th ed., The Babcock & Wilcox Company, New York, 1972.

3. Reynolds, H. B., et al.: "New Boiler Equipment at the Interborough Rapid Transit Co.'s Fifty-Ninth Street Power Station," *Trans. ASME,* vol. 48, 1926, p. 1369.

4. Greene, A. M., Jr.: "The Tale of Two City Stations: A Half Century of Progress in Steam-Power Generation," *Mechanical Engineering,* vol. 63, 1941, p. 109.

5. Keller, E. E., and F. Hodgkinson: "The Steam Turbine in the United States, I: Developments by the Westinghouse Machine Company," *Mechanical Engineering,* vol. 58, 1936, p. 683.

6. Christie, A. G.: "The Steam Turbine in the United States, II: Early Allis-Chalmers Steam Turbines," *Mechanical Engineering,* vol. 59, 1937, p. 71.

7. Robinson, E. L.: "The Steam Turbine in the United States, III: Developments by the General Electric Company," *Mechanical Engineering,* vol. 59, 1937, p. 239.

8. Kimball, D. S.: "The Century's Great Inventions," *Mechanical Engineering,* vol. 59, 1937, p. 507.

9. Blowney, W. E., and G. B. Warren: "The Increase in Thermal Efficiency Due to Resuperheating in Steam Turbines," *Trans. ASME,* vol. 46, 1924, p. 563.

10. Christie, A. G.: "Development and Performance of American Power Plants," *Mechanical Engineering,* vol. 58, 1936, p. 539.

11. Kerr, H. J.: "Once-Through Series Boiler for 1500 to 5000 lb. Pressure," *Trans. ASME,* RP-54-1a, vol. 54, 1932.

12. Gastpar, Jacques: "European Practice with Sulzer Monotube Steam Generators," *Trans. ASME,* vol. 75, 1953, p. 1345.

13. Rowland, W. H., and A. M. Frendberg: "First Commercial Supercritical-Pressure Steam Generator for Philo Plant," *Trans. ASME,* vol. 79, 1957, p. 409.

14. Dauber, C. A.: "Avon No. 8-A Supercritical-Pressure Plant," *Trans. ASME,* vol. 79, 1957, p. 727.

15. Campbell, C. B., C. C. Franck, Sr., and J. C. Spahr: "The Eddystone Super-Pressure Unit," *Trans. ASME,* vol. 79, 1957, p. 1431.

16. Wolfe, R. W.: *Energy Conversion Alternatives Study (ECAS), Westinghouse Phase I Final Report,* vol. XI: *Advanced Steam Systems,* NASA CR-134941, Feb. 12, 1976.

17. "The Bull Run Steam Plant," Tennessee Valley Authority Technical Report No. 38, 1967.

18. Elston, C. W.: "Design and Development Philosophy to Achieve High Reliability and Long Life in Large Turbine Generators," Paper No. AAS73-055, presented to the American Astronautical Society, September 1976.

19. Davis, C. M., et al: "Large Utility Boilers—Experience and Design Trends," *Proceedings of the American Power Conference,* vol. 38, 1976, p. 280.

20. Henke, W. G.: "The New 'Hot' Electrostatic Precipitation," *Combustion,* October 1970.

21. Bazelmans, C. L., et al.: "Study of Options for Control of Emissions from an Existing Coal-Fired Electric Power Station," Oak Ridge National Laboratory Report No. ORNL-TM-4298, September 1973.

22. Burchard, J. K., et al.: "Some General Economic Considerations of Flue Gas Scrubbing for Utilities," *Proceedings of Conference on Sulfur in Utility Fuels: The Growing Dilemma, Drake Hotel,* Chicago, Oct. 25-26, 1972.

23. Elliot, M. A.: *Chemistry of Coal Utilization,* 2d suppl. vol., Wiley, New York, 1981.

24. "Multi-Stream Coal Cleaning System Promises Help with Sulfur Problem," *Coal Age,* vol. 81, January 1976, pp. 86–88.

25. Anson, D.: "Fluidized Bed Combustion of Coal for Power Generation," *Progress in Energy and Combustion Science,* vol. 2, 1976, pp. 61–82.

26. Squires, A. M.: "Applications of Fluidized Beds in Coal Technology," *Alternative Energy Sources,* Academic Press, Inc., New York, 1976, chap. 4, p. 49.

27. Stringfellow, T. E., and J. G. Branam: "Start-Up and Initial Operation of the Rivesville 30 MW Fluid Bed Boiler," *Proceedings of Fifth International Conference on Fluidized Bed Combustion,* Washington, D.C., Dec. 12–14, 1977.

28. Spencer, D. F., O. D. Guildersleeve, and R. A. Loth: "Initial Comparative Analysis of the Market Penetration Potential of Coal and Coal Derived Fuels in the United States Utility Industry (1985–2005)," presented at the IEEE PES Summer Meeting, Mexico City, Mex., July 17–22, 1977.

29. Brown, D. H., et al.: *Energy Conversion Alternatives Study (ECAS), General Electric Phase II Final Report,* vol. II: *Advanced Energy Conversion Systems—Conceptual Designs,* part 2, "Closed Cycles," NASA-CR 134949, SRD-76-064-2, December 1976.

30. Becker, T. W.: "Application of Atmospheric Fluidized Bed Combustion for Electric Power Generation," *Proceedings of Fifth International Conference on Fluidized Bed Combustion,* Washington, D.C., Dec. 12-14, 1977.

31. Corell, R. B.: "Conceptual Design of a 570 MW Combustion Engineering, Inc. Atmospheric Fluidized Bed Steam Generator," *Proceedings of Fifth International Conference on Fluidized Bed Combustion,* Washington, D.C., Dec. 12-14, 1977.

32. Reed, K. A., and R. L. Gamble: "Conceptual Design of a 570 MW Foster-Wheeler Energy Corp. Atmospheric Fluidized Bed Steam Generator," *Proceedings of Fifth International Conference on Fluidized Bed Combustion,* Washington, D.C., Dec. 12-14, 1977.

33. Fraas, A. P., G. Samuels, and M. E. Lackey: "A New Approach to Fluidized Bed Steam Boiler," ASME Paper No. 76-WA/Pwr-8, December 1976.

34. Farmer, M.: "Application of Fluidized Bed Combustion Technology to Industrial Boilers," paper presented at the Fluidized Bed Combustion Technology Exchange Workshop, McLean, Va., Apr. 13-15, 1977.

35. Webb, R.: "Natural Versus Forced Circulation in Fluidized Bed Combustion," paper presented at the Fluidized Bed Combustion Technology Exchange Workshop, McLean, Va., Apr. 13-15, 1977.

36. Miller, W.: "Fluidized Bed Boiler at Georgetown University," paper presented at the Fluidized Bed Combustion Technology Exchange Workshop, McLean, Va., Apr. 13-15, 1977.

37. Iredale, A. J. F., and N. P. Grimm: "Ice Condenser Reactor System Containment," *Nuclear Engineering International,* vol. 16, October 1971, pp. 864-867.

38. Walker, R. E., and T. A. Johnston: "Fort Saint Vrain Nuclear Power Station," *Nuclear Engineering International,* vol. 14, December 1969, pp. 1069-1073.

39. "Construction of the World's First Full-Scale Fast Breeder Reactor," *Nuclear Engineering International,* vol. 23, June 1978, pp. 43-60.

40. Crowley, J. H.: "Power Plant Cost Estimates Put to the Test," *Nuclear Engineering International,* July 1978, pp. 39-43.

41. Leung, P., and K. A. Gulbrand: "Power System Economics: An Evaluation of Plant Auxiliary System Incremental Kilowatt Consumption," *J. Eng. Power,* July 1977, pp. 419-423.

42. Wood, R. A.: "Status of Titanium Blading for Low Pressure Steam Turbines," EPRI Report No. AF-445, February 1977.

43. Rush, R. E., and A. V. Slack: *Status Report on Flue Gas Desulfurization at Coal-Fired Power Plants,* Utility Air Regulatory Group, Jan. 15, 1979.

44. "Shifting SO_2 from the Stack," *EPRI Journal,* July/August 1979, pp. 15-19.

45. *Interagency Flue Gas Desulfurization Evaluation,* vol. 1, first draft report, Nov. 30, 1977.

46. Klett, M. G.: *Typical Costs for Electric Energy Generation and Environment Controls,* Interagency Energy/Environmental Protection Agency Report No. EPA-600/7-79-026, January 1979.

47. Bloom, S. G., et al.: *Analysis of Variations in Costs of FGD Systems,* Electric Power Research Institute Report No. FP-909, October 1978.

48. "Symposium on Health Effects of Sulfur Oxides and Related Particulates," *Bulletin of the New York Academy of Medicine,* vol. 54, no. 11, December 1978.

49. Fraas, A. P., et al.: "Assessment of the State of the Art of Pressurized Fluidized Bed Combustion Systems," Oak Ridge National Laboratory Report No. ORNL/TM-6633, June 1979.

50. Reed, K. A., and G. G. Cervenka: "Conceptual Design of a Foster Wheeler Energy Corporation Atmospheric Fluidized Bed Steam Generator," *Proceedings of Fifth International Conference on Fluidized Bed Combustion,* Dec. 12-14, 1977, vol. II, MITRE Corp., pp. 285-310.

51. Becker, R. W.: "The Application of Atmospheric Fluidized Bed Combustion for Electric Power Generation," *Proceedings of Fifth International Conference on Fluidized Bed Combustion*, Dec. 12-14, 1977, vol. II, MITRE Corp., pp. 267-282.
52. Covell, R.B.: "Conceptual Design of a 570-MW Combustion Engineering, Inc. Atmospheric Fluidized Bed Steam Generator," *Proceedings of Fifth International Conference on Fluidized Bed Combustion,* Dec. 12-14, 1977, vol. II, MITRE Corp., pp. 326-340.
53. Lackey, M. E.: "Design and Performance Testing of a Coal Feed and Metering System for the MIUS Fluidized Bed Combustor," Oak Ridge National Laboratory Report No. ORNL/HUD/MIUS-47, December 1977.
54. Reynolds, J. Z.: "Power Plant Cooling Systems: Policy Alternatives," *Science,* vol. 207, no. 4429, Jan. 25, 1980, pp. 367-372.
55. Haberski, R. J., and J. C. Bentz: *Conceptual Design and Cost Evaluation of a High Performance Dry Cooling System,* U.S. Department of Energy Report No. COO-4218-1, Curtiss-Wright Corp. Report CWC-WR-78-001, Mar. 1, 1978.

Section **3**

Power Plant Costs— Their Estimation and Use

Power plants are capital intensive—i.e. they require large amounts of capital for construction. For this reason, the selection of the type of plant and unit size are usually a function of cost. So every power-plant designer is faced with cost considerations from the moment of inception through the day the plant goes on line.

Costs of utility power plants are well defined by consulting engineering firms, Governmental agencies, and industry trade groups. Industrial power plants, however, are not as clearly analyzed since costs are often unavailable from private industry, trade-group compilations are unheard of, and plant design varies far more widely than in the utility field.

With the greater attention given cogeneration in recent years, more cost data for industrial power plants are becoming available. So this section combines information on both utility and industry plant costs. The data will be valuable in guiding any designer towards the most economic choices for either type of plant. Likewise, cogeneration considerations and costs are also included.

From *Engineering Evaluation of Energy Systems,* by Arthur P. Fraas. Copyright © 1982. Used by permission of McGraw-Hill, Inc. All rights reserved.

The prime criterion for the selection of a power plant type or unit size is minimum cost, if for no other reason than that there is usually a statutory requirement that the utility supply electric power to the consumer at the lowest practicable cost. The cost problems involve not only engineering complexities and a host of arbitrary (but generally reasonable) cost accounting formalities, but also a confusing tangle of laws and political considerations.[1] The latter stem in part from the fact that there are major reductions in cost associated with increasing scale. As a consequence, throughout the United States and the world the electric utilities have been granted a monopoly position: each utility organization is sufficiently large to operate at low costs and thus is given a monopoly in its area of operation. In some instances a local, state, or federal nonprofit organization operates the utility, but most electric power in the United States (~75 percent) is produced by investor-owned (i.e., privately financed) utility companies. The distinction between public and private ownership is important from the cost standpoint, first because the private utilities must pay both property and income taxes, and second because in this capital-intensive business there is a major difference in capital charges. Municipal, state, and Rural Electric Association utilities obtain their capital from bonds that are not only tax-free but also are backed by the tax base of the community; thus they involve almost no risk, and they bear a much lower interest rate than required for corporate bonds or for the dividend return rate for corporate stocks if these are to be competitive in the financial markets. (The ratio of bonded debt to equity capital for investor-owned utilities is about 65:35.) It is this difference in taxes and capital charges that is responsible for the lower rates charged by publicly owned utilities as compared to those charged by investor-owned utilities; e.g., average charges were $0.0241/kWh and $0.0416/kWh, respectively, for the United States as a whole in 1978.[2] As is the case in so many areas, superficial comparisons, however tempting politically, can be highly misleading; a sound analysis requires rigorous and tedious cost accounting.

MAJOR ELEMENTS OF POWER COSTS

The simplest and most widely used cost breakdown is shown graphically in the bar chart of Fig. 3-1, in which the costs of power generation at the plant output bus bars are divided into three categories, i.e., capital, fuel, and operation and maintenance. Note the large differences in these costs for the three different types of plant, the nuclear plant entailing the highest capital and lowest fuel costs whereas the reverse holds for the oil-fired plant.

A quite different cost breakdown is shown in Fig. 3-2, which gives the total expenditures of all United States electric utilities for the 1970–1978 period with the percentages in each of six major categories.[3] The category labeled "Production" corresponds to the costs represented in Fig. 3-1. Of particular interest is the item for "Transmission and Distribution," which represents only 5.6 percent of the total cost to the consumer, or about double the cost of handling customers' accounts. Those who claim that huge savings in distribution costs are possible through the use of many small, individual, widely distributed power plants apparently have not examined publicly audited actual costs for both public and private utilities. In this connection it should be mentioned that the power losses in long-distance electric-power transmission lines for most utilities run between 1 and 2 percent.[1] Thus transmission losses and costs are small compared to the economies of scale for large central stations.

CAPITAL COSTS

Each power plant, whether fossil-fuel steam or nuclear steam, usually consists of one to four units, though occasionally of as many as 10. In thermal power plants each unit is a tightly integrated heat-source–turbine–generator unit with its own controls and instrumentation. Units are usually not built simultaneously at a given plant site, but in sequence as the expanding load

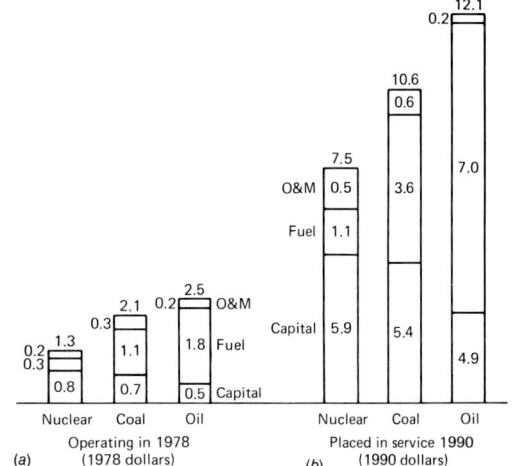

Figure 3-1 Principal cost components for electric power generation in cents per kilowatthour for nuclear, coal, and oil fuels for (a) plants operating in 1978 and (b) plants assumed to be placed in service in 1990.[1]

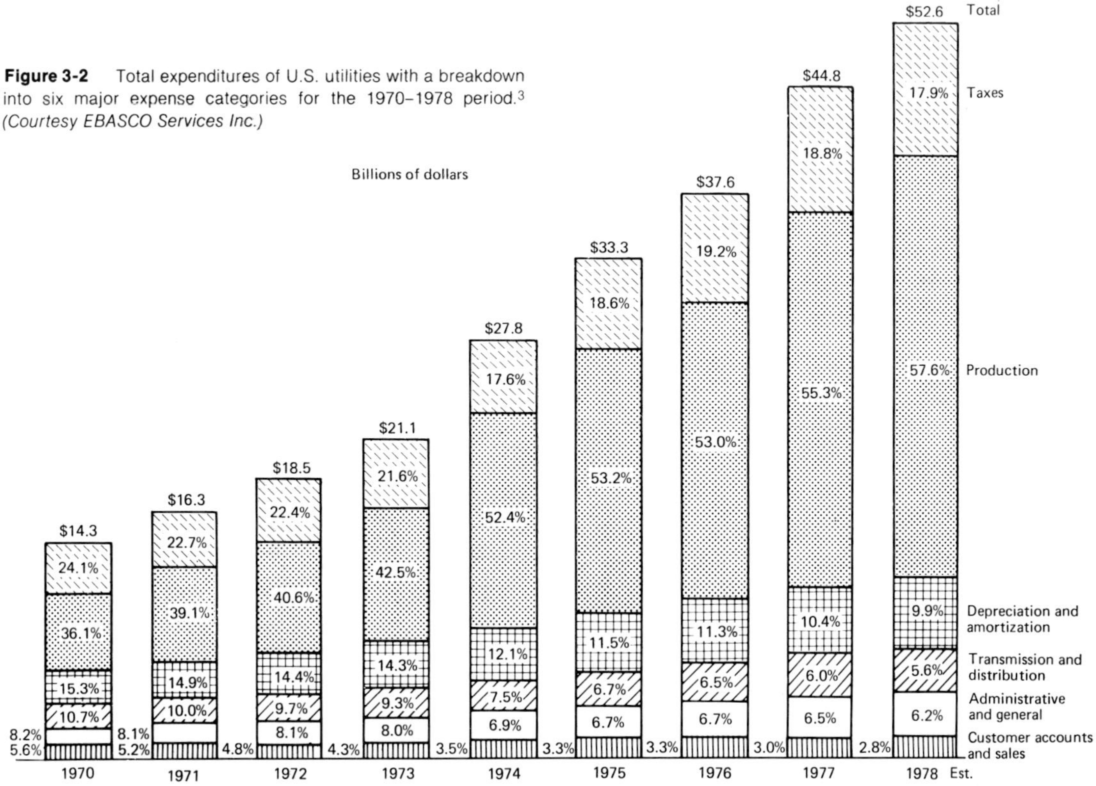

Figure 3-2 Total expenditures of U.S. utilities with a breakdown into six major expense categories for the 1970–1978 period.[3] *(Courtesy EBASCO Services Inc.)*

TABLE 3-1   Order-of-Magnitude Cost Estimates for 1000-MW Initial Units at New Sites[5]   (All Amounts in $1000—Mid-1975 Price Level)

Ebasco Account No.	Description	Nuclear Total	Nuclear Material	Nuclear Installation	Coal-Fired Total	Coal-Fired Material	Coal-Fired Installation
1.	Improvements to site	$ 1,690	$ 780	$ 910	$ 2,030	$ 1,000	$ 1,030
2.	Earthwork and piling	10,150	2,350	7,800	4,050	1,250	2,800
3.	Circulating water system	15,850	7,350	8,500	11,600	5,300	6,300
4.	Concrete	44,900	18,300	26,600	9,550	3,500	6,050
5.	Structural steel, lifting equipment, stacks	43,600	26,200	17,400	28,600	16,600	12,000
6.	Buildings	15,100	6,500	8,600	10,600	4,600	6,000
7.	Turbine generator	41,570	39,000	2,570	29,550	27,700	1,850
8.	Steam generator and accessories	—	—	—	128,900*	93,200	35,700
9.	Nuclear steam supply system	72,300	65,100	7,200	—	—	—
10.	Other mechanical equipment	20,900	18,300	2,600	14,700	12,600	2,100
11.	Coal- and ash-handling equipment	—	—	—	18,800	13,600	5,200
12.	Piping	45,900	24,700	21,200	17,300	9,600	7,700
13.	Insulation and lagging	3,700	1,040	2,660	10,900	3,300	7,600
14.	Instrumentation	2,870	2,000	870	2,680	2,300	380
15.	Electrical equipment	43,200	24,300	18,900	25,400	15,800	9,600
16.	Painting and finishing	2,100	600	1,500	1,790	500	1,290
17.	Off-site facilities	—	—	—	19,030	630	18,400
18.	Substation	2,290	1,820	470	2,290	1,820	470
	Total direct construction cost	$366,120	$238,340	$127,780	$337,770†	$213,300	$124,470
	Indirect construction cost	38,300	—	38,300	28,600	—	28,600
	Subtotal for contingencies	$404,420	$238,340	$166,080	$366,370	$213,300	$153,070
	Contingencies	77,300	35,750	41,550	51,900	21,300	30,600
	Total specific construction cost	$481,720	$274,090	$207,630	$418,270	$234,600	$183,670
	Home office services and fees	62,580			41,830		
	Total construction cost	$544,300			$460,100		
	Interest during construction and other client charges	Not included			Not included		
	Total project cost	$544,300			$460,100		

*Includes electrostatic precipitator and SO_2 removal equipment. Direct cost of SO_2 removal equipment included in Account No. 8 is $52,700,000.

†Total cost of SO_2 removal system including cost of required structures, foundations, off-site facilities, piping, electrical equipment, wiring, etc., for installation and operation of this system that are covered under the other appropriate accounts is $81,900,000.

Oil-Fired		
Total	Material	Installation
$ 1,620	$ 750	$ 870
1,820	550	1,270
11,600	5,300	6,300
4,060	1,380	2,680
11,800	6,700	5,100
6,930	2,930	4,000
29,550	27,700	1,850
38,800	26,300	12,500
—	—	—
16,380	13,800	2,580
—	—	—
14,800	8,200	6,600
2,190	600	1,530
2,150	1,850	300
16,400	10,100	6,300
1,320	370	950
—	—	—
2,290	1,820	470
$161,710	$108,410	$53,300
12,300	—	12,300
$174,010	$108,410	$65,600
23,900	10,800	13,100
$197,910	$119,210	$78,700
19,790		
$217,700		
Not included		
$217,700		

of a utility requires additional capacity. Both the unit output and the capital costs commonly increase from the first to the latest unit installed. To facilitate rate setting, the Federal Power Commission (FPC)—subsequently incorporated into the DOE as the Federal Energy Regulatory Commission (FERC)—standardized the major cost categories involved in the construction and operation of each power plant unit and organized a standard format for presentation of these costs.[4] Cost summaries of these data are presented in Table 3-1 for typical nuclear, coal-fired and oil-fired power plant units.[5] The complete detailed cost breakdowns from which these summaries were prepared commonly total around 50 pages for each unit with each detailed item bearing a standard FPC account number; e.g., account number 234.11 is for closed feed-water heaters and account number 232.151 is for the chlorine injection system for the condenser cooling-water system. Table 3-2 has been included both to show the standard format for summarizing the capital costs and to indicate the relative size of the principal cost items. Note particularly that although the heat source, steam generator, and turbine generator are the major cost items, both Table 3-1 and Table 3-2 indicate that their combined cost is still only about 25 percent of the total cost of the plant.

Major Factors Affecting Capital Costs

Capital costs vary widely from one unit to another. The extent of this variation is indicated by Fig. 3-3, which shows a scatter band of actual costs for coal-fired power plant units as a function of the year in which they began commercial operation.[6] A similar scatter band for light-water reactors is displaced upward 10 to 30 percent, that for oil-fired steam units is displaced downward about 40 percent, and that for gas-fired steam units is displaced downward at least 50 percent. In all cases the width of the scatter band is large, the lowest costs running about half of the highest. The reasons for these variations include the size of the unit, whether the unit is the first to be installed at the site or is an addition (i.e., whether site preparation costs have already been charged off), the soundness of foundation conditions, the amount and type of excavation required, and local labor rates.

Size of Unit

One of the biggest factors affecting the capital cost per kilowatt is the size of the unit, the unit cost of a 1000-MWe unit being only ~60 percent as much as that for a 200-MWe unit[1] (Fig. 3-4). This stems in large measure from the fact that many costs, such as those for engineering and for instrumentation and control equipment, increase relatively little with the size of the unit, and the cost of any particular item of equipment does not go up in

TABLE 3-2 Plant Capital Investment Summary

BASIC DATA

Name of plant	Seabrook Station Unit #1 + ½ Common	Cost basis:	at start of construction
Net capacity	1150 MWe		
Reactor type	Westinghouse PWR		**Type of cooling**
Location	Seabrook, N. H.	Run of river	
Design and construction period		Natural draft cooling towers	
Month, year NSSS order placed	January 1973	Mechanical draft cooling towers	
Month, year of commercial operation	April 1983	Other (describe)	Atlantic Ocean
Length of workweek	40 hours, 5 days hours		
Interest rate, interest during construction	9½ or compound?		

COST SUMMARY

Account number	Account title	Total cost (thousand dollars)
DIRECT COSTS		
20	Land and land rights	$ 1,250
	PHYSICAL PLANT	
21	Structures and site facilities	198,092
22	Reactor plant equipment	170,232
23	Turbine plant equipment	68,264
24	Electric plant equipment	48,930
325, 352, 353	Misc. plant equipment	20,950
	Subtotal	$ 507,718
	Spare parts allowance	3,236
	Contingency allowance	38,835
	Subtotal	$ 549,789
INDIRECT COSTS		
91	Construction facilities, equip't. and services	$ 40,502
92	Engineering and const. mg't. services	141,316
93	Other costs	46,341
94	Interest during construction	371,000
	Subtotal	$ 599,159
	Start of construction cost	$1,148,948
	*Escalation during construction (8%/yr)	88,673
	Total plant capital investment ($1077/kW)	$1,237,621

*Indicate separate escalation rates for site labor, site materials, and for purchased equipment, if applicable. Escalation rate is 8%/yr, simple.

Note: Cost data above are for Unit #1 plus ½ of the Common facilities. Date of latest construction cost estimate is January 1979. (Table submitted to the U.S. Nuclear Regulatory Commission June 22, 1979 in connection with a request for a license, Docket Nos. 50-443 and 50-444.)

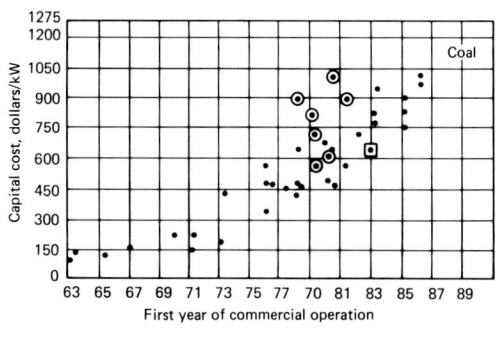

Figure 3-3 Capital cost in dollars per kilowatt for single- and twin-unit fossil power plants in the size range of 250 to 1300 MW.[6]

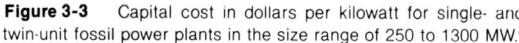

Figure 3-4 Effects of size on the 1964 cost and the performance of coal-fired steam-electric plants consisting of two units.[7]

proportion to its size. For example, the cost of the coal-handling equipment per ton of coal handled drops off rapidly as the capacity of the system is increased up to 1000 tons per hour (which is equivalent to ~ 2000 MWe), an effect shown in Fig. 3-5.[8]

Effects of Fuel Used

The amount and cost of the auxiliary equipment required varies widely with the type of fuel, coal requiring a large coal yard, conveyors, crushers, pulverizers, precipitators, and flue-gas desulfurization, soot-cleaning, and ash-handling equipment, as well as an ash-disposal area. Most of these costs can be avoided if oil is used, and all of them can be avoided if gas is the fuel. Further, the burners are simpler and less expensive for oil and even more so for gas. Costs also vary with the type of coal burned and its ash content. The furnace size must be increased for lignite, for example (Fig. 2-11). Nuclear plants require a

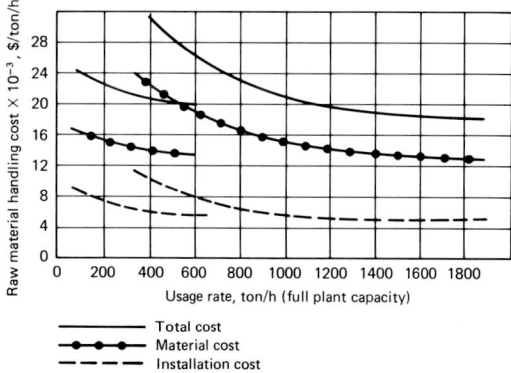

Figure 3-5 Effects of size on the unit price of systems for handling coal, limestone, and dolomite from the point of delivery by rail to silo storage at the furnace.[8]

large capital investment in the reactor plus much nuclear instrumentation and complex facilities for spent fuel and radioactive waste handling as dictated by nuclear safety regulations.

Plant Complexity

The thermal efficiency of a steam plant can be improved by refining the thermodynamic cycle through the inclusion of reheating and progressively more elaborate provisions for regenerative feed-water heating. The more elegant these details in the thermodynamic cycle, the greater the amount and complexity of the piping and equipment; increases in thermal efficiency are therefore obtained at the expense of an increase in the capital cost per kilowatt. Thus, in coal-fired plants designed for intermediate loads and peaking service the system design is simplified and compromised in the direction of reduced capital costs at the expense of a somewhat reduced efficiency.

Effects of Steam Temperature and Pressure on Steam-Turbine Costs

Increasing the steam pressure and temperature increases the cycle efficiency, and this reduces the capital charges per unit of electric output for auxiliaries such as the coal-handling, feed-water, and condenser-cooling systems. It also reduces the unit costs for the steam generator and turbine up to the point where an increase in the temperature requires the use of much more expensive materials. This effect is shown by Fig. 3-6, which gives a curve for the cost of the turbine, condenser, and feed-water system for a 100-MW unit as a function of the steam temperature into the turbine.[9] Although the curve of Fig. 10.6 was prepared in 1953, the same basic trends still hold at the time of writing, because the effects of design temperature on materials costs have not changed significantly since 1950. Recent data for a higher temperature range are given in the chapter on steam plants (Fig. 2-12).

The increases in efficiency and reductions in cost associated with the increasing temperature and pressure, along with increases in scale, were major factors in the big reduction in electric power rates in spite of inflation in other costs through the period from 1900 to 1970. For example, as can be seen in Fig. 10.7, the cost of electricity to consumers in then-current dollars dropped by a factor of 2 between 1930 to 1970, and by a factor of 5 after adjustment for inflation.[10] This is a good measure of the effectiveness of the continuing effort by the electric-utility industry to reduce costs to the consumer.

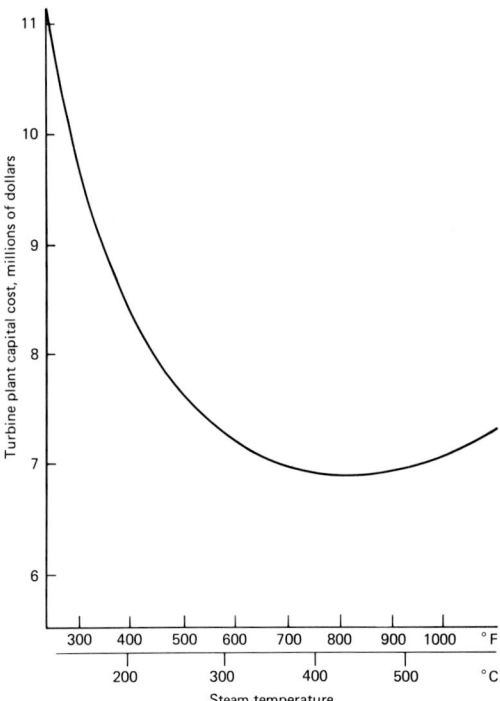

Figure 3-6 Effect of steam-turbine inlet temperature on the capital cost of the steam turbine, condenser, and feed-water system (but not the boiler, electric generator, or other elements of the plant).[9] Steam conditions chosen:

Steam cycle		Feed-water system	
Pressure, psig	Temperature, °F	Heaters	Temperature, °F
175	377	0	92
140	500	1	200
265	525	1	240
400	746	3	280
500	750	3	280
565	800	3	280
850	900	5	400
1250	950	5	435
1450	1000	5	450

Effects of Head on the Costs of Hydraulic Turbines

The head on a hydroelectric turbine generator is analogous to the temperature "head" on a steam turbine, and increasing the head reduces the cost per kilowatt of output. This stems from the fact that the physical size of the turbine is determined by the volumetric flow rate while the power output is directly propor-

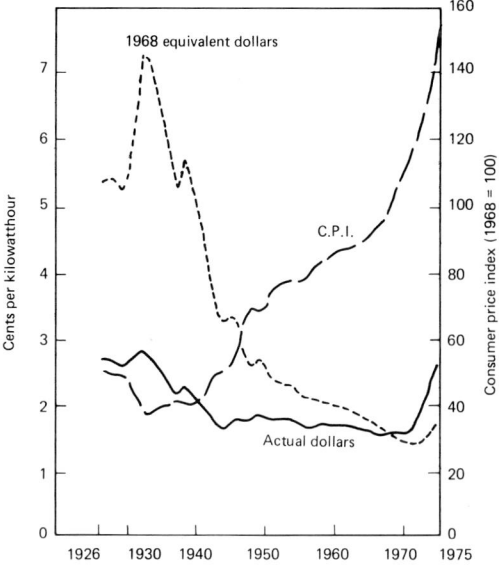

Figure 3-7 Price of electricity to ultimate consumers.[10]

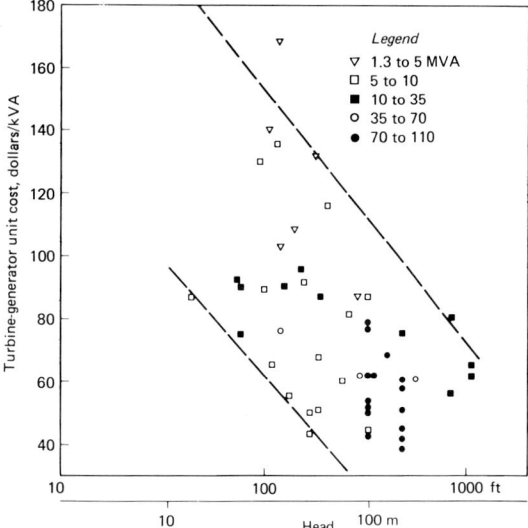

Figure 3-8 Effects of head and power capacity on the unit costs of 117 hydroelectric turbine-generator units. The data were taken from Ref. 11 for the 1950–1958 period, and the costs were converted to 1975 dollars by using the cost escalation factors of Fig. A10.1. A single point was plotted for multiple units purchased on the same contract.

tional to the head. Figure 3-8 shows a scatter band of cost data for 117 turbine-generator units purchased by the U.S. Bureau of Reclamation during the 1950–1958 period.[11] In preparing Fig. 3-8 the author corrected the original data for inflation to give equivalent 1975 dollars by using a suitable correction factor. Note that there are points in Fig. 3-8 for power capacities ranging from 1.3 to 110 MVA (megavoltamperes). Inspection of the scatter band indicates that the unit cost is inversely proportional to the head. For outputs above 6 MVA there is no significant effect of capacity on the unit cost, but all the points for outputs below 6 MVA lie in the upper part of the scatter band. Note, too, that only 6 of the 117 cases are for heads below 30 m (100 ft), and none are for a head below 12 m (40 ft) because in that region unit costs were too high to be attractive. Although the fact is not readily shown, a review of the basic data disclosed that the lower points in the scatter band were for orders placed during periods of economic recession—a condition to be expected, because fabricators will bid low during such periods in order to avoid laying off their employes and so will accept a loss on the job. While the scatter band of Fig. 3-8 may appear wide, even the bids for a particular contract commonly differ by as much as a factor of 2, a reflection of differences between manufacturers in the availability of designs, model test data, and tooling applicable to the particular specifications provided.

A sophisticated effort to relate the costs of hydraulic turbines to the design parameters was prepared for publication just as this book was being completed. The analysis was based on data for 29 Kaplan and 17 Francis turbine contracts (each involving from 1 to 14 turbines) let by the U.S. Army Corps of Engineers during the 1950–1979 period.[12] The cost of the generator was not included; the generators were procured under separate contracts. A multiple-regression computer program was employed to obtain empirical values for the constants in the equations used. Eight major cost items were considered, of which the turbine cost typically represented ~83 percent of the total, while the balance was for model tests, tools, spare parts, and installation. The resulting equations for the major cost item—the design, construction, and delivery of the turbine itself—for Kaplan and Francis units are as follows:

Kaplan: $W = 7.8215 \times 10^{-5} D^{3.3407} n^{0.064} H_s^{1.363} \text{hp}^{-0.7338}$ (3-1)

Cost (\$) = $(\text{HCIT})^{1.0101} (nW)^{0.9104}$ (3-2)

Francis: $W = 355 \times 10^{-5} D^{1.9566} H_s^{0.331}$ (3-3)

Cost (\$) = $38.1725 (\text{HCIT})^{1.0394} D^{2.0008} n^{0.8203}$ (3-4)

where W = weight, tons
D = throat diameter, in
n = number of units in the contract
H_s = static head, ft
hp = horsepower
HCIT = hydro cost index for turbines

This cost index is compiled by the Water and Power Resources Service (formerly the U.S. Bureau of Reclamation) and published quarterly in *Engineering News Record*. It is approximately proportional to the *Engineering News Record* Construction Cost Index running ~83 percent of the latter in 1979.

If one examines these equations to appraise the effects of the prime design variables (the output and head) on the unit cost in dollars per kilowatt, cost escalation and the number of units can be dropped and the cost for the turbines can be divided by the power output which is proportional to the head and the water flow rate. For Kaplan turbines

$$\text{hp} \sim D^2 V H_s \quad (3\text{-}5)$$

where V is the axial flow velocity through the throat, and

$$\frac{\text{Cost}}{\text{hp}} \sim \frac{D^{3.34} H_s^{1.36} \text{hp}^{-0.73}}{\text{hp}} = \frac{D^{3.34} H_s^{1.36} (D^2 V H_s)^{-0.73}}{D^2 V H_s} \quad (3\text{-}6)$$

$$\frac{\text{Cost}}{\text{hp}} \sim \frac{1}{D^{0.12} V^{1.73} H_s^{0.37}} \quad (3\text{-}7)$$

For Kaplan turbines which normally operate at relatively low heads (< 30 m, or < 100 ft), the rotational speed is usually not limited by cavitation considerations. Hence the wheel speed and the throat velocity vary as $H_s^{1/2}$ (Chap. 3), and the unit per cost per kilowatt becomes:

$$\text{Unit cost (\$/kW)} \sim \frac{1}{D^{0.12} H_s^{1.23}} \quad (3\text{-}8)$$

Thus the cost data show that the unit cost of Kaplan turbines is not very sensitive to the diameter but increases rapidly with a reduction in the head, both vital factors when considering the possibilities of exploiting small, low-head hydro sites.

A similar rationale for Francis turbines yields

Cost ~ D^2

$$\text{Unit cost (\$/kW)} \sim \frac{D^2}{D^2 V H_s} \sim \frac{1}{V H_s} \quad (3\text{-}9)$$

For the higher head region (above ~ 30 m) in which Francis turbines are normally employed, cavitation considerations limit the tip speed to around 36 m/s (120 ft/s) and hence the axial velocity through the throat. Thus V is essentially constant for the range of interest, and

$$\text{Unit cost (\$/kW)} \sim \frac{1}{H_s} \quad (3\text{-}10)$$

which indicates that the unit costs of Francis turbines are only a little less sensitive to the head than for Kaplan turbines, and are essentially independent of the size.

Cooling Towers

As discussed in the previous chapter on environmental effects, cooling towers are commonly required for new power plants on lake and river sites, and these are expensive.[13] The commonly used wet cooling towers entail incremental capital costs over those required for the simpler direct use of lake or river water of ~ \$30/kW in 1978 dollars. To save water in arid regions, dry cooling towers using water-to-air heat exchangers have been employed in a few cases, but their incremental cost ran ~ \$100/kWe over wet towers in 1975.[13] (See also the last section of Section 2.)

Time for Construction

The time required for the construction of a power plant unit is substantial, running as much as 10 years for a nuclear plant in the latter 1970s. The sequence of events and the time required for each phase of a project can be envisioned by examining Fig. 3-9, which outlines the major steps between the time a contract is awarded and the time commercial operation is commenced.[14] Of course, a substantial period of work precedes the contract award, including site selection, core drilling to establish foundation conditions if a new site is used, and conceptual design studies to establish firm design specifications. Note that a cost estimate is made at the time the design contract is awarded, and that Fig. 3-9 shows the cost estimate being revised at three subsequent points as the design and subsequent construction progress. Note, too, that construction is started before the

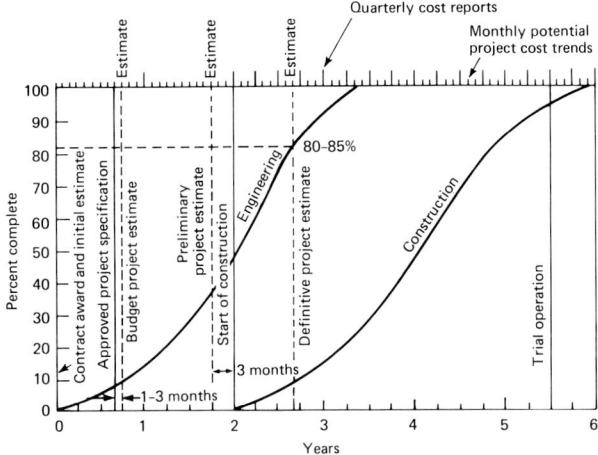

Figure 3-9 Typical schedule for estimates and cost reports—fossil project.[14]

detail design work is completed in order to expedite the project. However, this can lead to trouble if the design work falls behind schedule (usually because of changes in government regulations). In some instances construction workers are forced to stop work for lack of drawings, and in one instance in the author's experience, piping was installed without drawings and the drawings were subsequently corrected to conform to the actual installation! In any event such delays increase costs substantially.

A major step toward minimizing the time for construction has been the development of techniques for estimating the time required for preparing drawings, procuring materials and equipment, and carrying out the various construction and installation jobs at the site. The detailed estimates are organized in a chart that shows the flow of items into the site, and a computer program is commonly used to assist in detailed surveys and analyses. The chart and computer data are revised regularly as work progresses, mainly to highlight items that are falling behind schedule and may cause serious delays, but also to keep a close check on costs.

Costs of Interest During Construction and Escalation

In recent years the combined effects of construction delays stemming from government regulations, high interest rates, and inflation have increased the cost of interest on the investment during construction to as much as 35 percent of the total construction cost. This cost is called *Interest During Construction* (IDC), or *Allowance for Funds Used During Construction* (AFUDC). Figures 3-10 and 3-11 show that this cost plus escalation can total nearly double the cost of the plant estimated on the basis of the prevailing costs at the time the commitment for the plant construction was made.[1,15,16]

Additional costs to the consumer are likely to result from delays in construction if the utility finds itself short of power and must purchase power at a premium from another utility or operate high-cost oil-fired peaking units for intermediate load service.

The 1970s have seen a rate of increase in construction costs without precedent in history. Much of this has stemmed from inflation, but costs have also increased because of reduced labor productivity and increased regulation. Thus construction costs have escalated at a higher rate than inflation in spite of advances in technology, such as the widespread use of automatic welding equipment and larger earthmoving equipment.

Licensing

The direct costs associated with licensing a power plant were running roughly one-third of the cost of new plants in the latter 1970s. An incredible amount of paper work is required: from 10 to 20 tons for a new nuclear power plant! Many local, state, and

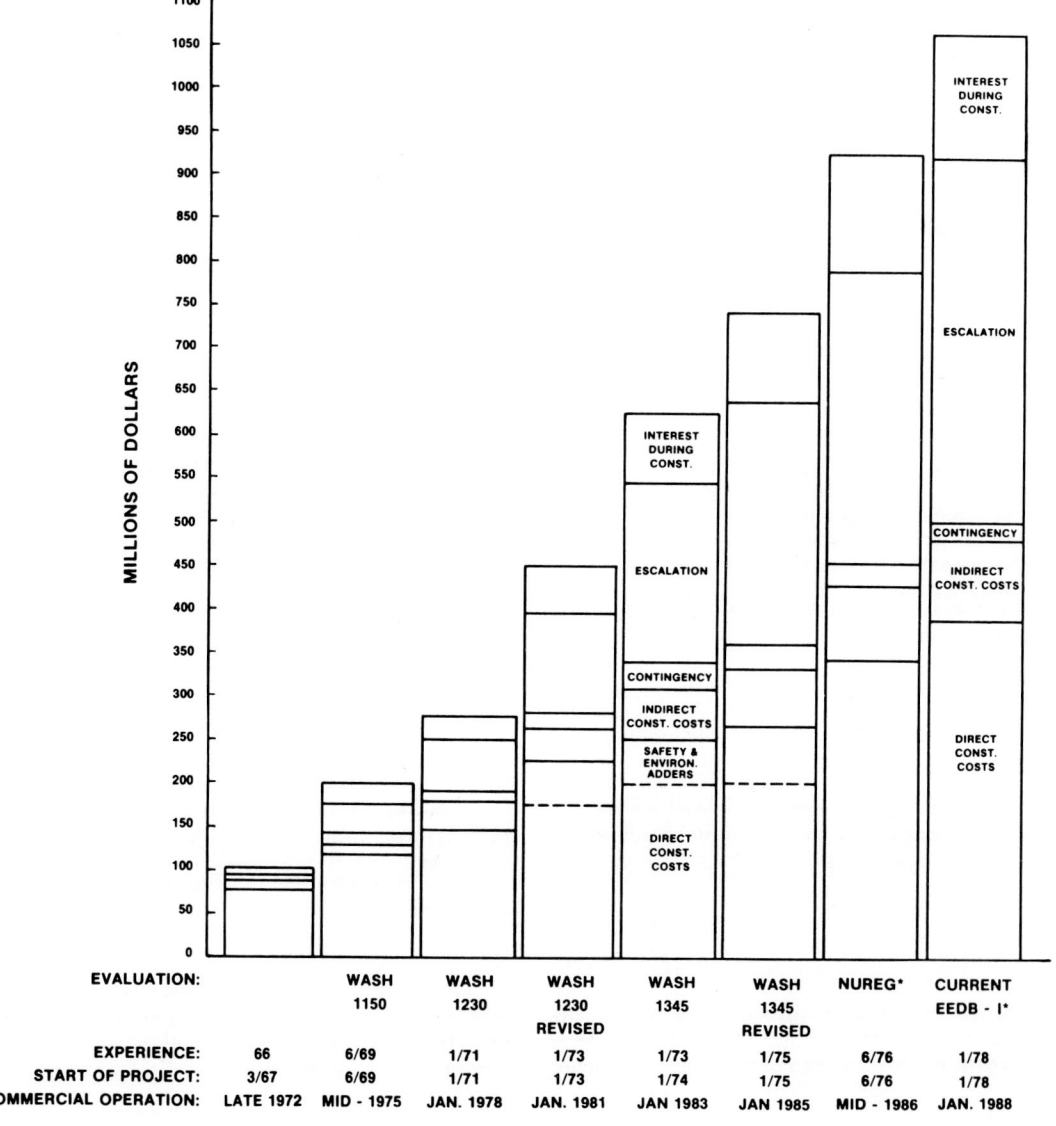

Figure 3-10 Comparison of coal-fired plant cost estimates (total investment cost for 1000-MWe units.).[15]

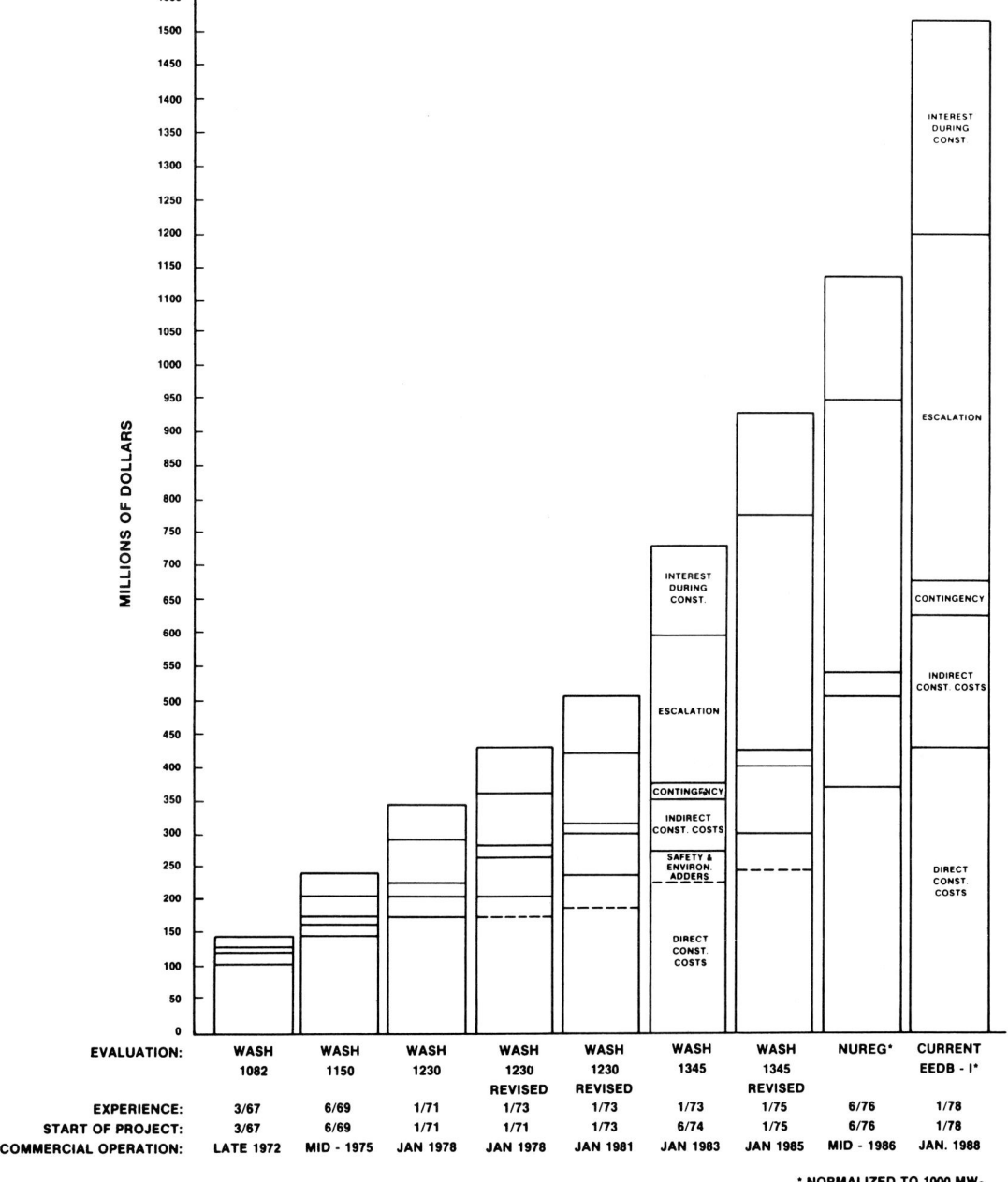

Figure 3-11 Comparison of nuclear plant cost estimates (total investment cost for 1000-MWe units).[15] (*Courtesy United Engineers and Constructors, Inc.*)

federal agencies (including, on the federal level, the FERC, EPA, OSHA, NRC) each have voluminous requirements that must be met with reports prepared according to their own particular formats and content specifications, often inconsistent with each other and sometimes actually contradictory. The resulting amount of engineering work and the attendant costs are staggering, commonly running over 30 percent of the direct cost of a nuclear plant. Some notion of the character and magnitude of this work is given by Figs. 3-10 and 3-11 and by Table 3-3, which also show the enormous increase in these requirements between the 1960s and 1970s. It is not clear that there has been any increase in public health or safety as a consequence of these requirements, but the increase in the cost of electric power to the consumer unquestionably has been high, especially when both direct and indirect costs are included.

Relative Amounts of Major Capital Cost Items

The relative amounts of the direct and indirect costs of new power plants have changed greatly in the 1970s, an effect shown graphically in Fig. 3-12.[15] Note that this bar chart also gives an excellent insight into the relative size of the various components of the direct and indirect costs.

Effects of Extended Workweek

Efforts to "make up time" and stay on schedule by going to an extended workweek lead to increased costs, not only because of the wage bonuses for overtime and Sunday work but also because of reduced worker productivity as a consequence of fatigue. This effect is indicated in Fig. 3-13 and applies to both craft labor and professional workers. The effect was first determined quantitatively in England during World War II, and essentially similar data have been obtained in the United States.

Capital Costs of New Types of Power Plants

Detailed cost breakdowns for conventional fossil fuel and nuclear power plants are available, and so good cost estimates can be made except for uncertainties stemming from possible delays in licensing and the amount of inflation. The situation is

TABLE 3-3 Changes in the Documentation Requirements for Nuclear Units from the 1960s to the 1970s[17]

Item	1960s	1970s
Engineering man-hours	500,000	3,500,000
Job duration, years	8	13
Engineering personnel	125	350
Correspondence		
Transmitted	12,400	55,000
Received (not including vendor dwgs. trans.)	8600	17,000
PSAR volumes	2	13
Environmental reports		
Construction permit stage	0	10 (volumes)
Operating license stage	1 (volume)	12 (volumes)
Public hearings	2 (days)	16 (days) (ACRS & ASLB)
Quality manuals	1	6
Calculation pages	1400	70,000
Hanger engineering calculation	0	12,000
Pipe stress computer calculations	500	500
Civil/structural seismic computer calculations	16	200
Drawings (total)	2200	45,000
Purchase orders	240	400
Subcontracts	25	60
Specifications	230	490
Vendor drawings	37,000	90,000

Source: AIF January 1978

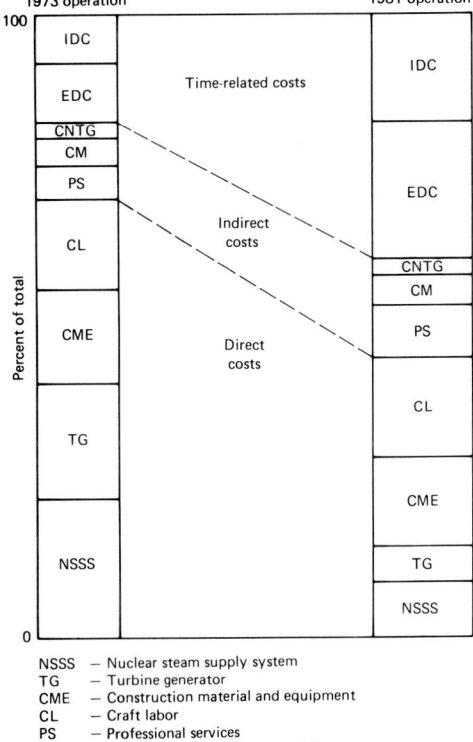

Figure 3-12 Shifts in distribution of nuclear power plant capital costs.[15] (*Courtesy United Engineers and Constructors, Inc.*)

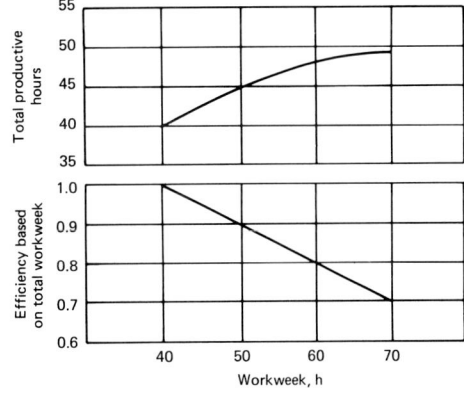

Figure 3-13 Effects of sustained overtime on productivity of site labor.[15] (*Courtesy United Engineers and Constructors, Inc.*)

much more difficult with respect to new types of power plants, and the uncertainties are inherently greater. Cost estimates must be made, however, and this is ordinarily done by estimating the cost of each individual piece of equipment required by using generic data for similar components for both power plants and the chemical process industries. Compilations of these cost data can be found in books, such as Ref. 17, where costs are given in terms of such parameters as the power for motors, the surface area for heat exchangers, the diameter and length for piping, the capacity and head for compressors, fans, and pumps. Cost estimates for unusual mechanical equipment, such as mountings for solar mirrors designed to track the sun, can be made on the basis of weight, type of material, and the amount and precision of the machining required by using costs for roughly similar machinery. In fact, component weight is probably the best single index of cost if the appropriate cost factors for the type of equipment and the material involved are employed. Under any circumstances, for unusual equipment it is best to make estimates on two or three different bases to provide some degree of assurance that the cost estimates are meaningful.

Heat-Exchanger Costs

Heat exchangers often constitute the largest item in the cost of a power plant, particularly in advanced types. Hence a generalized survey of factors determining their costs appears in order, particularly because they give a good illustration of the various design factors affecting costs. To keep the cost of equipment at as low a level as possible, it is desirable to employ as inexpensive a material as can be used after allowances are made for corrosion considerations and pressure stresses. This, in turn, implies that the peak temperature and pressure chosen for the thermodynamic cycle must represent a compromise between cycle efficiency and the pressure differentials across the walls of piping, tubing, heat-exchanger shells, etc., so that thick tube walls of an expensive alloy will not be required.

Example The effect of the choice of material and tube-wall thickness on the cost of tubing is given in Fig. 3-15 for 0.75-in- OD tubing made to a conventional ASTM specification for non-nuclear applications.[18] The basic data were obtained via conversations with tubing vendors in May 1975 assuming only modest production quantities of about 3048 m (10,000 ft). The effects of tube-wall thickness on costs were available only for Incoloy 800. As a consequence, in constructing Fig. 3-15, the writer assumed that the effects of wall thickness on cost would be similar for the other materials and simply drew in a set of parallel curves through the points obtained for a wall thickness of 1.65 mm (0.065 in). If the tubing must meet nuclear quality assurance standards, the extra cost for quality control and inspection in 1975 ran about $10/m ($3/ft), essentially irrespective of the alloy employed.

The tube-wall thickness required for any given installation depends in part on the pressure differential that must be accommodated across the tube wall and in part on the extent to which the wall must be thickened to allow for corrosion in the course of the life of the heat exchanger. In attempting to appraise the relative advantages of different materials after allowing for differences in strength at elevated temperatures, the curves of Fig. 3-14 were prepared using the data of Fig. 3-15 and assuming a heat exchanger that would have a pressure differential across the tube walls of 34 atm (500 psi) and a temperature difference sufficient to yield a heat flux of 15.8 kW/cm² [50,000 Btu/(h·ft²)]. Allowable stresses were taken from the ASME pressure vessel code for the three iron-chrome-nickel alloys, while the allowable stress for the niobium alloy was taken as 60 percent of the stress for 1 percent creep in 10,000 h, a value which is consistent with the relation between the ASME code stresses for the other three materials and their stresses for 1 percent creep in 10,000 h. In examining Fig. 3-14 and applying a factor of 2 to 2.5 to convert the tubing cost in dollars per kilowatt thermal to dollars per kilowatt of electric output, it is interesting to see that if the heat flux can be as high as 15.8 kW/cm² [50,000 Btu/(h·ft²)], the cost of the tubing is not a serious charge even for the expensive alloys. Note, too, that above 870°C (1600°F) the higher strength of Nb-1% Zr gives a lower cost than Hastelloy X, and that for an advanced nuclear plant employing a 982°C (1800°F) boiler for a potassium vapor cycle with a thermal efficiency of about 50 percent, if allowances are made for the lower pressure the cost of the Nb-1% Zr tubing would run about $18/kWe, a not unreasonable figure.

One of the most important steps that can be taken to reduce the cost of equipment is to design it so that it can be fabricated and assembled in the shop rather than in the field. This, in turn, implies that it is desirable to design components such as a large heat exchanger so that they consist of a multiplicity of basic modules, each of which can be fabricated and inspected independently before final assembly. If possible, these modules should be assembled into complete units before shipment to the

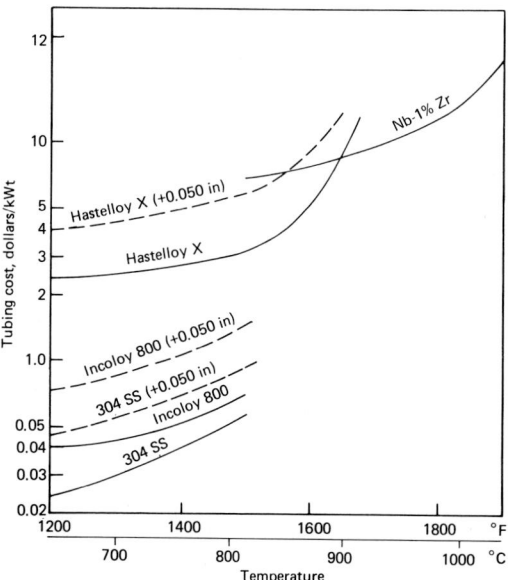

Figure 3-14 Effects of design temperature on the cost of tubing for a 500 psi pressure differential and a heat flux of 50,000 Btu/(h·ft²) assuming the estimated costs of Fig. A10.3 and ASME code allowable stresses (or equivalent). Two curves are given for the Fe-Cr-Ni alloys, one with and one without an extra thickness allowance of 0.050 in for corrosion in a coal-fired furnace.[18]

site, although if they are too large for shipment of assembled units to be feasible, they can at least be shipped as subassemblies.

A major factor in the cost of heat exchangers is the fabrication of the headers including making the tube-to-header joints. If a high degree of leak-tightness must be obtained, as in a helium or an alkali metal system, the joints must be welded and the detail design of the weld at the header sheet deserves much attention both to facilitate the welding operation and to ease inspection. For a nuclear quality heliarc weld in Fe-Cr-Ni alloys, the cost in 1975 was around $30 per tube-to-header joint including inspection, while for Nb-1% the cost was about double that because the welds must be made in a dry box.

It is usually necessary to make use of bent tubes. Where possible, all the bends should be kept in the same plane, because this not only simplifies the fabrication operation but also eases inspection. Springback problems make it difficult to maintain close tolerances on the shape of the tubes. Hence it is desirable to design the heat exchangers so that the tubes will be flexible enough for their shape to be determined by spacers and close tolerances on the bends will not be required.

FUEL COSTS

Fuel costs are a major fraction of the cost of power generation (Fig. 3-1), running about half the total cost for coal-fired plants. But the absolute and relative costs have changed markedly since 1970, with the price of oil in particular escalating rapidly. The world's fuel resource base is such that coal and nuclear fuel are clearly the lowest-cost fuels of the future; hence fuel cost trends can be expected to continue well into the twenty-first century.[19] There is little reduction in fuel costs gained by an increase in scale beyond ~600 MWe because the thermal efficiency increases relatively little with a further increase in the size of the plant (Fig. 3-4). The biggest uncertainties lie in the questions of changes in the relative costs associated with the environmental and safety regulations attending the use of these fuels.

Flue Gas Desulfurization

The environmental effects of sulfur emissions are treated in the previous chapter, while the equipment required and its performance characteristics are treated in the next chapter on steam power plants. The discussion here is thus limited to a brief look at the overall cost effects. A typical estimate of the incremental capital cost of flue gas desulfurization when incorporated in new coal-fired plants in the design stage has been made from actual installation costs by Battelle Columbus Laboratories.[20]

This study yielded an incremental capital cost in 1980 dollars of $101/kW for a 175-MW unit and $70/kW for a 300-MW unit on the basis of 1978 EPA requirements (subsequently made more stringent in July 1979). The study indicated that costs would be reduced ~10 percent if the design value for the coal sulfur content for which the system was designed were less than 1 percent, while capital costs would be increased by ~10 percent if the sulfur content were above 3.5 percent. In addition, there would be a capital cost for sludge disposal running from as little as $2/kW to as much as $66/kW, depending on the plant site. Retrofitting an FGD system to an existing power plant gives costs from 30 to 50 percent greater than for systems incorporated in a plant in the design stage.

The direct operating costs of FGD systems are substantial. A typical 1978 estimate was 3 to 8 mills per kilowatthour, or 40 to 60 percent of the total costs for operation and maintenance.[20] In addition, there have been major costs stemming from the increased plant downtime caused by corrosion and clogging of FGD system components.

Nuclear Plant Fuel Costs

The overall costs for the fuel cycle of nuclear plants are complicated by the large capital charge for the investment in the fuel and the expenses associated with its preparation, together with political uncertainties with respect to reprocessing and radioactive waste disposal. An excellent yet brief treatment of these cost factors for LWRs is given in Ref. 1, from which Table 3-4 has

TABLE 3-4 Projected Nuclear Fuel Cycle Cost for 1250-MW Plant Placed in Service in 1990 (Ref. 1)

Fuel cycle cost, cents/kWh	No recycle	Recycle
Levelized for first 5 years	1.3	1.1
Levelized over 30 years	2.0	1.5
Components (based on 30 years), %		
Uranium	68.0	58.0
Conversion	1.6	1.4
Enrichment	22.0	21.0
Fabrication	5.9	9.4
Transportation	0.6	1.2
Reprocessing	—	5.0
Waste	1.9	4.0

Note: Financing of fuel cycle is included in above. Costs are expressed in 1990 dollars. Recycle is believed more appropriate than nonrecycle for fuel cost predictions in 1990, and is used for prediction of generating cost.

TABLE 3-5   Staff Requirement for Coal-Fired Plants with FGD Systems [19]

	400-700 MWe unit				701-1300 MWe unit			
	Units per site				Units per site			
	1	2	3	4	1	2	3	4
Plant manager's office								
Manager	1	1	1	1	1	1	1	1
Assistant	1	2	3	4	1	2	3	4
Environmental control	1	1	1	1	1	1	1	1
Public relations	1	1	1	1	1	1	1	1
Training	1	1	1	1	1	1	1	1
Safety	1	1	1	1	1	1	1	1
Administrative services	13	14	15	16	13	14	15	16
Health services	1	1	1	2	1	1	1	2
Security	7	7	9	14	7	7	9	14
Subtotal	27	29	33	41	27	29	33	41
Operations								
Supervision (excluding shift)	3	3	5	5	3	3	5	5
Shifts	45	50	60	65	45	50	60	65
Fuel and limestone handling	12	12	12	18	12	12	12	18
Waste systems	15	30	45	60	15	30	45	60
Subtotal	75	95	122	148	75	95	122	148
Maintenance								
Supervision	8	8	10	12	8	8	10	12
Crafts	90	115	135	155	95	120	140	160
Peak maintenance annualized	33	66	99	132	35	70	105	140
Subtotal	131	189	244	299	138	198	255	312
Technical and engineering								
Waste	1	2	3	4	1	2	3	4
Radiochemical	2	2	3	4	2	2	3	4
Instrumentation and controls	2	2	3	4	2	2	3	4
Performance, reports, and technicians	14	17	21	24	14	17	21	24
Subtotal	19	23	30	36	19	23	30	36
Total	252	336	429	524	259	345	440	537

TABLE 3-6   Maintenance Materials Cost Factors as a Percentage of Maintenance Labor Cost* (Ref. 19)

	Fixed	Variable	Total
LWR	100	0	100
Coal	50	17	67
Coal with FGD	53	29	82

*Estimated at 80% plant capacity factor.

been taken. Two columns of numbers are given, one for "no recycle" (the practice up to the time of writing) and the other for "recycle" (the approach adopted by nearly every country having nuclear power plants). Kennedy in Ref. 1 addressed the charges by antinuclear activists that the costs of Table 3-4 do not include all the charges that should be made, and he quite effectively refutes these claims. The biggest questions are environmental, and these are treated in the previous chapter.

For breeder reactors, the fuel cycle costs would be much lower than for the LWRs of Table 3-4, because the cost of fresh uranium—the biggest item in Table 3-4—would be almost nil. The costs and advantages of the breeder reactor fuel cycle are treated in Ref. 21. This paper is representative of the views of those who have made a thorough engineering and economic study of the problems involved and who have generally concluded that basic economics and the relative availability of fuels of various types will eventually drive the world energy economy to the widespread use of breeder reactors early in the twenty-first century.

OPERATION AND MAINTENANCE

The operation and maintenance (O&M) costs of power plants vary substantially with both the size of the individual units and the number of units in a plant.[22] The biggest cost is for personnel. As shown in Table 3-5, the number of people required increases much less rapidly than the number of units or their size; hence here again there are major economies of scale that have contributed to the trend toward ever larger power plants.

Nuclear Versus Coal-Fired Plants

As one might expect, the manpower requirements for fossil fuel and nuclear plants are similar. However, nuclear plants require an additional 50 to 100 employees in the security force but only about half as many in the operating crew and roughly two-thirds as many in the maintenance force; they do not entail the extra work and the troubles with dirt and corrosion in the coal- and ash-handling equipment that coal-fired plants involve. Oil- and gas-fired plants require substantially smaller crews than coal-fired plants because the fuel- and ash-handling problems are greatly eased with fuel oil and eliminated with natural gas.

Utility systems normally have special crews that move from plant to plant to expedite maintenance work during major shutdowns, particularly during the annual shutdowns for general maintenance and overhaul. Allowances for these employees were included in Table 3-5 and the above discussion.

Hydroelectric Plants

Hydroelectric units and plants are generally much smaller than the 400-MW size that is the smallest in Table 3-5. An interesting comparison can be made, however, with the 24 plants totaling 3000 MW of hydro operated by TVA. These 24 plants have about the same number of operation and maintenance personnel as a single nuclear steam plant with four units each in the 400 to 700 MW range.

Materials and Supplies

The expenses for materials and supplies vary somewhat with the type of plant and its age, as well as the scale of the plant. It has been found empirically that a reasonable estimate of these costs can be obtained by multiplying the personnel costs of the plant by the factors given in Table 3-6.[22]

COST OF ELECTRICITY

In view of the substantial differences in the costs of the distribution of electricity as a consequence of the fixed costs for the lines and differences in the amount consumed by the customer, it seems best to confine the discussion here to the cost of electricity at the plant output bus.

Capital Charges and Taxes

Capital costs for new investor-owned electric utility plants that formerly ran around 6 percent increased to about 12 percent for either stock or bonds in the 1970s. The factors involved and their implications are complex[1] and consequently not generally understood or appreciated by the general public. Ad valorem taxes levied by state and municipal governments are also roughly proportional to the capital investment and commonly run 1 to 2 percent. Depreciation is ordinarily computed on a straight-line basis for an assumed 30-year life of the plant, and it is based on the initial rather than the replacement cost, giving a first-year capital charge of $3\frac{1}{3}$ percent. Federal corporate income taxes are normally about 50 percent of the net income before calculation of the income tax; hence if all the capital investment were in the form of corporate stock, to obtain a 12 percent net yield on this stock would require earning an additional 12 percent to cover the income tax. If the ratio of bonded debt to stockholders' equity is 2:1, the capital charge to cover the income tax would be 4 percent instead of 12 percent. (This shows the incentive to obtain as much capital through bonds as possible.) The total of these charges for a 2:1 ratio of bonded debt-to-stockholders' equity thus becomes about 21 percent of the initial capital investment (not including charges for Interest During Construction). These costs are essentially independent of the amount of operation and hence are inversely proportional to the capacity factor, usually ~60 percent.

Fuel Costs

The fuel cost component of the electricity produced depends on both the thermal efficiency and the cost of the fuel itself, together with the costs related to the fuel consumed—such as the transportation, reprocessing, and waste charges for nuclear fuel and the limestone required for flue gas scrubbing of sulfur from coal. These costs are roughly directly proportional to the amount of power produced rather than a function of the capacity factor.

Operation and Maintenance

The operating staff must be on duty whether the unit is running or not; hence the bulk of the operating costs are inversely proportional to the capacity factor, as is the case for the capital charges. Some of the maintenance costs may be reduced if the unit is not operating or is at low loads; the extent to which this affects the O&M charge depends on the type of unit, the fuel used, etc. Note that deterioration from corrosion in many components is likely to proceed more rapidly during shutdown when acidic moisture can collect than during hot operation when the parts are dry.

GENERAL PROBLEMS

The general problems associated with setting electric utility rates are both complex and vitally important to the nation. An enormous amount of capital—over 500 billion dollars—will be required in the 1980s for constructing new or replacement electric utility plants, and about two-thirds of this will have to be new capital (the balance will come from depreciation accounts for old plants).[1] Since this investment will represent ~25 percent of the new capital raised by private industry in financial markets, reasonable returns in the form of bond interest and stock dividends must be available. Probably the prime objection to these returns is that raised by no-growth advocates who also overlook the need for replacing worn-out plants. It is on the basis of this point of view that they approach such complex questions as whether to include in electric power rates a portion or all of the interest incurred during construction of new and replacement facilities. The inclusion of these charges in rates as a "Construction Work in Progress" (CWIP) item has been a major public issue that has involved a confusing mixture of statements and much misrepresentation. Unfortunately, the long-term costs to the consumer stemming from inadequate rates to finance the construction needed are likely to be much greater than those associated with adequate rates to finance an orderly construction program. It makes little difference whether this is done by providing an adequate profit for the utility so that its stock and bond offerings bring in all the required capital or whether a portion of it comes from inclusion in the rates of a charge for interest during construction of new facilities. Experience has shown that if these investments are not made, the consumer will eventually pay substantially more because of the high costs of power from gas turbines or premium power purchased from other utilities.

Cost Estimating

Estimating the costs for a projected plant has been rendered particularly difficult by the recent high rates of inflation with the escalation rate varying from one cost component to another. As a result, it is not surprising that estimates from different cost-estimating groups may differ substantially. A typical set made on the basis of similar ground rules is given in Table 3-7. It shows differences of ±20 percent in some items, and ±10 percent in the cost of the complete plant.[23]

TABLE 3-7 Three Independent Estimates of 500-MW Pulverized-Fuel Station Costs in Dollars per Kilowatt Using Similar Ground Rules

Plant item	Estimator		
	General Electric	Bechtel	United Engrs. and Constrs.
Coal handling	10.6	6.7	13.2
Ash handling	7.5	2.8	8.0
Boiler	108.4	120.3	78.1
Turbogenerator	43.7	54.9	47.2
Electrostatic pptr.	13.3	31.1	13.4
Scrubber	78.9	98.5	61.3
Cooling tower	10.3	*	13.7
Pumps, pipes, condenser, etc.	77.7	80.1	82.0
Electrical, instrumentation, etc.	51.9	59.3	74.0
Civil, yardwork, etc.	56.2	83.3	61.5
Total	458.5	543.4	452.5

*Not given separately.

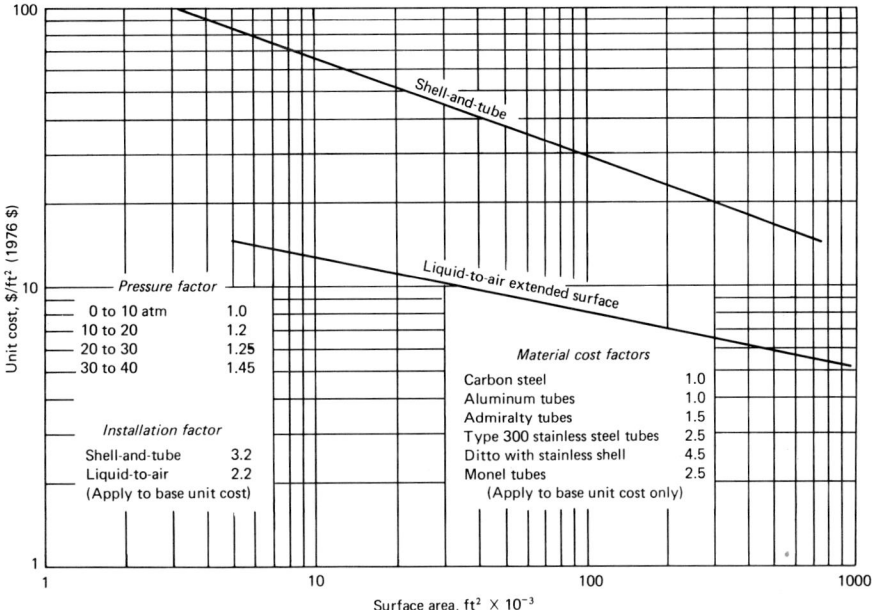

Figure 3-15 Unit costs of conventional shell- and tube- and liquid-to-air extended surface heat exchangers of carbon steel with aluminum fins for extended surfaces. (*Based on data in Chemical Engineering Progress, July 1976, p. 73.*)

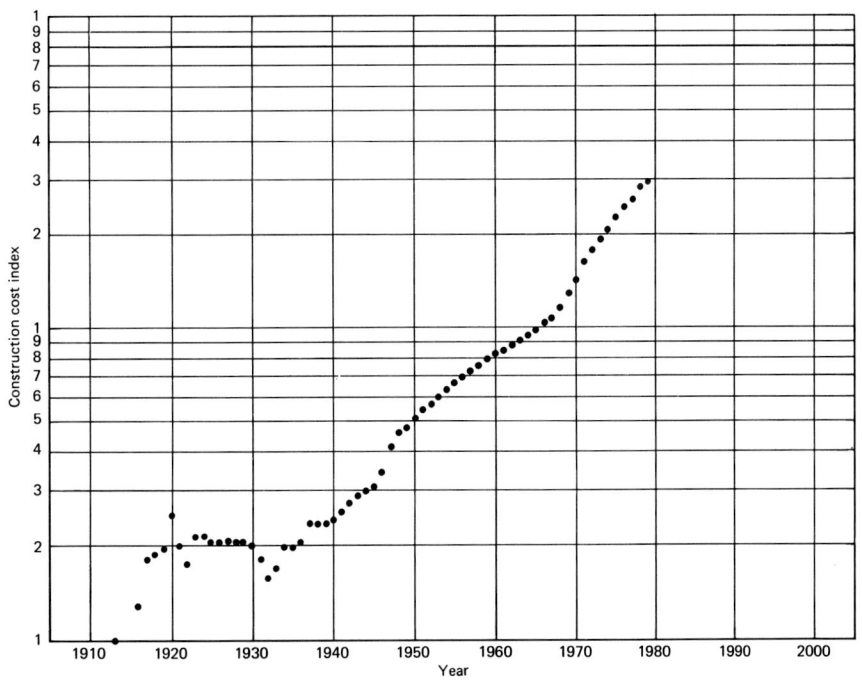

Figure 3-16 U.S. construction cost index as a function of year. (*Data plotted from Engineering News Record, Mar. 22, 1979, p. 73.*)

Probably the best and most convenient source of up-to-date cost information is the *Engineering News Record*, which gives quite detailed data for the costs of craft labor and construction materials for 20 cities in the United States, as well as some overall indices for the country as a whole. Data for one of these, *plant construction costs,* are plotted in Fig. 3-16 for the 1913-1979 period. Space has been provided so that the reader can fill in later data to provide an up-to-date basis for estimating the effects of inflation.

An important aid to cost estimators is a comprehensive set of eight reports of 200 to 500 pages each covering the capital, fuel, and operating costs of coal and nuclear plants prepared for the Nuclear Regulatory Commission (NRC).[24] Reference 25 is a similar but much less detailed study prepared for the Electric Power Research Institute (EPRI).

Relation between Energy Costs and Capital Costs

The rapid increase in the market prices of fuels in the 1970s led many to expect a shift in the relative importance of capital and fuel costs in the production of electricity. This has not occurred because, as pointed out by H. W. Parker,[26] energy is the driving force in our economy, and its cost directly affects the cost of everything including equipment and construction. A striking illustration of this is provided by the plot of Fig. 3-17 showing the industrial fuel cost index plotted against the process plant construction cost index for the 50-year period of 1926 to 1978. These data show a one-to-one relation between fuel and construction costs except in the 1958-1974 period, when a world oil glut temporarily depressed the price of petroleum and greatly stimulated the economy of the entire world.

The one-to-one relation between bulk energy costs and capital costs shown in Fig. 3-17 has a number of implications. First, it is apparent that increases in fuel cost apparently have not justified increasing the capital cost to get an increase in thermal efficiency and hence a lower fuel consumption. Secondly, while the effect is not obvious, it means that the cost of producing fuel oil and gasoline from coal increases about as rapidly as the cost

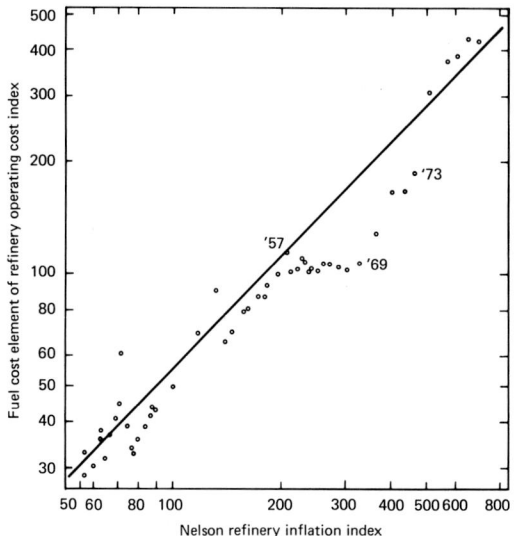

Figure 3-15 Relation between industrial fuel costs and process plant construction costs for the 1926 to 1978 period.[26] *(Courtesy ESCOE.)*

of bulk industrial fuel, because both the cost of the input material and the capital cost of the processing plant are roughly proportional to the industrial fuel cost. It is for this reason that the estimated cost of synfuel from coal seems to increase at almost the same rate as crude petroleum, thus somehow remaining substantially more expensive. Thirdly, the costs of "natural" sources of energy also continue to escalate at about the same rate as fossil fuels, because they require large capital investments; hence the expected "crossover" in costs does not materialize and the economical feasibility of "natural" energy sources continues to shift into the future.

REFERENCES

1. Kennedy, W. J. L.: "The Economics of Nuclear Power," paper presented to Atomic Industrial Forum, Inc., Conference on Nuclear Power and the Public, Feb. 26, 1979.

2. Weslowski, J. D.: "50 Locally-Owned Utilities in 1978 Outperformed Investor-Owned Counterparts," *Electric Light and Power,* vol. 57, no. 7, July 1979, pp. 10-12.

3. *1978 Business and Economic Charts,* Ebasco Services, Inc., 1979.

4. *Uniform Systems of Accounts Prescribed for Public Utilities and Licensees,* U.S. Federal Power Commission, April 1, 1973.

5. *Fossil and Nuclear 1000 MW Central Station Power Plants Investment Estimates,* prepared for EPRI by EBASCO Services, Inc., September 1975.

6. Budwani, R. N.: "Fossil-Fired Power Plants: What It Takes to Get Them Built," *Power Engineering,* May 1978, pp. 36-42.

7. *National Power Survey,* U.S. Federal Power Commission, 1964, p. 70.

8. Beecher, D. T., et al.: *Energy Conversion Alternatives Study—ECAS—Westinghouse Phase I Final Report,* vol. 1, *Introduction and Summary and General Conclusions,* NASA CR-134941, vol. 1, 1976.

9. *Nuclear Power Project Report,* Foster Wheeler Corp. and Pioneer Service and Engineering Co., October 1958.

10. *Factors Affecting the Electric Power Supply, 1980-85, Executive Summary and Recommendations,* U.S. Federal Power Commission, Dec. 1, 1976.

11. *Estimating Reclamation Instructions,* App. A, Estimating Data, Series 150, U.S. Bureau of Reclamation.

12. Sheldon, L. H.: "Cost Analysis of Hydraulic Turbines," *Water Power and Dam Construction,* June 1981.

13. Guyer, E. C., et al.: "Optimization-Simulation Methodology for Wet-Dry Cooling," EPRI Report No. FP-1096, May 1979.

14. Hallberg, L. K.: "Of Time and the Estimate," *Combustion,* September 1977, pp. 16–17.

15. "Power Plant Capital Investment Cost Estimates: Current Trends and Sensitivity to Economic Parameters," USDOE Report DOE/NE-0009, June 1980.

16. "1000 MWe Central Station Power Plant Investment Cost Study," vols. I, II, and III, USAEC Report WASH-1230, United Engineers and Constructors, Inc., June 1972.

17. Guthrie, K. M.: *Process Plant Estimating, Evaluation, and Control,* Craftsman Book Company of America, New York, 1974.

18. Fraas, A. P.: "Heat Exchangers for High Temperature Thermodynamic Cycles," ASME Paper No. 75-WA/HT-102, December 1975.

19. Brandfon, W. W.: "Comparative Costs for Central Station Electricity Generation," *Atomic Industrial Forum Conference on Energy for Central Station Electricity Generation,* Apr. 18, 1978.

20. Bloom, S. G., et al.: "Analysis of Variations in Costs of FGD Systems," prepared by Battelle Columbus Laboratories for EPRI, EPRI Report No. FP-909, October 1978.

21. Stauffer, T. R., et al.: "To Breed or Not to Breed," *Mechanical Engineering,* February 1977, pp. 32–41.

22. Myers, M. L.: "A Procedure for Estimating Nonfuel Operation and Maintenance Costs for Large Steam-Electric Power Plants," Oak Ridge National Laboratory Report No. ORNL/TM-6467, January 1979.

23. Johnston, R.: *The Economics of Coal-Based Electricity Generation by Conventional Methods and by Fluidized Bed Combustion,* International Energy Agency, Economic Assessment Service, Working Paper No. 33, September 1978.

24. *Commercial Electric Power Cost Studies,* prepared for the Nuclear Regulatory Commission and ERDA by United Engineers and Constructors, Inc., June 1977. Set of eight reports (200 to 500 pp. each):

   1. *Capital Cost: Pressurized Water Reactor Plant,* NUREG-0241, COO-2477-5
   2. *Capital Cost: Boiling Water Reactor Plant,* NUREG-0242, COO-2477-6
   3. *Capital Cost: High and Low Sulfur Coal Plants—1200 MWe,* NUREG-0243, COO-2477-7
   4. *Capital Cost: Low and High Sulfur Coal Plants—800 MWe,* NUREG-0244, COO-2477-8
   5. *Capital Cost Addendum: Multi-Unit Coal and Nuclear Stations,* NUREG-0245, COO-2477-9
   6. *Fuel Supply Investment Cost: Coal and Nuclear,* NUREG-0246, COO-2477-10
   7. *Cooling Systems Addendum: Capital Total Generating Cost Studies,* NUREG-0247, COO-2477-11
   8. *Total Generating Costs: Coal and Nuclear Plants,* NUREG-0248, COO-2477-12

25. *Technical Assessment Guide,* Electric Power Research Institute Report No. EPRI PP-877-SR, June 1978.

26. Parker, H. W.: "The Energy Component of Future Energy Costs," *ESCOE Echo,* The Engineering Societies Commission on Energy, vol. 3, no. 24, Nov. 19, 1979.

Section 4

Cogeneration Engineering and Design

Cogeneration Engineering and Design

Fuel-price increases (and decreases) have led plant designers to seek ways of reducing overall energy costs for a variety of services. One important solution has been cogeneration—defined as the sequential production of electricity and heat, steam, or useful work from the same fuel source.

Cogeneration was used for many years in early power plants but it had a different name or names—like by-product power, district heating, and process-steam utilization. Regardless of the name, the objective was the same-to wring the maximum energy from a unit of fuel, at an acceptable cost.

Facilities for cogeneration are operating in a variety of services—chemical, petrochemical, refinery, mining and metals, paper and pulp, food-processing, and utility. The three most significant factors affecting a cogeneration project are: (1) availability of financing, (2) acceptable economic payback period, and (3) ability to satisfy regulatory requirements. Most cogeneration projects providing their base process steam and/or heating and drying thermal loads, in addition to satisfying internal electric energy needs (except possibly during peak demand periods), are easily proved to have an economically attractive payback period. Each of these items is considered in the following discussion.

Pages 4-3 through 4-26 and references from *Engineering Evaluation of Energy Systems*, by Arthur P. Fraas. Copyright © 1982; pages 4-26 through 4-92 from *Cogeneration, Engineering, Design, Financing, and Regulatory Compliance*, by Charles H. Butler. Copyright © 1984. Used by permission of McGraw-Hill, Inc. All rights reserved.

COGENERATION AND BOTTOMING CYCLES

Inasmuch as all the systems used commercially for converting heat into electricity reject from 1.5 to 5 times as much energy to the environment in the form of heat as they yield in the form of electric energy, there are clearly incentives for utilizing this waste heat. As discussed in Section 2, steam cycles are fairly widely used to extract useful energy from the exhaust of gas turbines. As will be discussed in the latter part of this chapter, steam or other Rankine cycles can be used to recover the heat rejected from diesel engines or from chemical processes where the heat is usually available at substantially lower temperatures than the heat from gas turbines. However, a greater potential for conserving energy is through cogeneration, i.e., the design and operation of a power plant in such a way that it meets the needs for both heat and electricity in an industrial plant, a building complex, or an urban area. Cogeneration is widely used by industry in paper mills, chemical processing plants, etc., and for institutional heating systems; about 10,000 MW of electric power (~15 percent of U.S. industrial electricity consumption) was produced by U.S. industry in 1976 from cogeneration power plants. The concept of comprehensive urban area cogeneration offers the possibility of huge energy savings, but it is not widely used because it involves not only a formidable capital investment and a time span of decades for implementation, but also even more formidable institutional, legal, and political barriers. Thus this chapter begins with a brief look at cogeneration plants tailored to meet the needs of specific industrial or institutional applications and then turns to a comprehensive survey of the diverse problems of integrated urban energy systems in order to provide perspective on the possibilities, the technical requirements, and the institutional problems. The last section of the chapter surveys the possibilities of using working fluids other than water in Rankine bottoming cycles for special applications that are often alternatives to cogeneration.

COGENERATION PLANTS FOR INDUSTRIAL PROCESSES AND INSTITUTIONS

As soon as electric power began to come into use, it was evident that an industry that required both electricity and low-pressure steam for process heat could operate its boilers at a pressure substantially higher than the pressure at which steam was used for the process, and the steam could be expanded through an engine placed between the boiler and the process equipment. The incremental capital and fuel costs were small; hence the resulting cost of the electricity produced was low. As improvements in steam power plants made possible progressively higher boiler temperatures, the range of possibilities for cogeneration was broadened and the economic potential improved. A flowsheet for a typical modern plant is shown in Fig. 4-1 for an industry that requires steam at two pressure levels. In this case, 56 percent of the chemical energy in the fuel is converted into either electricity or process heat, and, in some plants, the fraction may be as high as 90 percent.

Cogeneration plants for institutions such as hospitals, universities, prisons, etc., are similar to those for industry except that most of the steam is used for building heating and hot water and so there is no large bleed-off of steam at moderate pressures, and the turbine discharges at a relatively low pressure—usually only a little above atmospheric pressure. This arrangement increases the fraction of the energy in the fuel converted into electricity and usually makes possible a somewhat higher overall thermal efficiency at the design point. However, the wide variations in the heat load with the weather lead to extensive periods of part-load operation in which there is a poor match between the requirements for heat and those for electricity. As a consequence, much or all of the electric power in mild weather is commonly purchased from a utility.

COGENERATION FOR URBAN AREAS

The extent to which cogeneration of heat and electricity can be employed depends on the degree of urbanization in a nation; hence a survey of this factor is in order. Urbanization, one of the most important trends of our century, has proceeded concurrently with industrialization.[1] Studies indicate that a city ordinarily has the full complement of urban advantages and problems when its population exceeds 100,000. One of the best indices of the degree of urbanization of a nation is the fraction of the population dwelling in cities of over 100,000; a nation may be classed as urban when this fraction exceeds 50 percent. Amazingly, by that definition no nation in the world could be classed as urban until England reached that point in 1850. Other nations passed that point much later, for example, the United States in 1915, Japan in 1955, and the U.S.S.R. in 1965. As a point of interest, only 2 percent of the population of Europe lived in cities over 100,000 in the year 1800, although the fraction in England was up to 10 percent. The remarkable thing is that, as time has gone on, the period of time required for the population distribution to shift from 10 percent urban to over 50 percent urban has gone from 79 years for England, 66 years for the United States, 48 years for Germany, 36 years for Japan, to 26 years for Australia. In the United States and Europe the movement to cities appears to have leveled off, with ~75 percent of the population in urban areas having populations of 100,000 or more. The accelerated rate of industrialization in underdeveloped countries will probably lead to even more rapid rates in their urbanization in the future. It is estimated that by the year 2000 half the population of the world will be urban, thus providing a huge market for cogeneration in central station plants.

As many studies show, the concentration of people in our large urban complexes has led to distressing air and water contamination problems, while thermal pollution of our streams, lakes, and estuaries has become a serious matter. Thus the cogeneration of heat and electricity for urban areas offers the possibility of reducing not only energy resource consumption but also environmental pollution. Inasmuch as over 75 percent of the population in developed countries lives in urban areas suited to comprehensive cogeneration systems, the potential application is enormous.

Total Energy Requirements

If one takes a comprehensive look at all these problems and considers long-term measures to alleviate them, a first logical step is to examine the total energy requirements of an urban complex. Cities currently require about twice as much energy for heating buildings as for generating electricity. Since World War II, air conditioning has come to represent an additional large energy requirement. Water shortages have begun to develop in some areas leading to a consideration of seawater distillation, which would require large amounts of heat energy. Although heretofore our attempts to meet these various requirements for energy have led to ad hoc solutions for each specific requirement, it has become evident that there is a strong incentive to obtain a single integrated public utility complex that would be a major element in the master plan for the city as a whole.

A point of departure is provided by the estimate of the United States energy requirement for the year 2000 presented in Table 4-1. Note that about 80 percent of the energy required by residences and over half of the energy employed by commercial operations is for heating buildings. The bulk of the energy used for transportation in 1979 was supplied by petroleum products, and it was assumed in preparing Table 4-1 that this would still be the case in 2000. Pressure for reductions in atmospheric pollution continues to build up; hence it may be that by the year 2000 a substantial fraction of automotive energy requirements might be met by using new types of electric storage battery. At the time of writing, over 10 percent of the energy used in industrial processes was in the form of electricity; the fraction had more than doubled in the 1960-1980 period. Note, too, that about one-third of the industrial process heat is employed at relatively low temperatures so that it could be supplied by extracting steam from a turbine at a temperature of 204°C (400°F) or less. In some industries (e.g., chemical) about 90 percent of the heat required is in this temperature range. For cities as a whole the percentage is large although it differs substantially from one city to another, depending on the local industrial operations. The balance of the heat required by industry is largely for high-temperature metallurgical or ceramics work. In any event, if one projects the current use of electricity to the year 2000, it appears that electric energy consumption may total about 25.6×10^{15} Btu/year, or about 15 percent of the total energy requirements of the nation at that time.

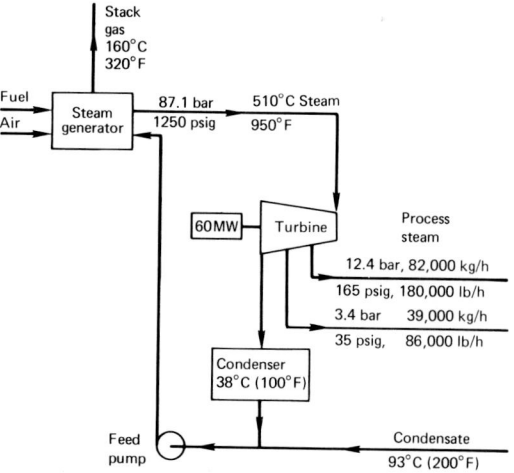

Figure 4-1 Typical cogeneration system for an industrial plant.

TABLE 4-1 A Typical Upper Limit Estimate of U.S. Energy Requirements in the Year 2000

Application	Electricity, 10^{15} Btu/year*	Heat, 10^{15} Btu/year	Total, 10^{15} Btu/year
Residential	5.0	19.5	24.5
Commercial	12.0	15.5	27.5
Industrial			
Food, paper, chemicals ($T \leq 400°F$)	2.7	19	21.7
Steel	1.3	12	13.3
Ceramics	0.6	5	5.6
Miscellaneous	4.0	21	25
Subtotal	8.6	57	65.6
Transportation		48	48
Total	25.6	140	165.6

*This column gives the electric energy; the thermal energy input to the plants generating electricity will be several times as great.

TABLE 4-2 Typical District Heat Consumption in 1952 (Ref. 2)

Location	Area served, mi^2	Steam distributed, 10^6 lb/(year·mi^2)
Baltimore	0.45	2,741,998
Boston	0.98	2,534,854
Chicago	0.20	3,361,990
Cleveland	0.88	3,239,226
Dayton	0.51	3,821,153
Detroit	1.58	2,553,496
Indianapolis	1.88	2,182,870
Milwaukee	0.87	2,190,074
New York	6.31	3,058,501
Philadelphia	1.13	1,883,422
Pittsburg	0.33	5,295,891
Rochester	0.86	3,070,672
St. Louis	0.82	2,253,461
AVERAGE	1.29	2,935,970

Heat for Buildings and Low-Temperature Industrial Processes

It is evident from Table 4-1 that heat for buildings and low-temperature industrial processes represents a large fraction of the total energy requirements, and that there is a strong incentive to employ the waste heat from the thermodynamic cycle of the electric power plants for these purposes. Fortunately, some valuable practical insights can be obtained from the more than 100 large district heating systems employed in cities in the United States, most of them providing the heat in the form of low-pressure steam.[2,3] In addition, there are also numerous smaller cogeneration systems that serve large institutions, shopping centers, and housing developments. Table 4-2 shows that nine of the larger district heating systems serve areas of 2 to 15 km^2 (0.8 mi^2 to over 6 mi^2) and that the steam distributed runs approximately 0.5 kg/(year·km^2) [3 × 10^9 lb/(year·mi^2)]. Because of the distorting effects of government rate setting for utilities, the steam for most of these district heating systems is usually generated in plants that are separate from those producing electric power. Where steam is extracted from steam-turbine-generator units, the method of computing costs imposed by government regulatory agencies usually requires that each Btu of extracted steam costs the same as a Btu of prime steam. No reduction in cost is made for the value derived from the steam expansion to produce electricity. This policy reduces the rate charged residential consumers for electricity and increases the rate charged commercial establishments for heat, a politically advantageous course. To avoid the legal hassles, district heating systems are usually operated by separate companies set up for that purpose, and most of the steam for district heating systems is generated in separate steam plants that do not generate electricity. Thus, irrespective of the source, the average price of steam sold by district heating systems in 1973 was $1.38/10^9 J ($1.45/10^6 Btu).[4] Of this amount, about half the cost could be attributed to the generation of the steam and about half to the distribution system. It is especially interesting to note that the large mains were accountable for only about one-quarter of the cost of the distribution system; most of the cost derived from the small branching mains and pipes.[2]

The real cost of low-temperature heat can be reduced drastically if, instead of generating steam for heating as a separate process, steam and electricity are produced simultaneously by bleeding steam from the lower turbine stages of an electric power plant. This solution is particularly cost-effective if the power plant is a nuclear plant having low fuel costs. A typical estimate in 1969 indicated that steam for a district heating system could be produced at $0.07 to $0.30/$10^9$ J ($0.07 to $0.30/$10^6$ Btu) with a nuclear plant if it were used simply for generating steam, whereas if it were a dual-purpose plant generating both electricity and steam, the cost would run from $0.024 to $0.092/$10^9$ J ($0.025 to $0.096/$10^6$ Btu), depending on the type of reactor and the type of financing.[3] (Inflation roughly tripled these costs by 1980.) As one might expect, the lower figures in these two cases are for a high-temperature reactor such as an LMFBR, whereas the higher figures are for low-temperature water reactors. For coal-fired plants the costs run about 50 percent higher because of higher fuel costs, even though a coal-fired plant could be in the city, while a nuclear plant would have to be ~30 km outside the city according to current regulatory practice.

Air Conditioning and Refrigeration

Many of the customers of existing district heating systems employ steam during the summer months to drive air-conditioning systems. In some instances this is done by expanding the steam through low-pressure turbines that drive compressors for the refrigerant, while in others an absorption type of refrigeration system is employed. The two methods are roughly comparable in overall efficiency. The consequent load on the district heating system in northern U.S. cities in 1975 was typically about half of that represented by the building heating load in the cold-weather months,[3] but it is expected to rise to equal the heating load.

The processes described above for air conditioning and refrigeration in large commercial buildings can also be employed in residential units. The American Gas Association has sponsored the marketing of gas-heated absorption refrigeration systems, and several gas organizations have sponsored the development of small Rankine cycle systems with gas-fired boilers that provide both electricity and heat. These absorption refrigeration units could be adapted to use high-temperature water from a district heating system preferably at a temperature of 120°C (250°F) or higher.

Industrial Process Heat

Detailed data on the temperature levels at which industries utilize thermal energy are not easy to find, and the sources available give values that differ substantially, in part because of differences in ways in which the raw data have been organized or interpreted.[5,6] Figure 4-2 gives a good indication of the distributiion of energy utilization by industry, while Table 4-3 gives the temperature level at which process heat is required by the six industries that have the highest energy consumptions.[6] The electric power requirements of these industries are also included in Table 4-3. In the 1970s only ~15 percent of this electric power was produced by on-site power plants owned by the industrial user and designed to supply both process heat and electricity. The percentage has not been larger for several reasons: A cogeneration plant involves a substantial extra capital investment over that required by a simple boiler for process steam; small-scale electric power plants entail higher operating and capital costs (Chap. 10); and in many cases the process steam requirements at a particular plant are much greater than the electric power requirements, and, until 1978, most utilities would not purchase the excess

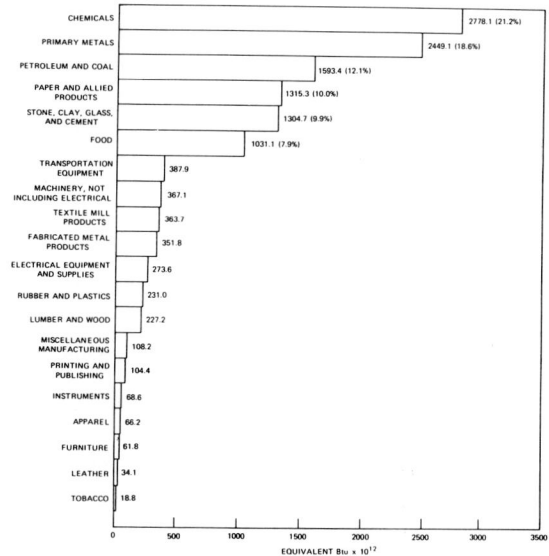

Figure 4-2 Fuels and electricity purchased by industry in 1971 (1 Btu = 1.055 kJ).[5]

TABLE 4-3 Annual Process Heat/Temperature and Electric Requirements for Six Major Industries (Ref. 6)

Industry	Process heat consumption, 10^{12} kJ (10^{12} Btu)				Electricity, 10^9 kWh	Date (reference)
	177°C (350°F)	177–593°C (350–1100°F)	593–816°C (1100–1500°F)	816°C (1500°F)		
Iron/steel	69 (65)	0 (0)	0 (0)	1806 (1712)	57.5*	1973 (6)
Petroleum	63 (60)	2958 (2804)	241 (230)	0 (0)	23.1	1973 (6)
Chemicals	322 (305)	195 (185)	0 (0)	37 (35)	80.0	1977 (5)
Paper/pulp	324 (307)	504 (478)	0 (0)	0 (0)	36.0	1973 (6)
Stone/clay/glass	77 (73)	19 (18)	12 (11)	1080 (1024)	24.0	1974 (6)
Food processing	271 (256)	64 (61)	0 (0)	0 (0)	36.0	1977 (5)

*20 percent generated in plant.

electric power that the industrial plant might produce. In fact, up until 1980, the only step taken by Congress to encourage cogeneration after the Yom Kippur War was a new law requiring electric utilities to buy such power where available. Another unfavorable factor is that the large capital investment required for electric power production can be justified only if a plant lifetime of 20 to 40 years can be expected, whereas market conditions and the technology for chemicals and many other industrial products changes so rapidly that the useful lifetime of a new processing plant may be only 3 to 5 years. Although the steam plant might be used to supply heat to another process after the first process became obsolete, the efficiency of the steam turbine is sensitive to its exhaust temperature and pressure, as well as the steam flow rate; hence a poor match is likely if the plant is used for another process. All these factors commonly favor an arrangement whereby the electric utility makes large blocks of process heat available to many industries—preferably in one or more industrial parks relatively close to the electric power plant. With a substantial number of users, the perturbations to the steam-electric plant are much reduced if a new plant process is started up or an old one discontinued.

Distillation of Seawater

Distillation of seawater to provide potable water for domestic use has been carried out in a few water-short areas for many years. At the time of writing, seawater distillation plants are in use at the U.S. naval base in Guantanamo, Cuba, and in Arabia. Extensive studies have been carried out since the early 1960s on the use of large nuclear power plants for desalting seawater for agricultural purposes with the waste heat from the generation of electric power that would be used for making aluminum, fertilizer, or other products requiring a large power input.[7,8] These agro-industrial complexes show fascinating possibilities for "making the desert bloom" in certain areas such as the upper Gulf of California, the Middle East, the Gujarat Peninsula on the northwest coast of India, and western Australia. The capital costs are huge, however, and the estimated cost-benefit ratio marginal, so that up to 1980 there had been no firm commitment to a definite project.

Distillation of Sewage

As a consequence of the growing concern over water pollution, many different efforts are being made to improve the methods of treatment used in sewage plants. Most of these efforts entail removal of organic solids. This is not sufficient for inland cities, however, because the mineral content of the water flowing through a city system is increased by about 250 ppm. Inasmuch as 500 ppm is considered to be the maximum tolerable mineral content of the water going into the city mains, it seems likely that many cities eventually must employ some means of reducing the content of the dissolved solids. If cheap low-temperature steam were available from a nuclear electric plant under off-peak conditions, one of the processes that might be developed for accomplishing this is distillation by means of multistage evaporators.[9] For the Philadelphia area with a population of 2 million people, the incremental capital cost of provisions for distilling all the sewage with heat from a centrally located nuclear electric power plant appeared in a 1969 study to be about 50 million dollars if the entire complex were properly integrated in the design stage.[9] This compared with a capital investment in the existing water and sewage systems of about 375 million dollars (in 1969 dollars). The requirements are about 237,000 J/L (850 Btu/gal) of water and about 378 L (100 gal) of water per day per person in the area served. This would amount to about 10^{19} J per year (10^{16} Btu per year) for the entire United States in the year 2000, and it would represent perhaps one-third of the waste heat available from the electric power plants.

It is believed that the costs of distilling sewage would not be greatly different from those for distilling seawater. The much lower content of dissolved inorganic solids would tend to reduce that type of fouling of heat-transfer surfaces, but, on the other hand, the presence of organic material might increase the fouling problem. In addition, it would probably be necessary to evaporate the waste solid content to dryness so that it could be burned, or it might be sterilized by heating and sold as fertilizer. Thus, with low-grade heat from a nuclear electric power plant, the entire sewage effluent from a large city could be distilled at a price equivalent to an increase in the operating cost of the water and sewage systems of about 50 percent. If waste heat from a coal-fired plant were employed, the incremental cost would be ~75 percent. In water-short areas the distillate could be recycled through the city water system. Preliminary estimates have indicated that, by using low-grade heat from the low-temperature end of the thermodynamic cycle of high-temperature nuclear power plants, it would be possible to distill all the sewage from the urban complexes in the United States by installing a set of systems whose incremental capital cost if designed into the set of nuclear utility systems would be only about 5 billion dollars (on the basis of 1969 costs). This would require a capital expenditure of 500 million dollars per year for 10 years in 1969 dollars, or roughly double that in 1980 dollars—a relatively modest cost. The cost of heat required for the distillation process would of course be substantial unless low-grade heat from nuclear power plants were available at a low cost under off-peak load conditions. If the value of that heat were $0.20/$10^9$ J ($0.20/$10^6$ Btu) (in 1969 dollars), the cost of the heat would be about 2 billion dollars per year—again a small fraction of the U.S. annual budget in the year 2000.

Municipal Solid Wastes as an Energy Source

Cities generate an enormous quantity of solid wastes—roughly 1 ton per year per capita, of which about one-third is from residences and the balance from commercial and industrial operations.[9-12] The heating value of municipal wastes commonly totals ~12 × 10^6 J/kg (5500 Btu/lb) so that in 1979 the chemical energy content approximated 10 percent that of the 1979 U.S. annual coal production of ~7 × 10^8 tons per year. Table 4-4 summarizes both the total amount of solid waste generated in the United States and that practically collectible from various sources. About 5 percent of the U.S. electric power consumption might be produced from these wastes if they could be burned efficiently in steam plants located within ~50 km of the source of the waste so that collection and transportation costs would be tolerable. Until the 1970s, although many municipal waste incinerators were in use and some produced useful heat and/or electric power, the extra costs of handling and burning solid wastes, together with the poor boiler efficiencies (60 to 70 percent) and potentially severe corrosion where heat recovery was attempted, led to solid-waste disposal in sanitary landfills wherever land was available for this purpose, which it usually was.[12,13] (In 1970 the on-site costs of landfill disposal ran only $1 per ton of wastes.) The rapidly rising cost of fuels in the 1970s changed the relative cost situation while the availability of sites for landfills declined. This situation led to many experimental investigations of the use of municipal solid wastes as fuels as well as the construction of some new power plants designed expressly for the use of solid wastes.[10,14,15] Even though these plants took advantage of extensive experience in Europe with the use of municipal wastes as fuel in power plants, their operation has presented a host of problems.

About half of the usable solid waste listed in Table 4-4 consists of municipal solid wastes. Although much of this consists of paper products, there are substantial amounts of metals and glass, and the value of the waste as fuel is degraded by a high moisture content (Fig. 4-3, Fig. 6-6). For use as fuel, the wastes are usually separated into light and heavy fractions by shredding and air flotation. These processes are usually preceded, however, by manual sorting to remove inner-spring mattresses, automobile batteries, massive metal parts, etc., that would foul or damage the shredders, and so perhaps one-third of the total must be diverted to a landfill disposal site.

TABLE 4-4 Dry, Ash-Free, Organic Wastes Generated and Potentially Collectible for 1971 (Ref. 10)

	Quantity, million tons/year	
Source	Total generated	Total available
Animal wastes	200	26.0
Urban refuse	129	71.0
Logging and other wood refuse	55	5.0
Agricultural and food wastes	390	22.6
Industrial wastes	44	5.2
Sewage solids	12	1.5
Miscellaneous	50	5.0
Total	880	136.3

TABLE 4-5 Expected Ranges in Mixed Municipal Refuse Composition

	Percent composition as received (dry weight basis)	
Component	Anticipated range	Nominal value
Paper	37–60	55
Newsprint	7–15	12
Cardboard	4–18	11
Other	26–37	32
Metallics	7–10	9
Ferrous	6–8	7.5
Nonferrous	1–2	1.5
Food	12–13	14
Yard	4–10	5
Wood	1–4	4
Glass	6–12	9
Plastic	1–3	1
Miscellaneous	< 5	3

Source: C. L. Wilson, "A Plan for Energy Independence," *Foreign Affairs,* July 1973, pp. 657–675.

Worse, attendants must watch for cans of gasoline, solvents, gunpowder, and even dynamite. Some 97 explosions in shredders in solid-waste disposal demonstration plants in the latter 1970s represented the equivalent of one serious explosion every few months in a plant handling 1000 tons per day, and at least two fatalities have resulted, as well as much expensive damage to facilities.[10] Yet another problem is that the plastic materials mixed with paper in the combustible light fraction from the air separation process include substantial amounts of chlorinated hydrocarbons that yield HCl in the stack gas and hence serious boiler and stack corrosion problems. Yet another difficulty is that the solid-waste composition varies widely from one truckload to another, so that furnace control is difficult if the wastes are burned alone. As a consequence, combustion of municipal solid wastes is usually carried out along with some other fuel, such as coal or oil.

Extensive operating experience with demonstration plants has generally been less than satisfactory and the capital costs of the facilities have been high: typically ~ $50,000 per ton per day of capacity (in 1979 dollars).[10] For example, two plants at St. Louis and Nashville[15] burning solid wastes in boiler furnaces designed for the purpose have proved financially burdensome to these cities even with substantial funding from the federal government, and they have experienced many technical difficulties, particularly in meeting EPA requirements on particulate emissions and with HCl corrosion. Both problems have been eased by burning the solid wastes as a fuel supplementary to coal having a substantial sulfur content.[16] Dilution of the coal with the solid wastes not only reduces the effective sulfur content but also somewhat increases the retention of sulfur compounds in the ash because of reactions involving the sulfur and the chlorine in the solid wastes. These reactions also keep down corrosion of the boiler tubes by HCl so that fire-side corrosion is about the same as for furnace operation on coal alone.[16]

In Baltimore a different approach employing a pyrolysis process presented so many problems that it was almost abandoned in 1977 after 2 years of shakedown tests. The problems were finally solved sufficiently so that by 1979 the plant was generating $39,000 per week in revenues for the city.[14] It handled 700 tons per day of refuse, shredding it and partially burning it first in a rotary kiln in an oxygen-deficient atmosphere at ~ 900 °C. The gases from the pyrolysis were then burned in a conventional boiler furnace to produce steam that

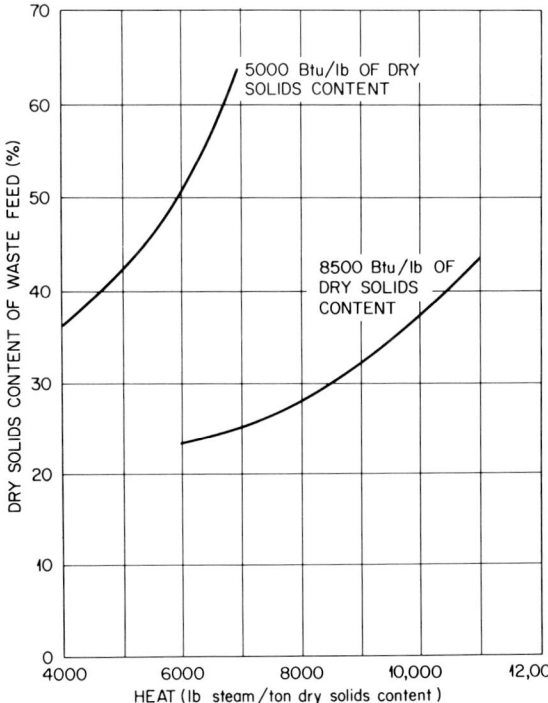

Figure 4-3 Relationship between the water content of waste feed and the low-pressure saturated steam generating capability of a Copeland fluidized-bed system at autogenous combustion.[12]

was sold to the Baltimore Gas and Electric Company for $3.90/1000 lb of steam in 1979, reducing the utility's fuel oil consumption by 5×10^6 gal per year.

Recycling metals and glass from the heavy fraction of the solid-waste separation process has been an important element in most of the projects.[10,13] After the shredding and air flotation operations extract the combustibles, the iron can be removed magnetically, nonferrous metals can be removed with an eddy current generator operated in conjunction with a magnet, and glass can be separated from the residue by a water flotation process. The products of these processes are of such low grade that, while it has proved possible to market the iron in at least some cases, it has generally not been possible to find buyers for the nonferrous metals and glass. The cost of their separation, therefore, is apparently not justified unless special markets can be developed. Possibilities include the use of glass as aggregate in concrete, in brickmaking, etc.

Combustion of municipal solid wastes greatly reduces their volume, but there is still ash to be disposed of in a landfill. The alkali content of this ash is substantial, and thus it presents a leaching and groundwater contamination problem similar to that of coal.

Matching Thermal and Electric Energy Requirements

Most of the process heat produced in the cogeneration plants of the U.S. manufacturing industry has been obtained from

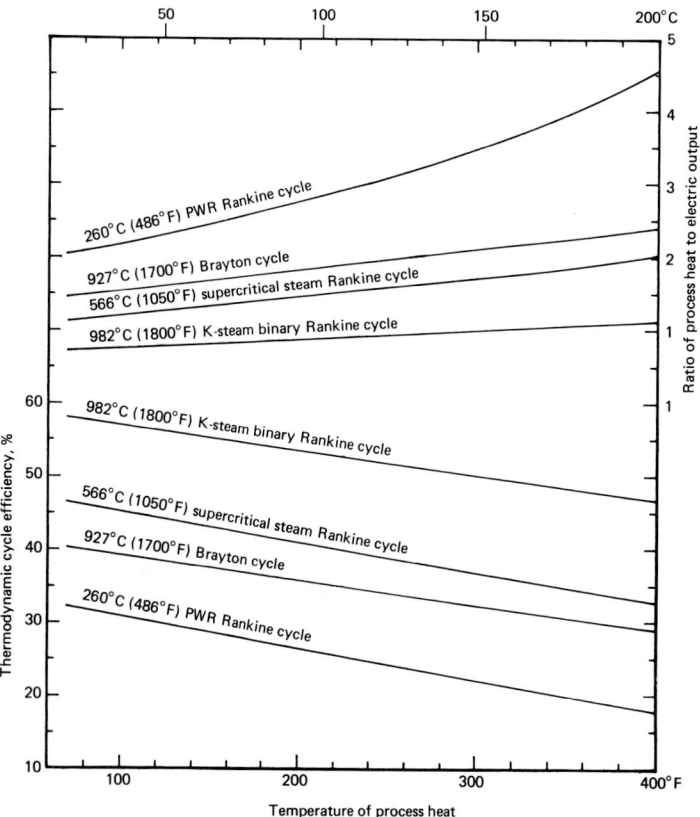

Figure 4-4 Effects of process heat-removal temperature on the overall thermal efficiency of the cycle for electric power generation and on the ratio of energy to the process heat system to the energy to electric power for some typical thermodynamic cycles.[19]

back-pressure steam turbines, i.e., turbines that discharge their steam against a substantial back pressure rather than expanding it to the low pressure obtainable with a condenser running at 38 °C (100 °F) or less.[17,18] In plants where steam is required at several temperatures, extraction turbines are used, and occasionally, where the requirement for electric power is large, the amount of steam extracted may be only a substantial fraction of the inlet flow. In this case, some steam is extracted and some is expanded to a conventional condenser.

The greater the temperature at which the steam is required for process purposes, the smaller the electric output per unit of energy input to the steam generator. The resulting fraction of the thermal energy input that is obtained in the form of electricity depends on both the steam extraction temperature and the peak temperature in the cycle. Figure 4-4 shows this effect for a water reactor plant, a conventional supercritical-pressure steam plant, and a concept for a plant of the twenty-first century employing a fusion reactor coupled to a potassium-vapor-steam binary vapor cycle plant. Figure 4-4 also shows a curve for a conventional gas turbine which has the inherent advantage that it rejects its heat at a relatively high temperature and for this reason is in widespread use for cogeneration in the chemical processing industry. In addition to the curves for the efficiency of the thermodynamic cycle for electric energy production, Fig. 4-4 gives a set of curves for the ratio of the energy obtained in the form of process heat to that in the form of electricity. This is also an important parameter, and may determine the choice of a gas turbine in preference to a steam turbine, or vice versa. Figure 17.5 shows a direct comparison of these two systems in this respect.[20] It can be seen that the overall utilization of the energy input for the conditions of Fig. 4-4 would be 100 percent except for stack losses if all of both the electricity and process heat could be utilized to good advantage. There is an important proviso in this respect relative to the gas turbine; Fig. 4-4 was prepared on the assumption that the process heat would be used in the form of superheated water and so the heat could still be extracted efficiently from the turbine exhaust up to fairly high temperatures. If the heat were required in the form of high-temperature steam, the pinch-point effect would make it

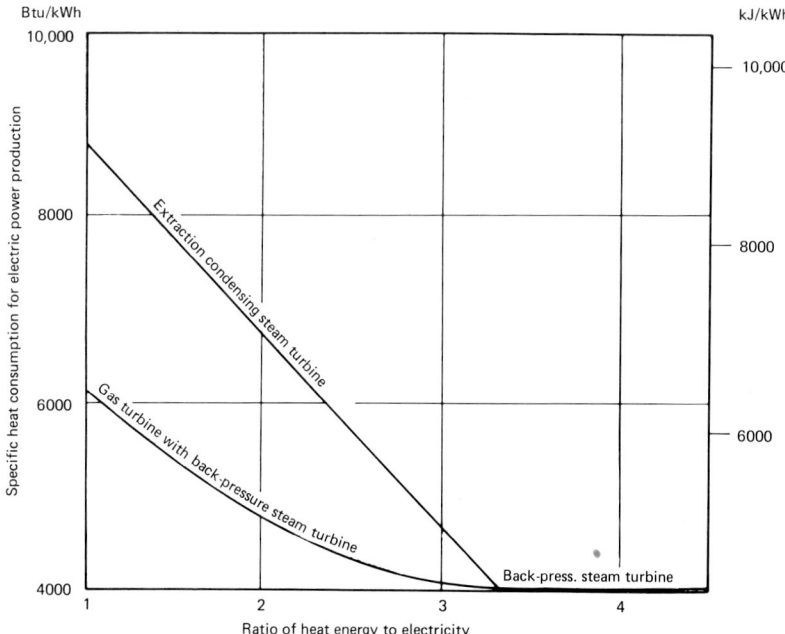

Figure 4-5 Replot of data from *Sulzer Technical Review* (April 1975) showing the heat rates for both gas and steam turbines where both process heat and electricity are required. (*Courtesy Oak Ridge National Laboratory.*)

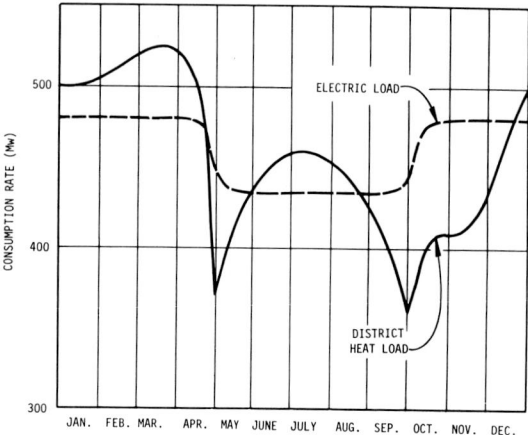

Figure 4-6 Annual electric and district heating system loads near Philadelphia.[3] (*Courtesy Oak Ridge National Laboratory.*)

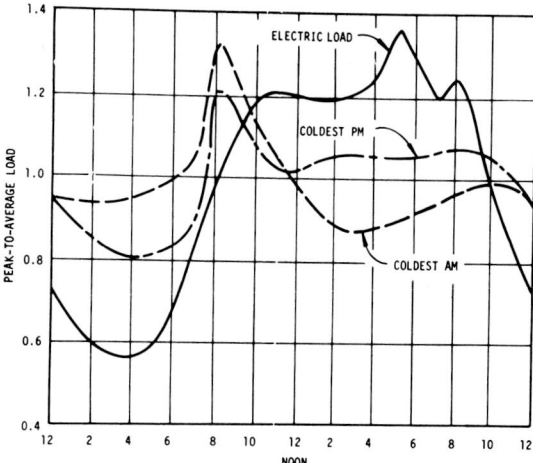

Figure 4-7 Variations in the winter diurnal electric and heating loads near Philadelphia.[3] (*Courtesy Oak Ridge National Laboratory.*)

impossible to get a high overall efficiency with the gas-turbine system.

Heat Storage

One element of the matching problem cited above is possible variation in the ratio of the heat required to the electricity produced. For district heating systems, both seasonal and diurnal variations in this quantity can be quite large; e.g., the building heating load tends to be maximum during a winter night, whereas the electric load is a maximum in the late afternoon (Figs. 4-6 and 4-7). Little can be done toward utilizing all the heat available during seasons of low heat demand, but the short-term storage of heat in the form of high-temperature water can be accomplished efficiently with relatively little capital investment to accommodate diurnal variations in the ratio of heat to electric energy demand, thus giving the plant operator a valuable degree of freedom in meeting both the heat and electric loads required. The flexibility and part-load performance characteristics of the power plant then determine the overall efficiency of energy utilization, a matter that will be discussed in a later section in this chapter in regard to a particular case, the Munich municipal system.

Plant Siting

Electric energy can be transmitted substantial distances efficiently for modest capital costs by using high voltages (Chap. 20), but heat transport is another matter, in part because of pumping and heat losses and in part because of much higher capital costs. For this reason it is important that a cogeneration type of power plant be located fairly close to the heat load, the distance acceptable being dependent on the heat load and prevailing costs. Thus these factors, as well as the usual plant siting criteria, must be considered.

Plant Site Criteria

Up until the 1960s the principal considerations in steam power plant siting were the availability of cooling water, a favorable geological structure for a firm foundation, the cost of land, and the relative cost of fuel transportation and electric power transmission if coal were the fuel. Most plants were built in or on the edge of urban areas, although coal transportation costs sometimes favored riverine sites close to coal mines. The siting problem became more complicated with the introduction of nuclear plants; the AEC imposed stringent requirements that

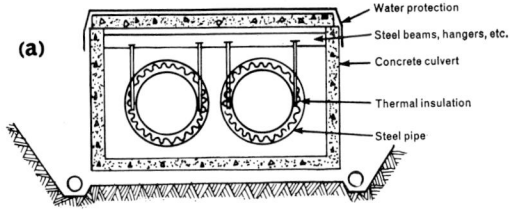

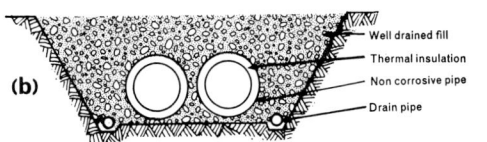

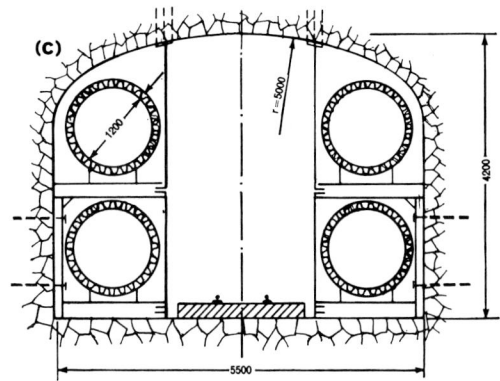

Figure 4-8 Types of construction used for the culverts and mains of district heating systems: (a) steel pipes in a concrete culvert, (b) armored plastic or prestressed concrete laid in the ground without a culvert, and (c) steel pipes in a tunnel.[27] (Larsson and El Mahgary, IAEA Report No. AG-62/6.)

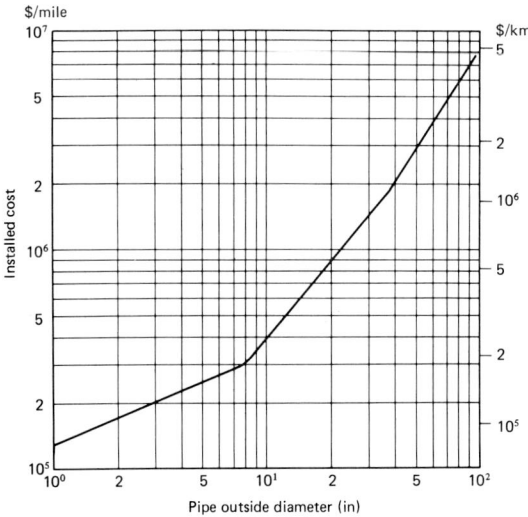

Figure 4-9 Cost of installing supply-and-return pipe system sealed with poured concrete for district heating system mains. Based on high-temperature water with an operating pressure of 28.6 bars (400 psig) and 1969 dollars.[3]

made it necessary to site plants in regions of low population density with large, fenced, exclusion areas around each plant. These factors, coupled with the problems of transporting radioactive spent fuel, have led to intensive studies of large nuclear energy centers, or *power parks,* that would include fuel reprocessing facilities.[21,22,23]

The siting problems for both fossil fuel and nuclear power plants have become so strongly dominated by environmental factors (see Section 9 on environmental considerations) and politics that at the time of writing it was not possible to delineate any sound logical basis for plant siting that could be applied for any actual site without being subject to legal attack by environmentalist groups that are basically opposed to the construction of any new plants. As a consequence, the discussion here on the important problems of plant siting is concerned mostly with the problems of transporting heat from the electric power plant to an urban load center. It will suffice to mention that nuclear fission plants could be located in mined caverns under urban areas, as suggested by Edward Teller, and thus protect the public from any conceivable accident.[24] There is extensive experience in Sweden with this type of construction for industrial manufacturing plants, chosen in that case as a means of protecting the facilities from aircraft and missile attack in the event of war. This Swedish experience indicates that where the local geology and terrain are favorable (which is usually the case), the incremental cost of the mined cavern construction might be as little as 10 percent of the total plant cost. However, for reactors, provisions for getting large pieces of equipment, such as pressure vessels, in and out of deep caverns through vertical shafts pose severe problems that appear to require expensive solutions. Note, too, that some designs for fusion reactors, if they prove practicable, offer the possibility of a vastly less serious hazard potential than do fission reactors, and so a good case can be made for siting fusion reactors in urban areas without the need to go underground.[25]

Superheated Water versus Steam for District Heating

Most of the district heating plants in the United States were built before 1929, served relatively small areas filled with tall buildings, and made use of steam as the heat-transport fluid. Both analyses and experience indicate that if the heat must be transported over distances greater than about ~ 1 km, the distribution system costs can be reduced by employing high-temperature water (100 to 200°C) rather than steam as the heat-transport medium.[4] The cost reduction stems from the fact that water has ~ 1000 times the density of low-pressure steam, and so, even though the heat content per unit weight is lower, the diameter of the mains is greatly reduced for a given pumping-power-heat-transport ratio. This is particularly important because the cost of the mains increases rapidly with diameter, in part because of the relatively high pressures and consequently large wall thicknesses required and in part because of the physical size of the trench in which the mains are installed, together with the amount of thermal insulation and concrete duct required for protecting them (Fig. 4-8).[26] Estimates in 1969 indicated that the cost of hot-water mains is largely a function of their diameter (Fig. 4-9 and Table 4-6), and the heat delivered through mains carrying high-temperature water is directly proportional to the temperature drop available.[3] For a temperature drop of $\sim 100°C$ the cost of heat transport was estimated in 1969 to run about $0.02/10^9$ J [$0.02/10^6$ Btu/(h · mi)] for large mains handling ~ 1000 MWt.[3]

The study of Ref. 3 indicated further that if steam were sold at the price cited above as the 1969 average for district heating systems in the United States—i.e. at $1.45/10^6$ Btu—and the steam were obtained from a nuclear plant at a cost of $0.10/10^9$ J ($0.10/10^6$ Btu) rather than $0.76/10^9$ J ($0.76/10^6$ Btu), and if half of the difference were considered as offsetting the additional cost of a distribution system, it would be possible to transport heat economically as much as 30 km (20 mi) from the electric power plant to the heat load center.[3]

The size of the pipe required to supply hot water for heating a single-family residence is actually less than that normally used for its conventional water supply. Even after allowing for the extra costs of thermal insulation, the heavier pipe required to withstand the higher pressure, and heat losses from the distribution system, preliminary calculations made by the writer indicate that heat could be distributed economically even to single-family residences in built-up areas on the outskirts of a city if the heat were supplied from a nuclear plant. European experience confirms this, and some examples will be cited later in this chapter.

Reliability Considerations

While not necessarily obvious, close integration of the various elements of the utility system for an urban complex carries with it the requirement for an exceptionally high degree of reliability. As serious as an electric power shortage may be, a failure of the heating system of an urban complex would be even more serious because there would be nothing comparable to the countrywide electric power grid to fall back on. Fortunately, there is a tremendous amount of heat capacity in the heating systems for an urban complex, so that an outage for an hour or two would probably not be serious. The heat required for the distillation of seawater or sewage need not be continuously available; in fact, with ample storage capacity for both

TABLE 4-6 Percentage of Total District Heating Pipe System Cost by Component

	Percentage of total cost						
	Pipe diameter (IPS)						
Component	4 in	8 in	12 in	16 in	24 in	30 in	36 in
Excavation	19.1	16.3	14.3	12.6	11.3	10.3	9.6
Concrete	29	26.1	24.5	22.4	21.8	20.6	20.2
Pipe	35.5	42.1	44.7	50.1	52.7	54.8	55.7
Insulation	16.4	15.5	16.5	14.9	14.2	14.3	14.5

TABLE 4-7 Estimated Size Distribution of the Electric Power Requirements of United States Metropolitan Areas in the Year 2000

Electric load, MW	Number of cities in range	Total power, MWe	Percent of total power
0–200	1	173	0.042
200–300	15	3,831	0.93
300–400	23	7,927	1.93
400–500	31	14,140	3.44
500–700	29	17,249	4.19
700–1000	38	32,055	7.79
1000–1500	24	28,416	6.90
1500–2000	13	22,341	5.43
2000–3000	18	44,163	10.73
3000–4000	10	36,387	8.84
4000–6000	7	32,279	7.84
6000–10000	7	51,386	12.49
10,000	6	121,189	29.45
Total	222	411,536	100.00

water and sewage, it appears possible in a cold spell to operate for as much as 30 days without any distillation.

In reviewing the above and considering all the fluctuations that are necessarily to be expected in both electric power demand and the demand for heat for buildings, together with the requirements for a very high degree of reliability, it seems likely that the heat load for a city should be shared by at least three, and preferably four, power plant units to assure the requisite reliability. To investigate this requirement, Table 4-7 was prepared to provide a basis for judging the size of power plant required. If most of the people in the U.S. urban complexes are to be served, it will be necessary to utilize power plant units with design outputs of no more than about 100 MWe for the smaller urban areas. Thus nuclear plants may not be well suited for use in the smaller demand areas because of their relatively high cost per kilowatt in sizes below ~500 MWe.

Typical Cogeneration Systems for Urban Areas

Although some U.S. electric utilities provide back-pressure steam for district heating systems, no U.S. city has a comprehensive integrated system supplying the greater part of the city's needs for both heat and electricity. There are a number of such examples abroad, however, such as Malmö and Västerås in Sweden, Munich in Germany, and Sapporo in northern Japan. The Sapporo case is interesting in that up until the latter 1960s the buildings were heated mainly with coal, and the release of coal smoke at low level from a multitude of small chimneys led to severe smogs under the frequent temperature inversion conditions prevailing in the winter because of the local topography. When the city was chosen as the host for the 1970 Winter Olympics, the city council decided that something had to be done, and a major program was launched to build a district heating system coupled to a large steam power plant with tall stacks.[27] The system was designed by a U.S. firm, built, and has proved to be eminently successful both economically and environmentally. The systems of the Swedish cities of Malmö and Västerås are particularly interesting in that they show that it is possible for small cities with populations of 100,000 to 250,000 to move in an orderly and economically attractive fashion from having no municipal district heating system to having one with cogeneration in a comprehensive integrated system.[26,28] In these instances the orderly buildup of the heat load (Fig. 4-10) was nicely paced with the construction of steam generation facilities especially designed for cogeneration—and consequently the central station is more complex and expensive than a conventional electric power plant. As in Sapporo, the systems have proved economically viable and environmentally attractive not only for commercial and multifamily building areas but also for areas of single-family residences. Further, as can be seen in Fig. 4-11, they have drastically reduced air pollution, in part because of the far better control of combustion conditions and better combustion efficiency in large power plants. They thus refute the Lovins hypothesis that "small is better."

Street Heating in Västerås

A special added feature in Västerås is the extensive use of street heating to free the streets of ice and snow in the winter.[30] By 1973 the street heating system had been extended to provide 1,300,000 m² of snow- and ice-free pavement heated by water entering 25-mm polyethylene tubes on 25-cm centers about 10 cm below the surface. The tubes are arranged in 200-m-long grids and supplied with water at 35°C which cools to 20°C in the circuit. Efforts to quantify the costs and benefits have yielded the bar chart of Fig. 4-12, which seems to give good justification for the Västerås street heating system.

The Munich Municipal Power System

The Munich Municipal Power System (München Stadtwerke) provides an excellent example of the application of cogeneration in a city of about 1 million people to supply both electricity

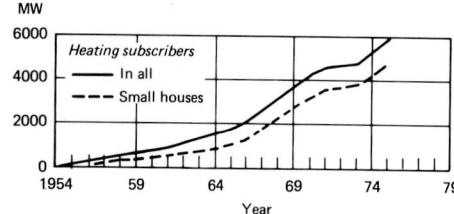

Figure 4-10 Growth of the heat load on the district heating system of Västerås, Sweden, from its inception in 1954, when the population was 80,000 to 1975, when the population reached 120,000.[29]

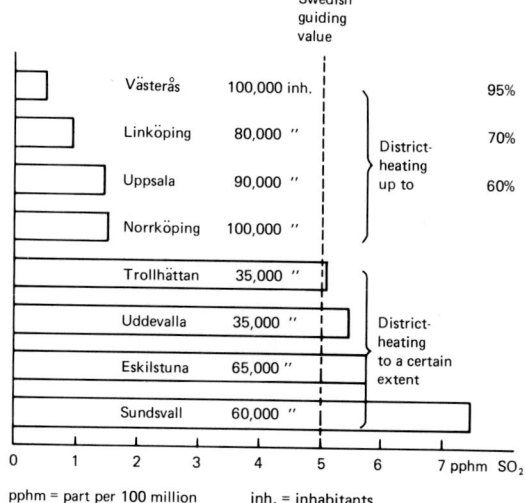

Figure 4-11 Effects of district heating on the SO₂ content in parts per hundred million of the air of typical Swedish cities in February 1971.[29]

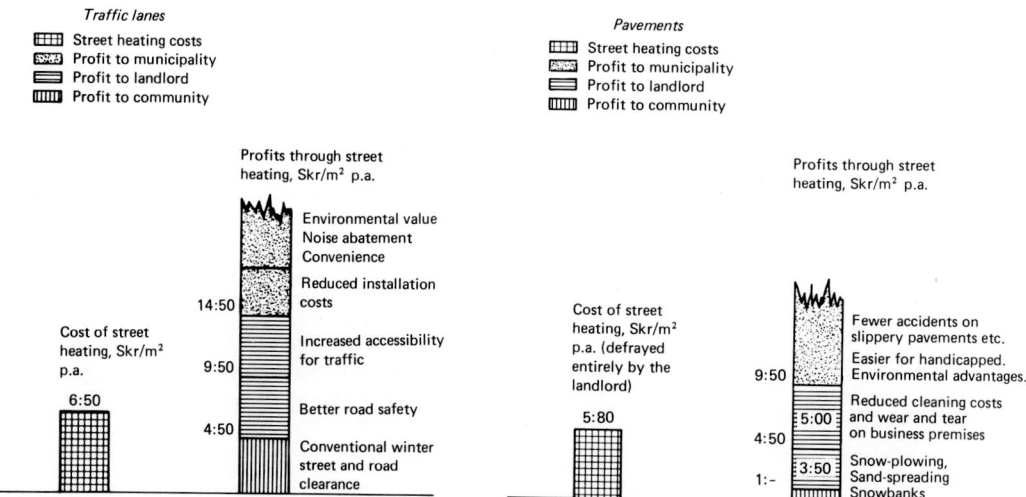

Figure 4-12 Estimated costs compared with the benefits from street heating in Västerås, Sweden. The bars at the left are for automotive traffic lanes, and those at the right are for sidewalks. At the time of the estimate in 1973 the exchange rate was 4.7 Skr per dollar.[30]

and heat.[31] The system is not allowed to supply heat or power to anyone outside the city limits, and since it is under serious governmental restrictions with respect to the purchase of electric power from private utilities, it is not connected into a utility grid. Municipally owned and operated, the electric power capacity was 600 MWe when the author visited the system in 1972. There were four steam plants and one gas-turbine plant, with a sixth plant, a 172-MWe gas-turbine plant, under construction and due for completion in 1974. There is no heavy industry in Munich, yet ~50 percent of their heat load is industrial and 50 percent is commercial and residential. The base heat load is 25 percent of the plant's full capacity and is mostly industrial. One of the steam plants is fired with rubbish and garbage with auxiliary fuel oil burners. Another of the steam plants—that on the upwind side of Munich—is gas-fired with gas from a Bavarian natural gas field. The system also includes a number of pumped-storage hydro units in the mountains just south of Munich.

Gas-Turbine Plant The first gas-turbine plant was intended as a peaking power unit when initially built in 1961. However, it proved to be not only more efficient than the design estimates had indicated, but also nicely flexible in handling varying ratios of heat to electric loads. Thus it has actually become essentially a base-load plant with one or both turbines in operation over 6000 h per year. The gas-turbine plant has demonstrated an excellent reliability and an availability of 98 percent while using a light distillate fuel oil and operating with a turbine inlet temperature of 720°C.

The gas-turbine plant design was intended to give a good overall efficiency for any of four modes of operation. If the electric load is dominant, the plant is operated with a regenerator extracting heat from the gases leaving the gas turbine and putting it into the air leaving the compressor before it reaches the burners. When the remaining heat in the exhaust gas leaves the regenerator, it is given up to water in the heat exchanger for the superheated-water distribution system. With this arrangement the thermal efficiency of the plant for generation of electricity alone is over 30 percent, and the overall thermal efficiency counting the heat given up to the superheated-water distribution system is about 90 percent. If the heat load on the hot-water distribution system becomes sufficiently high to justify it, the hot gases leaving the turbine may bypass the regenerator and go directly to the heat exchanger for the hot-water distribution system. A third mode of operation lies between the two just mentioned: it involves modulating from one mode to the other and operating with some regeneration. For unusually high heat loads the fourth mode of operation entails lighting an afterburner at the inlet to the heat exchanger for the hot-water distribution system. Note that the heat exchanger costs for this plant ran about four times the cost of the turbine-generator units.

Heat Storage An important element in the system both for the gas-turbine plant and for the steam power plants is the use of large tanks about 5 m in diameter and ~35 m tall for hot-water storage. These tanks are operated full of water all the time and function much like a domestic hot-water tank, with the hot water segregated at the top and cold water at the bottom. With these tanks it is possible to store heat during periods of low heat loads and then recover it and deliver it to the distribution system during high-peak-heat-load periods. Figure 4-13 indicates that the cost of storing heat in high-temperature water is low—definitely less than in fusible salts.[32]

Hot-Water Distribution System In 1972 the Munich heat distribution system covered an area of 16 km² and had a capacity of 500 gigacalories (Gcal) per hour (2×10^{12} J/h, or 2×10^9 Btu/h). This gave a heat-load density of approximately 30 Gcal/(km² · h) [120×10^6 Btu/(km² · h)]. In 1972 the heat was sold at 21 deutsche marks (DM) per gigacalorie (about $1.90/$10^6$ Btu). The capital cost for the water-main distribution system for a new, single-family residential area where the heat

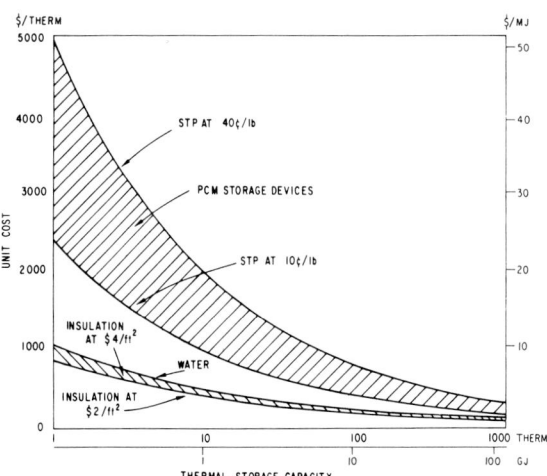

Figure 4-13 Comparative costs per therm (10^5 Btu) of thermal-energy storage devices using water versus phase-change material (PCM) for storage. (STP = sodium thiosulfate.)[32]

distribution system was installed before the construction of the streets and houses amounted to 10^5 DM/(Gcal · h) ($25/kWt).

The high-temperature water distribution system makes use of two basic water supply temperatures. The water for industry is maintained at 160 °C as it goes into the hot-water mains. It returns to the power house at a temperature of around 50 °C. The water supplied for building heating is normally at about 90 °C, but in cold weather the temperature is increased to as much as 110 °C. The building heating system is normally operated by taking the 50 °C water being returned from industry and adding to it 160 °C water to give the desired temperature level. The hot-water distribution system in Munich is an interconnected network, and so the outage of any one of the plants will not lead to a loss in supply of hot water to either the industrial or residential consumers.

Experience in Munich has shown that if a new residential section is built and all the houses are to be heated with hot water from a central system, a system using hot water from power plants gives the lowest overall capital costs and much lower fuel costs than any other system. Two large single-family residence areas of about 1 km² each have been built with spacious yards; the capital cost for the hot-water distribution system was less than $300 per residence for construction about 1963. This low cost was possible only if the hot-water mains were installed before the streets were paved and the houses built. Further, for low costs for single-family residences, it is essential that all the houses on each street served should use the hot-water system in order to give an adequate heat-load density.

The size of the mains is adjusted to the heat load of the area served. The smallest mains are 10 cm in diameter with a 5-cm-thick layer of thermal insulation. The hot-water mains are installed inside a small concrete duct with clearance around the main to reduce heat conduction losses to the earth (Fig. 4-8). The return water main is thermally insulated but is not surrounded with a concrete duct. The mains are usually installed at a depth of about 2 m. The design accommodates a velocity of about 4 m/s in 60-cm-diameter mains and about 2 m/s in the 10-cm-diameter mains. The capital cost of installing a set of mains where there are no obstacles to their installation (i.e., in a new area) in 1972 ran about 2300 DM/m ($250/ft) for a 50-cm-ID main and about 750 DM/m ($82/ft) for a 10-cm-ID main. The thermal insulation on the larger-diameter mains was about 10 cm, but it was only about 5 cm thick on the 10-cm-ID mains.

Plant Proposed for the Minneapolis–St. Paul Area

A major problem in getting new cogeneration projects underway in cities where there is no district heating system is that special turbines and higher capital costs are required, yet it will take some years to build up the heat load for the new steam plant. During this period the plant will be handicapped by both higher capital costs and the inherently poorer efficiency of the turbines designed for high steam-bleed rates. These problems have been explored in a set of design studies of 400- to 800-MWe turbine-generator units to give good efficiency for operation over a wide range of ratios of electric to heat outputs, e.g., from 550 MWe and zero heat output to the district heating system to 471 MWe and 705 MWt for district heating.[33] The corresponding thermal efficiencies based on the heat in the steam supplied to the turbine at 165 bars, 538 °C/538 °C (2400 psi, 1000 °F/1000 °F) would be 43.2 and 82.9 percent, respectively. Further, the loss in turbine efficiency at zero heat load would be only ~0.3 percent and the incremental capital cost would be only ~5 percent when compared with a conventional turbine. Thus the cost and efficiency penalties for providing the power plant to serve as the basis for a new cogeneration system appear to be small.

An excellent insight into the operation of the proposed system is given by Figs. 4-14 to 4-16. In warm weather, when there would be little heat load except for hot water, the water circulation rate in the district heating system would be maintained at a constant relatively low rate with a hot-water supply temperature of 71 °C (160 °F). In colder weather, the water circulation rate would be increased to the full design flow rate and the water temperature would be increased sufficiently to meet the heat demand (Fig. 4-14). Figure 4-15 shows the fraction of the time that one would expect any given heat load.

The thermal efficiency of the plant based on the heat in the steam supplied to the turbine as influenced by the heat and electric loads is delineated in Fig. 4-16. This shows the energy savings made possible by cogeneration.

BOTTOMING AND OTHER SPECIAL RANKINE CYCLES

In many cases, heat is available from an industrial process at moderate temperatures (e.g., 60 to 100 °C), together with a heat-sink temperature in the 10 to 20 °C range, but the use of steam turbines is not attractive because the excessively large machines needed to handle the low-density vapor make the unit cost high. In fact, in conventional central stations the limitations on steam-turbine size generally make it uneconomical to design for steam exhaust temperatures below ~32 °C (90 °F) because the density of steam drops rapidly at lower temperatures. In cold climates the use of a working fluid having a higher vapor pressure than water offers the possibility of extracting more work from the system via a low-temperature *bottoming cycle*. It happens that lower temperature cycles such as this require essentially the same working fluid characteristics as cycles for

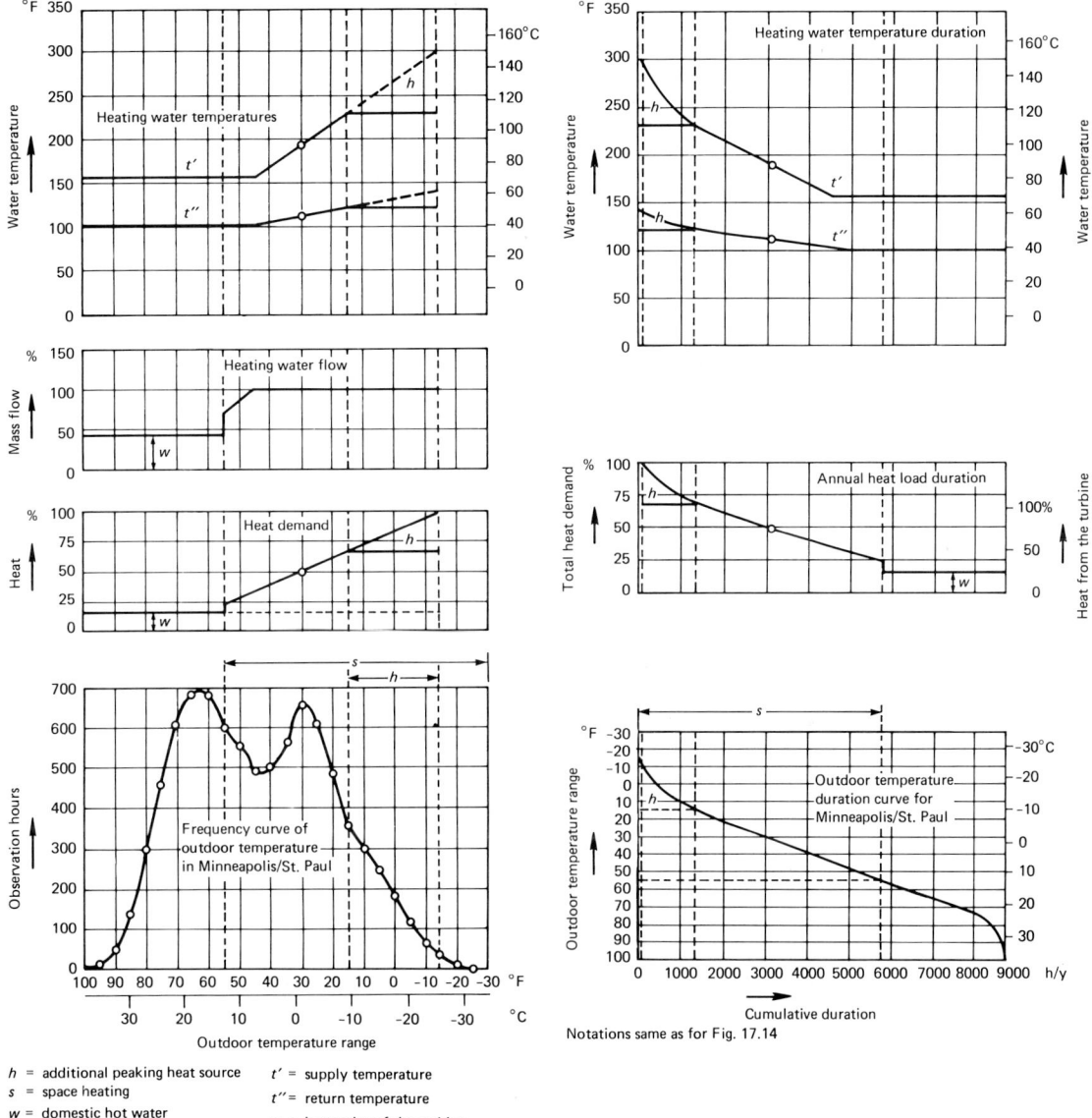

h = additional peaking heat source
s = space heating
w = domestic hot water

t' = supply temperature
t" = return temperature
o = best point of the turbine

Figure 4-14 Typical heat load data for a U.S.-district heating installation.[33]

Figure 4-15 Duration of heat load conditions for a U.S.-district heating installation.[33]

use with solar, geothermal, and ocean-thermal-difference energy systems (treated in Section 5); hence the material presented here is also applicable to those systems.

The use of fluids other than water in Rankine cycles has also received considerable attention for special applications in which the system output would be relatively small so that Reynolds number effects make possible a substantially higher turbine efficiency with a high-molecular-weight fluid than obtainable with steam in a small turbine. Examples include small nuclear electric power systems for space or undersea applications[34,35] and small systems for certain terrestrial applications.[36,37,38] Other fluids appear attractive where there are incentives to avoid high pressures at moderately high temperatures, as in certain chemical processes (Refs. 37, 38, and 39). Still other fluids are of interest to keep the turbine size down for large outputs at lower temperatures, as for geothermal and Ocean Thermal Energy Conversion (OTEC) applications (Section 5).[40,41]

Working Fluids

The many organic compounds, together with the fluorocarbons and chlorinated hydrocarbons, appear to offer an overwhelmingly wide range of candidate compounds for special Rankine cycles, but practical considerations narrow the field to a relatively small number. The first consideration is the vapor pressure in the temperature range of interest. For favorable proportions in the turbine, heat exchangers, and vapor passages, it is desirable that the vapor pressure at the lowest temperature in the cycle be ~0.3 bar or more, although a pressure as low as 0.03 bar may be acceptable. (The pressure in the condenser of a conventional steam power plant is about as low as is practicable, i.e., commonly 0.03 to 0.1 bar.) A second major consideration is the compatibility of the fluid with the other materials in the system, together with its long-term chemical stability, particularly freedom from a tendency to polymerize to form gums, resins, or tars. If a hermetically sealed system is employed, it is essential that the fluid not decompose to give noncondensable gases, such as H_2 or CH_4. It is desirable that the fluid be nontoxic and nonflammable, that the melting point be well below temperatures that might be reached when the system is shut down, and that the cost be reasonable. The fluids that have proved of greatest interest meet most of the above criteria; their principal properties are summarized in Table 4-8. Of these, NH_3 has the highest vapor pressure and is of interest mainly for the lowest temperature cycles, while Dowtherm has the lowest vapor pressure and is well suited to the higher temperature cycles. Water is included in Table 4-8 not only for purposes of comparison but also because, after all the con-

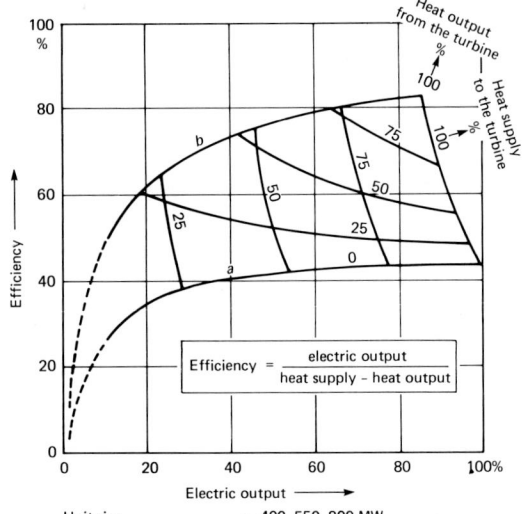

Figure 4-16 Net thermal efficiency of the cogeneration turbine as a function of heat and electric loads.[33]

siderations are weighed, it is usually the best candidate even for special applications.[39] A major reason for its suitability is that its heat-transfer characteristics, including a high heat of vaporization and relatively good thermal conductivity, are exceptionally good for both boiling and condensing conditions.

Heat-Transfer Considerations

The heat exchangers in a system ordinarily cost several times as much as the turbine. Hence the boiling and condensing heat-transfer characteristics are likely to prove to be the dominant factors in the choice of a working fluid. The column at the right end of Table 4-8 gives a parameter that is proportional to the forced-convection heat-transfer coefficient for the liquid—an important factor if a regenerator is employed.

The forced-convection heat-transfer parameter in the right-hand column of Table 4-8 also influences the condensing heat-transfer coefficient, because the film of liquid condensate on

TABLE 4-8 Properties of Typical Rankine Cycle Working Fluids at 25°C or 1 atm

	Properties of liquid							
Fluid	Molecular weight	Density, g/cm³	Specific heat, J/(g·°C)	Viscosity, cP	Thermal conductivity, W/(m·°C)	Heat of vaporization, J/g	Freezing point, °C	$\dfrac{c_p^{0.4} k^{0.6}}{\mu^{0.4}}$
Ammonia	17	0.602	4.81	0.124	0.50	1370	−78	2.85
Benzene	78.1	0.876	1.74	0.60	0.14	390	5	0.47
Dowtherm A	165.6	1.05	2.5	1.08	0.12	286	54	0.39
Ethanol	46	0.787	2.45	1.095	0.17	846	−114	0.48
Freon 12	121	1.315	0.97	0.27	0.07	165	−158	0.34
Methanol	32	0.789	2.55	0.56	0.22	1102	−98	0.54
Monochlorobenzene	113	0.876				323	−45	
Toluene	92.1	0.865	1.72	0.55	0.13	363	−95	0.46
Water	18	1.000	4.19	0.89	0.61	2262	0	1.47

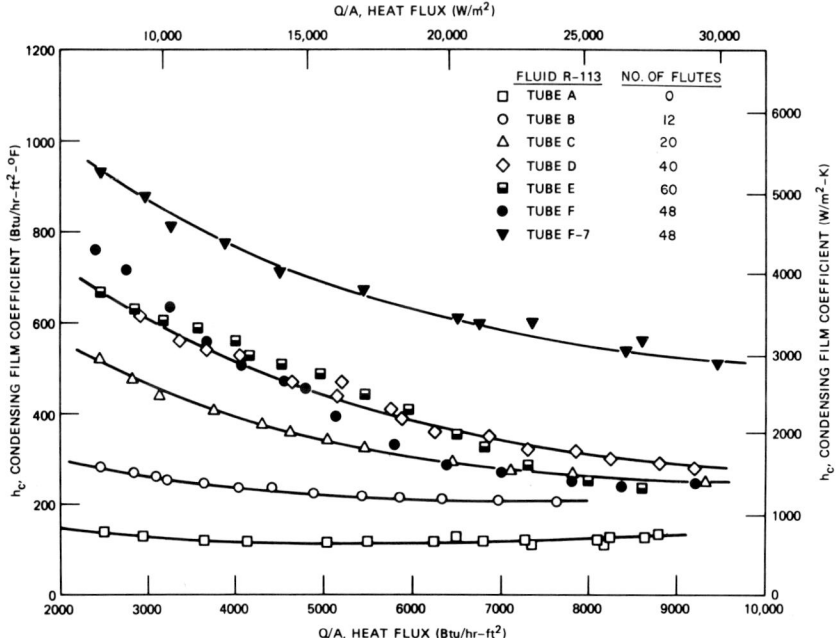

Figure 4-17 Effects of the flute geometry on the condensing heat-transfer performance of 1-in-OK, 4-ft-long, vertical condenser tubes operating with R-113 (CCl_2F-$CClF_2$). Tube F-7 was fitted with seven neoprene drain-off skirts (or collars) to reduce the thickness of the condensate film to a minimum.[43] (*Courtesy Oak Ridge National Laboratory.*)

the surface of a condenser acts as a barrier to heat transfer. This is such an important factor that the success of power plants designed to operate on ocean thermal differences and low-temperature geothermal heat sources depends in large measure on minimizing this temperature loss. One method that has been found effective is to employ fluted vertical condenser tubes. The liquid condensate film is concentrated in the flutes by surface tension and runs down the tiny channels formed by the grooves, leaving only an extremely thin film covering the ridges between the grooves. Figure 4-17 shows that such a geometry can increase the condensing heat-transfer coefficient by as much as a factor of 7 over that obtainable with a plain tube. As shown in Fig. 4-18, the magnitude of the enhancement depends on the physical properties of the condensed liquid, particularly the thermal conductivity, surface tension, and viscosity. Note the clear superiority of ammonia from this standpoint.

The high condensing heat-transfer coefficient obtainable with fluted condenser tubes gives a situation in which the heat-transfer coefficient on the cooling water side becomes controlling so that there is little incentive for further increases in the condensing coefficient. This effect is shown in Fig. 4-19, which gives the overall heat-transfer coefficient as a function of the condensing (or boiling) coefficient for units operating with seawater flowing through the tubes.[43]

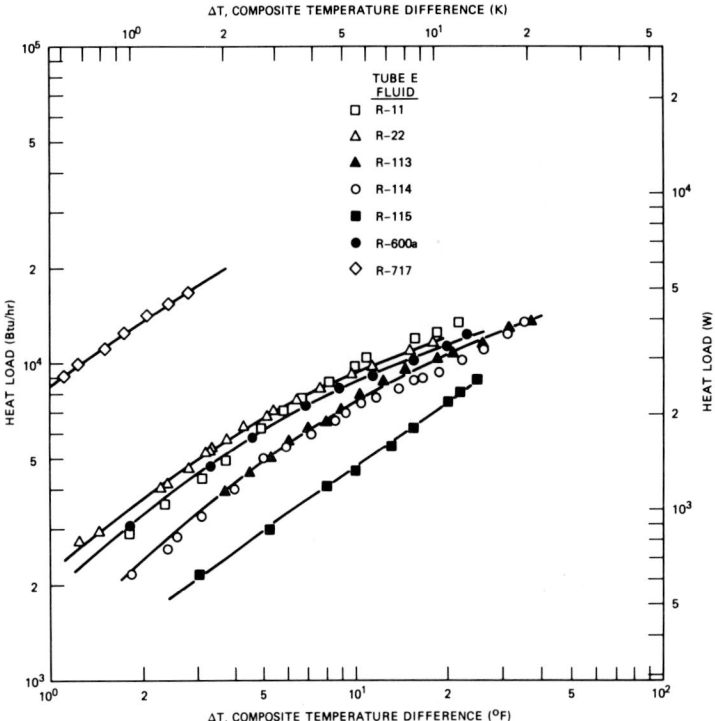

Figure 4-18 Effects of the physical properties of the working fluid on the condensing heat transfer rate for Tube E of Fig. 17.17 (surface area = 1.05 ft²). Data are plotted for five Freons, isobutane (R-600a), and NH_3 (R-717). The "composite temperature difference" includes the temperature drop through the tube wall.[43] (*Courtesy Oak Ridge National Laboratory.*)

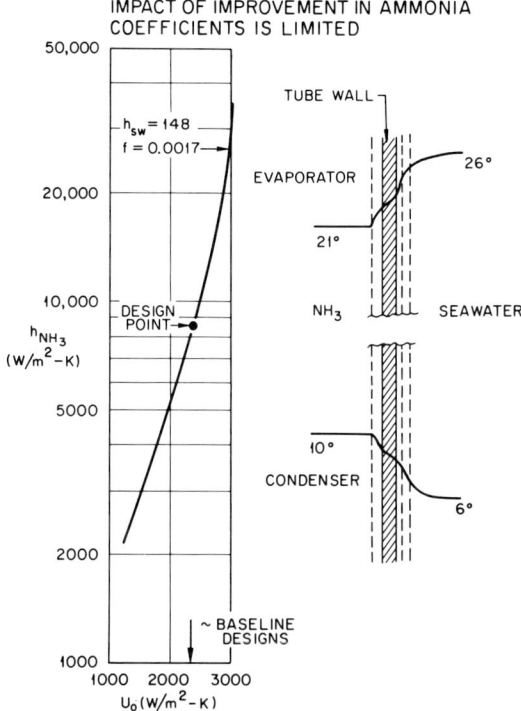

Figure 4-19 Effects of the boiling or condensing heat-transfer coefficient on the overall heat-transfer coefficient for operation with seawater as the tube-side fluid giving a heat-transfer coefficient of 812 W/(m² · K) [143 Btu/(h · ft² · °F)].[43] (*Courtesy Oak Ridge National Laboratory.*)

Turbine Design Considerations

The principal proponents of working fluids other than water for special low- and moderate-temperature Rankine cycles are turbine designers who are mainly interested in improving the proportions and/or increasing the Reynolds number in the turbine through the use of a high-molecular-weight fluid.[34,37,38] Table 4-9 shows the effects of the choice of working fluid on some of the principal design parameters for a small turbine. Tip leakage losses associated with a 0.16-mm blade height for the full admission steam turbine would be so severe that it would be better to go to partial admission, though this would still give a lower turbine efficiency than for full admission with the greater blade heights of the units designed for operation with chlorinated hydrocarbons.

Cycle Efficiency Effects

Table 4-8 does not include a column indicating differences in cycle efficiency between the working fluids. The reason for this is that, fundamentally, there is no difference in the efficiency of the ideal cycles operating between the same temperature limits; differences in efficiency for the actual cycles stem from differences in component efficiencies attributable to Reynolds number effects, design compromises, and the usable temperature range which is limited by factors such as pinch point effects.[40] An indication of the magnitude of the latter consideration is given by Fig. 4-20, which shows the effectiveness of energy utilization as a function of the temperature at which a geothermal fluid is available.[43] It can be seen from these curves that R-115, a fluorocarbon, appears to be the best choice from the purely thermodynamic standpoint for a source temperature of 140 °C, whereas NH_3 would be best for a source temperature of 300 °C.

TABLE 4-9 Effects of the Choice of Working Fluid on the Principal Design Parameters of a Small Turbine Designed to Produce 285 W (Ref. 37)

	Boiler temp., °C	Condenser temp., °C	Rpm for 140-mm wheel	Total throat cross section, mm²	Blade height (mm) for full admission, 20 deg injection angle, impulse
Steam	110	45	62,500	5.8	0.16
Monochlorobenzene	110	45	26,500	13.2	0.61
Dichlorobenzene	110	45	25,000	144.0	5.00

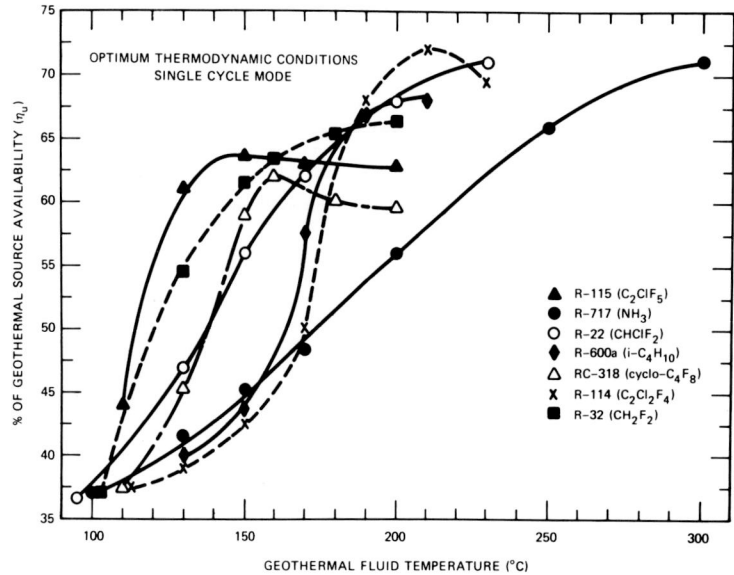

Figure 4-20 Effect of working fluid characteristics on the fraction of the thermal energy available that can be put into the thermodynamic cycle as a function of the geothermal source temperature.[43] (*Courtesy Oak Ridge National Laboratory.*)

Bottoming Cycles for Conventional Steam Plants at Dry, Cold Sites

The size and cost of steam turbines and condensers increase rapidly if one tries to increase the cycle efficiency by reducing the condenser temperature. At 120 °C (250 °F), for example, the specific volume of steam is 0.86 m³/kg (13.8 ft³/lb), whereas at 38 °C (100 °F) it is 21.8 m³/kg (350 ft³/lb) and at 15 °C (60 °F) it is 75 m³/kg (1200 ft³/lb). For an 1800-rpm steam turbine, it is difficult to make use of a diameter greater than 4.2 m (14 ft) because of tip speed effects on both the stresses in the turbine blades and compressibility losses in the steam. As a consequence, the velocity of the steam leaving the last stage of the turbine must be so high that it represents a large fraction of the energy that one might expect to be available from expansion from 38 °C (100 °F) down to 15 °C (60 °F). One way to avoid the losses in cycle efficiency that this entails and also effect a marked reduction in the turbine cost, would be to make use of a working fluid with a much higher vapor pressure than steam in this temperature range. Hydrocarbons such as butane, some of the Freons, and particularly ammonia have looked attractive for this purpose.[44]

The principal barrier to the use of a bottoming cycle in steam plants at favorable sites is that it inherently requires a heat exchanger between the lower end of the steam cycle and the upper portion of the bottoming cycle. Analyses indicate that the temperature loss across the heat exchanger can be kept to as little as 7 to 10 °C with acceptable costs. Even this low temperature difference inevitably leads to a loss in cycle efficiency unless the heat-sink temperature will run below around 10 °C for most of the year. For such conditions the bottoming cycle has the additional major advantage that it could be used with dry cooling towers without danger of freezing of the working fluid. This is especially important in cold areas where there is not an adequate supply of cooling water.

Small Systems in the Intermediate-Temperature Range

A typical special situation favoring the use of an organic working fluid is offered by an industrial operation in which sufficient heat to produce ~100 kWe is available at ~350 °C (662 °F) and there is a need for process heat at ~100 °C (212 °F). One of the best of the working fluids in Table 4-8 for this application is Dowtherm because its chemical stability is the best among the fluids listed,[35,45] its vapor pressure is in the right range, and it yields a more efficient turbine than steam in this size range.

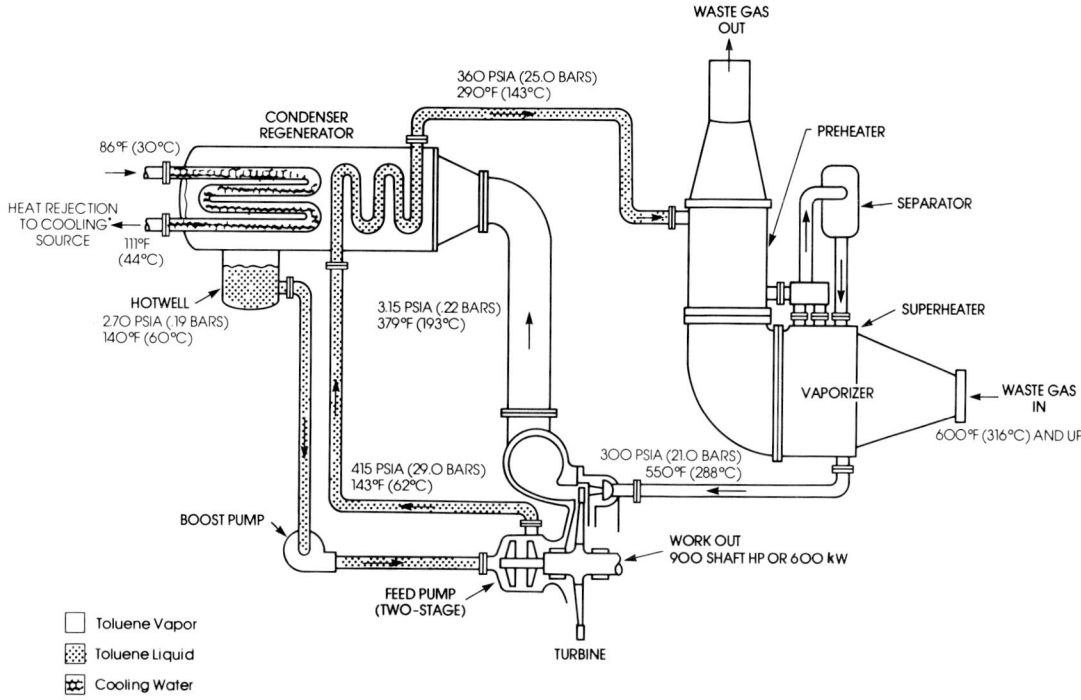

Figure 4-21 Flowsheet for an organic Rankine cycle system utilizing toluene as the working fluid for recovering heat from diesel engine exhaust gas or industrial process fluid streams to produce 600 kWe. (*Courtesy Sundstrand Corp.*)

To obtain a good efficiency from a Dowtherm cycle, it is necessary to cope with a special problem introduced by the shape of the temperature-entropy diagram. The vapor expansion from saturated conditions at the turbine inlet entails a path into the strongly superheated region. This is advantageous in that it avoids moisture formulation in the turbine and the consequent problems with moisture churning losses and possible blade erosion, but for good cycle efficiency it is necessary that the heat removed from the superheated vapor in cooling it to the saturated condition must be employed for regenerative heating of the liquid feed to the boiler with a counterflow heat exchanger. This makes it necessary to remove the heat by forced convection with a much poorer heat-transfer coefficient than for condensing conditions; hence the amount of heat-transfer surface area required is much greater than for a condenser.[35]

Commercial Applications

Although there have been many studies of potential applications of working fluids other than water for use in low- and intermediate-temperature Rankine cycle systems, none of these has gone beyond the experimental and demonstration stages to any substantial commercial application. Thus these systems are likely to continue to be of interest only for quite special heat-recovery applications,[42] and for low-temperature heat sources such as in geothermal, solar, and OTEC applications. An excellent example of such a system for obtaining electric power where heat is available at intermediate temperatures is a 600-kWe unit built by Sundstrand for heat recovery from diesel exhaust and industrial processes.[38] As of the latter part of 1979, three of the systems were in commercial use and three more were under construction with some support funding from the conservation division of DOE. A flowsheet giving the operating conditions for this system is shown in Fig. 4-21. The system is being used where the thermal energy is in effect free, and so the cost of power depends mainly on the capital and maintenance costs. Utilizing a simple system with a fluid having good thermal stability so that corrosion and deposits are not a problem helps reduce both capital and operating costs. In this instance toluene was chosen as the working fluid because it made possible a good efficiency in a single-stage turbine, whereas steam would have required a multiplicity of stages and hence a higher cost.[38] This basic system is also suited to solar thermal power systems in the 500- to 1000-kWe range.[38]

The thermal efficiency of the system of Fig. 4-21 is modest, about 9 percent, because of the low peak temperature and the high heat-rejection temperature for a process heat application; these conditions give an ideal Carnot cycle efficiency of 25 percent. More significant is the ratio of the overall cycle efficiency to that of the Carnot cycle, a ratio that in this case is 37 percent. Part of the loss stems from the fact that small, single-stage turbines of the sort used here have a turbine efficiency of 70 to 75 percent and generators in this size range have efficiencies of 93 to 95 percent. The rest of the losses stem from differences between the actual cycle and the ideal Carnot cycle.

PRELIMINARY STUDIES, ENGINEERING AND CONSTRUCTION PLANNING, AND DESIGN DOCUMENTATION

This segment outlines the tasks required to successfully perform studies, engineering, design, and construction of cogeneration projects.

PRELIMINARY STUDIES AND CONCEPTUAL ENGINEERING

Cogeneration studies are initiated to evaluate cost-effective alternatives and to select the most appropriate options. This is achieved by performing a technical feasibility and economic cost-benefit study to rank and recommend the alternatives. Determination of technical feasibility includes a realistic assessment of each application with respect to environmental impact, regulatory compliance, and interfaces with a utility. An economic payback analysis determines the payback period which is then compared to client financial goals. The selected design can be further developed during a subsequent detailed feasibility study to obtain a more comprehensive technical definition of the project and improved accuracy of cost estimates. Many clients elect to proceed directly to the engineering and design phase upon comparison of the initial technical and economic evaluations. After obtaining client authorization to proceed with a project, detailed engineering, design, procurement, construction, and start-up activities can be initiated.

The various types of analyses and their objectives are listed below.

- TECHNICAL FEASIBILITY STUDY

    Provide a realistic assessment of cogeneration potential.

    Identify available cogeneration alternatives.

    Establish fuel availability and priority classifications.

    Prepare concept design criteria.

    Determine modes of operation.

Optimize the power cycle.

Utilize existing proven cogeneration equipment and extensive system data base experience.

Match systems and equipment to the generation of electricity and process steam or exhaust heat demands.

Plan for the efficient utilization of fuel.

- ECONOMIC PAYBACK ANALYSIS

    Compare economically competitive cogeneration options with noncogeneration alternatives.

- ENVIRONMENTAL ASSESSMENT

    Select the site.

    Assist in the preparation of permit applications.

    Examine air quality problems and compliance with regulations.

    Select appropriate emission control equipment.

- IDENTIFICATION OF REGULATORY REQUIREMENTS

    Clarify applicable regulatory requirements.

    Determine the exemption status.

    Advise the client regarding regulatory applications.

    Assist the client in dealing with regulatory agencies.

    Identify permit requirements.

- ESTABLISHMENT OF ELECTRIC UTILITY INTERFACE

    Clarify the avoided cost rate.

    Define the buy-and-sell contract and ownership options.

    Determine the technical electrical interconnection interface.

- IDENTIFICATION OF FUNDING SOURCES

ENGINEERING AND CONSTRUCTION PLANNING

Preliminary or detailed engineering is initiated by establishing the scope of the project. Meetings are held with the client to identify the design and operating requirements for a new or an existing facility. In developing the design criteria the following should be considered.

- Specific information regarding the site and conditions must be obtained.
- The system must be clearly defined.
- Individual components or combinations of components must be identified.
- The specific mode or modes of operation must be determined.
- The operating modes must be satisfied by selection of the proper cogeneration equipment, systems components, and controls.
- A thorough review of the existing facility operating modes and the existing systems interconnection interface points with the cogeneration facility must be conducted.
- The assumptions and design criteria to be used in the analysis or study must be defined.
- Detailed description of the system design, installation, operations, and environment must be obtained.
- Complete system and equipment data must be identified.
- All the alternatives or combinations of cycles should be examined.
- The systems and equipment must be sized.

Steps listed below should be followed in choosing engineering systems, selecting equipment, and utilizing the latest cogeneration technology to match electrical and process steam or heating applications to provide the highest possible efficiency and lowest total installed cost.

Technical evaluation of a project should

- Establish design criteria.
- Identify fuel considerations.
- Verify the fuel supply.
- Match varying client requirements to the cogeneration application: process steam, exhaust heat utilization or drying, hot water, generation of electricity.
- Select the most suitable plant, systems, equipment, and control systems for the facility.
- Optimize the power cycle.
- Size the plant equipment and systems to match the application.

- Apply experience with the development of compact and dependable heat recovery equipment.

Major equipment and component applications should be reviewed to optimize overall system efficiency and operating characteristics in a comparison of technical performance with cost-effectiveness. Feedwater quality has a direct affect on plant design and cost. The makeup and feedwater for existing facilities should be reviewed relative to equipment water quality requirements.

Efficient cogeneration energy systems design, equipment selection, and construction have become increasingly important because of the rising price and scarcity of fuel. More efficient cogeneration systems conserve energy. The following effective energy conservation and heat recovery practices should be integrated into the power cycle and system design.

- PROJECT MANAGEMENT

    Planning.

    Scheduling.

    Cost control.

- ENERGY CONSERVATION

    Improve energy efficiency by implementing an energy conservation plan.

    Obtain more efficient utilization of fuel by

        Increasing the overall efficiency of electrical generation.

        Providing energy at the lowest possible cost.

    Economically recover heat rejected to exhaust and/or water systems.

        Utilize lowest technically practical temperature relative to ambient conditions.

    Obtain maximum economical heat recovery while ensuring reliability and operability of the equipment.

    Reduce exhaust losses.

    Reduce cooling system losses.

    Minimize the energy demand.

    Utilize available energy for internal process applications and displace less efficient sources.

    Convert existing standby power generator sets to continuous profit producers.

- **ENGINEERING ASSURANCE**

    Assure client of technical quality and economical project.

    Preclude cost of nonconformance and schedule delays.

    Review project for compliance with specifications and industry standards.

- **FINANCIAL CONSIDERATIONS**

Performing the activities listed below will help to identify the economic impact of the project.

- Establish economic evaluation criteria.
- Prepare and evaluate economic factors:

    Determine fuel costs and consumption.

    List buy or sell prices for electricity.

    Estimate electric energy and capacity purchase prices.

    Assist client in preparing the power agreement.

    Prepare an equipment cost estimate.

    Determine operating and maintenance expenditures.

- Perform a cost-benefit and cost-effectiveness analysis.
- Estimate the total installed cost of the project.
- Prepare estimated cash flow requirements.
- Provide liaison services with local economic development officials.
- Advise client on financial assistance programs.

Cogeneration projects should be designed to satisfy applicable environmental requirements to ensure that the facility will be permitted by regulatory agencies. The following factors should be taken into account.

- **ENVIRONMENTAL CONSIDERATIONS**

    Indentify applicable air quality regulations.

    Coordinate the activities of local, state, and federal regulatory agencies.

    Evaluate tradeoffs for air quality.

    Apply best available control technology (BACT).

Determine the requirements of the Environmental Protection Agency and the Air Resource Board concerning emission offsets.

- **REGULATORY REQUIREMENTS**

    Identify applicable governmental regulatory policies.

    Determine the Fuel Use Act exemption status of project.

- **PROJECT PROCUREMENT**

- **CONSTRUCTION SERVICES**

    Preconstruction planning.

    Construction management.

    Construction.

    Start-up.

- **SUPPORT SERVICES**

    Quality assurance.

    Maintenance.

    Training.

DESIGN DOCUMENTATION

Design documentation includes the preparation of project flow diagrams, piping and instrument diagrams, general arrangement drawings, equipment layouts, building and structural drawings, foundation drawings, and electrical one-line diagrams.

Integrating cogeneration projects into existing facilities requires a review of the site areas available and considerations of planned expansion. This area is sensitive to the type of fuel selected. Solid fuels require more space for delivery, storage, handling, and ash removal facilities. The following objectives should be considered in designing a project.

- Design for a specific base fuel.
- Design the facility to integrate with existing client plant operating characteristics:

    Define system and control philosophy.

    Satisfy the electric utility interconnection requirement.

- Design for existing space allocations or limitations.
- Select equipment suitable for the application.
- Arrange equipment for easy access and maintenance.
- Assist in finding a site for the facility.

Fuel availability and alternative fuel possibilities should be analyzed. Potential alternate fuels to be considered for cogeneration applications include natural gas, fuel oil, petroleum-based fuels, wood waste, biomass, coal, lignite, and refuse-derived fuel. Fuel selection should be based on the following considerations.

- AVAILABILITY
    Reliability of equipment and system
    Delivery time

- SITE LOCATION
    Fuel delivery alternatives available
    On-site versus off-site storage
    Delivery costs

- SITE AREA REQUIREMENTS
- AIR EMISSIONS
- EQUIPMENT DESIGN, SIZE, AND COST

- OPERATIONS
    Maintenance
    Training of employees
    Personnel staffing
    Control response time

- GOVERNMENT REGULATIONS
    Exemptions
    Air quality standards
    Tax credits

The equipment and systems listed below are generally considered major factors in the design of a cogeneration facility.

- Combustion turbine generators
- Waste heat recovery steam generators
- Steam turbine generators
- Condensers
- Cooling towers
- Chiller-heaters
- Fuel systems
- Instrumentation and control systems
- Electric switchgear and motor control centers
- Condensate and feedwater systems
- Water treatment systems
- Heat recovery components

POWER CYCLES AND APPLICATIONS

POWER CYCLES

Cogeneration projects generally employ topping or bottoming power cycles. The topping cycle has the widest industrial application. The bottoming cycle is of limited application because of its high cost and the probability that more efficient use of the waste energy stream can be made.

> TOPPING CYCLE The primary energy source is used to produce useful electric or mechanical power output. Reject heat from power production is then used to provide useful thermal energy.

> BOTTOMING CYCLE The primary energy source is applied to a useful heating process, and the reject heat emerging from the process is then used for power production.

These power cycles are shown in Figures 4-22 to 4-25. The efficiency and heat utilization standards established by the Public Utility Regulatory Policy Act and the Federal Energy Regulatory Commission are indicated. A topping cycle for a simplified combined cycle cogeneration facility is shown in Figure 4-22. A topping cycle for a simplified combustion turbine exhaust heat cogeneration facility is shown in Figure 4-23. A topping cycle for a simplified internal-combustion diesel or gas exhaust heat cogeneration facility is shown in Figure 4-24. A bottoming cycle for a cogeneration facility is shown in Figure 4-25.

Total overall thermal efficiencies of up to 80 percent are possible for energy cogeneration systems. This efficiency is approximately twice that of a conventional power plant application. Conventional power plants are approximately 30 to 40 percent efficient. Boilers for conventional plants are approximately 75 to 85 percent efficient, as compared to unfired waste heat recovery boilers for cogeneration applications which can be 98+ percent efficient. Because cogeneration cycles are more efficient they use 30 to 40 percent less total energy. The gross heat rate (Btu

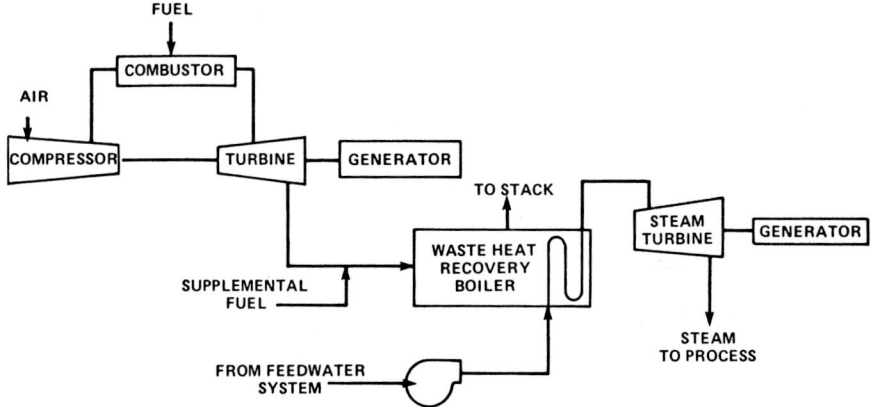

Figure 4-22 Simplified combined cycle cogeneration. Topping cycle: fuel → electricity → process.

PURPA and FERC Efficiency and Heat Utilization Standards

1. Useful thermal energy is the steam to process which must be greater than 5 percent of the total energy.

2. Electricity generated plus one-half the useful thermal energy must be equal to or greater than 42.5 percent of the fuel energy input.

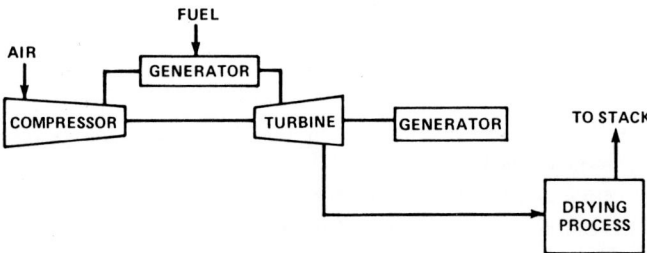

Figure 4-23 Simplified combustion turbine exhaust heat cogeneration facility. Topping cycle: fuel → electricity → process.

PURPA and FERC Efficiency and Heat Utilization Standards

1. Useful thermal energy is the exhaust heat to drying process which must be greater than 5 percent of the total energy.

2. Electricity generated plus one-half the useful thermal energy must be equal to or greater than 42.5 percent of the fuel energy input.

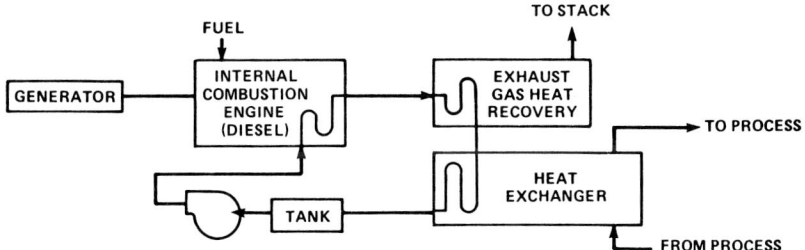

Figure 4-24 Simplified internal-combustion diesel exhaust heat cogeneration facility cycle. Topping cycle: fuel → electricity → process.

PURPA and FERC Efficiency and Heat Utilization Standards

1. Useful thermal energy is the exhaust heat recovered which must be greater than 5 percent of total energy.
2. Electricity generated plus one-half the useful thermal energy must be equal to or greater than 42.5 percent of the fuel energy input.

Process	Engine	Exhaust to stack	Engine cooling system	Radiation to ambient air
Overall engine thermal efficiency without				
Heat recovery 33%	33			
Losses 67%		−30	−30	−7
Total fuel input 100%				

Process	Radiation to ambient air	Engine cooling system	Exhaust to stack	Engine
Overall engine thermal efficiency with				
Heat recovery 75%		30	+12	+33
Losses 25%	+7		+18	+0
Total fuel input 100%				

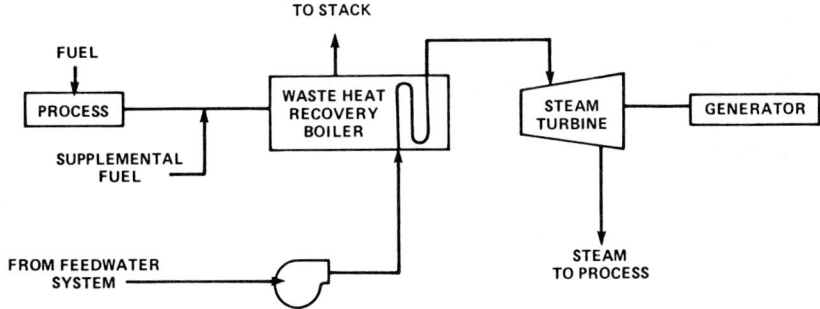

Figure 4-25 Cogeneration bottoming cycle: fuel → process → electricity.

PURPA and FERC Efficiency Standards

Electricity generated must be equal to or greater than 45 percent of the fuel energy input.

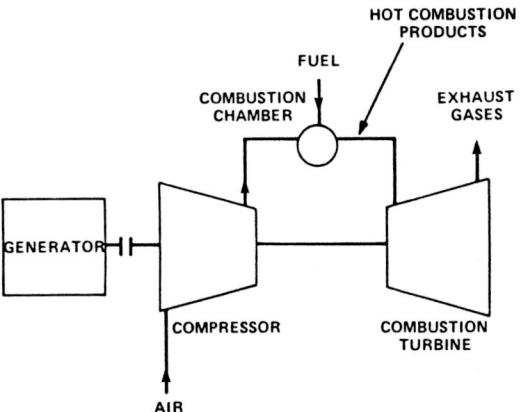

Figure 4-26 Simple cycle gas turbine generator for power generation.

per kilowatthour) is the energy required by the prime mover without considering the use of its exhaust heat. The net heat rate is the gross heat rate less the heat used in the process. The lower the heat rate, the more power produced for a given amount of heat and the more efficient the facility.

A BASIC COMBUSTION TURBINE GENERATOR FOR POWER GENERATION

The basic combustion turbine generator configuration shown in Figure 4-26 is simple and compact. It has the following advantages.

- A minimum cost investment in a combustion turbine generator plant can be made when emergency or peak power demand periods exist.

- The combustion turbine generator plant can be installed at any site and remote-controlled to reduce personnel requirements.

- The installation time is shorter than for other thermal plants. It is therefore possible to sequence the installation cost to planned energy requirements by purchasing completely independent units.

- Combustion turbines can be fired by various fuels such as natural gas, no. 2 fuel oil, kerosene, and low-cost residual fuels.

A combustion turbine generator power package plant is shown in Figure 4-27. Engineering data and standard performance are given in Table 4-10. Figure 4-28 illustrates a typical combustion turbine cross-sectional view.

Figure 4-27 W191 Gas turbine power package plant. (*Westinghouse Standard Proposal for the W191 Econopac Gas Turbine Power Plant*)

Figure 4-28 Combustion turbine cross section. (*Westinghouse Combustion Turbines, "The Ready Source of Power Worldwide," Bulletin SA 11152*)

Table 4-10 W191G Gas Turbine Generator Econo-Pac Power Plant: Engineering Data*

Type of operation	Plant net rating, kW	Minimum exhaust flow, lb/hr	Minimum exhaust temperature, °F	Plant performance Heat rate, Btu/kWh	Efficiency, %
Fuel—Standard natural gas: Heating value 1,000 (HHV) and 900 (LHV) Btu/scf					
Base	17,700	985,000	780†	13,390	25.49
Peak	19,000	985,000	813†	13,210	25.84
Fuel—Standard no. 2 distillate oil: Heating value 19,300 (HHV) and 18,300 (LHV) Btu/lb					
Base	17,300	985,000	781†	13,680	24.95
Peak	18,600	985,000	815†	13,510	25.26

	60 Hz	50 Hz
Gas turbine:		
Axial compressor stages	15	15
Turbine stages	5	5
Unit pressure ratio	7.47–1	7.37–1
Combustor baskets	6	6
Rated speed, rpm	4912	4830
Equipment weight (approximate), lb:		
Turbine (erected including piping and lube oil)	169,500	169,500
Gear	43,500	46,770
Generator and exciter	89,300	101,000
Auxiliary assembly	20,100	20,100
Complete turbine generator	322,400	337,370
Heaviest piece to be handled		
During installation (turbine)	145,000	145,000
After installation (generator rotor)	40,800	46,000
Turbine rotor	26,000	26,000
Turbine cylinder cover	29,000	29,000
Starting diesel	12,000	12,000
WR^2 (turbine generator unit referred to turbine speed), lb ft²	26,600	27,500
Lubrication system:		
Oil reservoir capacity, gal (U.S.)	1,750	1,750
Lube system heat load, Btu/min	30,000	30,000

Table 4-10 (Cont.)

	60 Hz	50 Hz
Lubrication system (cont.):		
Oil pump, capacities, gal/min (U.S.)		
Shaft-driven, main	370	364
Motor (ac)-driven, primary auxiliary	300	300
Motor (dc)-driven, secondary auxiliary	150	150
Starting device:		
Diesel engine (rating at 100% speed), hp	430	430
Fuel oil tank capacity (U.S.)	50	50

	60 Hz		50 Hz	
	Generator	Exciter	Generator	Exciter
Type of unit: salient pole				
Rated				
Kilovoltamperes	20,820	111	20,850	111
Power factor	0.85	0.90	0.85	0.90
Kilowatts	17,700	100	17,720	100
Air flow, cfm	40,000	1500	37,500	1500
Ambient air, °C	15	15	15	15
Rpm	900	900	750	750
Phase	3	3	3	3
Cycle	60	75	50	62½
Voltage	13,800	125	11,500	125
Short circuit ratio	0.645		0.60	
F.L. amps	870		1050	
Low ambient maximum capability, KVA	22,940		22,940	
Class insulation	B	B	F	B
Speed of response		0.5		0.5
Reactances				
Transient (sat., Xd')	0.363		0.40	
Subtransient (sat. Xd'')	0.259		0.27	
Synchronous (Xd)	2.03		2.06	
Zero sequence (X_0)	0.159		0.13	
Negative sequence (X_2)	0.248		0.26	
Time constant, sec	6.8		8.0	
Generator and exciter WR^2 at generator speed, lb/ft^2		80,000		94,100
Cooling water required, gal/min at °C	None		None	

Table 4-10 (Cont.)

	Horsepower at 50 Hz	Horsepower at 60 Hz	Operation Category§
Plant auxiliaries			
Ac drives—440-V, three-phase, 60-Hz or 380-V, three-phase, 50-Hz			
Primary auxiliary lube oil pump	30	25	BA
Control and atomizing compressor	5	5	AR
Lube oil cooler fan	2–20	2–20	C
Enclosure exhaust fans (total)	20	20	AR
Optional arrangements, ac motors			
Inlet air evaporative cooler pump	20	20	C
Inlet air filter screen (approximate total)	½	½	C
Fuel oil (main) pump	50	40	BC
Fuel oil (tank-to-unit) transfer pump	3	3	BC
Air conditioner	16	8	AR
Fuel oil filter	⅛	⅛	BC
Glycol pump (option)	20	20	AR
Dc drives—125-V			
Secondary auxiliary lube oil pump	3	3	D
Turning gear	5	5	BD
Clutch air compressor	1–5	1–5	B
Controls, kW	1	1	ABCD
Heaters LV, three-phase			
Generator space heaters, kW	5.6	5.6	AR
Building unit heaters, kW	60	60	AR
Lube oil heater, kW	18	15	AR
Exciter space heater, kW	0.6	0.6	AR
Controls and lighting, kV	32	32	AR

*ISO ambient conditions: Temperature 59°F (15°C); Barometric pressure 14.7 psia (sea level)
†Exhaust temperatures shown are for a 50-Hz plant or 776 base and 809 peak at 60 Hz.
‡Exhaust temperatures shown are for a 50-Hz plant or 777 base and 810 peak at 60 Hz.
Performance based on lower heating value of the fuel. Performance based on unfiltered inlet and exhaust systems with level A silencers (4-in. H_2O pressure loss).
§A, Standby; B, start-up, C, unit operating normally; D, emergency unit shutdown; AR, as required.
SOURCE: Westinghouse Standard Proposal for the W191 Econopac Gas Turbine Power Plant.

Fig. 4-29 shows a cutaway section of a typical gas turbine generator power plant. A compact arrangement requiring minimum platform space for offshore combustion turbine generator applications is shown in Figure 4-31. A cutaway view of a boiler and a condensing steam turbine generator marine propulsion power system is given in Figure 4-30. A modular combustion turbine designed for high availability, good serviceability, and easy maintenance and overhaul at the site is shown in Figure 4-32. Materials and temperatures for industrial applications are indicated on Figure 4-33. A cutaway drawing of a combustion turbine generator power station is shown in Figure 4-34.

Figure 4-29 Combustion turbine cross section. (*Stal-Laval, Inc.*, "*Turbine Power,*" *Bulletin 572E9.79.4.000*)

Figure 4-30 Marine propulsion power system. (*Stal-Laval, Inc.*, "*Turbine Power,*" *Bulletin 572E.9.79.4.000*)

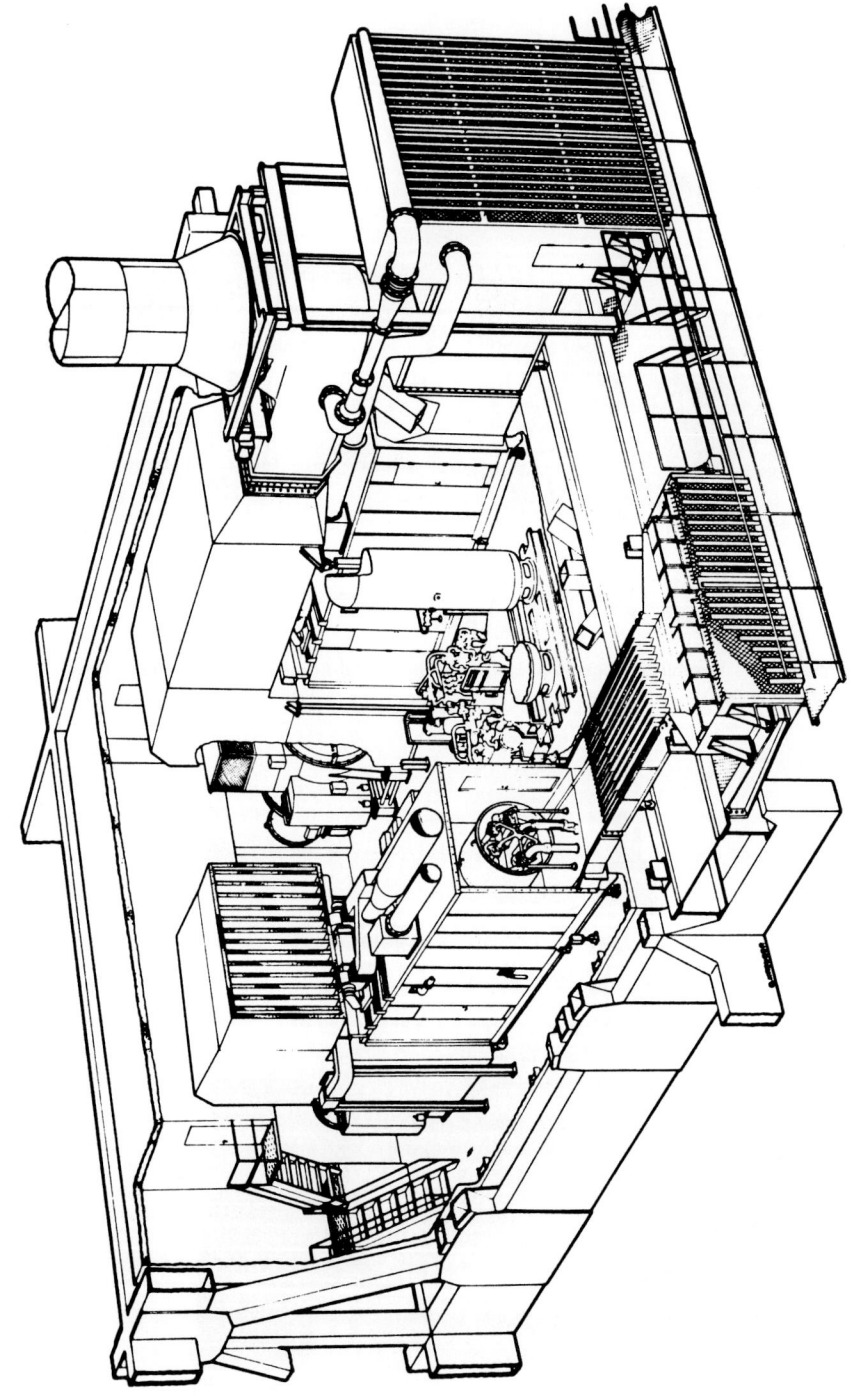

Figure 4-31. Offshore platform combustion turbine generator application. (Stal-Laval, Inc., "The GT 35 Gas Turbine 10-15MW," Bulletin 588E02.801.000)

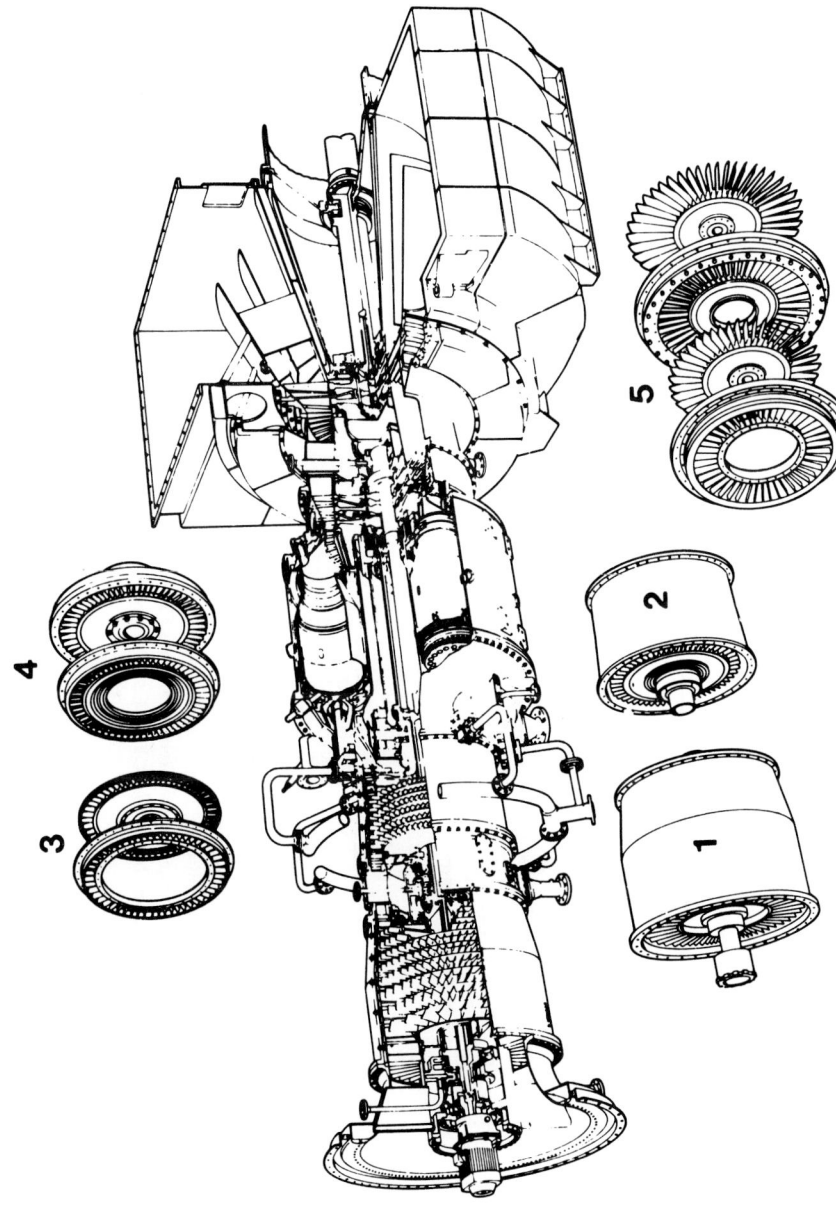

Figure 4-32 Modular combustion turbine. GT 35 and modules: (1) LP compressor, (2) HP compressor, (3) HP turbine, (4) LP turbine, (5) power turbine. (*Stal-Laval, Inc., "The GT 35 Gas Turbine 10-15MW," Bulletin 588E02.80.1.000*)

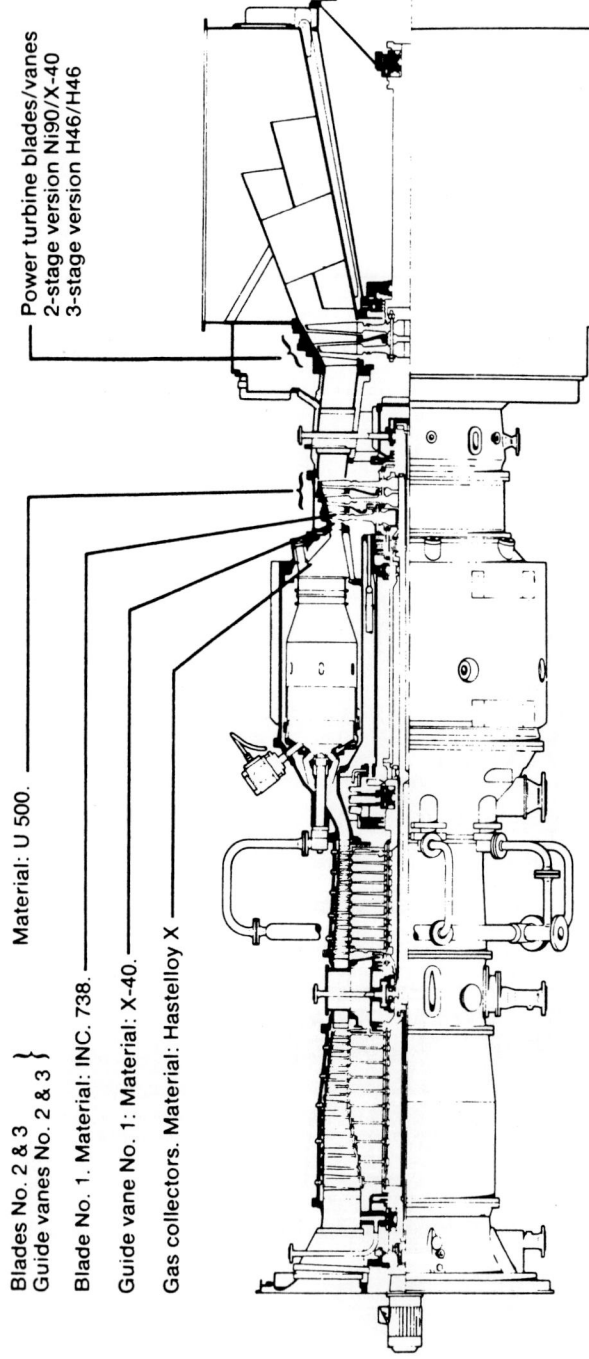

Figure 4-33 Industrial combustion turbine material and temperature. Temperatures, materials, and coatings: At base load, ISO conditions, turbine inlet temperature is 810°C (1490°F). At maximum continuous load, ISO conditions, turbine inlet temperature is 850°C (1562°F). Coatings are available for hostile environments. (*Stal-Laval, Inc., "The GT 35 Gas Turbine 10-15MW," Bulletin 588E02.801.000*)

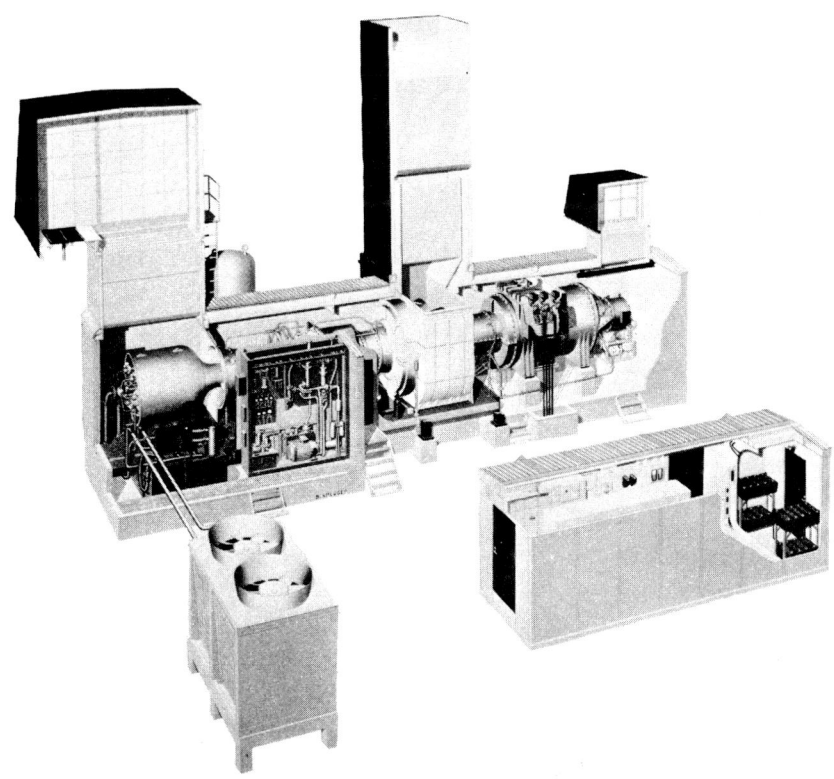

Figure 4-34 Combustion turbine generator power station. (*Stal-Laval, Inc.,* "*The GT 35 Gas Turbine 10-15MW,*" *Bulletin 588E02.801.000*)

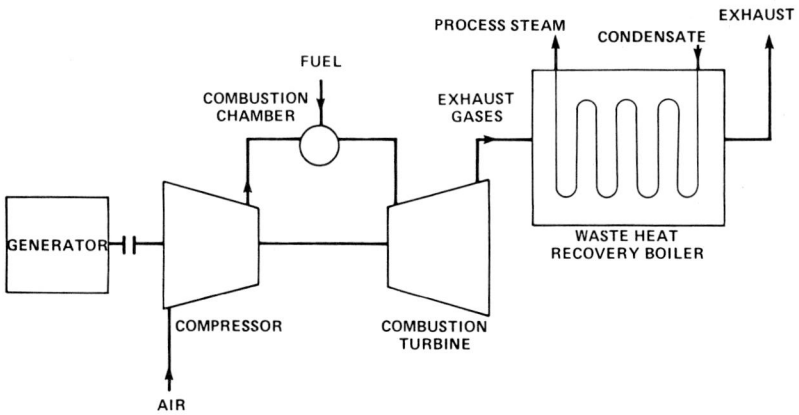

Figure 4-35 Simple cycle gas turbine generator for power and process heat generation.

A HEAT RECOVERY COMBUSTION TURBINE GENERATOR CYCLE FOR GENERATION OF BOTH POWER AND PROCESS HEAT

The heat content of the turbine exhaust gas is recovered through a waste heat recovery steam generator as shown in Figure 4-35. Typical combustion turbine generator and waste heat recovery steam applications are shown in Figure 4-36. The high temperature of the turbine exhaust gas and the heat transferred to the environment make the combustion turbine more suitable than other thermal engines because of the ratio between the useful power, both as electricity and hot

Figure 4-36 Typical cogeneration installations. (*a*) Repowering application in Canada. Two W251 (35-megawatt) combustion turbines with waste heat making steam for existing steam plant. (*b*) Cogeneration in the United States. W501D combustion turbine generator and waste heat boiler making steam for process. (*c*) Combined cycle in Mexico. A Westinghouse preengineered, packaged 520-megawatt PACE plant. (*Westinghouse Combustion Turbines, "The Ready Source of Power Worldwide," Bulletin SA 11152*)

fluid, and the power supplied to the plant by the fuel. This heat utilization ratio can be as high as 80 percent, a value that can be achieved with installations of no particular complexity. Heat utilization varies from process steam for industry requirements to hot water or steam for district heating. Installation of a combustion turbine with a waste heat recovery boiler is particularly simple. Installation time and the number of operating personnel are considerably reduced.

A COMBINED CYCLE FOR POWER GENERATION

The combined cycle shown in Figure 4-37 utilizes the steam produced from the exhaust gas heat recovery steam generator to drive a steam turbine generator set. The steam turbine is the condensing type. It can either be manufactured for the application or be an existing unit repowered by replacing an existing boiler with a waste heat recovery boiler. In plants of this kind the output of the steam turbine may be approximately half that of the gas turbine, and efficiency can reach 45 percent, exceeding that of all other thermal plants.

Combining a gas turbine generator and a waste heat recovery boiler with an existing steam turbine-generator gives new life to and increases the output of

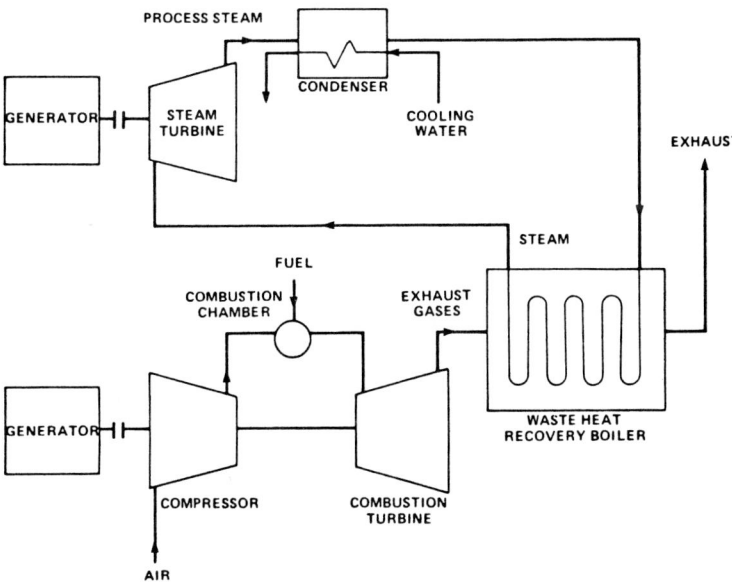

Figure 4-37 Combined cycle for power generation.

plants that, because of obsolescence of the conventional boiler, would have been dismantled.

A cross-sectional illustration of a gas turbine generator, steam turbine genera-

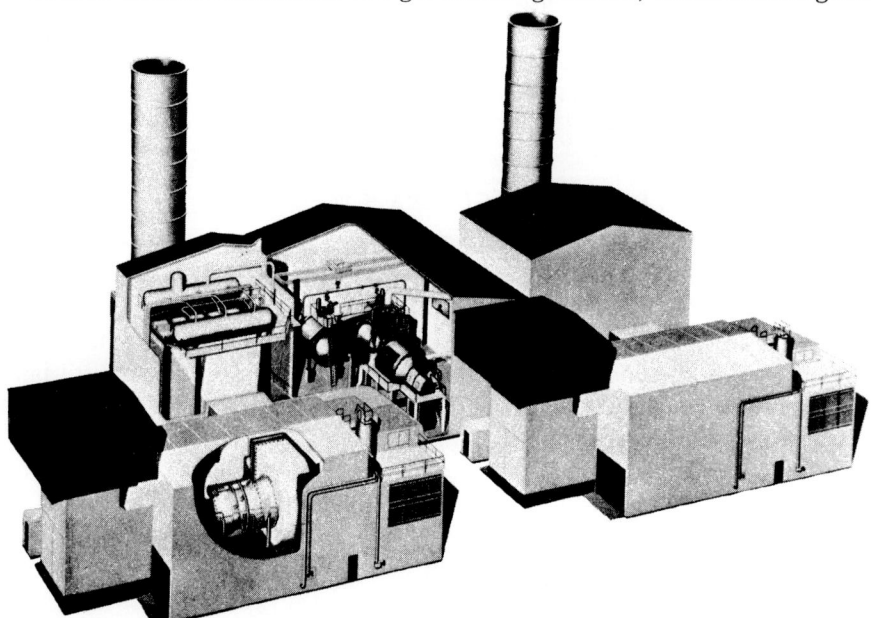

Figure 4-38 Combined cycle power plant. (*Stal-Laval, Inc., "Turbine Power," Bulletin 572 E.9.79.4.000*)

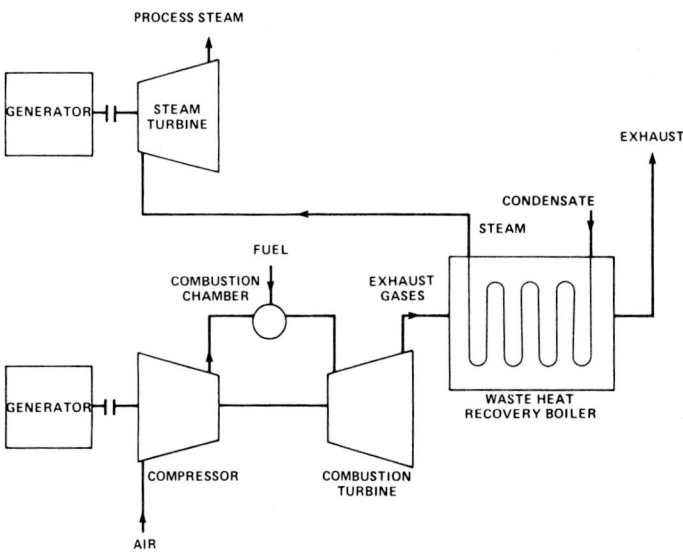

Figure 4-39 Combined cycle for power and process heat generation.

tor, and waste heat recovery steam generator combined cycle facility is shown in Figure 4-38.

A COMBINED CYCLE FOR GENERATION OF BOTH POWER AND PROCESS HEAT

A gas turbine generator and waste heat recovery boiler can be combined with a condensing or extracting or a back-pressure steam turbine generator, as shown in Figure 4-39, to provide electric power and process steam. Multidisk radial steam turbine cross sections and steam flow paths are shown in Figures 4-40 and 4-41.

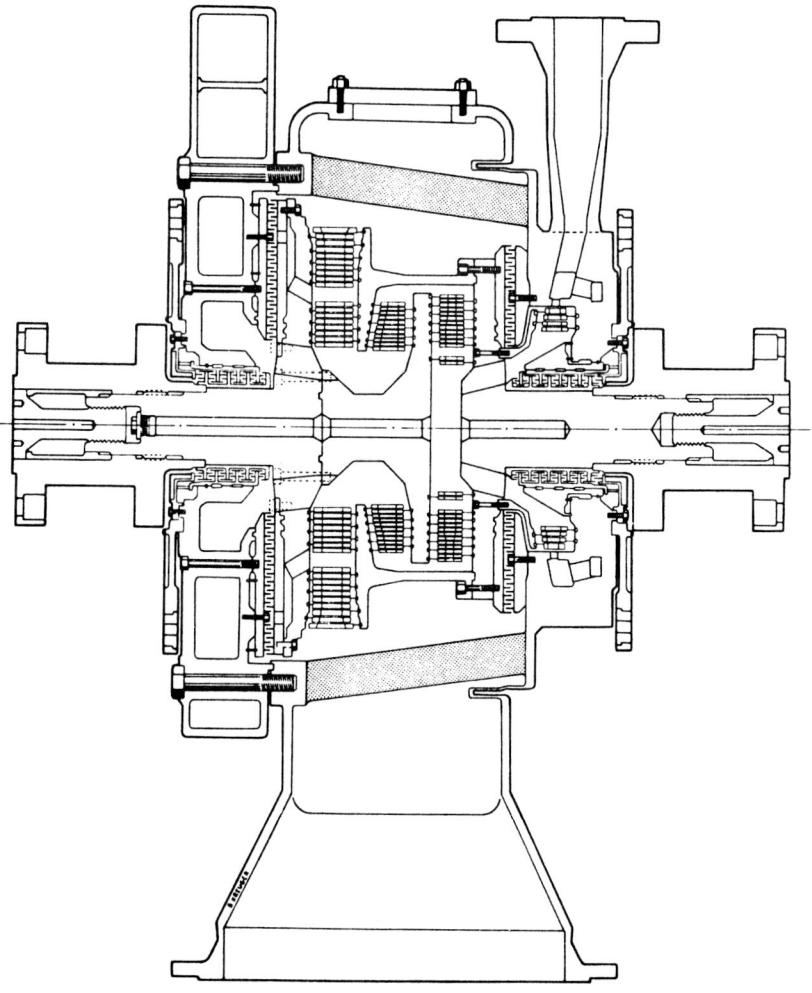

Figure 4-40 Steam turbine cross-sectional view. (*Stal-Laval, Inc.*, "*Turbine Power*," *Bulletin 572E.9.79.4.000*)

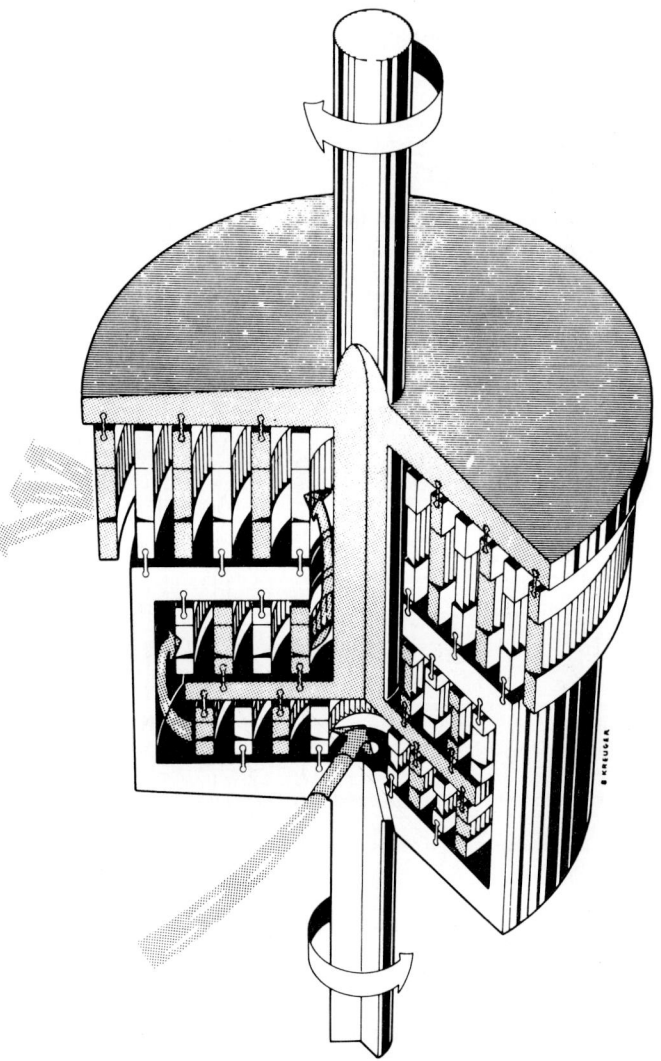

Figure 4-41 Radial steam turbine steam flow path. (*Stal-Laval, Inc., "Turbine Power," Bulletin 572E9.79.4.000*)

APPLICATIONS

Cogeneration is most feasible and financially rewarding for the following types of applications requiring a large consumption of natural gas and energy.

- INDUSTRIAL FACILITIES

    Paper and pulp.

    Food processing.

Chemical manufacturing.

Petroleum refining.

Oil processing.

Secondary recovery in low-gravity oil fields. Extracting heavy crude oil from the ground is difficult and expensive. For every 4 barrels of oil produced, more than 1 barrel is burned to generate steam. The steam is injected into wells to heat the oil in the ground so it can be pumped.

Lumber.

Canning.

- HOSPITALS
- HOTELS
- INSTITUTIONS

    Correctional facilities

    Educational facilities and schools

- OFFICE BUILDINGS
- APARTMENT COMPLEXES
- SHOPPING CENTERS
- UNIVERSITIES AND COLLEGES

Some simple cycle diagrams are included here to help define cogeneration.

A process steam plant consisting simply of a boiler providing steam to a process is shown in Figure 4-42. An ideal temperature-entropy diagram is shown for this cycle. 1' is the boiler inlet, and 2' is the superheater outlet. Both the work out (2' to 3') and the heat of condensation (3' to 4') are received by the process. Ideally this is an efficient cycle with an actual plant efficiency in the low 70 percent range.

The conventional power plant cycle, shown in Figure 4-43, has additive losses in the mechanical equipment plus heat loss to the condenser. The station heat rate is approximately 10,000 Btu per kilowatthour. The ideal temperature-entropy diagram for this cycle shows boiler work in from 1 to 2, assuming no loss in the main steam piping, and turbine work out from 2 to 3. The heat removed to condense the turbine exhaust steam (3 to 4) is wasted in this case. This heat is proportional to the area under line 3-4. Typical enthalpy values are shown, resulting in a calculated theoretical efficiency of 43.5 percent which converts to an actual efficiency of 34.8 percent.

The cogeneration cycle depicted in Figure 4-44 is very much like the conventional power cycle. It can vary in numerous ways, but for this example a single process steam requirement of 60 pounds per square inch gauge was chosen. A noncondensing turbine is used exhausting at 60 pounds per square inch gauge. The temperature-entropy diagram is much like the one for power, except that the bottom line is higher because less energy was removed from the steam in the turbine. The energy passed on to the process via the exhaust steam is 1112-262 Btu per pound. There is less turbine workout and the work-in by the boiler is slightly reduced, but the large loss in the condenser is eliminated, giving an actual plant efficiency in the 72 percent range. Normal heat rates of 4500 to 5000 Btu per kilowatthour are possible.

Various options must be studied when optimizing cogeneration cycles. The cycle must be designed to fit the requirements of the process. Examples of various methods of providing steam for process requirements are given in Figures 4-45 to 4-47, and the power cycle optimization and evaluation are shown in Figure 4-48. The following factors must be considered.

- Are the steam or power supplies to be continuous? What is their ratio?
- Additional types of turbines (i.e., back pressure, condensing).
- Process steam conditions.
- Initial steam conditions.
- Number of feedwater heaters (if any).
- Quantity and quality of condensate.

The evaluation should

- List all the operating condition options.
- Select the system best suited for each operating condition.
- Apply the systems chosen to each operating condition for evaluation of the best overall system.

Higher temperature and pressure steam can be produced at the turbine throttle. This additional temperature and pressure produce an amount of electricity many times greater in value than the additional fuel cost. The cycles of some typical cogeneration projects are depicted in Figures 4-49 to 4-52.

SUPPLEMENTAL FIRING

A high oxygen content in the exhaust gases permits supplemental firing in the steam generator furnace or the ducting between the combustion turbine exhaust and the waste heat recovery boiler, as shown in Figure 4-53. Supplemental firing can be used if the exhaust gas heat from the combustion turbine or cycle is insufficient for the desired operating mode. Supplemental firing is one approach to meeting peak steam or hot water demands while allowing wide flexibility of application. Temperature limitations must not be exceeded during supplemental firing.

Supplemental firing increases the steam generation capacity through higher gas turbine exhaust temperatures. The steam generation capacity can also be maintained through supplemental firing when combustion turbine generator loads are reduced. Where standby boilers do not exist, a 100 percent back-up burner system should be considered for the waste heat recovery boiler. Environmental considerations limit a 100 percent back-up burner system to 35×10^6 Btu per hour. Selective catalytic reduction and low nitrogen oxide burners are required for waste heat recovery boiler auxiliary firing systems larger than 35×10^6 Btu per hour.

Supplemental fuel-fired waste heat recovery boilers require less fuel consumption than conventional boilers providing the same steam generation.

Supplemental firing is usually done in the inlet ducting, as shown in Figure 4-54. A low–nitrogen oxide duct burner system normally fires either natural gas or no. 2 fuel oil. Ducting from the combustion turbine to the duct burner may be fabricated of alloy steel or carbon steel plate internally lined with a ceramic fiber blanket and covered with a stainless steel sheet to prevent erosion of the insulation. The ducting from supplemental firing burners and the transition duct from the superheater to the boiler may be internally lined with ceramic fiber blanket thermal insulation covered with a rigidized wet felt ceramic liner to withstand temperatures up to 2000°F resulting from supplemental firing. This duct is sized for uniform temperature distribution into the waste heat recovery boiler.

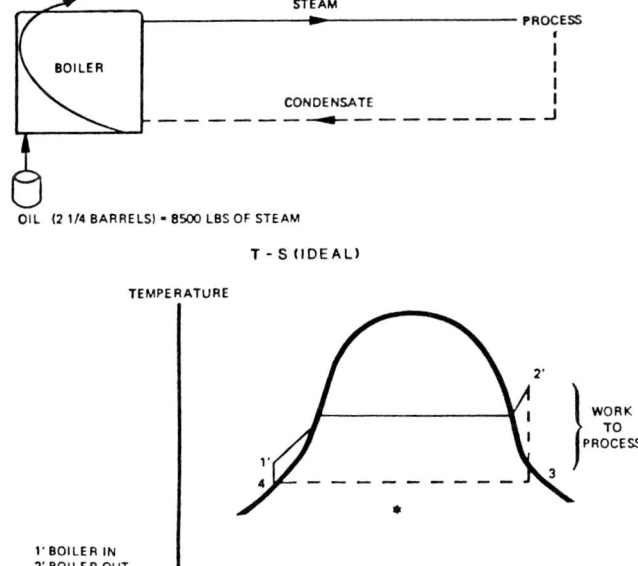

Figure 4-42 Process steam plant.

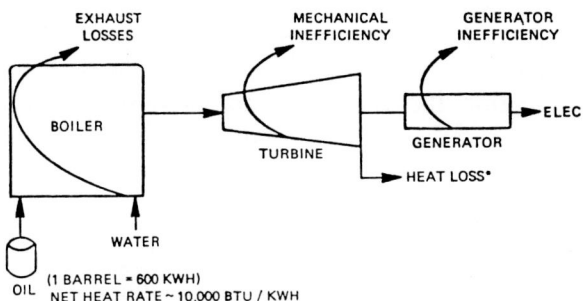

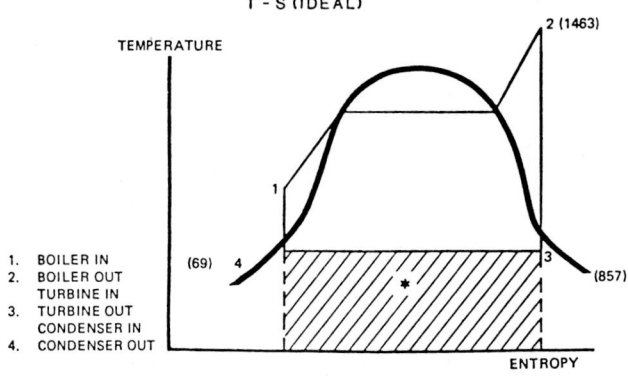

*REJECTED HEAT IS PROPORTIONAL TO THIS AREA
( ) TYPICAL ENTHALPYS

$$EFF_{TH} = \frac{WORK\ OUT}{WORK\ IN} = \frac{1463 - 857}{1463 - 69} \times 100 = 43.5\%$$

$$EFF_{(ACT)} = 0.80 \times 43.5 = 34.8\%$$

Figure 4-43 Conventional plant.

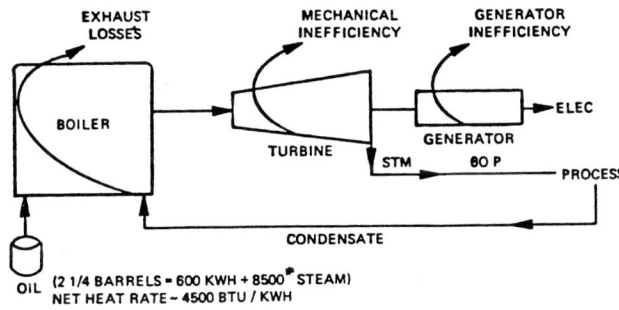

T - S (IDEAL)

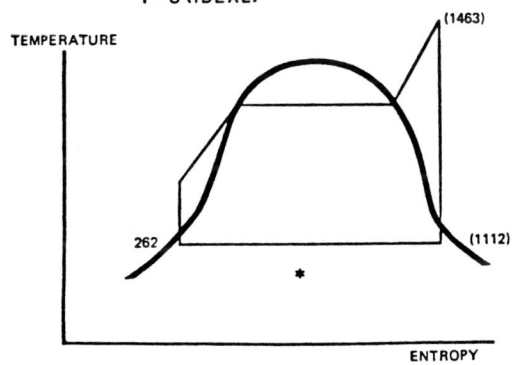

*HEAT LOSS CHARGED TO PROCESS
( ) TYPICAL ENTHALPYS
TURBINE WORKOUT = 1463 - 1112 = 351 BTU/lb
STEAM WORKOUT = 1112 - 262 = 850 BTU/lb
PLANT EFF$_{(ACT)}$ ~ 72%

Figure 4-44 Cogeneration plant.

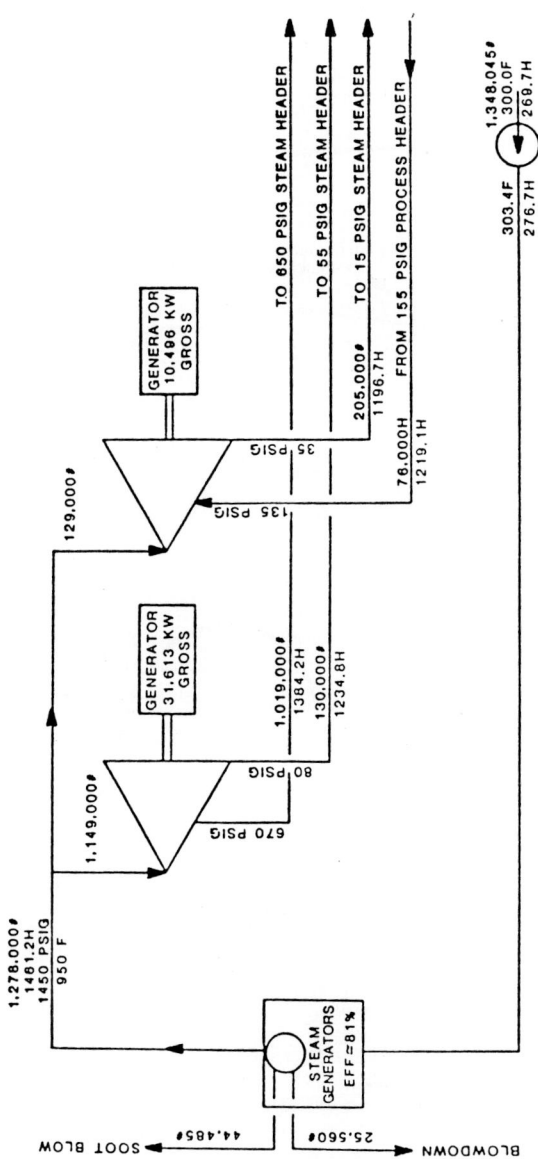

Figure 4-45 Steam generator and extraction steam turbine generator heat balance.

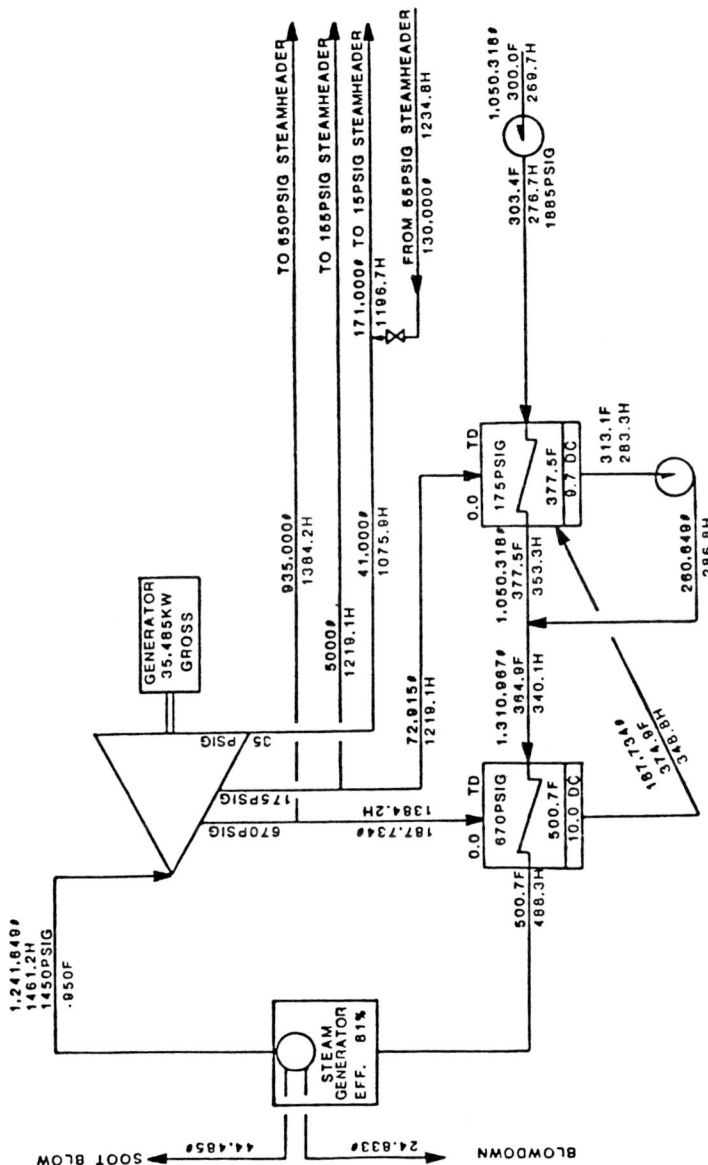

Figure 4-46 Regeneration cycle heat balance.

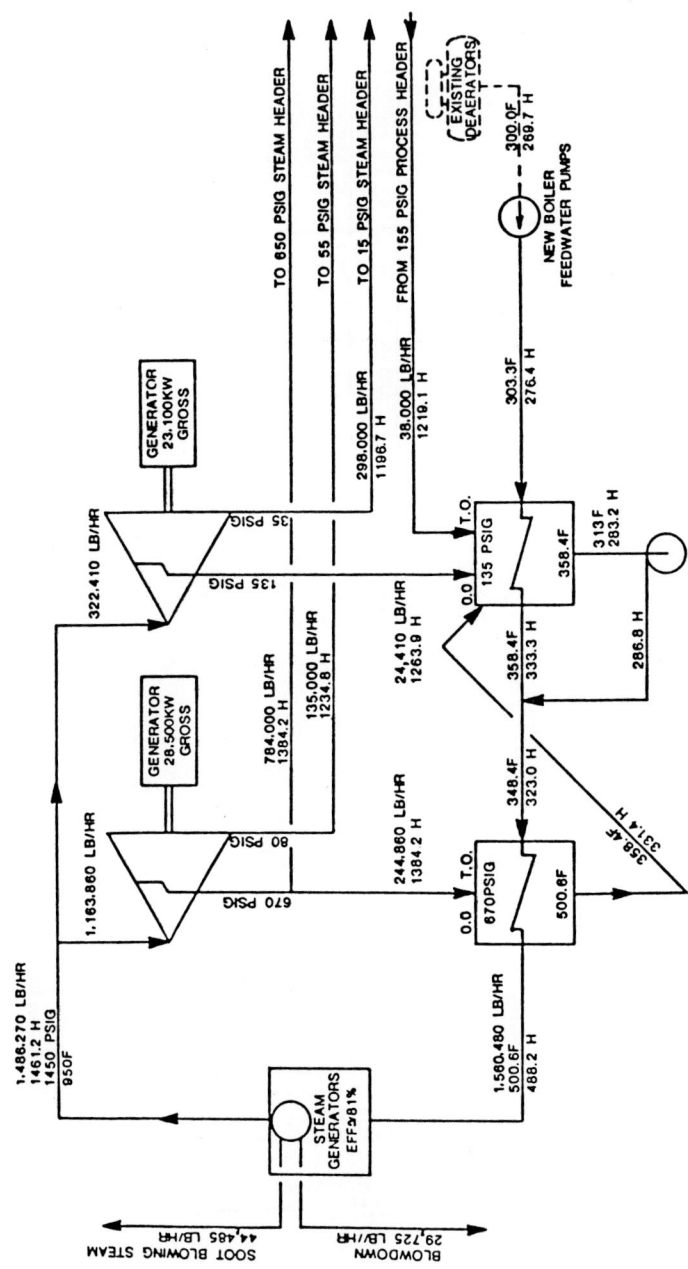

Figure 4-47 Dual steam turbine generator regenerative cycle heat balance.

OPTIMIZATION CONSIDERATIONS

- Ratio of power to process steam
- Type of turbine
- Process steam temperature and pressure
- Initial steam temperature and pressure
- Number of heaters
- Quantity and quality of condensate from process

Figure 4-48 Turbine thermal cycle.

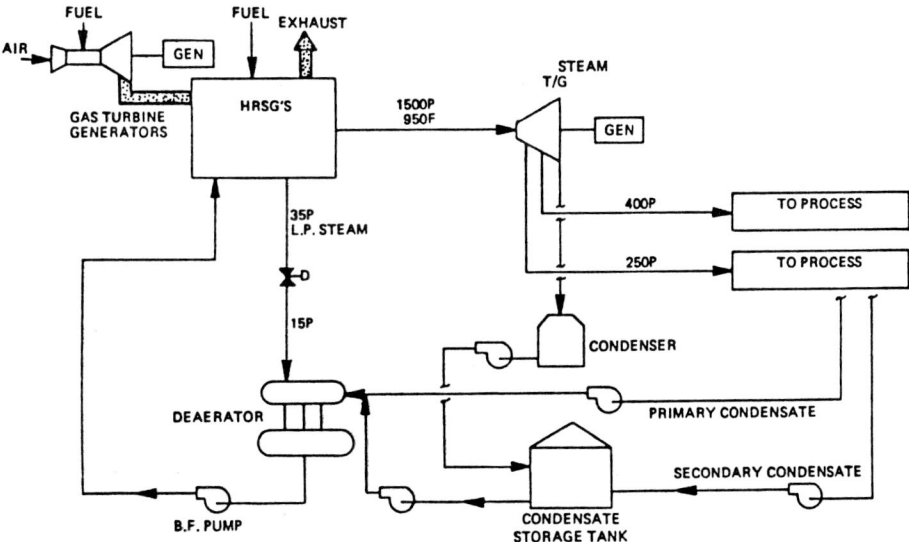

Figure 4-49 Extraction condensing steam turbine cycle.

4-59

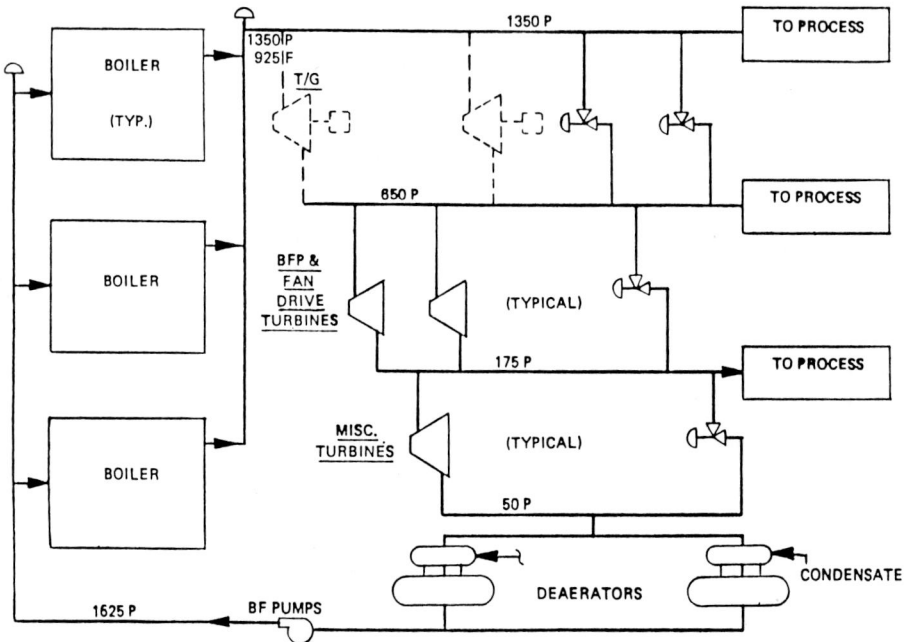

Figure 4-50 Topping turbine power cycle.

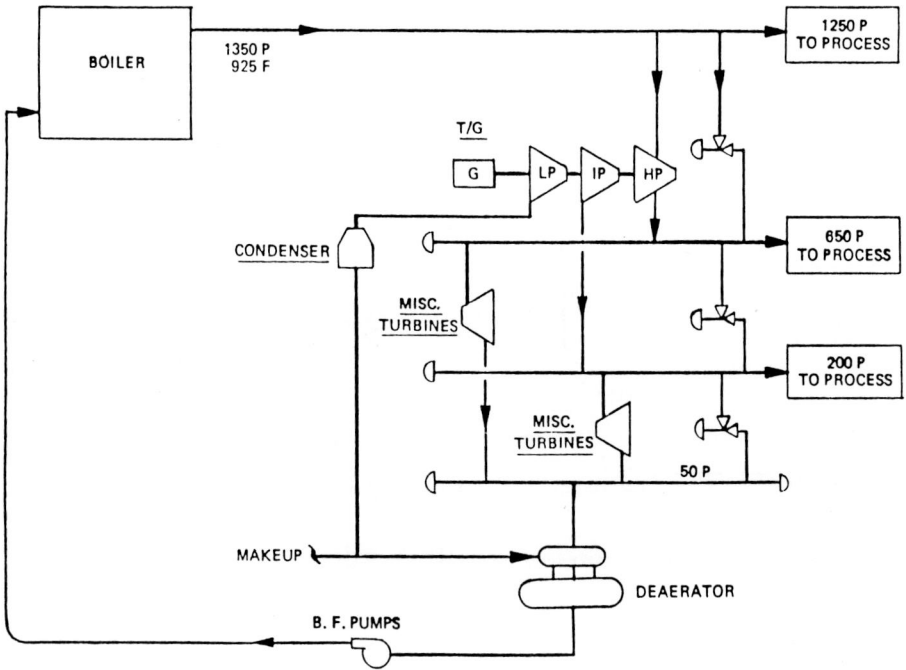

Figure 4-51 Tandem steam turbine generator cycle.

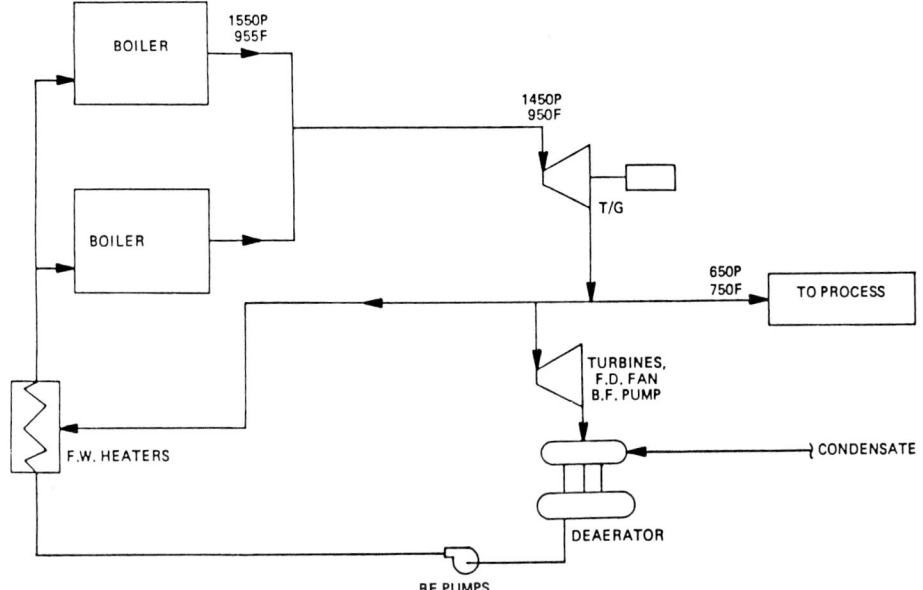

Figure 4-52 Power and process steam regenerative power cycle.

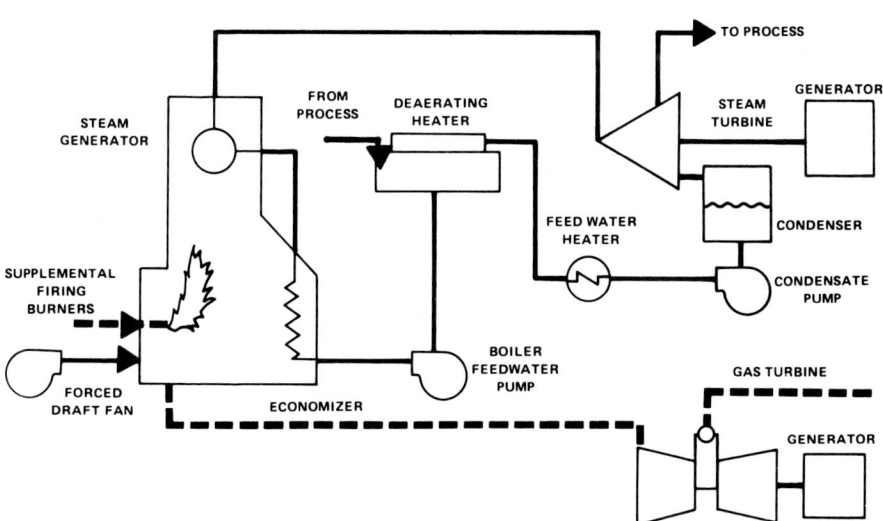

Figure 4-53 Supplemental fired heat recovery power cycle.

4-61

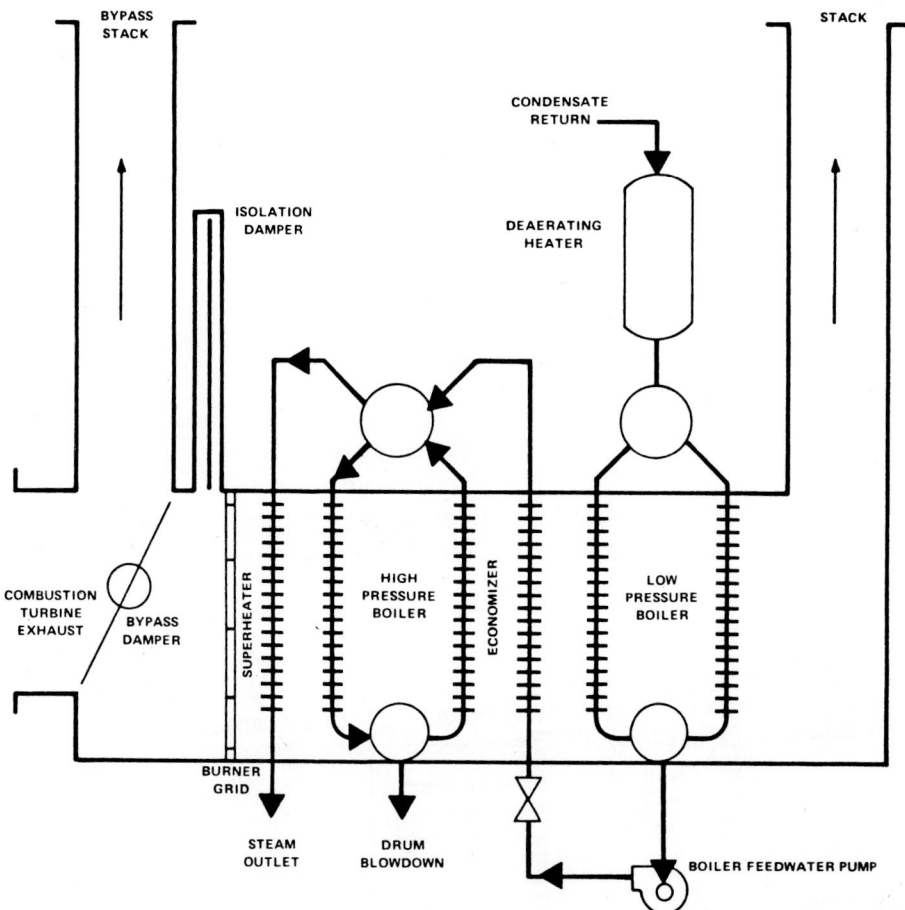

Figure 4-54 Combustion turbine exhaust heat recovery multipressure steam generator system.

COGENERATION IN HEATING AND COOLING SYSTEMS FOR BUILDINGS AND DISTRICTS

Cogeneration for building and district space heating and cooling purposes consists of producing electricity and sequentially utilizing useful energy. Useful energy in the form of steam, hot water, or direct exhaust gases is provided for building or district heating and air-conditioning loads as shown in Figure 4-55. The benefits from combined district heating and production of electric power are illustrated in Figure 4-56. A district heating power station and control room are shown in Figures 4-57 and 4-58. An integrated district heating and power-producing complex flowchart is given in Figure 4-59.

The two most common heating, ventilation, and air conditioning (HVAC) cycles are the vapor compression cycle and the absorption cycle.

VAPOR COMPRESSION CYCLE

The vapor compression cycle shown in Figure 4-60 consists of a compressor, a condenser, an expansion valve, and an evaporator. The compressor driver can be an electric motor, steam turbine, or combustion turbine. A single fluid used as the refrigerant vaporizes in the evaporator as a result of heat transfer from the refrigerated space, is compressed by applying work to the compressor, and condenses in the condenser as a result of heat transfer to cooling water or to the surroundings. The refrigerant leaves the condenser as a high-pressure liquid. The pressure of the liquid is decreased as it flows through the expansion valve, and some of the liquid flashes into vapor.

Centrifugal chillers are the most commonly used chillers in the United States, especially for office building air-conditioning systems. They are simple and clean. The chiller compressor drive is usually an electric motor, although either a gas turbine or a steam turbine drive can be used. The centrifugal chiller HVAC application is shown in Figure 4-61. This arrangement is essentially the same as the cogeneration arrangement shown in Figure 4-55 for the absorption chiller cycle, except that the chiller-equipment-related pumps can be driven by electric motors, combustion turbines, diesel or gas engines, or steam turbines. The cost of power or energy consumption for an electric-motor-driven centrifugal chiller is approximately 30 to 50 percent more than that for an absorption chiller system.

ABSORPTION REFRIGERATION CYCLE

The absorption refrigeration cycles for cogeneration air-conditioning applications primarily utilize water as the refrigerant and lithium bromide solution as the absorbent. Lithium bromide solution has a strong affinity for water vapor, thus giving the absorbent the ability to compress the refrigerant vapor.

The absorption cycle shown in Figure 4-55 consists of an absorber generator (concentrator), condenser, expansion valve (pressure-reducing device), evap-

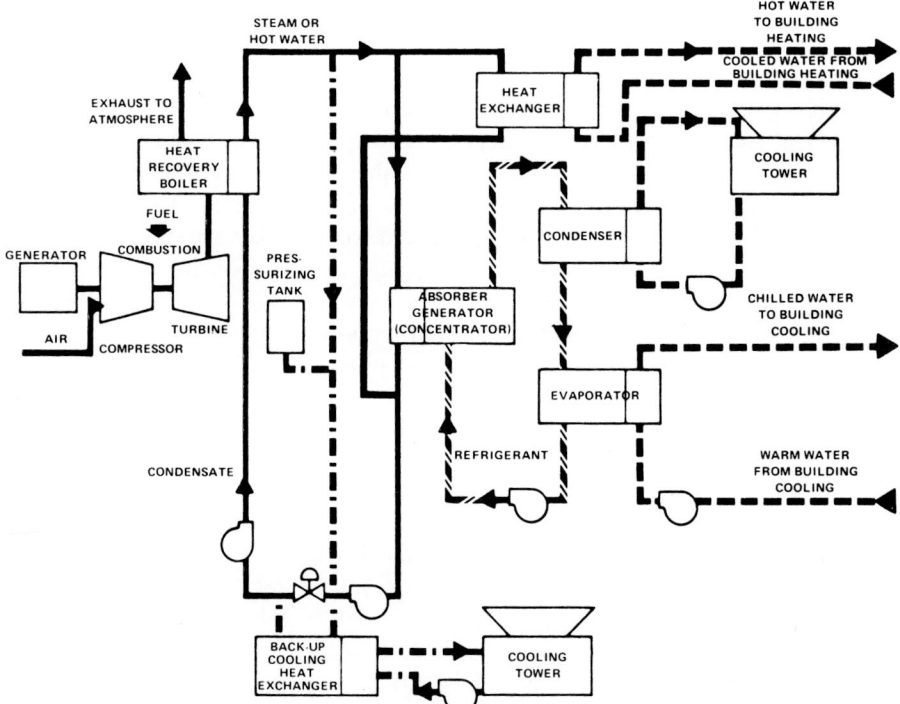

Figure 4-55 Cogeneration absorption district heating and cooling cycle: cogeneration → electricity → process topping.

PURPA and FERC Efficiency and Heat Utilization Standards

1. Useful thermal energy is the steam to process which must be greater than 5 percent of the total energy.

2. Electricity generated plus one-half the useful thermal energy must be equal to or greater than 42.5 percent of the fuel energy input.

orator, and pump. The absorber functions as a compressor. The pump circulating the absorbent requires less energy than that to drive the compressor in the vapor compression cycle.

This cycle, combined with one or more combustion turbine generators or internal-combustion engines exhausting to one or more waste heat recovery steam or hot water generators, provides a common cogeneration system.

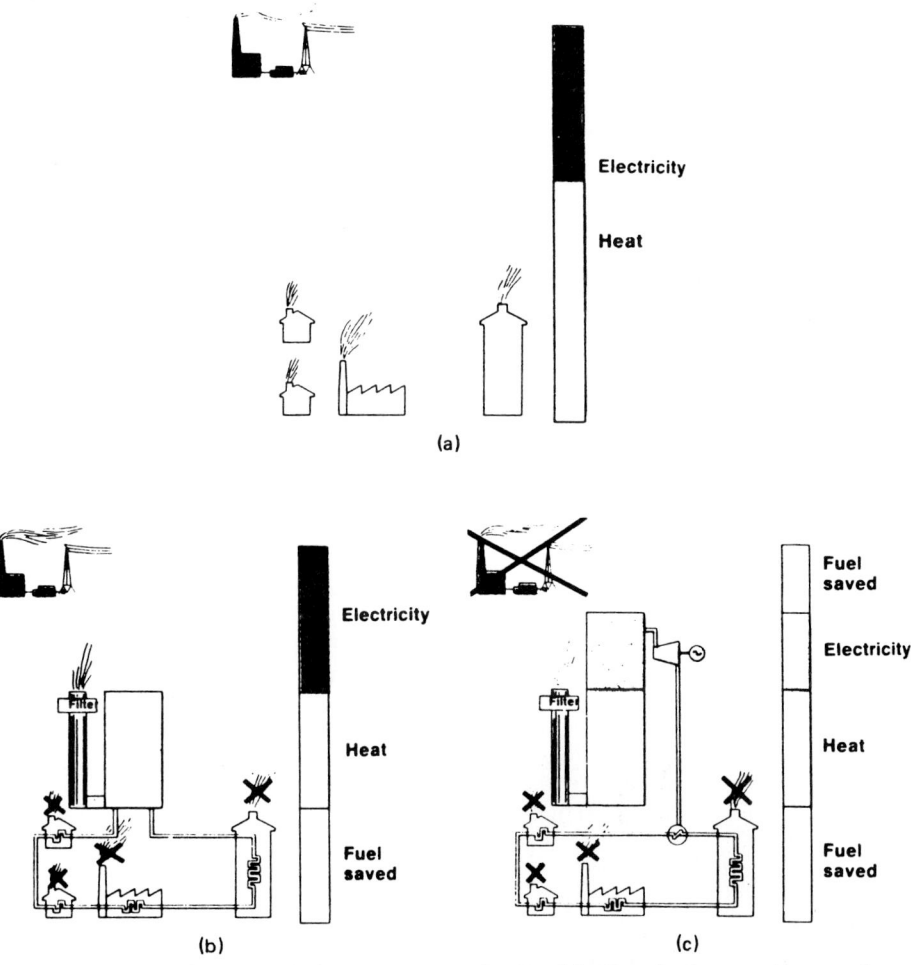

Figure 4-56 District heating and electric power production. (a) When the owners of houses, flats, and industrial premises make their own arrangements for heating and buy their electric power, there is a high rate of fuel consumption. (b) If there is a district heating boiler plant which can supply hot water for heating, a great deal of fuel can be saved and, in addition, there is less atmospheric pollution. (c) If a district heating power station is built, fuel is used more effectively and large amounts of electric power are obtained at the same time. (*Stal-Laval, Inc.,* "*Turbine Power,*" *Bulletin 572E9.79.4.000*)

BUILDING HEATING AND COOLING SYSTEMS

Typical cogeneration systems supply power and steam or hot water for heating and air-conditioning loads. The following systems are used.

- Combustion turbine generator exhausting into a heat recovery steam generator (HRSG) supplying steam to a back-pressure turbine generator exhausting steam to an air-conditioning absorption chiller and a steam heating system
- Combustion turbine generator exhausting directly into a HVAC absorption chiller-heater
- Combustion turbine generator exhausting into a HRSG supplying hot water for HVAC applications
- Combustion turbine generator exhausting directly into a HRSG supplying steam for HVAC applications

Figure 4-57 District heating and power station. (*Stal-Laval, Inc., "Turbine Power," Bulletin 572E9.79.4.000*)

Figure 4-58 District heating and power station control room. (*Stal-Laval, Inc.*, "*Turbine Power*," *Bulletin 572E.9.79.4.000*)

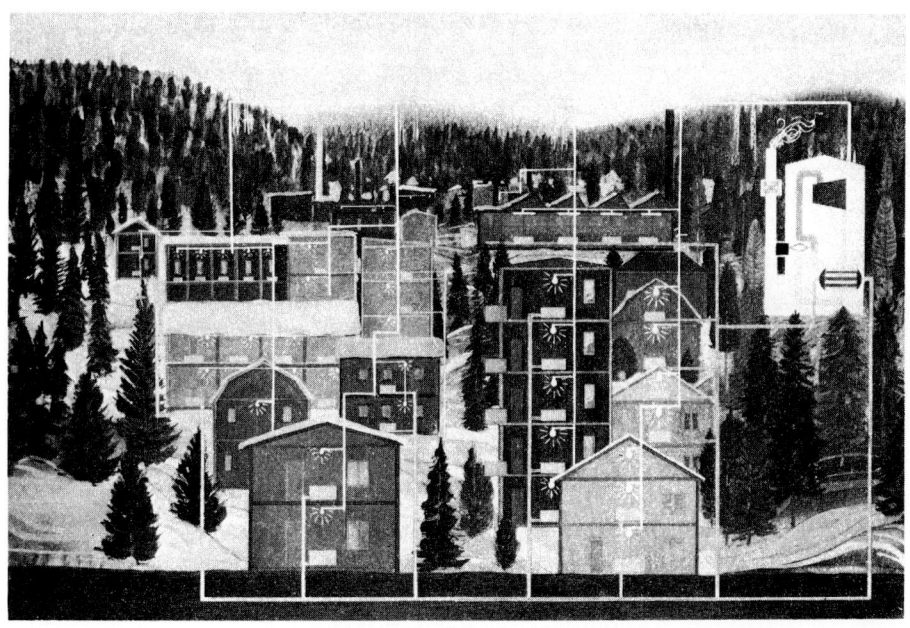

Figure 4-59 District heating and power simplex flowchart. (*Stal-Laval, Inc.*, "*Turbine Power*," *Bulletin 572E9.79.4.000*)

- Internal-combustion engine generator set exhausting into HVAC absorption chiller-heaters
- Internal-combustion engine generator set exhausting into a HRSG supplying steam or hot water for HVAC applications

The above-mentioned system, consisting of a combustion turbine generator exhausting directly into an absorption chiller-heater, offers benefits such as a minimum number of components, lower associated capital costs, and higher efficiency associated with direct energy conversion of exhaust heat, lower fuel consumption per HVAC output, lower maintenance and spare parts costs, a requirement for less space, flexibility to be used for heating, and simplicity of operation. Absorption chiller-heaters can be supplied with hot water or steam from a HRSG or provided with optional direct gas-fired burners to supplement or use when the combustion turbine is not in service.

AIR-CONDITIONING SYSTEMS

Typical air-conditioning system alternatives are steam absorption chiller-heaters, direct-fired absorption chiller-heaters, and centrifugal chillers. (Drivers that may be used include combustion turbines, steam turbines, electric motors, and diesel or gas engines.)

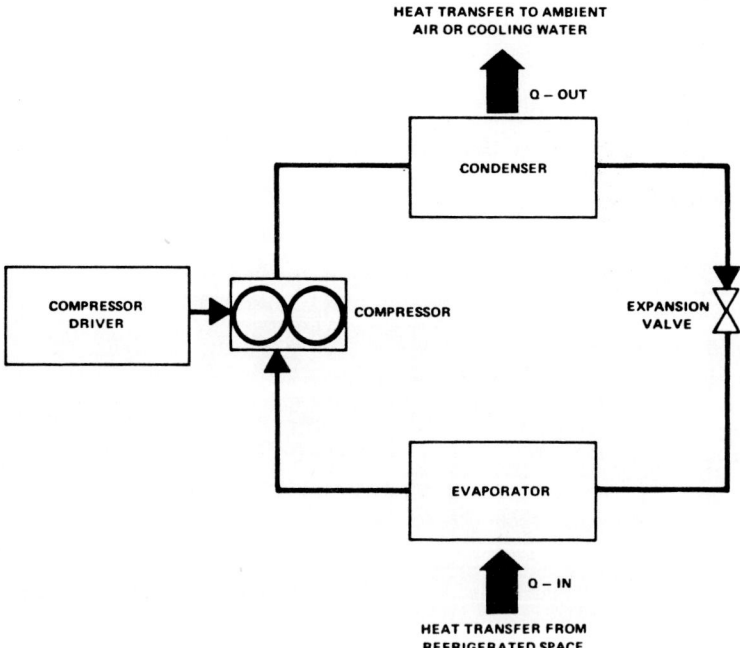

Figure 4-60 Basic vapor compressor cycle.

HEATING SYSTEMS

The two heating systems commonly used are steam heating systems and hot-water heating systems. Hot-water heating systems are widely used in medium to large centralized heating systems because of their higher reliability, lower maintenance requirements, lower operating costs, and longer component life, as compared with steam heating systems. The initial cost of a hot-water heating system, however, is slightly higher than that of a steam heating system.

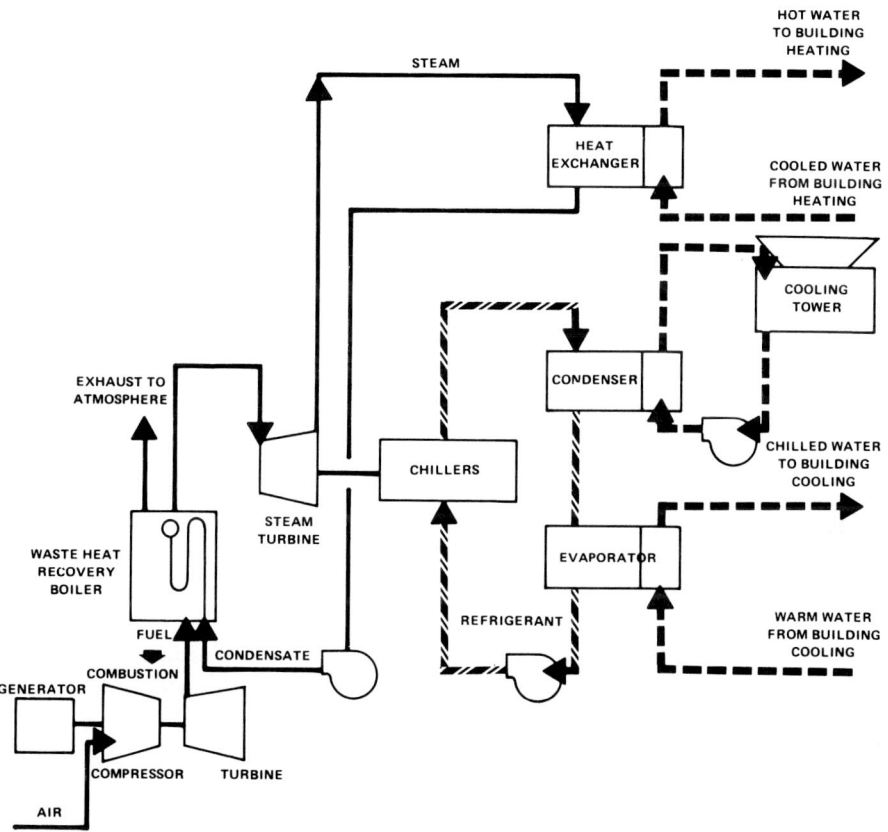

Figure 4-61 Cogeneration centrifugal chiller district heating and cooling cycle: Cogeneration fuel → electricity → process topping.

PURPA and FERC Efficiency and Heat Utilization Standards

1. Useful thermal energy is the steam to process which must be greater than 5 percent of the total energy.

2. Electricity generated plus one-half the useful thermal energy must be equal to or greater than 42.5 percent of the fuel energy input.

PLANT DESIGN

The cogeneration facility should be thoroughly examined to confirm the design basis for the plant. The list below includes guidelines for the major steps that must be covered in the preliminary study.

- Determine the suitability of the identified site for the cogeneration facility. Factors to consider include the physical site available, legal matters, and other pertinent features.
- Determine the energy utilization factors.
- Determine the nature of the heating and cooling loads.
- Determine the technical and economic benefits of practical alternative cogeneration thermal cycles and their associated costs.
- Match alternative thermal cycles with heating and cooling loads and with thermal heating and cooling heat-sink options.
- Plan the integration of the building thermal load cycling with additional off-peak accessible heat-sink and cooling sources, such as

    Municipal swimming pools, park ponds, and so forth

    Cooling capacity derived from fire protection storage water

    Architectural enhancements such as fountains and reflecting pools (which can also be utilized for building cooling)

- Prepare design criteria for the cogeneration facility and its proper integration with the electric system and the building or district heating and cooling development.
- Prepare an economic and financial analysis of the proposed cogeneration facility.
- Identify environmental constraints and determine viable mitigation measures such as nitrogen oxide, sulfur dioxide, and carbon dioxide control applications.
- Prepare a project implementation schedule containing the analyses conducted, the design criteria selected, and a summary of the results, together with conclusions and recommendations for implementation.

**INTEGRATION OF COGENERATION WITH
BUILDING HEATING AND COOLING SYSTEMS**

Efficiently designed cogeneration projects provide technical and economic features which can be optimized to meet both electrical and thermal heating and cooling requirements.

It is important to assess load requirements on a seasonal basis so that heat-utilization factors for the cogeneration plant can be accurately determined. A comprehensive knowledge of the nature of the plant heating and cooling requirements, as well as the cogeneration cycle characteristics, is essential to creation of an optimal overall design for both the building and the cogeneration facility.

The engineer or project manager should work closely with the client to assess present and future electrical and thermal load requirements. The steps involved in performing this assessment include

- Identify and quantify the overall energy requirements.
- Identify the peak heating and cooling requirements.
- Establish daily and seasonal variations.
- Develop daily load duration curves to characterize heating and cooling requirements.

The most cost-effective integrated system possible for the particular facility may be designed by

- Integrating cogeneration energy systems into the entire pattern of energy consumption associated with the building heating and cooling systems
- Designing the system to compensate for either heat loss or heat gain in a building, and taking into consideration factors such as infiltration of outside air, thermal transmission, ventilation, lighting, solar heat gain, equipment for heat dissipation, and number of occupants and type of activity.

The cogeneration concept—combining heating and cooling in one energy system—offers tenants

- Cost-saving incentives relative to energy consumption
- Satisfactory lighting levels
- Satisfactory heating and cooling

The heating-and-cooling system must be properly balanced to achieve the airflow rates necessary for the safety and comfort of building occupants. Listed below are some factors which must be considered in the design phase.

- Systems must be adjusted and balanced to minimize improper control causing

    Overcooling

    Overheating

Poor zoning

Poor distribution

- For proper air balance and to retard infiltration-caused heat losses and heat gains, a slight positive pressure should be maintained.
- Both positioning of dampers and selection and maintenance of filters are important in creation of a properly balanced system.
- Damper sequencing of outdoor air and fan control adjustments should be programmed to take into account hours of occupancy and equipment utilization.
- Total energy management can substantially reduce energy consumption and provide dramatic savings. These ends can be achieved by combining programmed start-stop and night setback adjustments. Demand forecasting and load shedding can be combined with more sophisticated computerized control techniques, such as

    Enthalpy switchover and supply air reset control

    Optional morning start-up

    Chiller plant control

- Centralized control also means that fewer personnel are needed, and yet constant surveillance over all building systems can be maintained.

Cogeneration savings go hand in hand with building temperature control and energy management systems. All the factors listed below should be evaluated in relation to possible energy-saving improvements which can lead to improved cost-effectiveness and increased return on investment.

- The climatic conditions of the building
- The working environment
- Business or other equipment required by tenants in addition to basic heating, cooling, lighting, and ventilation
- Building complex envelope or boundaries consisting of light, insulation, glazing, alternate fuels, thermal storage, spot cooling or heating, and the power factor correction

RECIPROCATING INTERNAL-COMBUSTION ENGINES

A reciprocating internal-combustion diesel engine and a gas Otto cycle engine can be used as the prime mover to drive an electric generator, as shown in Figure 4-62. Waste heat can be recovered from the engine exhaust gases, the engine cooling water, and the engine lubricating oil to produce steam, hot water, or both, to be used for building heating and cooling applications, as shown in Figure 4-63.

At reduced loads, the heat rates of gas engines and combustion turbine engines increase more significantly than do the heat rates of diesel engines.

Cogeneration facilities which use a reciprocating engine operate at lower temperatures than alternative cogeneration systems because of engine temperature limits. Overall engine thermal efficiency can be increased up to approximately 75 percent by waste-heat recovery.

OTTO CYCLE ENGINES

With minor modifications, gas engines can burn fuels such as gasoline, propane, butane, natural gas, sewage gas, alcohol, and gasoline and alcohol mixtures. Gas engine fuel-to-air ignition is achieved by a spark plug or magneto-type ignition system.

The Otto cycle gas engine can be used to drive an electric generator. Engine heat rejected to the cooling water, lubricating oil, and engine exhaust can be recovered for project heating. Cooling water can be used to generate low-pressure steam or hot water; however, lubricating oil heat recovery is sufficient only to produce hot water. The engine exhaust represents about 30 percent of the recoverable heat at temperatures sufficient to produce steam. The conversion efficiency for fuel to electricity is approximately 30 to 33 percent. Total system efficiencies of 75 to 85 percent can be achieved by rejected heat recovery in the conversion of fuel energy to usable energy.

The recovered heat temperature is limited by the permissible temperature of the engine and is in the range of approximately 180 to 250°F. Low-temperature

Figure 4-62 Engine generator sets. (*Colt Industries, Fairbanks Morse Engine Division*)

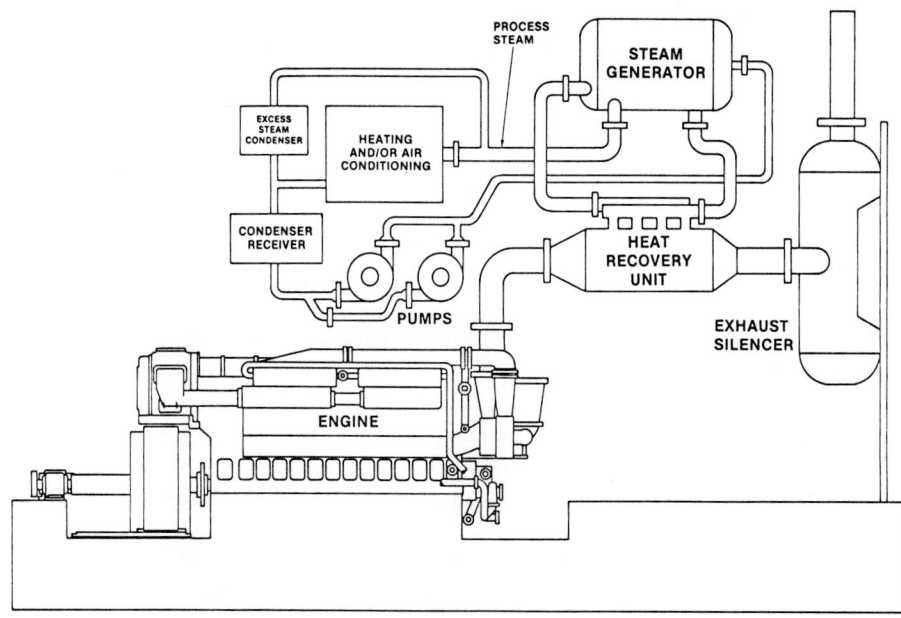

Figure 4-63 Internal combustion engine heat recovery system. (*Colt Industries, Fairbanks Morse Engine Division*)

single-stage absorption chiller machines should be used because of the low temperature of the recovered heat. These chillers are approximately 65 percent as efficient as steam turbine centrifugal chillers and high-temperature two-stage absorption machines.

The maximum hot-water temperature from the heat recovery system is limited to approximately 250°F. Each engine can have its own generator and heat recovery silencer, and heat can be recovered as hot water or as low-pressure steam.

Cooling is produced by means of low-temperature (250°F) absorption chillers with cooling towers. These towers can also be used to cool the engines when the load requires less cooling or less heat than is being produced and operation of the system for electricity generation alone is desired.

The Otto cogeneration cycle is essentially identical to the diesel cycle shown in Figure 4-24.

DIESEL GENERATOR SETS

Diesel generator sets consist of small, high-speed and large, slow-speed two-stroke engines. Diesel engine fuel-to-air spontaneous ignition is achieved from the heat of compression.

Large, slow-speed units are more efficient and have a longer operating period between maintenance and overhauls than small, high-speed engines. Large engines generally cost more and require more space per unit electrical output.

Waste heat is recovered from diesel exhaust and cooling system. The thermal energy recovered from exhaust gases ranging from 1000 to 1400°F can be used to generate hot water or steam through a HRSG. Hot water can be recovered from the diesel cooling water and lubricating oil systems with chiller-heater equipment.

Diesel engine generator sets have a thermal efficiency of about 33 percent, as indicated in Figure 4-23. This means that about one-third of the energy content of the fuel is converted to work. Recovering the exhaust gas heat can increase the efficiency to about 45 percent. Additional capital equipment investment to recover the cooling water and lubricating oil heat content can increase the efficiency to about 75 percent.

Diesel engines have a higher thermal efficiency than gas turbines and steam turbines and thus greater electrical conversion efficiency. The diesel engine also has a higher fuel-to-electricity conversion ratio than the gas engine because of the diesel's higher compression ratio.

Engine performance remains essentially constant down to a 50 percent load. Engine reject heat can be recovered effectively from the exhaust gas, the jacket, and the turbocharger air coolers.

The recommended downtime for engine inspection and maintenance is 300 to 450 hours/year, which is equivalent to a plant availability of about 95 percent.

A diverting valve in the exhaust system passes the exhaust to the heat recovery boiler, or, if required, directly to the adjoining stack.

Diesels emit less sulfur dioxide but more oxides of nitrogen per kilowatthour of electricity than conventional oil-fired electric utility power plants. Nitrogen oxide emissions are a concern that may affect compliance with environmental regulations.

HEAT RECOVERY

The heat requirements of a facility are usually the determining factor favoring one system of heat recovery over another. Heat recovery from engines may utilize the heat transferred from an engine radiator to a flow of air at temperatures ranging from 100 to 150°F. Air can be delivered essentially free of contaminants. The effect is a relatively efficient system converting approximately 33 percent of input fuel energy to work or power and 30 percent to recovered heat energy, resulting in an overall efficiency of approximately 63 percent. This percentage can be increased by approximately 12 percent by recovering a portion of the exhaust heat. Exhaust heat recovery equipment adds to the cost, except in processes where the exhaust can be used directly.

The primary responsibility of heat recovery equipment is to cool the engine. Secondary functions are to recover heat and to silence engine exhaust. These units include a heat recovery muffler, steam separator, and safety devices to protect the entire cooling system. Heat recovery units are constructed either as water tube (water inside a tube) or fire tube units. Pressure drops on both the water and gas sides must be considered when sizing the heat recovery unit to the engine.

Typical heat recovery systems include hot-water systems, steam systems, and ebullient systems, as shown in Figure 4-64.

Hot-Water System

A hot-water system utilizes a jacket water temperature of approximately 190 to 250°F recorded at the engine outlet. A shell-and-tube heat exchanger transfers rejected engine heat to a secondary loop. An exhaust heat recovery boiler or heater may also be included in the system. The primary coolant loop serving the engine jacket must be a closed system.

Pressure control provided in the engine coolant loop ensures sufficient pressure to preclude steaming or steam formation in the engine coolant loop. This pressure source may be designed as a static head from an elevated expansion tank, or controlled gas or air pressure in the expansion tank. Water-circulating pumps must be designed for the elevated temperatures and pressures.

Steam System

A steam system incorporates features of a hot-water system plus a boiler for generating low-pressure steam. Steam is generated in the boiler and in the piping to the boiler as a result of the pressure differential existing by design between the engine outlet and the boiler. A lower pressure prevails in the boiler than at the

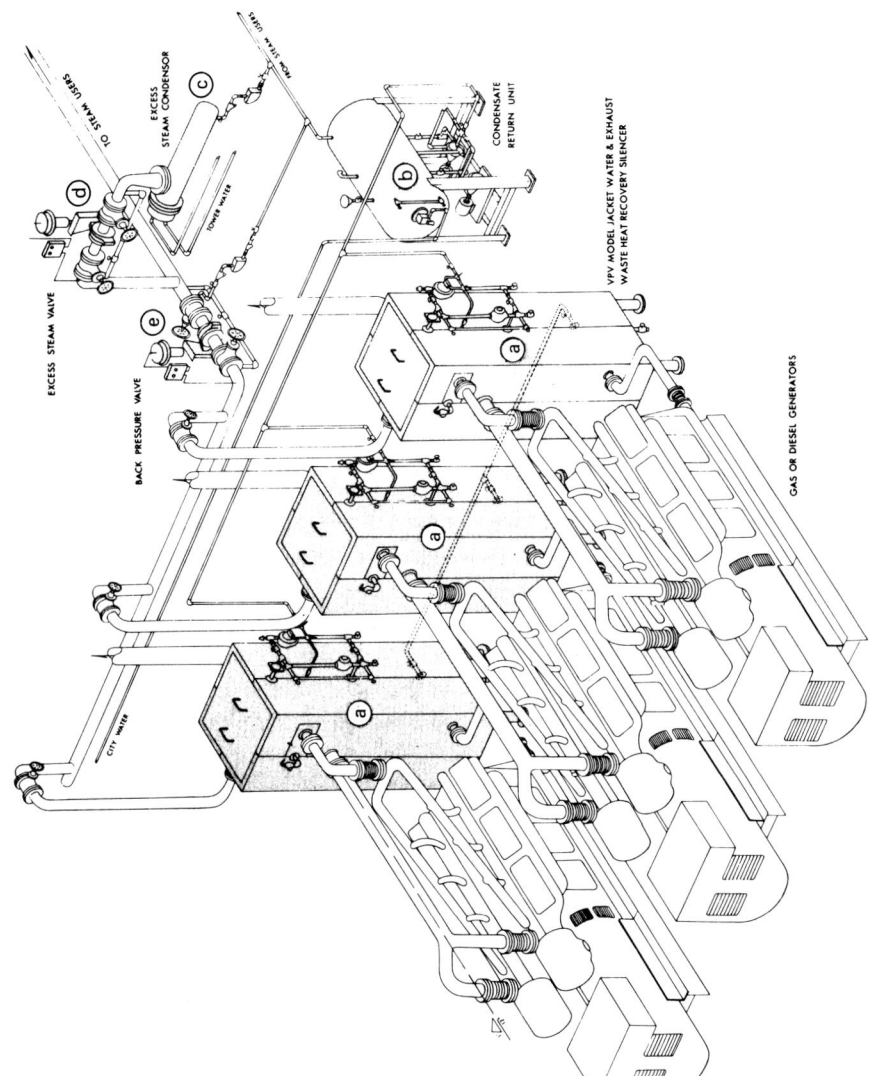

Figure 4-64 Internal-combustion engine. (*a*) Vaporphase Model VPV packaged jacket water and exhaust heat recovery silencers. (*b*) Vaporphase Model ECCR condensate return unit. (*c*) Vaporphase Model ST excess steam condensor. (*d*) Vaporphase Model ESB pneumatic excess steam valve and pilot. (*e*) Vaporphase Model BPB pneumatic back-pressure valve and pilot. (*Vaporphase packaged heat recovery units*)

engine outlet. As the high-temperature water from the engine approaches the boiler, the static head is reduced, as is the heat content of the liquid corresponding to the lower pressure. Heat of vaporization converts part of the water to steam. The temperature of both the steam and the remaining water adjust to the temperature corresponding to the controlled pressure in the boiler. The steam is supplied to the system and recycles through the process.

The total pressure in the engine cooling loop, consisting of the combined steam pressure and the static head, must be adequate to prevent boiling or flashing within the engine. For controlled boiler pressures, the static head required may be reduced but should always be adequate to prevent flashing in the engine cooling loop under normal pressure fluctuations.

A properly installed and operated pressure-reducing valve or orifice at the inlet to the boiler, instead of a static head and a controlled steam pressure system, is more susceptible to a pressure imbalance in the system and potential steam formation in the engine jacket water pump and in the engine, which can result in equipment damage.

Ebullient System

An ebullient system utilizes the heat of vaporization to cool the engine. The engine must be protected at all times by maintaining adequate circulation of the coolant to absorb rejected heat. Steam is moved through the water passages, along with the high-temperature water, by natural circulation to a steam separator located above the engine, as shown in Figure 4-65. Circulation of the coolant is thermally induced and depends on the static head or finite density difference between the coolant water and the steam-water mixture. This is the simplest and least costly form of waste heat recovery, and a jacket water-circulating pump is not required. The temperature differential between the water inlet and outlet of the engine is normally 2 to 3°F. Flow through the engine is due to the change in coolant density as it gains heat from the engine. Since the higher-temperature coolant is lighter, a pressure differential is created between the water inlet and water outlet connections to the engine. Almost all the heat gain in the coolant is added in the form of heat of vaporization.

The exhaust gas boiler and the steam separator can be combined into a single unit, and a direct-fired section added to the exhaust boiler, eliminating the need for an auxiliary boiler.

CONTROLS

The switchgear and controls used with gas and diesel engine generator sets vary in sophistication and cost. The type of facility and load served by the plant dictate the basic engineering sizing requirements for the switchgear.

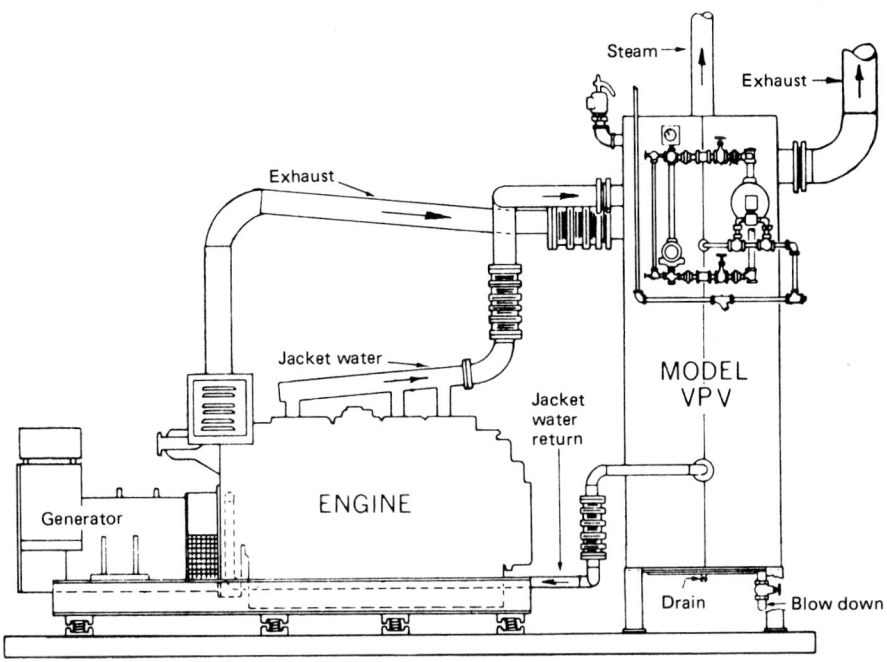

Figure 4-65 Ebullient reciprocating internal-combustion engine and waste heat recovery system. The Vaporphase Model VP packaged heat recovery silencer is specifically designed to provide maximum economical recovery of waste heat from engine jacket water and exhaust while satisfying the many problems encountered in existing total energy and other waste heat recovery installations. Sizes available are standard 100 through 1500 horsepower. For larger sizes consult factory. (*Vaporphase packaged heat recovery units*)

Cogeneration plants can be grouped into two basic categories:

1. MANUALLY CONTROLLED POWER PLANTS In manually controlled power plants, the operator sequences units in or out of service as the load profile demands. The units are generally paralleled and the load divided manually.

2. AUTOMATED POWER PLANTS An unattended or totally automated system utilizes an electronic governor to maintain control. Governors enable the units to be automatically paralleled and then divide the load proportionally. A signal-initiated device places units in operation or out of operation as the load profile demands. The electronic governor allows a group of engine-driven generator sets to be paralleled with reliability and ease. These units can be operated unattended by designing adequate monitoring and indicating devices to notify a distant or roving operator when abnormal operating conditions exist. Routine servicing and major maintenance is required.

TYPICAL SUPPLIERS

The following is a listing of some representatives and suppliers of reciprocating internal-combustion engine generator set equipment.

Alco Power, Inc.	Fairbanks Morse
Caterpillar Tractor Company	International Harvester
Cummins Engine Company Inc.	Sultzer Brothers, Inc.
Detroit Diesel Allison, A Division of General Motors Co.	Waukesha Engine Division, Dresser Industries, Inc.

TECHNICAL AND ECONOMIC FEASIBILITY

The objective of the hypothetical project described in this segment is to provide a cogeneration system having

- The least complexity
- Redundancy or backup features
- The highest reliability
- Low capital installed costs
- Minimum maintenance
- Ease of construction
- Simplicity of operation and interconnection with building heating and cooling systems

This facility is for the cogeneration of electricity and thermal energy for heating and cooling systems. The cogenerator sells all electric power produced to a local electric utility and buys back all site power. Revenues are generated from the sale of electricity, heat, and cooling thermal energy. Heating and cooling revenues are received from district heating consumers.

The financial, design, and operating criteria for this cogeneration project must provide

- Maximum efficiency and energy conservation features
- Satisfaction of environmental and regulatory requirements
- The least environmental impact
- The highest return on the investment at the least possible risk

This project is thermally matched to the system demand to optimize efficiency. The unit loading tracks the thermal demand for process steam and hot water while generating electricity, thus requiring the least fuel use. The project is evaluated on the basis of owner financing for economic comparisons.

Equipment and sizing selection criteria include heat rates and exhaust temperatures for achieving the greatest operating efficiency. This sizing approach fully utilizes the combustion turbine generator exhaust heat and meets the FERC topping-cycle efficiency standard of 42.5 percent, thus permitting the facility to benefit from the regulatory and tax advantages available to qualifying facilities.

The economic implication is that this thermally optimized unit has the operating flexibility to satisfy thermal load energy requirements—peak, seasonal, and daily load variations—at an optimum capacity factor, resulting in attractive overall operating revenues.

The design life of this plant is expected to be in excess of 20 years. This facility has been specified, engineered, and designed to meet best available control technology (BACT) standards and to satisfy emission control requirements.

This example consists of two alternative equipment and system arrangements. Alternate 1 is shown in Figure 4-55 and consists of one combustion turbine generator exhausting to one waste heat recovery boiler. Alternate 2 consists of two combustion turbine generators each separately exhausting to one of two waste heat recovery boilers.

Each combustion turbine generator has a rating of 3000 kilowatts net output, a lower heating valve heat rate of 12,970 Btu per kilowatthour, and a heat input of 40.2×10^6 Btu per hour with an exhaust temperature of 930°F.

The waste heat recovery boiler converts the combustion turbine generator exhaust gas mass flow rate of 126,000 pounds per hour at 930°F to hot water for heating the building. High-temperature hot water is pumped and distributed to heating units to supply the heating load.

Absorption chillers convert the combustion turbine exhaust heat to chilled water to supply the building cooling load, and chilled water is pumped through the cooling units. Mechanical draft cooling towers provide cooling water for the heat exchangers. Hot- and cold-water storage heat sinks are provided. The overall project schedule is 24 months.

An economic comparison of the engineering aspects of the two alternatives includes estimated equipment, materials, engineering, and construction costs for the 24-month construction period. At the top of the next page is a summary of the costs incurred.

Alternative 1 is based on one 3000-kilowatt combustion turbine generator and one waste-heat recovery steam generator (WHRSG). An economic comparison and cash-flow tabulation are given in Table 4-11. Alternative 2 is based on two 3000-kilowatt combustion turbine generators and two WHRSGs. An economic comparison and cash-flow tabulation are given in Table 4-12.

An interest rate of 11.5 percent is assumed during the 24 months of construction.

	Alternative 1	Alternative 2
Equipment and materials	$4,600,000	$ 5,800,000
Engineering and construction	3,100,000	3,950,000
Total installed cost before construction	$7,700,000	$ 9,750,000
Interest during construction	886,000	1,122,000
Total cost	$8,586,000	$10,870,000

FINANCIAL AND CASH-FLOW ANALYSIS

The project is financed with 30 percent equity and 70 percent debt. Construction costs are financed 30 percent from equity and 70 percent from short-term debt. This short-term debt is refinanced at the end of the 24-month construction period with an 8-year, 16 percent loan to be repaid in eight equal annual payments.

Operating Expenses

Operating expenses consist of fuel costs, operating and maintenance costs, property taxes and insurance, and interest.

Fuel Costs Fuel costs are based on the plant operating at 75 percent of its capacity and utilizing natural gas, with an annual base fuel consumption of $264{,}114 \times 10^6$ Btu for alternate 1. In the first year of operation the cost per therm (100,000 Btu per therm) is $0.6441. The cost of natural gas fuel escalates at 13.55 percent per year. (See Table 4-13.)

$$\text{Annual fuel consumption} = \text{(combustion turbine fuel consumption, Btu/hr)} \text{(hr/yr) (average capacity factor)}$$

$$\text{Alternative 1 annual fuel consumption} = (40.2 \times 10^6 \text{ Btu/hr})(8760 \text{ hr/yr})(75\% \text{ capacity factor})$$
$$= 264{,}114 \times 10^6 \text{ Btu/yr}$$

$$\text{Alternative 2 annual fuel consumption} = (80.4 \times 10^6 \text{ Btu/hr})(8760 \text{ hr/yr})(75\% \text{ capacity factor})$$
$$= 528{,}228 \times 10^6 \text{ Btu/yr}$$

The fuel gas cost assumes 13.55 percent escalation per year after the base year of initial operation.

$$\text{The annual fuel gas cost, \$/yr} = \text{(annual fuel consumption, Btu/yr)(fuel gas cost, \$/10}^6 \text{ Btu)}$$

$$\text{Fuel gas cost, \$/10}^6 \text{ Btu} = \text{(fuel gas cost, \$/therm)} \div (100{,}000 \text{ Btu/therm})$$

Table 4-11 Alternative 1
Economic Comparison and Cash Flow

	\multicolumn{10}{c}{Year}									
	1	2	3	4	5	6	7	8	9	10
Total plant cost	—	—	8,586,000	—	—	—	—	—	—	—
30% Equity	—	—	2,575,800	—	—	—	—	—	—	—
70% Long-term debt	—	—	6,010,200	—	—	—	—	—	—	—
Principal	—	—	751,275	751,275	751,275	751,275	751,275	751,275	751,275	751,275
Operating expenses										
Fuel costs	—	—	1,701,158	1,931,730	2,193,467	2,490,595	2,828,133	3,211,362	3,646,358	4,140,515
Operating and maintenance costs	—	—	270,000	288,900	309,129	330,762	353,915	378,689	405,197	433,561
Property taxes and insurance	—	—	92,000	92,000	92,000	92,000	92,000	92,000	92,000	92,000
Interest	—	—	961,632	841,428	721,224	601,020	480,816	360,612	240,408	120,204
Total operating expenses	—	—	3,024,790	3,154,058	3,315,820	3,514,377	3,754,864	4,042,663	4,383,963	4,786,280
Revenues										
Electricity	—	—	1,675,155	1,825,062	1,981,461	2,200,932	2,359,647	2,574,363	2,819,568	3,089,583
Thermal heating	—	—	763,360	872,130	951,480	1,075,770	1,191,100	1,317,090	1,473,500	1,643,670
Depreciation, federal	—	—	1,287,900	1,888,920	1,803,060	1,803,060	1,803,060	—	—	—
Depreciation, state	—	—	1,717,200	1,717,200	1,717,200	1,717,200	1,717,200	—	—	—
Investment tax credit	—	—	772,740							
Total revenues	—	—	6,216,355	6,303,312	6,453,201	6,796,962	7,071,007	3,891,453	4,293,068	4,733,253
Cash-flow summary										
Total payment to principal	—	—	751,275	751,275	751,275	751,275	751,275	751,275	751,275	751,275
Total operating expenses	—	—	3,024,790	3,154,058	3,315,820	3,514,377	3,754,864	4,042,663	4,383,963	4,786,280
Total revenues	—	—	6,216,355	6,303,312	6,453,201	6,796,962	7,071,007	3,891,453	4,293,068	4,733,253
Net income	—	—	2,440,290	2,397,979	2,386,106	2,531,310	2,564,868	(902,485)	(842,170)	(804,302)

Table 4-12 Alternative 2
Economic Comparison and Cash Flow

	\multicolumn{10}{c}{Year}									
	1	2	3	4	5	6	7	8	9	10
Total plant cost	—	—	10,870,000	—	—	—	—	—	—	—
30% Equity	—	—	3,261,000	—	—	—	—	—	—	—
70% Long-term debt	—	—	7,609,000	—	—	—	—	—	—	—
Principal	—	—	951,125	951,125	951,125	951,125	951,125	951,125	951,125	951,125
Operating expenses										
Fuel costs	—	—	3,402,317	3,863,460	4,386,334	4,981,190	5,656,265	6,422,724	7,292,716	8,281,030
Operating and maintenance costs	—	—	180,000	192,600	206,082	220,508	235,943	252,459	270,131	289,041
Property taxes and insurance	—	—	116,000	116,000	116,000	116,000	116,000	116,000	116,000	116,000
Interest	—	—	1,217,440	1,065,260	913,080	760,900	608,720	456,540	304,360	152,180
Total operating expenses	—	—	4,915,757	5,237,320	5,621,496	6,078,598	6,616,928	7,247,723	7,983,207	8,838,251
Revenues										
Electricity	—	—	3,350,310	3,650,310	3,962,222	4,401,864	4,719,294	5,148,726	5,639,136	6,179,166
Thermal heating	—	—	1,526,720	1,744,260	1,902,960	2,151,540	2,382,220	2,634,180	2,947,000	3,287,340
Depreciation, federal	—	—	1,630,500	2,391,400	2,282,700	2,282,700	2,282,700	—	—	—
Depreciation, state	—	—	2,174,000	2,174,000	2,174,000	2,174,000	2,174,000	—	—	—
Investment tax credit	—	—	978,300	—	—	—	—	—	—	—
Total	—	—	9,659,830	9,959,784	10,322,582	11,010,104	11,558,214	7,782,906	8,586,136	9,466,506
Cash-flow summary										
Total payment to principal	—	—	951,125	951,125	951,125	951,125	951,125	951,125	951,125	951,125
Total operating expenses	—	—	4,915,757	5,237,320	5,621,496	6,078,598	6,616,928	7,247,723	7,983,207	8,838,251
Total revenues	—	—	9,659,830	9,959,784	10,322,582	11,010,104	11,558,214	7,782,906	8,586,136	9,466,506
Net income	—	—	3,792,948	3,771,339	3,749,961	3,980,381	3,990,161	(415,942)	(348,196)	(322,870)

Table 4-13 Fuel Costs

Year	Fuel gas cost		Annual fuel cost	
	Dollars per therm	Dollars per 10^6 Btu	Alternative 1, $	Alternative 2, $
1	—	—	—	—
2	—	—	—	—
3	0.6441	6.441	1,701,158	3,402,317
4	0.7314	7.314	1,931,730	3,863,460
5	0.8305	8.305	2,193,467	4,386,334
6	0.9430	9.430	2,490,595	4,981,190
7	1.0708	10.708	2,828,133	5,656,265
8	1.2159	12.159	3,211,362	6,422,724
9	1.3806	13.806	3,646,358	7,292,716
10	1.5677	15.677	4,140,515	8,281,030

OPERATING AND MAINTENANCE COSTS

The operating staff provides supervision, administration, and technical support for the plant and is also involved in billing and collecting revenues. Technical maintenance includes repair and overhaul for the facility. Other maintenance costs include those for materials used in the maintenance of nontechnical items. The work is performed by the operating staff.

Operating staff and maintenance costs (Table 4-14) are assumed to be $45 per kilowatt times the net plant electricity output for alternative 2, or ($45 per kilowatt) (6000 kilowatts) = $270,000, based on two combustion turbine generators and two WHRSGs.

For alternative 1 $0.003 per kilowatthour is deducted for one combustion turbine generator and $25,000 per year for the controls and one WHRSG. The deduction for alternative 1 is therefore

(3000 kW) (8760 hr/yr) (85% capacity) ($0.003/kWhr)	= $67,014/yr
Deduction for controls and WHRSG	= $25,000/yr
Combined deduction for combustion turbine generator and WHRSG	= $92,014/yr
Deduction approximately	= $90,000/yr

Costs escalate at 7 percent per year after the first year of operation.

Property Taxes and Insurance   Property taxes and insurance are assumed to be 2 percent of the equipment and materials capital cost per year. The sale of steam, heat, or electricity by a qualifying cogeneration facility is exempt from state sales taxes.

For alternative 1 the annual property taxes and insurance are equal to 2 percent

Table 4-14 Annual Operating, Staffing, and Maintenance Costs

Year	Alternative 1, $	Alternative 2, $
1	—	—
2	—	—
3	180,000	270,000
4	192,600	288,900
5	206,082	309,139
6	220,508	330,762
7	235,943	353,915
8	252,459	378,689
9	270,131	405,197
10	289,041	433,561

of the $4,600,000 equipment and materials capital cost, or $92,000. For alternative 2 the annual property taxes and insurance are equal to 2 percent of the $5,800,000 equipment and materials capital cost, or $116,000.

Interest Interest on the long-term 70 percent 8-year debt is 16 percent (Table 11-5). The annual principal amount is equal to the long-term debt divided by the 8-year period.

For alternative 1 this is

$$(\$6,010,200) \div (8 \text{ yr}) = \$751,275/\text{yr}$$

The interest for year 10 is

$$(16\%)(\text{principal}) = (0.16)(\$751,275) = \$120,204$$

For alternative 2 this is

$$(\$7,609,000) \div (8 \text{ yr}) = \$951,125/\text{yr}$$

The interest for year 10 is

$$(16\%)(\text{principal}) = (0.16)(\$951,125) = \$152,180$$

REVENUES

Revenues consist of electricity revenues, thermal revenues, depreciation, and energy investment tax credits.

Electricity Revenue

The electricity revenue consists of a capacity price and an energy price (Table 4-16). The capacity price is based on a power purchase contract price from the electric utility of $112 per kilowatt on firm capacity for an 8-year contract life. The firm

Table 4-15 Interest on Debt

Year	Alternative 1, $	Alternative 2, $
1	—	—
2	—	—
3	961,632	1,217,440
4	841,428	1,065,260
5	721,224	913,080
6	601,020	760,900
7	480,816	608,720
8	360,612	456,540
9	240,408	304,360
10	120,204	152,180

capacity is 85 percent of the total capacity. The energy price is based on the average guaranteed rate for the first 8 years of operation. The following energy and capacity price projections assume that the cogeneration facility will operate at an average of 75 percent of its capacity. The guaranteed rate is the minimum price. The electric revenues estimated for these examples are therefore conservative and may be higher should the cogenerator negotiate a higher price from the electric utility.

The capacity price for alternative 1 is

$$(\$112/\text{kw yr}) (3000 \text{ kW}) (85\% \text{ firm capacity}) = \$285,600$$

The energy price for alternative 1 is

$$(\$0.0705/\text{kW yr}) (8760 \text{ hr/yr}) (75\% \text{ capacity factor}) (3000 \text{ kW}) = \$1,389,555$$

Thermal Revenue The thermal revenue for heating demand is assumed to be sold to the customer at a cost equivalent to producing heat in a natural-gas-fired boiler having an efficiency of 65 percent.

Alternative 1 is based on producing 40×10^9 Btu of heat, whereas alternative 2 is based on producing 80×10^9 Btu of heat. The heating revenues for alternatives 1 and 2 are computed in Table 4-17.

The thermal revenue for a cooling demand of $125,257.1 \times 10^6$ Btu for alternative 2 is converted to kilowatthours by dividing the cooling demand of $125,257.1 \times 10^6$ Btu by 3413 Btu per kilowatthour to obtain 36.7×10^6 kilowatthours. The thermal revenue is based on the peak period electric power cost adjusted for a coefficient of performance equal to 4.3. The cooling revenues for alternatives 1 and 2 are computed in Table 4-18 and the combined thermal revenues are shown in Table 4-19.

Federal Depreciation Allowances Should the cogenerator be able to take advantage of accelerated depreciation benefits, then it may choose a 5-year term, except for buildings which may have a 15-year term.

Table 4-16 Electricity Revenue

Year	Energy price, $/kWh	Capacity price, $/kW yr	Capacity price at 85% capacity factor		Price of energy		Total electricity revenue	
			Alternative 1, $	Alternative 2, $	Alternative 1, $	Alternative 2, $	Alternative 1, $	Alternative 2, $
1	—	—	—	—	—	—	—	—
2	—	—	—	—	—	—	—	—
3	0.0705	112	285,600	571,200	1,389,555	2,779,110	1,675,155	3,350,310
4	0.0772	119	303,450	606,900	1,521,612	3,043,224	1,825,062	3,650,124
5	0.0841	127	323,850	647,700	1,657,611	3,315,222	1,981,461	3,962,922
6	0.0942	135	344,250	688,500	1,856,682	3,713,364	2,200,932	4,401,864
7	0.1007	147	374,850	749,700	1,984,797	3,969,594	2,359,647	4,719,294
8	0.1103	157	400,350	800,700	2,174,013	4,348,026	2,574,363	5,148,726
9	0.1208	172	438,600	877,200	2,380,968	4,761,936	2,819,568	5,639,136
10	0.1323	189	481,950	963,900	2,607,633	5,215,266	3,089,583	6,179,166

Table 4-17 Thermal Revenues for Heating

Year	Fuel cost, $/10^6$ Btu	Adjusted fuel cost for 65% efficient boiler, $/10^6$ Btu	Annual heating demand Alternative 1, $\times 10^6$ Btu	Annual heating demand Alternative 2, $\times 10^6$ Btu	Heating revenue Alternative 1, $/yr	Heating revenue Alternative 2, $/yr
1	—	—	—	—	—	—
2	—	—	—	—	—	—
3	6.441	9.909	40,000	80,000	396,360	792,720
4	7.314	11.252	40,000	80,000	450,080	900,160
5	8.305	12.777	40,000	80,000	511,080	1,022,160
6	9.430	14.508	40,000	80,000	580,320	1,160,640
7	10.708	16.474	40,000	80,000	658,960	1,317,920
8	12.159	18.706	40,000	80,000	748,240	1,496,480
9	13.806	21.240	40,000	80,000	849,600	1,699,200
10	15.677	24.118	40,000	80,000	964,720	1,929,440

Table 4-18 Thermal Revenues for Cooling

Year	Peak period electricity cost, $/kWh	Adjusted electricity cost, for chiller COP* = 4.3 is $/kWh ÷ COP $/kWh	Annual cooling demand Alternative 1, ×10⁶ kWh	Annual cooling demand Alternative 2, ×10⁶ kWh	Cooling revenue Alternative 1, $/yr	Cooling revenue Alternative 2, $/yr
1	—	—	—	—	—	—
2	—	—	—	—	—	—
3	0.087	0.020	18.35	36.7	367,000	734,000
4	0.098	0.023	18.35	36.7	422,050	844,100
5	0.104	0.024	18.35	36.7	440,400	880,800
6	0.117	0.027	18.35	36.7	495,450	990,900
7	0.123	0.029	18.35	36.7	532,150	1,064,300
8	0.134	0.031	18.35	36.7	568,850	1,137,700
9	0.146	0.034	18.35	36.7	623,900	1,247,800
10	0.159	0.037	18.35	36.7	678,950	1,357,900

*Coefficient of performance.

Table 4-19 Combined Revenues

	3000 kilowatts			6000 kilowatts		
Year	Heating revenue, $/yr	Cooling revenue, $/yr	Combined thermal revenue, $/yr	Heating revenue, $/yr	Cooling revenue, $/yr	Combined thermal revenue, $/yr
1	—	—	—	—	—	—
2	—	—	—	—	—	—
3	396,360	367,000	763,360	792,720	734,000	1,526,720
4	450,080	422,050	872,130	900,160	844,100	1,744,260
5	511,080	440,400	951,480	1,022,160	880,800	1,902,960
6	580,320	495,450	1,075,770	1,160,640	990,900	2,151,540
7	658,960	532,150	1,191,110	1,317,920	1,064,300	2,382,220
8	748,240	568,850	1,317,090	1,496,480	1,137,700	2,634,180
9	849,600	623,900	1,473,500	1,699,200	1,247,800	2,947,000
10	964,720	678,950	1,643,670	1,929,440	1,357,900	3,287,340

Depreciation for this example is based on the accelerated cost recovery system (ACRS) of the Economic Recovery Tax Act of 1981. The recovery period for depreciable cogeneration property is 5 years. Depreciation for year 1 is 15 percent, for year 2 it is 22 percent, and for years 3 through 5 it is 21 percent per year for federal tax purposes.

Depreciation is based on the total plant cost of $8,586,000 for alternative 1 and $10,870,000 for alternative 2 (Table 4-20).

State Depreciation Allowances The state is assumed to have not adopted the federal ACRS depreciation provision. This issue is dependent on state policies. For state tax purposes the straight-line method and a 60-month term are assumed. The straight-line method of depreciation yields an annual amount obtained by dividing the total plant cost by the 60-month amortization period.

Table 4-20 Depreciation Rates

Year	Depreciation rate, %	Alternative 1, $	Alternative 2, $
1	15	1,287,900	1,630,500
2	22	1,888,920	2,391,400
3	21	1,803,060	2,282,700
4	21	1,803,060	2,282,700
5	21	1,803,060	2,282,700
Total	100	8,586,000	10,870,000

The straight-line depreciation annual amount for alternative 1 is

$$\$8,586,000 \div 5 \text{ yr} = \$1,717,200$$

The straight-line depreciation annual amount for alternative 2 is

$$\$10,870,000 \div 5 \text{ yr} = \$2,174,000$$

Energy Investment Tax Credit The project will use the same fuel source to produce both qualifying thermal energy and electric power. It is estimated that 90 percent of the project's cost will qualify for the 10 percent energy investment tax credit for cogeneration equipment. The remaining 10 percent of the project cost is for buildings which do not qualify for the regular investment credit.

For alternative 1 the investment tax credit is

$$(90\%)(\$8,586,000)(10\%) = \$772,740$$

For alternative 2 the investment tax credit is

$$(90\%)(\$10,870,000)(10\%) = \$978,300$$

REFERENCES

1. *Cities,* special issue of *Scientific American,* vol. 213, no. 3, September 1965.

2. *District Heating Handbook,* 3d ed., National District Heating Association, 1951.

3. Miller, A. J., et al.: "Use of Steam-Electric Power Plants to Provide Thermal Energy to Urban Areas," Oak Ridge National Laboratory Report No. ORNL-HUD-14, January 1971.

4. Anderson, T. D., et al.: "An Assessment of Energy Options Based on Coal and Nuclear Systems," Oak Ridge National Laboratory Report No. ORNL-4995, July 1975.

5. Graves, R. L., et al.: "Assessment of an Atmospheric Fluidized Bed Coal Combustion Gas Turbine Cogeneration System for Industrial Applications," Oak Ridge National Laboratory Report No. TM-6626, vol. 1, October 1979.

6. Sindt, H. A., et al.: "Costs of Power from Nuclear Desalting Plants," *Chemical Engineering Progress,* vol. 63, no. 4, April 1967, pp. 41–45.

7. Probstein, R. F.: "Desalination: Some Fluid Mechanical Properties," *J. Basic Eng., Trans. ASME,* vol. 94, 1972, pp. 286–313.

8. Spiewak, I.: "Investigation of the Feasibility of Purifying Municipal Waste by Distillation," Oak Ridge National Laboratory Report No. ORNL-TM-2547, April 1969.

9. Anderson, L. L.: "Energy Potential of Organic Wastes: A Review of the Quantities and Sources," U.S. Bureau of Mines Information Circular IC-8549, 1972.

10. Sherwin, E. T., and A. R. Nollet: "Solid Waste Resource Recovery: Technology Assessment," *Mechanical Engineering,* vol. 102, no. 5, May 1980.

11. Fife, J. A.: "Incineration: Steam Generation from Solid Wastes," *District Heating,* vol. LVI, no. 2, Fall 1970, pp. 18–24.

12. Boegly, W. J., et al: "MIUS Technology Evaluation—Solid Waste Collection and Disposal," Oak Ridge National Laboratory Report No. ORNL-HUD-MIUS-9, September 1973.

13. Stabenow, G.: "The Chicago Northwest and Harrisburg Incinerators: A Proven Method of Energy Recovery and Recycling of Ferrous Metals," *Proceedings of the 1976 National Waste Processing Conference, Boston, Mass., ASME,* 1976, pp. 81–96.

14. "Baltimore's Resource Recovery Plant now Working Well City Official Says," *Environmental Reporter,* Oct. 12, 1979, pp. 1348–1349.

15. McEwen, L., and S. J. Levy: "Can Nashville's Story Be Placed in Perspective?" *Solid Waste Management,* vol. 19, no. 8, August 1976, pp. 24–60.

16. Krause, H. H., et al.: *Corrosion and Deposits from Combustion of Solid Wastes,* part VI: *Processed Refuse as a Supplementary Fuel in a Stoker-Fired Boiler,* ASME Paper No. 78-WA/Fu-4, December 1978.

17. Kovacik, J. M.: "Guidelines for Future Cogeneration," *American Power Conference,* Apr. 23–25, 1969.

18. Wilson, W. B., and W. J. Hefner: "Economic Selection of Plant Cycles and Fuels for Gas Turbines," *Combustion,* April 1974, pp. 7–16.

19. Fraas, A. P., and G. Samuels: "Power Conversion Systems of the 21st Century," *Journal of the Power Division, ASCE,* vol. 104, no. PO1, Proceedings Paper 13545, February 1978, pp. 83–97.

20. Frei, D.: "Gas Turbines for the Process Improvement of Industrial Thermal Power Plants," *Sulzer Technical Review,* vol. 4, 1975, pp. 195–200.

21. *Considerations Affecting Steam Power Plant Site Selection, A Report Sponsored by the Energy Policy Staff,* Office of Science and Technology, U.S. Government Printing Office, 1969.

22. Piper, H.G., and G. L. West, Jr.: "Siting of Nuclear Reactors," Oak Ridge National Laboratory Report No. ORNL-HUD-11, 1971.

23. "Nuclear Energy Center Site Survey—1975, Practical Issues of Implementation," U.S. Nuclear Regulatory Commission, NUREG—0001, part IV, January 1976.

24. Beck, C.: "Engineering Study on Underground Construction of Nuclear Power Reactors," USAEC Report AECU-3779, Apr. 15, 1958.

25. Fraas, A. P.: "Environmental Aspects of Fusion Power Plants," *Proceedings of the Plenary Sessions International Conference on Nuclear Solutions to World Energy Problems,* Nov. 13–17, 1972, pp. 261–274.

26. *Mälmo Kraflvärmeverk,* The City of Malmö, Technical Division, Spring 1973.

27. Raisic, N.: "The Most Promising Area of Future Nuclear Heat Utilization?" *Nuclear Engineering International,* vol. 22, no. 260, August 1977.

28. Malfitani, L.: "District Heating System Warms Heart of Japanese City," *Power,* April 1970, pp. 50–53.

29. Some articles about townplanning in Västerås, City of Västerås, 1973.

30. Västerås District Heating Power Station, Springfeldt Annonsbyrå AB, Västerås, Sweden, 1975.

31. Heizkraftwerk Sendling, Stadtwerke München, Munich, West Germany, November 1972.

32. Segasser, C. S., and J. E. Christian: "Low Temperature Thermal Energy Storage," Oak Ridge National Laboratory Report No. ANL/CES/TE 79-3, March 1978.

33. Oliker, I., and H. J. Muhlhauser: "Technical and Economic Aspects of Coal-Fired District Heating Power Plants in U.S.A." *American Power Conference,* April 1980.

34. Wigmore, D. B., and R. E. Niggemann: "The Specification of an Optimum Working Fluid for a Small Rankine Cycle Turboelectric Power System," *Proceedings of the Seventh Intersociety Energy Conversion Engineering Conference,* September 1972, pp. 303–314.

35. Samuels, G., and R. S. Holcomb: "Reference Design for an Isotope Power Unit Employing an Organic Rankine Cycle," Oak Ridge National Laboratory Report No. TM-2960, March 1971.

36. Morgan, D. J., and J. P. Davis: "High Efficiency Gas Turbine/Organic Rankine Cycle Combined Power Plant," ASME Paper 74-GT-35, March 1974.

37. Tabor, H., and L. Bronicki: "Establishing Criteria for Fluids for Small Vapor Turbines," *SEA Trans.,* vol. 73, 1965, pp. 561–575.

38. Niggemann, R. E., et al.: "Fluid Selection and Optimization of an Organic Rankine Cycle Waste Heat Power Conversion System," ASME Paper No. 78-WA/Ener-6, Dec. 10, 1978.

39. *The Dowtherm Heat Transfer Fluids,* Dow Chemical Co., 1967.

40. Milora, S. L., and J. W. Tester: *Geothermal Energy as a Source of Electric Power,* The M.I.T. Press, Cambridge, Mass., 1976.

41. McGowan, J. G., et al.: "Conceptual Design of a Rankine Cycle Powered by the Ocean Thermal Difference," *Proceedings of the Intersociety Energy Conversion Engineering Conference,* Aug. 13–16, 1973, pp. 420–427.

42. Luchter, S.: "Power Recovery from Gas Turbines—A Review of the Limitations, and an Evaluation of the Use of Organic Working Fluids," ASME Paper No. 70-GT-113, May 1970.

43. Michel, J. W.: "Heat Transfer Considerations in Utilizing Solar and Geothermal Energy," paper presented at the Miami International Conference on Alternative Energy Sources, Dec. 5–7, 1977, Miami Beach, Fla.

44. Set, R. G. and W. Steigelmann: *Binary-Cycle Power Plants Using Dry Cooling Systems,* part 1: *Technical and Economic Evaluation, Final Report F-C3023,* The Franklin Institute Research Laboratories, January 1972.

45. Pummer, J. W.; *The Kinetics and Mechanism of the Pyrolytic Decomposition of Aromatic Heat Transfer Fluids, Final Report NBS Project No. 3110541,* National Bureau of Standards, April 1970.

Section 5

Hydroelectric Power Systems

Long a source of both central-station and industrial power, hydroelectric systems gained new popularity with the rise in price of fossil fuels. Sites which once appeared to be unsuitable for economic power generation are now being utilized as low-head installations in many areas.

With increased interest in hydroelectric power, designers are offered more choices as to the power source they might choose for a given load. Small, abandoned dams, once thought noncompetitive because of their high cost relative to thermal plants, now appear to be most promising sources of hydroelectric power. The economics of such sites are discussed in this section.

Other topics important to the power-plant designer covered here include hydroelectric power station costs, types of sites suitable for these plants, dams of various types used for hydro stations, turbines for power generation, estimating the flow potential of a site, tidal power, and standardized vs custom-designed units.

With the possibility of continued uncertainty in fossil-fuel supplies and prices, hydroelectric power will continue to be a possible alternative in many plant designs. For this reason every designer should be equipped with enough data to make intelligent decisions as to the most economic source of power for a given load or demand.

From *Engineering Evaluation of Energy Systems,* by Arthur P. Fraas. Copyright © 1982. Used by permission of McGraw-Hill, Inc. All rights reserved.

Just as the highly developed coal-fired steam plant has served as a convenient standard against which other thermal power plants can be compared, so conventional, reliable hydroelectric plants provide a basis for evaluating the potential of "natural" sources of energy such as solar, wind, tides, ocean thermal differences, etc., because in each case one is confronted with high capital costs which are site-dependent and are sensitive to the head and energy flux available and to cyclic variations characteristic of the site. As indicated in Ref. 1, approximately 11 percent of the electric power produced in the United States in 1977 was provided by hydroelectric installations. However, this fraction varies widely with the region of the country as indicated by Fig. 5-1, which shows the relative amounts of hydroelectric and thermal power plant capacities for the principal geographical regions of the country, indicating a much greater availability of hydroelectric power in mountainous areas.[1]

HISTORICAL BACKGROUND

As mentioned in the first chapter, the oldest documented use of a machine to replace human or animal power was a waterwheel described in a Byzantine manuscript written a little after 600 A.D. This machine employed a paddle wheel similar to that of a Mississippi stern-wheeler, with the wheel mounted above a fast-flowing stream. The outboard end of the rotor was supported on a pontoon that was positioned by a system of booms and guys. The power was transmitted through pulleys with rope belts to drive a millstone. Returning crusaders apparently carried the concept to Europe, where waterwheels gradually came into use during late medieval times. These machines provided the basis for the beginning of the industrial revolution and were used not only for grinding grain but also for a wide variety of industrial operations, such as driving the bellows for ironworks. For example, the first ironworks in the United States was built by Governor Bradford's son just north of Boston in 1648 and required eight large waterwheels for driving the bellows for the furnaces and for driving rolls and slitting shears. Thus, an important consideration in finding a suitable location for a plant was the availability of waterpower.

Undershot and overshot waterwheels, operating with heads of 2 to 4 m, were the source of waterpower up until the latter part of the eighteenth century, when small, vertical-shaft, turbine-type machines began to come into use not only in Spain, France, and Italy but also in the United States. These machines attracted the attention of men such as Euler, who evolved a theoretical basis for their design, so that by 1830 a number of vertical-shaft turbines were in use with efficiencies of over 80 percent.[2] The use of these machines increased rapidly, but they did not fully displace the overshot and undershot waterwheels until the end of the century. Incidentally, the largest of these machines was an 1839 overshot waterwheel, 60 ft in diameter and 22 ft wide, employed in an ironworks on the Hudson River just above Troy, New York. This machine developed 1200 hp.

The use of waterpower for generating electricity began in the United States on the Fox River in Wisconsin in 1882, a year after the first steam-driven electric utility plant began operating in Philadelphia.[3] The use of hydroelectric generators grew rapidly so that by the 1930s some 30 percent of the nation's generating capacity and 40 percent of the electric energy produced was from hydro units.

HYDROELECTRIC POTENTIAL OF THE UNITED STATES

The United States is blessed with one of the largest sets of hydroelectric power sources in the world; only China, the U.S.S.R., the Congo, and Brazil have greater national hydroelectric potential. The United States has developed a larger fraction of its hydroelectric potential than almost any other major nation and hence gives an excellent indication of the extent to which the hydroelectric potential both here and abroad can be utilized.

The total hydroelectric potential of the continental United States is estimated by the Federal Power Commission[1] to be about 168×10^6 kW of capacity, of which the principal undeveloped potentials are in the north Pacific and Alaskan areas (Fig. 5-2). The capacity that has been developed or is under construction amounts to 66×10^6 kWe, roughly half of the total potential excluding the 33×10^6 kWe undeveloped potential of Alaska. At first thought one might construe this to indicate that a huge amount of hydroelectric capacity remains to be developed, but in point of fact most of the attractive sites have already been developed, and the balance would entail progressively more expensive installations. Not only are the sites naturally less well suited to development, but also the cost of land has been rising steeply and environmental considerations are leading to vociferous opposition to construction of new facilities at almost any of the remaining sites. An indication of these effects is given by Fig. 5-3, which shows the developed capacity as a function of year. It is evident from Fig. 5-3 that the construction of new facilities has already tapered off, and, in light of the above considerations, it is unlikely to increase rapidly in the future.[4]

TYPES OF SITE FOR HYDROELECTRIC POWER PLANTS

Some insight into the effects of a site on the cost of a hydroelectric power plant can be obtained by considering the half dozen different types of site commonly developed. The principal costs involve the amount of land that must be acquired, replacement of assets submerged by the reservoir, and the quantity of material required for the construction of the dam and powerhouse. The useful power output depends on the head and water flow available and the extent to which these may vary with the season of the year. Seasonal variations in water flow constitute a particularly important factor in determining the size of reservoir best suited to the site.

Falls and Short, Steep Rapids

One of the best and lowest-cost hydroelectric sites in the world is Niagara Falls. Land requirements are minimal; a relatively short canal around the falls serves to supply a huge volume of water with a high head to the powerhouse located below the falls with little seasonal variation in the flow rate. The site is close to a major load center, and so no long transmission lines are required. Interestingly, it provided one of the first cases in which a conflict developed between environmentalists and the need for low-cost power. In this case, the conflict was resolved sensibly by compromising on the percentage of the water flow used for power, which was kept below the point at which it detracted noticeably from the aesthetic value of the falls. There is no other comparable site in the United States, but the great falls on the Parana River at Sete Quedas in Brazil provides a roughly similar situation. The water flow at Sete Quedas is much greater but varies more widely through the course of the year and the head is only half as great. More importantly, it is hundreds of miles from the nearest load centers, and so it was not until the 1970s that construction of a power plant was undertaken at this site. When fully developed, the Itaipu plant will be the largest hydroelectric plant in the world with an output roughly six times that at Niagara.

Low-Dam, Run-of-the-River Sites

The oldest and still one of the most common types of hydroelectric site entails the construction of a relatively low dam in a

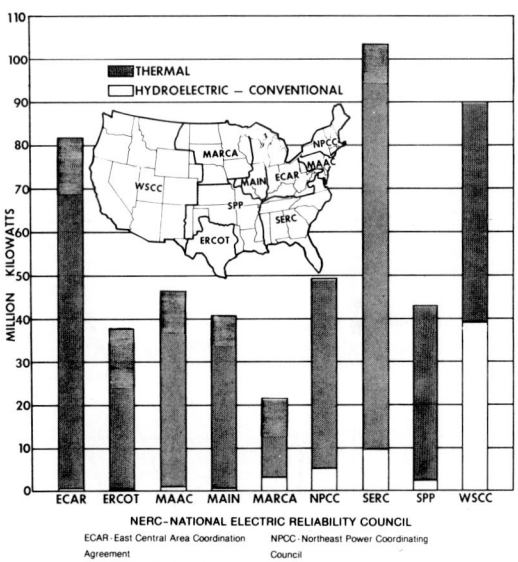

Figure 5-1 The power capacity of U.S. thermal and hydroelectric power plants in 1978.[1] *(Courtesy Federal Energy Regulatory Commission.)*

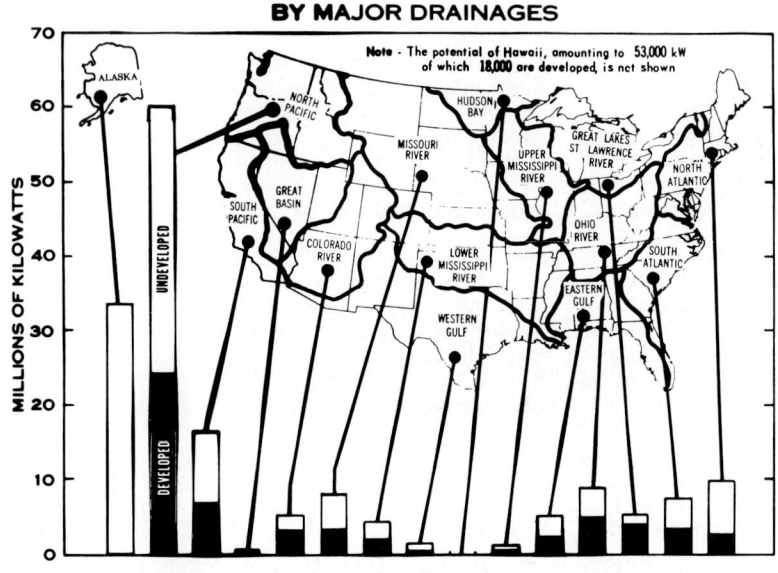

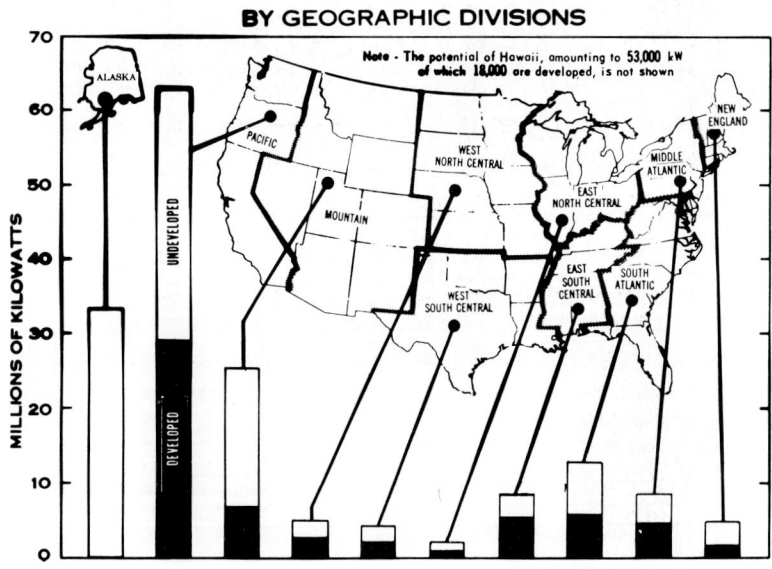

Figure 5-2 Developed and undeveloped hydroelectric power capacity of the United States in 1978 by drainage areas.[1] (*Courtesy Federal Energy Regulatory Commission.*)

stretch of river in which the stream level drops fairly rapidly. Such dams commonly yield heads of 6 to 18 m (20 to 60 ft), are often relatively small, and have little or no storage capacity. A relatively large example of this type is Wilson Dam at Muscle Shoals on the Tennessee River. Note that dams of this type are often multipurpose dams in that they include locks to improve the navigability of the river in which they are located, as is the case at Wilson Dam. When locks are included, a substantial fraction of the cost of the dam can be charged to navigation of the waterway.

High Dams

In mountainous areas where rivers drop fairly rapidly in long gorges, it is often advantageous to place a dam at a particularly narrow spot in the gorge, in part to give a high head and in part to provide a large storage capacity. Inasmuch as most rivers are characterized by large seasonal variations in flow, these dams are often multipurpose units, providing both flood control and a means of giving a more nearly uniform water flow rate downstream. The latter point is particularly important in ensuring adequate water supplies to downstream cities during seasons of low rainfall. Typical examples of such dams are TVA's Norris Dam in Tennessee and Shasta Dam in California.

Low Dams in Mountainous Areas

In mountainous areas it is sometimes possible to build a low dam in the upper reaches of a river with a fairly rapid drop around a U-bend or between one river and another so that a tunnel can be bored laterally through the ridge separating the two stretches of river and a much higher head can be delivered to the powerhouse than developed at the dam. An example of such a dam is the Santeetla Dam of Alcoa located in North Carolina. This dam has a maximum height of 66 m (216 ft) but yields a head at the powerhouse of 202 m (665 ft). A truly impressive possibility for a dam of this sort is offered by the great bend in the Brahmaputra River in India where the river flows out of the Himalaya Mountains. It is estimated that this project, if developed, would yield 20,000 MW.

Occasionally a low dam may be built on a river near the edge of an escarpment to give a high head for a plant at the foot of the escarpment. An example of this sort is the Alcoa Thorpe Dam (originally called Glenville) in North Carolina, which has a maximum dam height of 46 m (150 ft) but gives a head at the powerhouse at the foot of the escarpment of 368 m (1207 ft). Probably the most spectacular example of this type of site is

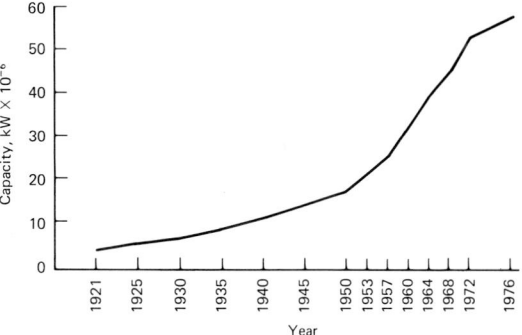

Figure 5-3 Developed hydroelectric capacity of the United States as a function of time.[4]

given by Cubatão, a plant at Santos in Brazil. In this area the Brazilian landmass is characterized by an escarpment ~ 1000 m high which rises up within a few kilometers of the Atlantic coast. Water drains inland, with the rivers draining southwestward into the Parana-Plata, which empties into the Atlantic near Buenos Aires. The prevailing winds carry the warm moist air from the South Atlantic over the escarpment, giving a rainfall of over 5 m per year (17 ft per year). In the early 1920s Billings, an American engineer, realized that a river up on the plateau could be dammed so that water would be backed up to a point near the edge of the escarpment and penstocks could be run down the very steep slope to sea level to drive Pelton wheels with a head of 713 m (2340 ft). This brilliant concept was kept secret, and options on the land required were obtained on the pretext of finding a good site for a British golf and country club. The options were all exercised, the plant was built, and the first power produced in the early 1920s. São Paulo was then a city of only 50,000 people engaged mainly in the coffee trade. The cheap power revolutionized the economy. São Paulo developed rapidly into an industrial metropolis which, at the time of writing, was a city of around 7 million people with a huge appetite for electric power. By adding small dams downstream on the rivers flowing inland and pumping water back to the edge of the escarpment, the power capacity of the Cubatão site has been greatly increased, and other, roughly similar, though less remarkably favorable, sites along the Brazilian coasts have been developed.

Similar but much smaller-capacity sites have been developed in Scandanavia, where—as in Brazil—the escarpment consists mainly of granite batholiths which make possible a special type

TABLE 5-1 Summary of Data on Hydroelectric Installations in the TVA System

				\multicolumn{5}{c}{Dam and appurtenances}	\multicolumn{2}{c}{Power plant}
Project	Owner	Year of closure	River	Max. height, ft	Overall crest length, ft	Max. spillway capacity, cfs	Volume of concrete, yd³	Volume earth and/or rock fill, yd³	No. units	Rated capacity, kW
Tennessee River Basin										
Kentucky	TVA	1944	Tennessee	206	8422	1,050,000	1,356,000	5,582,100	5	175,000
Pickwick Landing	TVA	1938	Tennessee	113	7715	650,000	679,100	3,081,000	5	218,000
Wilson	TVA	1924	Tennessee	137	4535	671,000	1,766,200	0	21	629,840
Wheeler	TVA	1936	Tennessee	72	6342	542,000	1,099,400	0	11	356,400
Guntersville	TVA	1939	Tennessee	94	3979	478,000	514,100	874,900	4	97,200
Nickajack	TVA	1967	Tennessee	81	3767	360,000	546,900	989,200	4	97,200
Chickamauga	TVA	1940	Tennessee	112	2960	470,000	506,400	2,793,500	4	108,000
Watts Bar	TVA	1942	Tennessee	112	2960	560,000	480,200	1,210,000	5	150,000
Fort Loudoun	TVA	1943	Tennessee	122	4190	390,000	586,700	3,594,000	4	131,190
Tims Ford	TVA	1970	Elk	175	1484	108,000	85,400	2,530,000	1	45,000
Appalachia	TVA	1943	Hiwassee	150	1308	136,000	237,800	0	2	75,000
Hiwassee	TVA	1940	Hiwassee	307	1376	112,000	800,600	0	2	117,100
Chatuge	TVA	1942	Hiwassee	144	2850	11,500	25,700	2,348,000	1	10,000
Ocoee No. 1	TVA	1911	Ocoee	135	840	45,000	160,000	0	5	18,000
Ocoee No. 2	TVA	1913	Ocoee	30	450	—	0	0	2	21,000
Ocoee No. 3	TVA	1942	Ocoee	110	612	95,000	82,500	82,000	1	27,000
Blue Ridge	TVA	1930	Toccoa	167	1000	55,000	—	1,500,000	1	20,000
Nottely	TVA	1942	Nottely	184	2300	11,500	21,700	1,552,300	1	15,000
Melton Hill	TVA	1963	Clinch	103	1020	122,000	246,800	0	2	72,000
Norris	TVA	1936	Clinch	265	1860	93,400	1,002,300	181,700	2	100,800
Tallico	TVA	1975	Little Tenn.	129	3238	135,000	78,000	1,883,000	—	—
Chilhowee	Alcoa	1957	Little Tenn.	91	1373	182,000	91,500	307,000	3	50,000
Calderwood	Alcoa	1930	Little Tenn.	232	916	260,000	—	0	3	121,500
Cheoah	Alcoa	1919	Little Tenn.	225	750	200,000	—	0	5	110,000
Fontana	TVA	1944	Little Tenn.	480	2365	134,300	2,815,500	760,600	3	225,000
Santeetlah	Alcoa	1928	Cheoah	212	1054	76,100	—	0	2	45,000
Nantahala	Alcoa	1942	Nantahala	250	1042	59,000	—	1,829,000	1	43,200
Thorpe	Alcoa	1941	Tuckasegee	150	900	56,000	—	1,060,000	1	21,600
Douglas	TVA	1943	French Broad	202	1705	342,000	556,400	127,900	4	113,500
Nolichucky	TVA	1913	Nolichucky	94	480	—	—	0	4	10,640
Cherokee	TVA	1941	Holston	175	6760	694,200	894,200	3,304,100	4	120,000
Fort Patrick Henry	TVA	1952	S Fork Holston	95	737	141,000	72,500	0	2	36,000
Boone	TVA	1952	S Fork Holston	160	1532	137,000	198,400	714,000	3	75,000
South Holston	TVA	1950	S Fork Holston	285	1600	116,200	97,500	5,897,400	1	35,000
Wilbur	TVA	1912	Watauga	77	375	34,000	—	0	4	10,700
Watauga	TVA	1948	Watauga	318	900	73,200	80,400	3,497,800	2	50,000
Pumped-Storage Project										
Raccoon Mt.	TVA	1974	Tennessee	230	8500	none	135,000	9,400,000	4	1,530,000
Cumberland River Basin										
Great Falls	TVA	1916	Caney Fork	92	800	150,000	—	0	2	31,860
Barkley	C of E	1963	Cumberland	157	10,180	570,000	1,258,400	3,335,600	4	130,000
Center Hill	C of E	1948	Caney Fork	250	2,160	349,600	995,000	3,609,900	3	135,000
Cheatham	C of E	1953	Cumberland	75	981	90,000	290,400	136,000	3	36,000
Cordell Hull	C of E	1967	Cumberland	93	1,306	155,000	338,800	280,600	3	100,000
Dale Hollow	C of E	1943	Obey	200	1,717	59,450	581,710	—	3	54,000
J. Percy Priest	C of E	1967	Stones	147	2,718	182,000	226,400	1,788,100	1	28,000
Laurel	C of E	—	Laurel	282	1420	86,000	32,500	3,225,800	1	61,000
Old Hickory	C of E	1954	Cumberland	98	3,750	200,000	460,800	451,400	4	100,000
Wolf Creek	C of E	1950	Cumberland	258	5,735	434,800	1,422,000	10,319,200	6	270,000

				Reservoir data and operating levels							
Area at full pool El., acres	Total volume below top of gates, acre-ft	Useful controlled storage, acre-ft	Length of shore line, miles	Back-water length, miles	Full pool (El.)	Top of gates (El.)	Min. expected pool level (El.)	Avg. tail-water level (El.)	Head, ft	Cost	Project
											Tennessee River Basin
160,300	6,129,000	4,008,000	2380	164.3	359	375	354	310	47	$117,984,000	Kentucky
43,100	1,105,000	417,000	496	52.7	414	418	408	362	50	45,605,000	Pickwick Landing
15,500	641,000	59,000	154	15.5	507	507	504.5	414	92	107,585,000	Wilson
67,100	1,071,000	351,000	1083	74.1	556	556	550	507	48	87,655,000	Wheeler
67,900	1,052,000	172,300	949	75.7	595	595	593	557	37	51,054,000	Guntersville
10,730	252,400	32,300	192	46.3	634	635	632	596	34	74,942,000	Nickajack
35,400	739,000	347,000	810	58.9	682	685	675	634	45	42,065,000	Chickamauga
39,000	1,175,000	379,000	783	72.4	741	745	735	682	56	36,065,000	Watts Bar
14,600	393,000	111,000	360	55	813	815	807	740	70	42,374,000	Fort Loudoun
10,700	617,000	323,000	246	34	888	895	860	752	138	50,900,000	Tims Ford
1,100	57,800	8,800	31	9.8	1290	1280	1272	840	380	24,051,000	Apalachia
6,090	434,000	362,200	180	22	1524	1526	1415	1275	254	24,440,000	Hiwassee
7,050	240,500	222,100	132	13	1927	1928	1860	1804	126	9,122,000	Chatuge
1,890	86,500	33,800	18	7.5	837	837	816	724	113	2,963,000	Ocoee No. 1
—	—	silted	—	—	1115	1115	—	843	252	3,035,000	Ocoee No. 2
621	4,040	3,770	24	7	1435	1435	1413	1119	313	8,997,000	Ocoee No. 3
3,290	195,900	183,900	60	10	1690	1691	1590	1543	147	5,507,000	Blue Ridge
4,180	174,300	161,600	106	20	1779	1780	1690	1612	174	8,081,000	Nottely
5,690	126,000	31,900	144	44	795	796	790	742	51	36,250,000	Melton Hill
34,200	2,549,000	2,260,000	800	72	1020	1034	930	826	196	33,368,000	Norris
16,500	447,300	126,000	310	33.2	813	815	807	740	70	69,000,000	Tellico
1,890	49,250	6,564	30	8.9	874	874	870	812	80	—	Chilhowee
536	41,180	1,570	—	8	1087	1087	1084	869	209	—	Calderwood
595	35,030	1,850	—	10	1276	1276	1273	1087	187	—	Cheoah
10,640	1,443,000	1,145,000	248	29	1708	1710	1525	1276	429	78,448,000	Fontana
2,863	158,250	133,300	85	7.5	1939	1939	1863	1275	597	—	Santeetlah
1,605	138,730	126,000	—	4.6	3012	3012	2881	2007	944	—	Nantahala
1,462	70,810	67,100	—	4.5	3491	3491	3415	2284	1200	—	Thorpe
30,400	1,475,000	1,394,700	555	43.1	1000	1002	920	873	129	47,029,000	Douglas
797	9,850	—	—	—	1245	1245	—	—	68	1,747,000	Nolichucky
30,300	1,544,000	1,460,000	463	59	1073	1075	980	925	149	36,805,000	Cherokee
872	26,900	4,200	37	10.3	1263	1263	1258	1195	75	12,289,000	Fort Patrick Henry
4,400	193,400	148,400	130	17.3	1385	1385	1330	1264	123	27,766,000	Boone
7,580	764,000	642,600	168	24.3	1729	1742	1616	1490	239	31,428,000	South Holston
72	—	—	3	1.7	1650	1650	1645	1585	62	2,503,000	Wilbur
6,430	677,000	624,700	106	16.7	1959	1975	1815	1850	309	32,590,000	Watauga
											Pumped-Storage Project
528	37,930	36,340	—	—	1672	—	1530	633	1040	155,000,000	Raccoon Mt.
											Cumberland River Basin
2,110	51,300	48,300	120	22	805	805	762	655	150	9,030,000	Great Falls
57,920	2,082,000	1,472,000	1417	118.1	359	375	354	312	44	143,750,000	Barkley
18,220	2,092,000	1,254,000	422	64	648	685	618	486	162	44,491,000	Center Hill
7,450	111,000	26,800	320	67.5	385	386	382	361	22	30,185,000	Cheatham
11,990	280,500	75,700	381	71.9	504	505	499	457	47	71,700,000	Cordell Hull
30,990	1,706,000	849,000	620	51	651	663	631	514	137	25,949,000	Dale Hollow
22,720	652,000	384,000	265	31.9	490	504	480	396	94	51,625,000	J. Percy Priest
6,060	435,000	185,000	206	19.2	1018	1018	1018	765	253	34,600,000	Laurel
22,500	467,000	110,000	440	97.3	445	447	442	398	45	48,684,000	Old Hickory
50,250	6,089,000	4,236,000	1255	101.3	723	750	573	561	137	79,082,500	Wolf Creek

of construction. That is, the expense of penstocks can be avoided by boring tunnels downward through the granite and lining them with concrete to give a smoothly finished surface. In fact, the casings for the turbines themselves can be made in the same way by carving appropriately shaped cavities in the solid rock. Another area favored with this type of site is the central portion of the west coast of India. Bombay derives a large fraction of its power from installations such as the Khapoli plant at the edge of the escarpment, which rises up about 300 m (1000 ft) some 50 km east of Bombay.

Strings of Power Plants

In some instances a series of dams can be located one after another along a river to give a system such as that built by Alcoa and TVA on the Little Tennessee River in eastern Tennessee and western North Carolina. The 146-m-high (480-ft-high) Fontana Dam in the headwaters provides a large storage capacity and a steady water flow into a series of four dams and powerhouses downstream which have heads of 20 to 60 m (67 to 200 ft).

DATA FOR TYPICAL HYDROELECTRIC SITES

At the time of writing there were over 1400 hydroelectric power plants in the United States. Data for a typical set of these is presented in Table 19.1 to give some useful insights into the principal parameters. This set, which is for the TVA-Alcoa hydroelectric power system, includes examples of almost every type of site discussed above. Most of the TVA units are multi-purpose dams, those on the main river being designed to provide navigation and flood control as well as hydroelectric power, while most of the tributary projects are designed to provide both flood control and hydroelectric power. Additional benefits include provision of adequate mainstream flow during the dry summer months to assure satisfactory water supplies to cities along the river system and a wealth of recreational opportunities for fishing, boating, and swimming. The Alcoa dams, on the other hand, were built primarily to provide electric power with little or no effort to provide flood control, although they provide fine recreational opportunities.

One of the most significant points implicit in the table is the fact that in no instance has it appeared appropriate to build a dam in the TVA system that yielded a head of less than 12 m (40 ft), and in only three instances is the power capacity less than 20 MW. The reasons of course were economic in spite of the fact that the dams have been built with federal funds, which have been available at low interest rates, and substantial portions of the cost of the dams have been written off against flood control or navigation. These and other points implicit in the table will be referred to frequently in the subsequent sections.

Storage Capacity

One of the major advantages of hydroelectric units is the rapidity with which they can pick up load. If a unit is kept on the line but is delivering essentially no power while operating in the condensing mode (to improve the power factor in the system), it can accept load in a matter of seconds, far more rapidly than any thermal unit. In addition, it is very important for the system to be able to pick up peak loads for a few hours a day. As a consequence, electric utility systems tend to have at least 10 percent of their power capacity in hydroelectric units. However, in order to take advantage of these characteristics of hydroelectric turbine generators, it is essential to have storage capacity in the dam. This should be at least enough to take care of diurnal load swings in power, and preferably sufficient to take care of weekly variations. In addition, it is highly desirable to be able to handle seasonal variations such as heavy air-conditioning loads in the summer and heavy heating and heat-pump electric loads in the winter. A convenient chart for relating the number of kilowatthours of electric power production obtainable to the water storage capacity of a reservoir is provided by Fig. 5-4. The enormous water storage capacity required to accommodate seasonal variations is indicated by taking data from Table 5-1 for the Norris power plant, which has a useful controlled storage of 2.7×10^8 m^3 (2.2×10^6 acre-ft). Neglecting the loss in head associated with dropping from full pool to the ordinary minimum level [i.e., from a lake elevation of 311 m (1020 ft) above sea level to 283 m (930 ft)], this storage capacity with a head of 61 m (200 ft) will give a total output of about 4×10^8 kWh. This would amount to about 4000 h at the full-power output of around 100 MW from the Norris turbines. Thus the Norris Dam serves to collect the heavy rainfall in the winter and spring months and has sufficient capacity to provide power through the summer and autumn months, when the river flow would otherwise be relatively small. (There are 8760 h in a year.) Note also in Table 5-1 that there is only one dam in the TVA system with a higher storage capacity than that of Norris, and only five of the dams of the system have useful controlled storage capacities in excess of 8×10^7 m^3 (650,000 acre-ft), i.e., the equivalent of 500 h at 100 MW with a head of 61 m (200 ft).

Neglecting the loss in head with reservoir drawdown, as was done above, involves much less of an error than one might at first think. Most reservoirs have volume characteristics similar to those of an inverted pyramid, so that the volume of water stored varies as the cube of the water height above the tail water. As a consequence, a 20 percent reduction in the level from full pool means that half of the volume at full pool will have been removed.

Geological Considerations

One of the important cost factors in constructing a dam, aside from the volume of material required, is the geological structure underlying both the dam site and the lake to be impounded. Extensive test borings must be made to assure a good foundation for the dam and freedom from cracks or permeable strata that might lead to leaks and a possible weakening of the foundation. For example, poor foundation conditions were a factor in the catastrophic failure of the Teton Dam in 1976. Several failures caused by poor foundation conditions have occurred, and this is one reason why many otherwise superficially attractive hydroelectric sites are not suitable for development. Aside from the possibility of catastrophic failure, porosity in the underlying rock strata can lead to serious leakage and a marked loss in the expected hydroelectric power capacity. One example of this sort of problem is given by the Aswan Dam: leakage through rock strata and losses to evaporation have reduced its hydroelectric power capacity by about one-third of the value originally expected.

Another major factor affecting the cost of the plant is the availability at the site of suitable material for construction. For a concrete dam this means the availability of a quarry for obtaining suitable aggregate and sand pits to provide the principal materials for the concrete. For an earthen dam or an earth-and-rock-filled dam, there must be an adequate supply of suitable earth and/or rock close to the dam site.

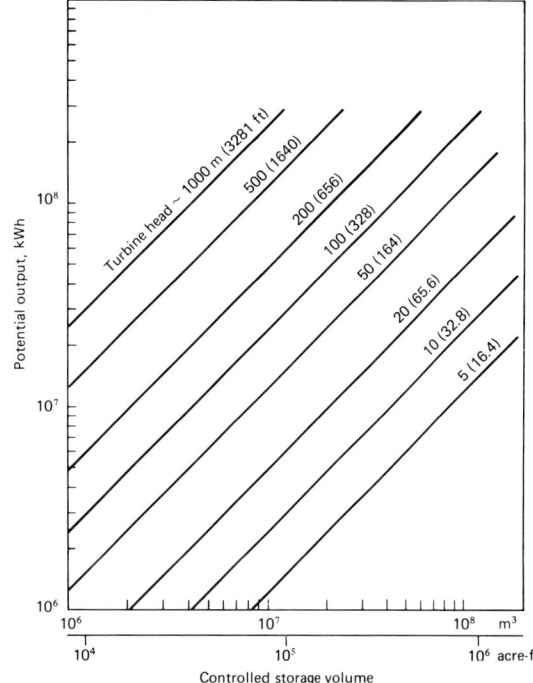

Figure 5-4 Potential electric energy output from a water storage reservoir for a 90 percent turbine-generator efficiency.

TYPES OF DAMS

The fifth and sixth columns under "Dams and Appurtenances" in Table 5-1 indicate the type of construction used in the TVA and Alcoa dams. One of the most common types is the concrete gravity dam with earth embankments. Note that this type of dam has a roughly triangular cross section with the broad base of the triangle resting on and keyed into bedrock so that it is held in place by its own weight. The construction is simple but the volume of concrete required per unit of length is large. The concrete volume required can be reduced by employing a concrete arch dam, the shoulders of which can rest against sound rock. The Calderwood and Hungry Horse dams are examples of this construction. The forces generated by the hydrostatic pressure of the impounded water act against the convex face of the dam and are carried in compression along the arc of the dam into the rock shoulders at either end. This type of structure can be used only in deep, narrow gorges with sound rock foundations on either side. The most common type is the earth dam—which may have several types of earth in the cross section—with rock riprap facing. The facing may be heavy, as in an earth-rock dam, such as the TVA Cherokee Dam, or the dam may be made largely of rock.

TYPES OF TURBINES

Four types of turbines are employed in hydroelectric power stations. For low heads—up to as much as 55 m (180 ft)—propeller turbines are employed, primarily because they give relatively high rotor velocities for relatively low water through-flow velocities. These are usually Kaplan turbines, i.e., propeller-type turbines with adjustable blades. The blade pitch can be varied to give a good efficiency over a wide range of loads,[5,6] a second and important reason for using them. For higher heads—30 m (100 ft) to as much as 300 m (1000 ft)—Francis turbines are employed. These are radial inflow units in which the water enters the rotor through a set of variable angle inlet guide vanes and flows radially inward and axially downward, with a substantial pressure drop taking place within the turbine wheel itself; i.e., the unit is a reaction turbine. The Deriaz turbine is a little-used intermediate variety with adjustable rotor vanes whose axes lie on the surface of a cone. For heads in excess of ~305 m (1000 ft), Pelton wheels are ordinarily employed. These are impulse turbines in which all the static head available in the turbine is converted into velocity in needle valve-controlled nozzles, and all this velocity energy is absorbed in the wheel so the water leaves at a negligible absolute velocity (Chap. 3). The ranges of application of the principal types of turbine are indicated graphically in Fig. 5-5.[7]

For electric power generation, the turbine speed must be kept constant to hold a constant output frequency. This poses problems for part-load operation. The changes in the angles with which the flow enters and leaves the turbine can be accommodated in the Kaplan and Francis turbines by varying either the rotor blade or the inlet guide vane angles, but this cannot be done in the Pelton wheel. As a consequence, the efficiency at part load suffers. Figure 5-6 shows typical curves for the efficiency of these three types of turbine as a function of load. Note that the Kaplan turbine with its very low pitch propeller blades yields the best performance over a wide range, the adjustable inlet guide vane Francis turbine is next best, and the Pelton wheels have a wider range for good performance than one might expect.

It is advantageous to increase the size of the turbine generator, partly because this reduces the hydrodynamic losses but mainly because it reduces the cost per kilowatt of the complete installation. However, increasing the size also increases the weight, the problem of handling the components for shipment and installation, and the bearing loads. Continuing improvements in design have made it possible to produce progressively larger units; Fig. 5-7 indicates the progress in this area.[8] Note that head is an important factor in making possible high out-

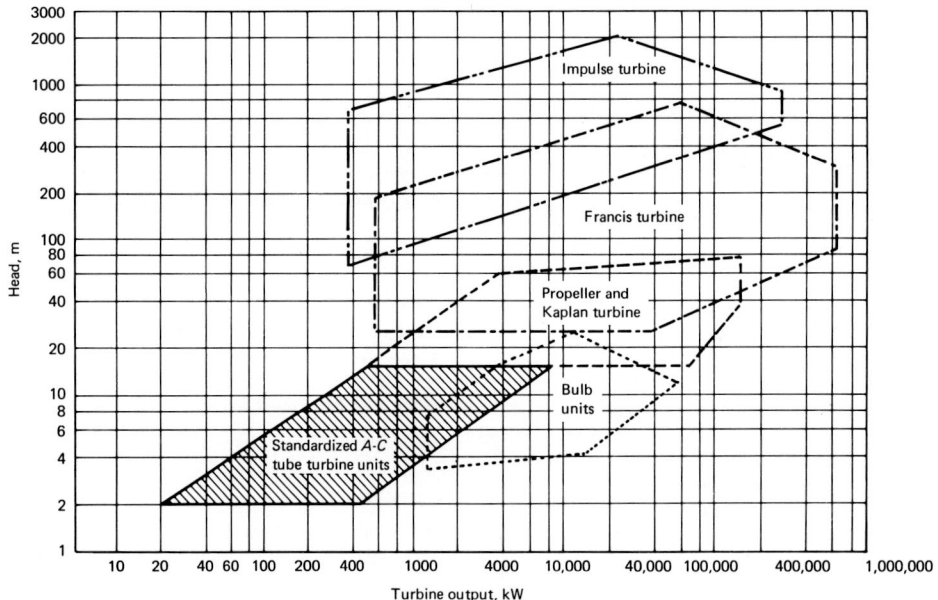

Figure 5-5 Ranges of application of the various types of hydraulic turbine units. Tube turbines and bulb units are specialized forms of propeller turbine in which the shaft is horizontal to reduce inlet and exit losses for low head sites. (*Courtesy Allis-Chalmers.*)

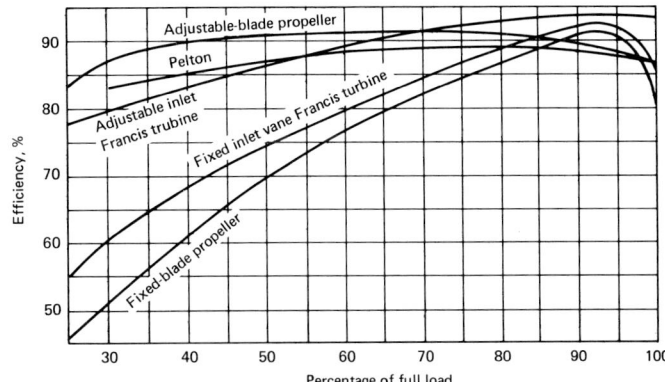

Figure 5-6 Effects of load on the efficiency of typical turbines.

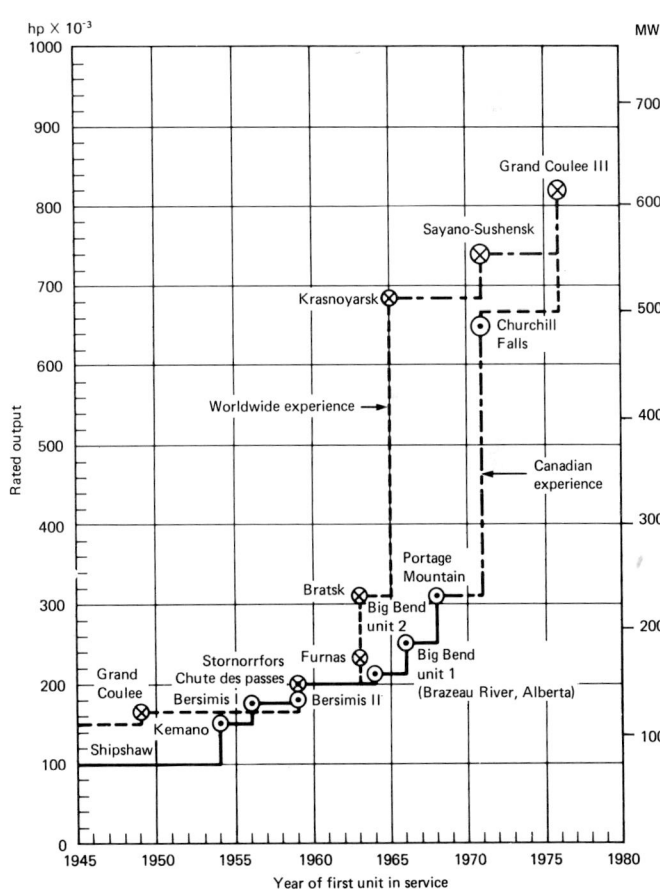

Figure 5-7 Increase in the maximum rating of hydraulic turbines in the 1945–1976 period.[8]

5–11

puts. The 600-MW Grand Coulee turbine in Fig. 5-7 was designed for a head of about 107 m (350 ft). Because of their inherently lower head, propeller-type turbines have been limited to outputs of about 100 MW, and, while they operate with very high heads, Pelton wheels have been limited to outputs of about 30 MW because of the restricted inlet nozzle area available in the wheels and the need to have a free exit for the water leaving the wheel.

To minimize fluid losses in the inlet and exit passages, propeller turbines are usually mounted within the dam itself. Francis turbines may be mounted in the dam or in a separate powerhouse. Figure 5-8 shows a typical installation for a Francis turbine. For low-head installations the inlet and exit losses can be reduced by inclining the shaft from the vertical. This is particularly true for propeller turbines, which are often mounted with the shaft horizontal.

To minimize exit losses, particularly in low-head installations, a diffuser or draft tube is commonly employed between the turbine wheel discharge and the tailrace. This improves the efficiency of the turbine by recovering much of the velocity energy at the turbine outlet, but is likely to reduce the static pressure at the turbine to a level below atmospheric. The reduction may lead to difficulties with air in-leakage and possibly to cavitation. The latter has proved to be a problem in some of the higher-speed Kaplan turbine installations. Difficulties with cavitation can be avoided by installing the turbine at a level well below the level of the tailrace so that the turbine is operating under a substantial static head. This approach has the disadvantage of preventing easy draining of the turbine during maintenance. The turbine can be readily drained if it is mounted above the level of the tailrace, one of the original advantages to the use of a draft tube.

HYDROELECTRIC PLANT COSTS

As mentioned at several points in this chapter, the cost of hydroelectric plants varies widely, depending on the site. Some indication of this is given by Table 5-1, which summarizes data for the TVA and Alcoa hydroelectric plants. Much additional data can be obtained from tables published by the Federal Energy Regulatory Commission (FERC). Summaries of data for four typical plants are presented in Table 5-2. Note the wide variation in the cost of both land and the structure, mainly the dam. However, a reasonably consistent pattern emerges with respect to the cost of the turbine-generator unit and its associated equipment. A plot of cost data for turbine-generator units built for the Bureau of Reclamation was presented in Fig. 3-8, which showed that the principal factor affecting the cost of the turbine-generator is the head. The cost of the complete in-

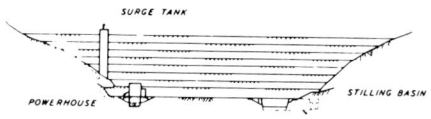

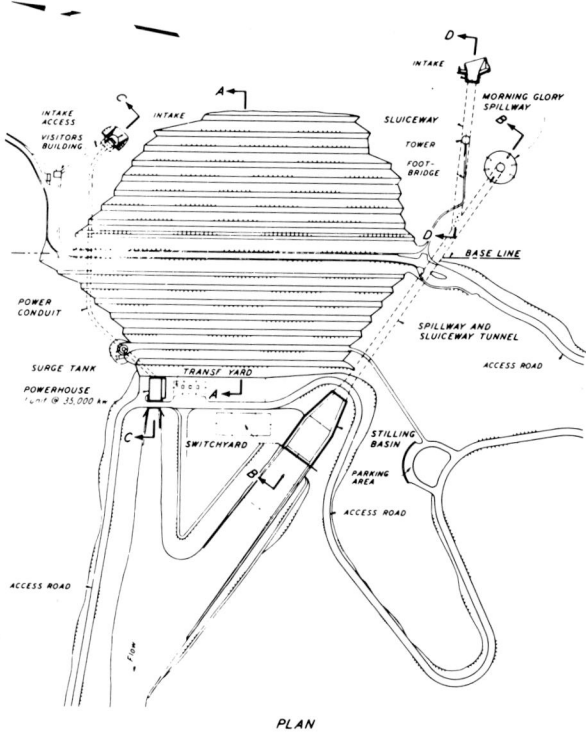

Figure 5-8 General plant elevations and sections through a typical hydroelectric power plant, the TVA South Holston project (*Courtesy TVA.*)

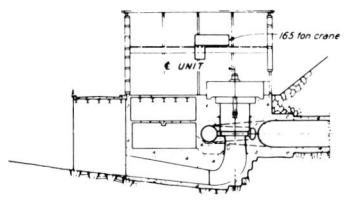

SECTION THRU UNIT

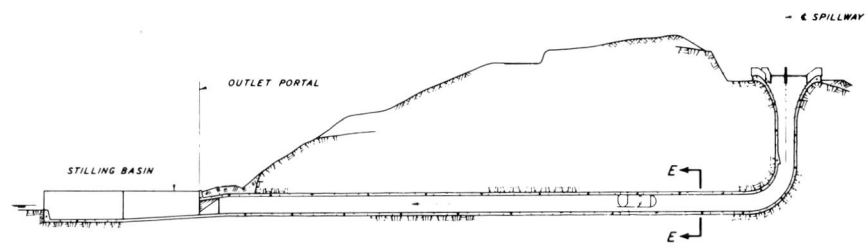

SECTION A-A

SECTION B-B

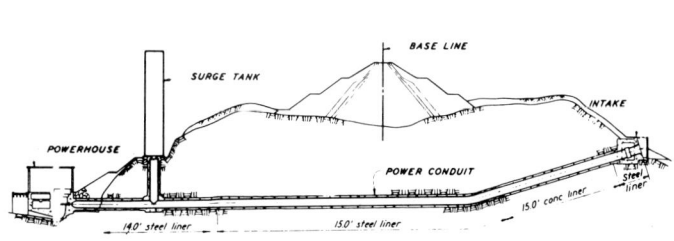

SECTION C-C

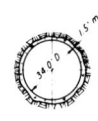

SECTION E-E

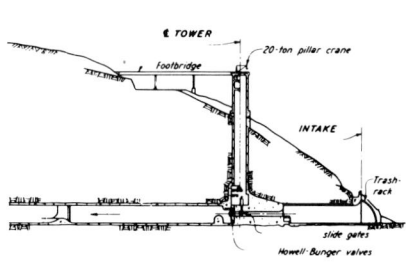

SECTION D-D

SITE PLAN

5-13

TABLE 5-2 Hydroelectric Plant Construction Cost and Annual Production Expenses, 1973[12]

Line No.		Name of Utility	U. S. DEPARTMENT OF THE ARMY, CORPS OF ENGINEERS [1/]							
		Name of Plant Project	Albeni Falls		Dworshak [3/]		Barkley		Wolf Creek	
		Post Office	Newport, Wash.		North Fork, Idaho		Grand River, Ky.		Jamestown, Ky.	
		County and State	Bonner, Idaho		Idaho		Lyon, Ky.		Russell, Ky.	
		River	Pend Oreille		N. F. Clearwater		Cumberland		Cumberland	
		Region and Power Supply Area	VII-41		VII-41		III-20		III-20	
		Licensed Project No.	—		—		—		—	
1		Installed Generating Capacity—Generator Nameplate - MW	42.6		400.0		130.0		271.2	
2		Pumping energy, Million kWh (Pumped storage)	—		—		—		—	
3		Net Generation, Million kWh	228.1		163.1		889.9		1,271.1	
4		Plant Factor, Percent, Based on Nameplate Rating	61		—		78		54	
5		Net Peak Demand on Plant, MW (60 Minutes)	43.0		NR		164.0		344.0	
6		Capability Under Most Favorable Operating Conditions - MW	49.0		460.0		166.0		300.0	
7		Capability Under Most Adverse Operating Conditions - MW	0		38.5		0		180.0	
8		Hours Connected to Load	7,998		2,998		8,597		8,447	
9		Planned Ultimate Generating Capacity - MW	42.6		—		130.0		271.2	
10		COST OF PLANT (Thousands of Dollars)								
11		Land and Land Rights	3,421		30,421		5,163		11,113	
12		Structures and Improvements	11,966		18,574		18,835		9,690	
13		Reservoirs, Dams, and Waterways	7,795		206,326		2,932		24,911	
14		Equipment Costs	8,733		20,077		18,588		13,552	
15		Roads, Railroads, and Bridges	—		5,954		38		682	
16										
17		Total Cost	31,915 2/		281,352 2/		45,556 2/		59,849 2/	
18		Cost per kW, Installed Capacity (Nameplate) $	749		703		350		222	
19		PRODUCTION EXPENSES	$1000	Mills kWh	$1000	Mills kWh	$1000	Mills kWh	$1000	Mills kWh
20		Operation Supervision and Engineering	39	.17	7	.04	32	.04	36	.03
21		Water for Power	—	—	—	—	—	—	—	—
22		Hydraulic Expenses	26	.11	—	—	2	—	—	—
23		Electric Expenses	116	.51	28	.17	123	.14	122	.10
24		Misc. Hydraulic Power Generation Expenses	2	.01	—	—	14	.02	26	.02
25		Rents								
26		Joint Operating Expenses (Allocated)	92	.40	107	.66	19	.02	88	.07
27		Maintenance Supervision and Engineering	29	.13	—	—	39	.04	40	.03
28		Maintenance of Structures	91	.40	—	—	38	.04	6	—
29		Maintenence of Reservoirs, Dams, and Waterways	—	—	—	—	—	—	—	—
30		Maintenance of Electric Plant	40	.18	—	—	74	.08	177	.14
31		Maintenance of Misc. Hydraulic Plant	20	.09	—	—	6	.01	8	.01
32		Joint Maintenance Expenses (Allocated)	133	.58	81	.50	4	—	5	—
33		Total Production Expenses	588	2.58	223	1.37	351	.39	508	.40
34		Production Expenses per kW (Nameplate)	13.80		.56		2.70		1.87	
35		Cost of Pumping Energy (Pumped Storage) $	—		—		—		—	

TABLE 5-2 Hydroelectric Plant Construction Cost and Annual Production Expenses, 1973[12] (*Continued*)

Line No.	Name of Utility	U. S. DEPARTMENT OF THE ARMY, CORPS OF ENGINEERS 1/			
	Name of Plant Project	Albeni Falls	Dworshak 3/	Barkley	Wolf Creek
	Post Office	Newport, Wash.	North Fork, Idaho	Grand River, Ky.	Jamestown, Ky.
	County and State	Bonner, Idaho	Idaho	Lyon, Ky.	Russell, Ky.
	River	Pend Oreille	N. F. Clearwater	Cumberland	Cumberland
	Region and Power Supply Area	VII-41	VII-41	III-20	III-20
	Licensed Project No.	—	—	—	—
36	Average Number of Employees	15	12	17	22
37	Type of Operation	Manual		Manual	Manual
38	Initial Year of Plant Operation	1955	1973	1966	1952
39	**HYDRAULIC DATA**				
40	Drainage Area, Square Miles	24,200		17,598	5,789
41	Area of Pond at Normal Full Pond Level, Acres	94,000		57,920	56,250
42	Storage or Pondage from Maximum Draw-down, Acre-Feet	1,153,000		258,900	2,142,000
43	Gross Head (Pond Elevation Minus Tailwater Elevation), ft	28.3		46	163
44	(CFS per kW)—Full Station Load	NR		0.298	0.0815
45	**GENERAL DATA—FOR PLANTS ADDED IN 1973**				
46	Kind of Development				
47	Type of Dam				
48	☐ Pumped Storage				
49	Type of Powerhouse Construction:	☐ Conventional	☐ Outdoor	☐ Semi-Outdoor	
50	Project Use:	☐ Single Purpose	☐ Multiple Purpose		

EQUIPMENT CHARACTERISTICS—FOR PLANTS OR UNITS ADDED IN 1973										
WATERWHEELS						GENERATION				
No. Units	Design Head (Feet)	Max. H P* (1,000)	Type (Hor. or Vert.)	Type** Runner	Year Installation	No. Units	Nameplate Rating (MW)	P F %	Voltage k V	Year Installed
2	560	142.0	Vert.	F	1973	2	90.0		13.8	1973
1	560	346.0	Vert.	F	1973	1	220.0		13.8	1973

3/ New Plant-Commercial Operation: 1973

*At Design Head;   **F-Francis;   FP-Fixed Propeller;   AP-Automatic Adjusted Propeller;   I-Impulse.

NOTES: 2/ Includes tentative multiple purpose plant allocation.

Fiscal Year Ending: 1/   June 30th.

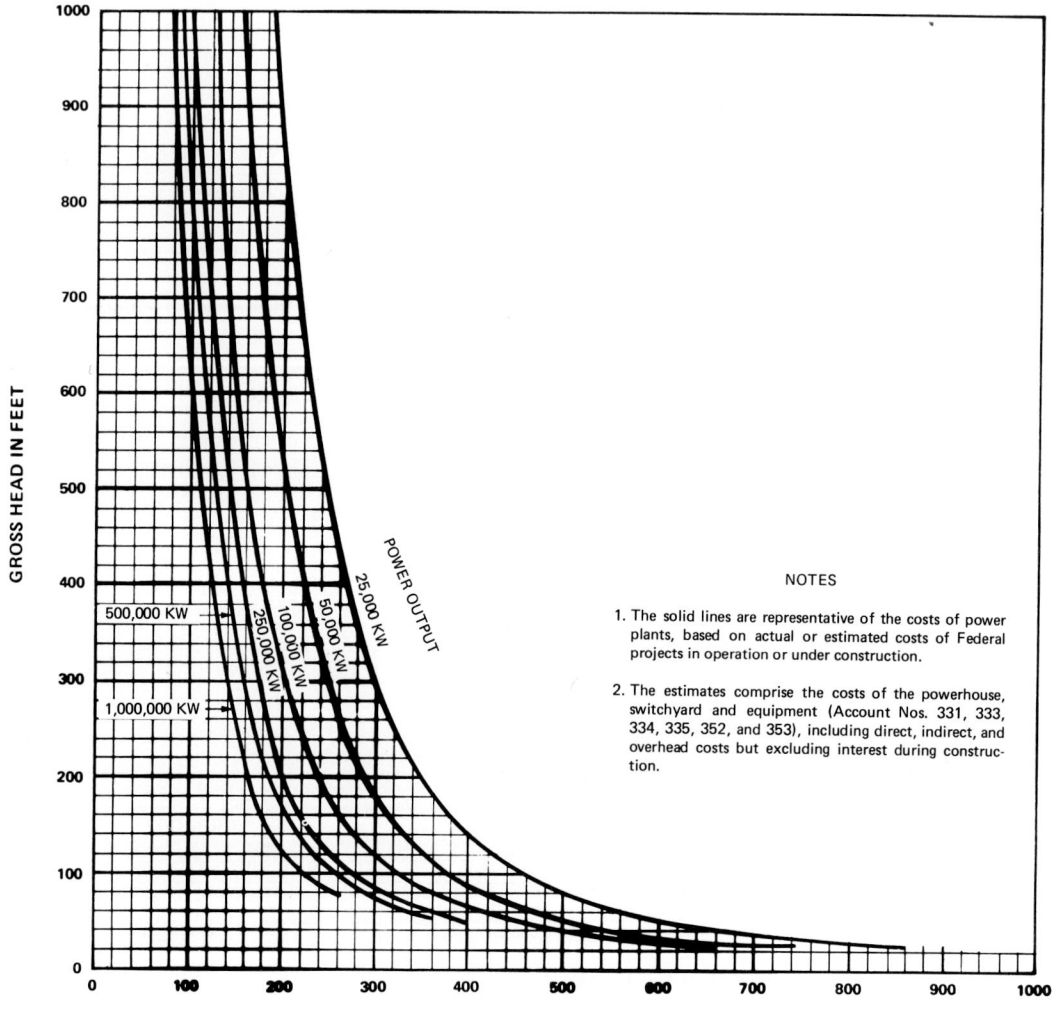

Figure 5-9 Federal Power Commission guide for estimating the cost of hydroelectric power plants (excluding the dam and reservoir) as of 1978.[1] *(Courtesy Federal Energy Regulatory Commission.)*

stallation of the turbine-generator unit with its associated control equipment and electric switchgear goes up less rapidly than linearly with power output; hence the cost per kilowatt of installed capacity falls off somewhat at a constant head with an increase in capacity. This effect is shown in Fig. 5-9.

UTILIZATION OF SITES FOR SMALL HYDROELECTRIC UNITS

The drastic increase in the cost of petroleum, natural gas, and coal following the 1973 Yom Kippur War has led to a reexamination of the possibility of utilizing small hydroelectric sites. In a study of this problem submitted to the President by the Corps of Engineers,[4] it was concluded that the most promising approach would be to install turbine-generator units in relatively small dams that had been abandoned because they were not competitive costwise with thermal power plants while the supplies of fossil fuels could be obtained inexpensively. The findings are summarized in Table 5-3, which indicates that rehabilitating existing hydroelectric dam sites could provide an additional hydroelectric power capacity of about 5000 MW, expansion of the capacity of existing dams in service could provide an additional 16,000 MW, the potential at existing nonhydroelectric dams (for irrigation, city water supplies, etc.) might be as much as 7000 MW, and the potential at existing nonhydro dams capable of less than 5000 kW of output might total as much as 26,000 MW. The total hydroelectric potential thus obtainable amounts to almost as much as that already developed and in service. Most of this would be in the form of low-head, low-power-capacity units, and hence the turbine-generator and associated equipment would inherently have a relatively high unit cost (Figs. 3-8 and 5-9). Further, the cost of maintenance and of coupling these units into a grid would be relatively high. The question is whether these high unit cost components would still yield an attractive system in view of the fact that the cost of the dam and associated structures should be low. In addition, other problems arise that could entail large legal and oher costs. One example cited was the Occoquan reservoir in northern Virginia, which supplies 220,000 m³ per day (58 × 10⁶ gal per day) of water to the Washington metropolitan area, yielding $12 × 10⁶ per year in gross revenues. If this dam were rebuilt to increase the hydraulic head available by a factor of 4 so that it would yield a capacity of 17,500 kW, it could produce 47 × 10⁶ kWh of electricity per year, which would have a sales value of $1.8 × 10⁶. Clearly the existing $12 × 10⁶ per year in revenues from supplying water to the urban market is roughly seven times the revenues that could be obtained from electric power production after a large new investment both in improving the dam and installing the hydroelectric unit. Further, there would be a serious conflict regarding the operation of the dam for water supply purposes as opposed to operation for hydroelectric power production. There is also the question as to whether the reliability of the urban water supply would be compromised by less than perfect reliability of the turbine-generator system.

The problems are much more complex than they appear on the surface. Certainly some of the small hydroelectric sites that have been abandoned because they were uneconomic can be rehabilitated to give economically attractive power, and this is being done in some cases. However, the sixfold increase in fuel costs from 1973 to 1979 has been accompanied by nearly as great an increase in capital costs (Fig. 3-14), so that the economics of hydroelectric plants relative to thermal plants has not changed as much as one might first think. Further, it seems clear that the construction of numerous new low-head dams on small streams does not offer a promising source of electric power at a competitive cost. A specific example that illustrates the problem quite well is given by an article that appeared in *The New York Times* in May 1978 giving an enthusiastic report on a small hydroelectric turbine-generator unit being marketed by a French concern. A 5-kW unit selling for $7800 was designed to operate on a head of about 3 m (10 ft). This gives a capital cost of $1500/kW, to which must be added the cost of its installation. It is difficult to estimate operating costs; certainly a substantial amount of maintenance would be required just to keep the trash racks at the turbine inlet clear of debris. Such a system would make a nice hobby for someone so inclined, but would probably not be attractive to most people.

A typical case in which changing economics have justified the redevelopment of an old hydro plant that had fallen into disuse is given by the Cornell plant in Michigan. In this instance the dam originally supplied power to a paper pulp mill, with 10 of the 12 turbines used for direct drives of pulp mills while two were used for generating electricity. The dam, which yields a head of 11 m (36 ft), has been modified to take three hydroelectric turbines of 10^4 kW each at a cost of $500/kW. Contracts for the bulk of the equipment were let in 1972, and the installation

TABLE 5-3 Conventional Hydroelectric Capacity (Constructed and Potential) at Existing Dams[4]

	Capacity, millions of kW	Generation, billions of kWh
Developed	57.0	271.0
Under construction	8.2	16.8
Total installed	65.2	287.8
Potential rehabilitation of existing hydro dams	5.1	24.4
Potential expansion of existing hydro dams	15.9	29.8
Potential at existing nonhydro dams greater than 5000 kW	7.0	20.4
Potential at existing nonhydro dams less than 5000 kW	26.6	84.7
Total potential	54.6	159.3
TOTAL (developed and undeveloped)	119.8	447.1

TABLE 5-4   Principal Cost Items for the Installation of a New 1-MW Turbine-Generator Unit in a Typical Existing Dam Giving a Head of 15 m (49 ft) (Ref. 9)

	Custom	Percent of total	Standardized	Percent of total
Equipment	$ 530,000	44	$450,000	45
Engineering	180,000	15	85,000	9
Civil Construction	500,000	41	460,000	46
Total	$1,210,000	100	$995,000	100

was completed in 1977, hence the $500/kW cost is probably in 1973 or 1974 dollars.[8]

Costs of Standardized versus Custom-Designed Installations

The major cost components in low-head, low-power hydro installations in existing dams are equipment, engineering, and construction in the field. All three items can be minimized by employing standardized turbine-generator units that include the controls.[7] Table 5-4 compares these costs for a typical case, first for a custom-designed, and secondly for a standardized, turbine-generator package unit.[7]

Estimating the Flow Potential of a Site

In attempting to estimate the hydroelectric potential of a site on a stream whose flow has not been well monitored, one can make a good estimate of the flow from the area of the drainage basin upstream of the site and the annual rainfall. A chart that makes it easier to carry out such estimates for the New England area is shown in Fig. 5-10. Note that the same chart can be used for other areas by applying a factor for the difference in rainfall between the other area and New England. If this is done, an additional factor to allow for differences in evaporation losses between the New England area and the area in question may be necessary. These evaporation losses may be quite high; they represent about two-thirds of the total rainfall in the TVA area, for example.

TIDAL POWER

Tidal energy has been used since the eleventh century to drive waterwheels, and "tidal mills" were fairly common along the French and English coasts up to the early twentieth century. The possibility of using the ebb and flow of tides to drive hydroelectric units has been an intriguing subject of discussion since 1890, and seems again to be a live issue at the time of writing. The range of sea level variation varies widely from relatively small values of about 1 m in regions near the equator to values of as much as 20 m along the northern coast of North America and Europe. There are normally two high and two low tides a day, depending on the phase of the moon. Maximum tides, called the *spring* tides, occur when the sun and the moon are both on the same side of the earth, while the *neap* tides occur when the sun and moon are at 90 degrees with respect to the earth. The neap tides are commonly about half as great as the spring tides.

The characteristics and problems of a tidal power station are well illustrated by examining the experience with the tidal power station built in the estuary of the river La Rance in Brittany, France, completed in 1966.[12] This plant has an installed capacity of 240 MW and produces power about 6 h per day, utilizing both the inflow and the outflow of water with changes in the tide. Figure 5-11 shows the variation in the sea level for a typical day, together with the variation in level of the water in the tidal basin behind the dam across the mouth of the estuary. In this instance, the turbines are operated as pumps for a short period, when the level in the basin and sea level are the same, to accelerate filling or emptying the reservoir. In that way they derive some extra power during the period when the difference in level is great enough to obtain useful power from the tidal flow. The turbines are large—their diameter is 5.35 m—and 24 units are required. These are mounted in the central part of the 750-m-long dam, in which the power station occupies about half of the length.

The principal characteristics of the power plant are summarized in Table 5-5, together with the corresponding data for the TVA Nickajack Dam, which was completed about the same time.[13] Note that the peak difference in water level in spring tides for the La Rance dam is about the same as the head available from the Nickajack dam, and the length of both dams is about the same. The turbines are designed for a lower head, and the total installed power capacity is more than $2\frac{1}{2}$ times as great in the LaRance power plant. As a consequence, it is not surprising that the cost of the La Rance power plant was about three times that of the TVA Nickajack Dam and powerhouse. In a sense, both projects are multipurpose, because the Nickajack dam also provides improved navigation and some flood control capability, while the La Rance dam serves as a bridge across the mouth of the estuary and is heavily traveled. Note that in the table two values for the La Rance power level are given: 60 and 240 MW. This is because the turbines develop power for only 6 h per day; the equivalent output on a sustained basis is only one-

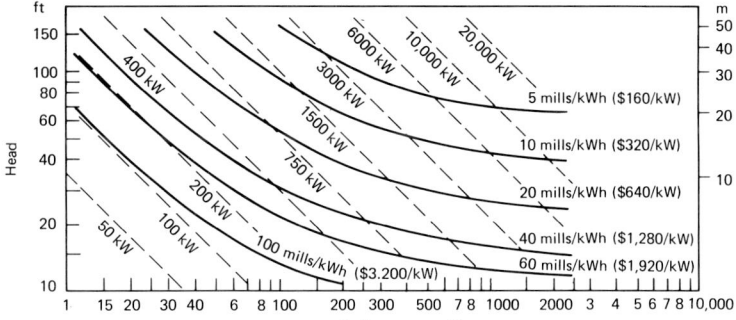

Figure 5-10 Typical capacities and minimal capital charges for small hydroelectric plants as estimated for rivers in the northeastern United States.[10]

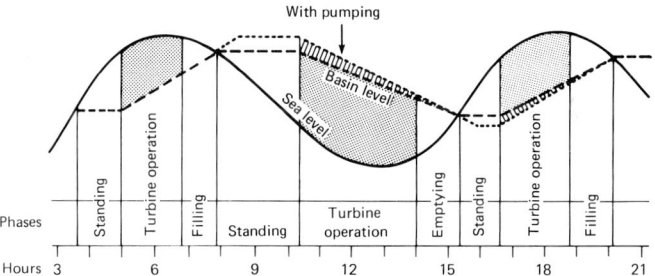

Figure 5-11 Typical operating regimes of the pump-turbines at La Rance. Note the additional head that can be gained by pumping, an option sometimes taken at off-peak times.[11]

TABLE 5-5 Comparison of Tidal Power Projects with a Conventional Low-Head Riverine Hydroelectric Project Including Actual Costs

Type of project	Riverine	Tidal	Tidal
Name of project	TVA—Nickajack[11]	La Rance[10]	Passamaquoddy[12]
Location	Tennessee River Tenn., U.S.A.	La Rance Estuary Brittany, France	Passamaquoddy Bay Maine, U.S.A.
Year of completion	1968	1966	
Design power output, MW	96	240 (60*)	166 (80*)
Dam length, m	1135	750	3660
Dam maximum height above foundation, m	24.7		45.7
Area of pool, km²	43.5	22	96
Operating head, m	13.6	5.5	3
Mean range for spring tides, m		13.5	7
Project cost, $	42,100,000	120,000,000†	
Cost per kilowatt of average output, $/kW	343‡	2000†	

*Output averaged over 24 h.
†Cost does not include the cost of the site survey and design or credit for value of the dam as a bridge.
‡Cost does not include credit for flood control and navigation.

quarter of the peak output, or about 60 MW. Although the total water flow available at Nickajack is not sufficient to permit full-power output throughout the year, the availability of the power capacity at any time is worth a great deal. Thus the 60-MW value for La Rance is reasonable for comparative purposes. Using this 60-MW value yields a project cost of about $2000/kW for the La Rance project, as opposed to about $435/kW for the Nickajack dam, if no allowance is made in either case for the benefits of the other uses of the dam. This large difference in cost is not surprising in view of the much greater turbine-generator capacity required for a tidal project. Further, many proposed projects, such as Passamaquoddy, entail exposure of dams to ocean storms, and the much more rugged construction required of such a dam would entail much higher costs.

In the United States the most promising site for a tidal power project has been the Passamaquoddy Bay in Maine. The project was originally proposed about 1920, and quite thorough design studies were carried out in the depression of the 1930s with the expectation that it would be constructed as a means of providing work for over 40,000 unemployed in the state of Maine.[13] Cost estimates proved to be so high that the project was shelved. The reason for this can be appreciated by examining the last column in Table 5-5, in which the design data for the Passamaquoddy project have been summarized. Note that the head available would have been about half that for the La Rance project, the effective output would have been only a little greater than for the La Rance plant, the dam length would have been about five times as great, and some of that would have been exposed to wave damage in ocean storms. No data were available for the maximum height above the foundation for the La Rance dam, but note that this parameter for Passamaquoddy is 45.6 m (150 ft), while the corresponding value for the Nickajack dam is 24.7 m (81 ft). Thus the volume of material per unit of length would be four times as great for Passamaquoddy as for Nickajack, and hence one would expect the cost of the three-times-longer Passamaquoddy dam to run at least 12 times that for the Nickajack dam. The average head on the turbines when operating would be less than one-fourth as great, and they would operate only about half the time; hence their capital cost would run roughly eight times as much. Further, the availability of any tidal power unit is a function of the timing of the tides, and since this would commonly not be in phase with the load demand on the power system, the value of the power capacity would be reduced by an additional factor. Thus it appears that electric power from the most attractive tidal power project in the United States would cost roughly 10 times as much as that from a typical hydroelectric installation on a river. There would be no other benefits, such as improved navigation or flood control, and there might be serious adverse ecological effects on estuarine marine life.

REFERENCES

1. *Hydroelectric Power Evaluation,* Federal Energy Regulatory Commission, DOE/FERC-0031, August 1979.

2. Keator, F. W.: "Benoit Fourneyron (1802–1867)," *Mechanical Engineering,* vol. 61, no. 4, 1939, pp. 295–301.

3. *Hydroelectric Power Systems,* Energy Research and Development Administration, Argonne National Laboratory, Report No. 107, 1977.

4. *Estimate of National Hydroelectric Power Potential at Existing Dams,* U.S. Army Corps of Engineers, Institute for Water Resources, July 20, 1977.

5. Hackert, H.: "Victor Kaplan: His Life and Work," *Water Power and Dam Construction,* November 1976, p. 39.

6. Terry, R. V.: "Development of Automatic Adjustable-Blade-Type Propeller Turbine," *Trans. ASME,* vol. 63, 1941, p. 395.

7. Haydock, J. L., and J. G. Warnock: "Towards 2,000,000 Horsepower for Giant Turbines," *Energy International,* vol. 7, no. 4, April 1970, p. 28.

8. Eberhardt, A.: "Cornell Hydro Plant Redevelopment," *Water Power and Dam Construction,* June 1976, p. 35.

9. Mayo, H. A., Jr.: "Modern Low Head Hydro Equipment," *1980 Joint Power Generation Conference,* Sept. 30, 1980.

10. O'Brien, E.: "Small Hydroplants for the Northeast," *Electrical World,* Aug. 15, 1977, p. 61.

11. "Ten Years of Tidal Power," *Water Power and Dam Construction,* December 1976, p. 55.

12. *Hydroelectric Plant Construction Cost and Annual Production Expenses,* Federal Power Commission, FPC S-256, 17th Annual Supplement, 1973, May 1976.

13. Casey, H. J.: "The Passamaquoddy Tidal-Power Project," *Mechanical Engineering,* vol. 57, 1935, p. 580.

Section **6**

Geothermal, Solar, Wind, Wave, and Ocean Thermal Difference Energy Systems

Nearly everyone in the world likes to get "something for nothing." And the users of power are no different from the rest of humanity. So there is an ongoing interest and desire to get power for nothing—wherever possible.

Free energy *is* available from the sun, the wind, and the ocean. The key question, of course, is the relative cost of this "free" energy when harnessed for productive use. Further, such "free" energy is often available at the wrong time in the wrong place. So its utilization can require a large capital investment for both storage and transmission of the energy.

Power-plant designers must be ready and able to analyze the costs of these alternative "free" energy sources. The reason for this is that people in charge of a project may be convinced that the "free" energy route is the way to go for a particular plant or load. While such may be true, "free" electricity may cost from two to twenty times as much as fossil-fuel generated power. When shown such relative costs, the proponents of "free" energy may revise their views and choose conventional power sources. But the designer must be able to prove the relative costs. This section helps in that task.

From *Engineering Evaluation of Energy Systems,* by Arthur P. Fraas. Copyright © 1982. Used by permission of McGraw-Hill, Inc. All rights reserved.

People are always eager to get something for nothing, and hence are much intrigued by the prospect of getting free energy from the sun or the wind or ocean waves. However, all these are low-head systems comparable to low-head hydroelectric systems, and thus would entail high capital investments. Often, as is the case with energy in tides as discussed in Section 5, these forms of energy are available in the wrong place at the wrong time hence would require large investments both for storage and for transmission of the electric produced to areas in which it could be used to advantage. There is no question but that electric energy can be produced from these various "natural" energy sources. The question is one of cost. If the cost of electricity is to run 2 to 20 times that of electricity from fossil fuel, it seems highly doubtful that most people will prefer expensive electricity from a source of free, low-grade energy. It is significant that efforts have been under way for many years to produce electricity from the "free" energy sources, but only two of these sources—geothermal and "solar biomass"—are currently yielding commercial electric power anywhere in the world. Only a few plants are presently utilizing this energy, and they are close to electric load centers.

Capital costs are the heart of the problem in each case. Although they are difficult to estimate in a definitive fashion, it is possible to get some significant insights by examining the proposed systems and estimating the costs of major components that are essentially similar to those in widespread commercial use, and this approach is followed here.

GEOTHERMAL ENERGY

There were five commercial power plants producing geothermal energy in 1980. The first of these—that at Larderello in Italy—began operation in 1904; that in New Zealand in 1957; the Geysers near San Francisco, California, in 1960; the 150-MWe Cerro Prieto plant in Baja California in 1972; and the most recent—the one in El Salvador—in 1975.[1-4] The fields in Italy and New Zealand appear to be limited in scope, and so these plants have not been expanded in recent years. The Geysers plant in California, however, has been steadily increased in size to a capacity of 900 MW at the time of writing, and further expansion to around 1500 MW is under way.[5] The new fields in El Salvador and Baja California also appear capable of providing for further expansion. The principal characteristics of these plants are summarized in Table 6-1.

Types of Geothermal Heat Sources

All the geothermal heat sources of interest have been created by the intrusion of hot magma from deep in the earth up into rock strata close to the surface.[6] Geologically recent intrusions of this sort have been responsible for extensive volcanic activity in the region stretching up through California, Oregon, and Washington. Note that the most recent volcanic eruptions in this area occurred between 1914 and 1921 at Mount Lassen, about 200 km north of Sacramento, and at Mount St. Helens in Washington in 1980. (The Geysers power plant is about 75 km west-northwest of Sacramento.) The thermal conductivity of rock is low and the depth of these masses of magma is usually of the order of many kilometers, and so the time required for cooling is of the order of 1 million years. A good idea of the location of these potential sources of heat is provided by extensive measurement of the temperature gradient as a function of depth as measured in wells drilled in efforts to find oil and gas deposits. These temperature measurements yield the data from which the vertical thermal gradient as a function of depth can be computed. The U.S. Coast and Geodetic Service has organized these data to provide the map shown in Fig. 6-1, which gives geothermal temperature gradient contours as determined from the average temperature gradient to a depth of 6 km.[7] The regions of interest for power production are those in which the temperature gradient exceeds about 20°C/km; these regions have been cross-hatched in Fig. 6-1.

TABLE 6-1 Engineering Data for Commercial Geothermal Power Plants Operating in 1979
(Data from Refs. 1-4)

	Larderello	Wairakei	Geysers	Ahuachapan	Cerro Prieto
Country	Italy	New Zealand	U.S.A.	El Salvador	Mexico
Year of initial operation	1904	1957	1960	1975	1972
Normal peak power output, MWe	400	180	908	60	150
Total installed power capacity, MWe	440	240	908	60	150
Steam temperature, °C	140–190	260 max.	172–240	250	167
Steam pressure, bars	7–40	12, 4.6, 1.14	6.5–7.5	14.6	5
Type of geothermal source	Dry steam	Steam @ 20% quality	Dry steam	Dry steam	Hot water
Number of wells					40
Well depth, m	<1000	171–1220	1200–3000		1100–1400
Well bore, cm	34	20			18 (liner)
Size of well field, km			4 × 11		
Weight fraction noncondensables, %	5–30	0.26	0.54		

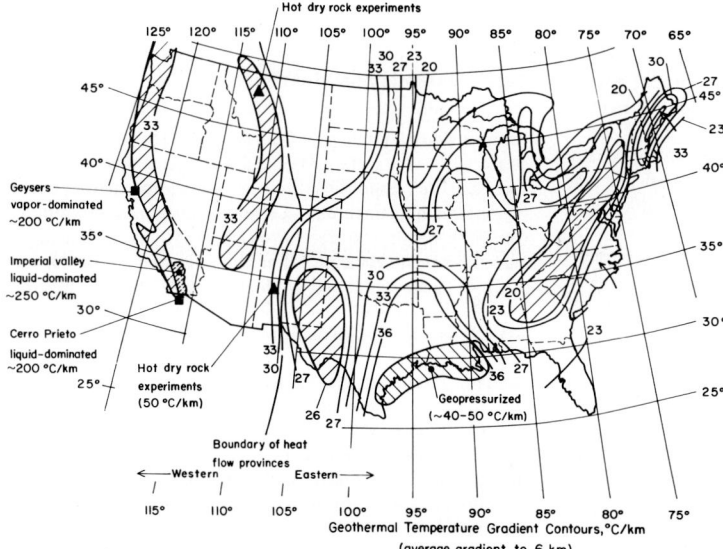

Figure 6-1 Map of the United States showing lines of constant average geothermal gradient.[6] *(Courtesy Oak Ridge National Laboratory.)*

The problems of extracting heat from a geothermal heat source vary widely with the particular type of rock strata in the area. In a few regions, porous rock is overlain by a low permeability stratum and above that an aquifer which allows water to trickle into the hot porous rock at a rate such that a steady flow of dry steam is generated. It is this type of formation that has been tapped at Larderello, the Geysers, and in El Salvador. A similar formation exists under the power plant in New Zealand except that there the aquifer feeds in an excess of water, so that only about 20 percent by weight of the steam-water mixture emerging from the wells is in the form of steam and the balance is in the form of superheated water. Favorable rock formations of this sort are rare, which is the reason that there are so few commercial geothermal power plants.

A second major geothermal source is the hot brines in porous strata. These brines sometimes contain as much as 30 percent salt by weight and hence are both highly corrosive and difficult to handle, because salt precipitates out rapidly as heat is removed from them. The geothermal heat sources in the Imperial Valley area of California are of this type. The hot water in the nearby Baja California field, on the other hand, has a low salt content. The geopressured hot-brine deposits found along the Texas-Louisiana gulf coast are not at as high a temperature as the brines in the Imperial Valley, but they contain a substantial amount of dissolved methane. In fact, the first well drilled into these brines to explore their potential as an energy resource disclosed that the amount of methane obtained from the hot brine emerging from the well was roughly double the amount that would have been obtained from a brine saturated with methane. Thus it appears that there are pockets of methane gas in these deposits, so that small bubbles of methane are entrained in the brine drawn from the strata. Analyses indicate that neither the thermal energy in the brine nor the contained methane would be sufficient by itself to make commercial exploitation of these deposits economically attractive, but by exploiting both the thermal energy and the methane, the economics may prove to be favorable. Unfortunately, the rock has been under such great pressure that its compressibility is low; hence only ~ 4 or 5 percent of the fluid is recoverable.[8]

A third type of geothermal heat source is the pools of molten magma beneath active volcanoes, such as those in Hawaii and Central America. No good way of exploiting this type of geothermal source has been proposed.

The fourth and largest type of geothermal heat source is hot, dry rock, and a substantial amount of effort was under way at the time of writing to develop means for exploiting it. The approach being followed is to drill into hot, dry rock in regions with a high thermal gradient and then hydrofracture the rock by pumping water in at pressures of the order of 200 atm.[6] This produces a system of vertical cracks radiating from the bore of the well. In principle, wells could be drilled in a checkerboard pattern and water injected into alternate wells to yield hot water and/or steam flowing out of the intermediate wells. Extensive analytical efforts indicate that percolation of water from the feed wells through the cracks in the rock could yield attractive rates of steam formation and collection from the steam wells. As the rock is cooled, further fracturing should occur as a consequence of thermal stresses, and so the yield of steam from a given set of wells may actually improve. On the other hand, the water flow through some cracks may tend to open them up to a greater degree than others, so that channeling will occur and the effective heat-transfer surface will be greatly reduced. Extensive and expensive tests will be required to determine what can actually be accomplished.

The extent of these various types of geothermal resources in the United States is indicated by Table 6-2,[5] which gives the amount of thermal energy available in terms of quads (10^{15} Btu) and in terms of units of 10^{18} J (virtually the same quantity of energy).[5] As shown in Table 6-13, the U.S. energy consumption in 1977 totalled about 76 quads per year of which about 10 percent was in the form of electricity. It must be remembered that the thermal efficiency in converting relatively low-temperature geothermal heat energy into electricity is likely to be around 10 percent, and the U.S. consumption of electricity in 1978, if derived from geothermal heat, would have required 75 quads of heat. Thus, the estimates of Table 6-2 indicate that, if economical ways to exploit them completely can be found, the geohydrothermal resources of the United States might provide our electric power needs at the 1978 level of consumption for roughly 160 years, and the geopressurized resources might serve for an equal period if one allows for the low recovery rate of ~ 5 percent that appears likely. However, the hot, dry rock resource represents an amount of energy roughly 1000 times that available from the geohydrothermal deposits.

Geothermal Wells

One hundred years of experience in drilling oil and gas wells have produced a mature technology for drilling and a good basis for estimating the costs of wells. The generally used drilling technique employs a drill bit consisting of three toothed rotors set 120 degrees apart, each tangent to the bore of the hole being drilled. These are made of very hard, wear-resistant material. The assembly is rotated at slow speed in a bath of "mud," which is a thick suspension of clay particles. The mud acts as a lubricant and coolant as well as the vehicle for removing the debris from the drilling. A heavy axial load of about 10,000 kg imposed on the drill bit acts to crush the rock under the points of contact

TABLE 6-2   Estimated U.S. Geothermal Resources[5]

Type of source	Thermal energy	
	Quads (10^{15} Btu)	10^{18} J
Geohydrothermal	12,000	12,600
Geopressurized	190,000	200,000
Magma	200,000	210,000
Hot, dry rock	13,000,000	13,700,000

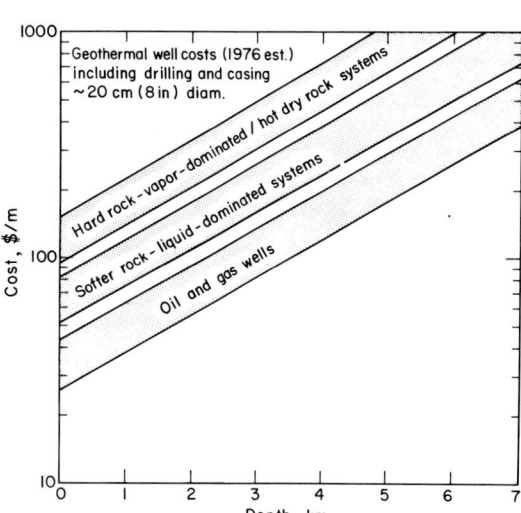

Figure 6-2 Estimated costs for drilling deep wells.[6] *(Courtesy Oak Ridge National Laboratory.)*

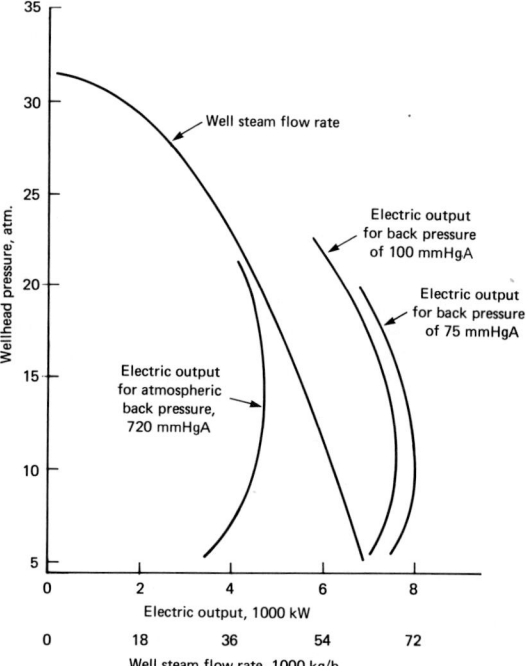

Figure 6-3 Relation of wellhead pressure to steam flow rate and electric output.[3]

between the teeth of the rotors and the rock. The penetration rate depends on the type of rock but is commonly of the order of 3 to 7 m/h in the sedimentary rock strata encountered in oil and gas well drilling. After being drilled, the well is commonly lined with a casing whose OD is 3.8 cm (1.5 in) less than the bore of the well to facilitate its insertion. For deep wells a larger diameter is used for the upper portion and the diameter is stepped down at depth.

Experience in drilling wells in the Geysers area has shown that the conventional technique for lubricating and cooling the drill bit with mud is not satisfactory for geothermal wells because the high temperature leads to both caking of the mud and clogging of the porous wall of the well so that the effectiveness of the well for producing steam is severely reduced.[9] This has necessitated the use of compressed air for blowing out the debris from the drilling operation. When the drill begins to enter the steam stratum, the mass of high-temperature steam escaping up the bore of the hole is substantial so that relatively little cooling can be accomplished with this compressed air. The combination of high temperatures and lack of lubrication leads to a drastically shortened life for the drill bits, and this greatly increases the expense of drilling the wells.

The cost of drilling deep wells in the search for new deposits of oil and gas is so great that new techniques for drilling have been investigated. One of these makes use of tiny water jets employing pressures of the order of 1500 to 3000 atm.[10,11] These jets enter cleavage planes between the rock crystals and act to break crystallites loose from the matrix. Rapid cutting action in porous rock has been obtained, but penetration rates are low for the hard, dense rocks, such as basalt, encountered in drilling geothermal wells. To date there has been no commercial application of this drilling technique. Another technique under investigation is designed to penetrate hard rock with a piercing tool having a heated nose made of a material, such as tungsten or molybdenum, that can withstand a high temperature and melt its way through the rock.[12] While intriguing, the results of analyses and experiments have not served to demonstrate the value of the system for penetrating hard rock. The heated piercing tool has also been suggested for use in drilling through salt formations, but these pose another problem; namely, the plastic character of the salt tends to close the hole and crush the casing. This problem stems from the fact that, aside from the nominal compressive stresses, bending stresses are also induced in the casing as a consequence of deviations from perfect circularity. Bending stresses are likely to lead to the phenomenon of creep buckling of cylindrical shells under external pressure. A much thicker and hence more expensive casing is therefore required.

Cost and Output of Wells

A good indication of the cost of wells as a function of depth is given in Fig. 6-2, taken from Ref. 6. Note the much higher cost associated with drilling through hard, porous rock permeated by hot steam. It must be remembered that the permeability of the rock is not high, so that, in order to get a large yield from a well, it is necessary to drill into the hot-rock region quite some distance to get sufficient wall surface area to yield a high steam flow rate.

An important factor limiting the energy obtainable from a well is the pressure drop in the steam or hot water flowing to the surface from the region tapped. An indication of the importance of this effect is given by Fig. 6-3, which was obtained for a well in the Geysers area. In this instance the most economical operating condition is at a high steam flow rate that yields a wellhead pressure only about one-third of that for zero flow. This loss stems from the pressure drop within the porous rock along with the pressure drop within the casing bringing the steam to the surface. The cost data plotted in Fig. 6-2 were for wells having diameters in the range of 15 to 30 cm, in part because drill rigs are not equipped for drilling larger wells and in part because the limitations on the flow rate imposed by rock permeability ordinarily limit the output of the well to a value within the capacity of a 15- to 30-cm bore.

The limitations on well output imposed by the pressure drop for the flow to the surface are indicated by Fig. 6-4 for some typical water and steam conditions for three sizes of casing. The electric output obtainable from the heat energy in the hot water or steam is indicated by Fig. 6-5. These charts can be used for the rough estimation of the useful output of a well if the pressure drops through the porous rock strata and the surface piping are small.

The value of a well is also a function of its operating life. This is determinable only from operation in each particular type of rock strata. In New Zealand, for example, the output of the wells fell off within the first year to a value not much more than half of that expected from the original test wells, but apparently the output has stabilized at that reduced value. It remains to be seen what the useful life of these wells will prove to be. In the Geysers area the average life span of a well appears to be about 8 years. Thus it is evident that the cost of the well must be written off over a period of the order of 10 years.[13] The relatively short life of the wells as compared to the operating life of the power plant indicates that the power plant should be situated within a reasonable radius of a number of wells. In any case, the steam obtained is not free: The capital charges are high and in the Geysers wells have led to a 1976 price for the steam at the wellhead equivalent to about 7 mills per kilowatthour.[13]

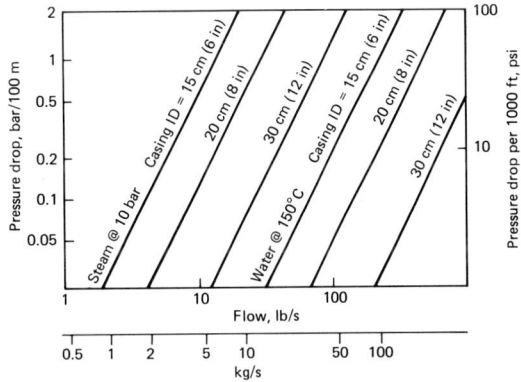

Figure 6-4 Chart for rough estimates of pressure gradients in geothermal wells. (For rough approximations, the pressure drop for a given flow varies inversely as the square root of the steam pressure, the electric output obtainable from steam at 10 bars is ~ 800 kW per 1.0 kg/s, and the electric output obtainable from 180° C water is ~120 kW per 1.0 kg/s. See Fig. 6-5.)

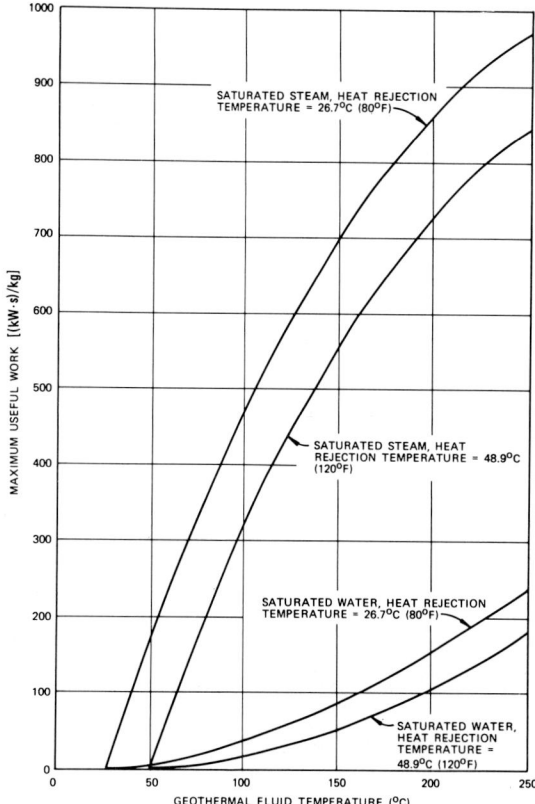

Figure 6-5 Maximum useful work or availability plotted as a function of geothermal fluid temperature for saturated-steam and saturated-water sources.[6] *(Courtesy Oak Ridge National Laboratory.)*

Thermodynamic Cycles

For wells such as those at the Geysers, the obvious course is to use the steam directly in the turbine after filtering it to remove particulates. In New Zealand, where 80 percent by weight of the fluid emerging from the wells is water, or in hot-brine fields such as that in the Imperial Valley, it is advantageous to consider more complex systems. Steam can be obtained from superheated water by flashing it in one or more stages, with some of the sensible heat in the water employed to vaporize a portion of the water in each stage. In the New Zealand plant this has led to a system in which steam is fed to the turbine at three pressures: 12 bars for the steam that is drawn from the well, 4.6 bars for the steam from the first-stage flash boiler, and 1.14 bars for the steam from the second-stage flash boiler. The thermal efficiency with which the heat can be utilized from each of these three pressure levels, of course, drops with the pressure, thus making the performance of the plant particularly sensitive to the temperature of the cooling water available for the condenser.

The high mineral content of hot brine makes it advantageous to have a heat exchanger between the brine and the thermodynamic working fluid. Further, in order to reduce the size of the turbine for these low-temperature vapor conditions, it may be advantageous to employ a working fluid with a higher vapor pressure than water, e.g., a Freon or an organic fluid. Systems of this sort are discussed in Section 4. The principal additional problem for hot-brine systems is the tendency toward heavy scaling and severe corrosion in the boiler. It has been suggested that the scaling problem could be handled by passing the hot brine through a liquid fluidized bed containing the boiler heat-transfer surfaces so that the particles in the bed will scrub off any scale that forms. A preliminary test indicates that this approach can be effective.[14]

Special Problems

Problems peculiar to geothermal power plants include many different types of corrosion stemming from impurities in the hot fluid, air pollution by noncondensables such as H_2S and NH_3, and surface water pollution by materials such as bromine and boron that may be present in the condensate from a dry steam well. An indication of the seriousness of these problems is that 20 percent of the electric output of the Larderello plant in Italy must be used for scavenging noncondensable gases from the condensers, and about 7 percent of the cost for the steam in the Geysers field is for disposing of the condensate by reinjecting it into the field because of the substantial boron and bromine contents. These problems are all very much site-dependent so will not be discussed further here.

SOLAR-ENERGY UTILIZATION

An enormous amount of energy flows to the earth from the sun: a total of over 10^{11} MW around the clock, or 1.39 kW/m² of area normal to the sun's rays.[15] (Eccentricity in the earth's orbit causes the latter value to vary from 1.35 to 1.44 kW/m².) A substantial fraction of the energy incident on the earth is absorbed in the earth's atmosphere or reflected by clouds or particulates so that the energy reaching the earth's surface on a clear day when the sun is directly overhead is commonly 0.9 kW/m². This energy serves to heat the earth and keep its temperature high enough to make it habitable. A small fraction of the solar energy incident on foliage induces chemical reactions that yield stored chemical energy in biomass, which can be burned to produce electricity. This commonly represents ~2 percent (in a few plants as much as 5 percent) of the incident light energy. At first glance it appears that some sort of system devised by humans ought to yield a higher conversion efficiency for the production of electricity. There are, however, practical problems, which are treated in this section.

Although some people include wind and waves and even tides and hydroelectric power units under the heading of solar energy, for purposes of this section only three basic types of energy are classified as solar energy: biomass, direct thermal conversion of solar energy into electric energy by a thermodynamic cycle, and direct conversion of solar energy into electric energy by photovoltaic cells. Of these three, the only one that has been and is currently being used to produce electricity commercially is biomass; a substantial amount of wood is being employed as fuel in steam plants.

Biomass

Although less than 1 percent of the wood produced by the U.S. forest products industry is employed as fuel, nearly half of the forest products harvested in the world are employed as fuel, largely in the underdeveloped countries.[16] The reason for this disparity is that harvesting the timber is a labor-intensive operation, and so the cost of the product is too high for it to be competitive with other fuels unless labor costs are very low. Even then, commercial operations on a substantial scale require tree farming, i.e., the planting of trees of highly productive species in uniform rows on relatively level land so that they can be harvested by cutting over large areas at a time. In traveling thousands of miles in rural areas in India and Brazil, the writer has seen numerous tree farms dedicated to the growth of firewood. Fast-growing evergreen species are employed: banyan trees in India and eucalyptus trees in Brazil. It is interesting to note that in Minas Gerais, Brazil, over 400,000 ha (10^6 acres) are producing ~ 10^7 tons of wood per year for fuel, much of it for making charcoal to smelt iron.

In the United States, wood has been getting some attention recently as a possible fuel for use by electric utilities. One example of a U.S. electric utility fueled by wood is a small, 10-MW, coal-burning plant of the Burlington Electric Department in Vermont.[17] In this instance the wood is harvested in the wake of timbering operations by removing the culls—stunted or otherwise undesirable trees not suitable for lumber. (As is evident in Fig. 6-6, the heating value of wood per unit weight is independent of the specie of tree; it depends only on the moisture content.[18] The density of wood varies widely; e.g., a cord of light wood such as poplar has a heating value only about half that for dense woods such as hickory or oak.) The trees harvested in this way are reduced to chips, hauled to the mill, and burned with oil, the proportions depending on the supply of wood chips. Inasmuch as wood chips are bulky, they are expensive to haul and store. Hence they ordinarily are brought in from a radius of not more than 80 km (50 mi), but usually less than 50 km (30 mi), and are burned within a day or two of their arrival at the plant so that the high cost of large storage bins can be avoided.[19]

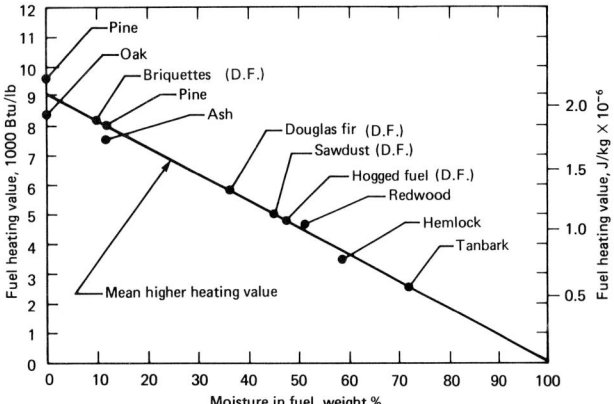

Figure 6-6 Higher heating value of wood fuel.[18]

TABLE 6-3 U.S. Commercial Forest Production and Its Energy Equivalent (Data from Refs. 16 and 20)

U.S. forestland area, ha	305×10^6 (750×10^6 acres)
U.S. forestland suitable for timber, ha	202×10^6 (500×10^6 acres)
U.S. forest industry wood production, tonnes/year (63% lumber, 35% fiber products)	226×10^6 (250×10^6 tons)
Wood waste fuel used by paper industry, J	0.96×10^{12} (0.91×10^{15} Btu)
Energy equivalent of U.S. 1977 forest products, J (assuming HHV = 5000 Btu/lb for green wood)	2.6×10^{18} (2.5×10^{15} Btu)
Total U.S. energy consumption in 1977, J	80×10^{18} (76×10^{15} Btu)

Forest Resources of the United States

In view of Burlington Electric's successful use of wood fuel obtained by following up timber clear-cutting operations, a survey of the U.S. forest products industry from the energy standpoint appears in order. Table 6-3 was compiled to indicate the amount of energy potentially available. If all the trees cut for timber and paper in the United States in 1977 had been employed instead as fuel, the energy available would have amounted to only about 3 percent of the total U.S. energy consumption. This utilization, of course, would not have been practical, because both lumber and paper represent higher value uses than fuel. Note that even the low-grade wood used for pulp was bringing ~$28 per cord, or about ~$20 per ton in Tennessee in 1980. This amounts to about $2/$10^9$ J ($2/$10^6$ Btu). It is interesting to note that although the paper industry is well integrated, with the paper mill facilities including huge tracts of land on which much of their own pulpwood is grown and harvested, the economics are such that only about 30 percent of the fuel used in the paper mills is in the form of wood waste; the balance is either coal or fuel oil. While it might be argued that a much larger fraction of the fuel could be obtained from the small branches and even the leaves of the trees that are cut, it is common practice to leave this material at the site to provide nutrients and organic matter for the soil. In fact, to improve the production rate from the forest areas, commercial fertilizer is commonly applied, usually by dispersal from an airplane.

It is often suggested that dead and wind- or lightning-damaged trees in forested areas could be culled and used for fuel. However, even when these areas are readily accessible by truck, it is difficult for a skilled lumberman to take out more than about one cord per day (128 ft³ of firewood or 160 ft³ of pulpwood). A cord of pulpwood weighs about 1.5 tons, hence even if labor were paid the minimum wage and there were no charges for overhead or transportation, the fuel cost per 10^6 Btu would be about double that for coal. For such operations it is not practicable to bring in heavy equipment because this not only damages the standing timber but also deeply ruts the forest floor and is likely to lead to serious gullying. Finally, it is evident from Table 6-3 that, at most, this approach would yield only a small fraction of the U.S. energy requirements.

Agricultural and Municipal Solid Wastes

Other forms of biomass have been suggested as fuel. A crop such as hay, for example, might be grown expressly for fuel. Again, a higher use of the land, i.e., for agricultural production, is clearly in order, particularly in view of the world food shortage. Agricultural wastes such as cornstalks might be employed, but again they have an important use as fertilizer and to renew the organic material in the soil. Even if all the agricultural and forest wastes together with all the municipal solid wastes were employed, estimates indicate that the total would at most be equivalent to about 7 percent of the 1976 U.S. energy consumption. (Ref. 10 of Section 4).

Harvesting biomass from the sea offers interesting possibilities. To reduce harvesting costs, extensive sea farms have been suggested, possibly including some deliberate arrangements to bring nutrients from deep waters to the surface to promote the rapid growth of kelp. (See the last section of this chapter.) After harvesting, the kelp would be sun-dried and then could be used as fuel. A number of experiments to investigate the possibilities of this approach are under way at the time of writing; results of this work should give a good indication as to its commercial feasibility. The ecological effects of massive harvesting of kelp will be more difficult to determine.

Conversion to Methane or Alcohol Fuels

Conversion of organic wastes such as manure into methane by anaerobic fermentation has received some attention.[21] This is technically feasible but expensive, and the heating value of the methane obtained is only about 60 percent that of the dried cattle dung widely used as fuel in India and other underdeveloped countries.

Conversion of crops such as grain or sugar cane to ethyl alcohol has been a popular possibility for many years. The technology is well established and the concept is strongly supported by political pressure groups. However, the costs are high and the energy efficiency is poor.

The basic cost of the raw material—a food crop—is inherently high as compared to the cost of fossil fuel. Even more important, the energy conversion efficiency is poor. Extensive studies have shown that the lowest cost and the highest energy efficiency are obtained with sugar cane; grains, nuts, cassava, etc., give a less favorable ratio of the energy available in the alcohol produced to the energy required for their production (tractor fuel, insecticides, refining, etc.).[21-24] Inasmuch as virtually all the water must be removed if the alcohol is to be mixed with gasoline, it is necessary to follow the distillation process by treatment with a dessicant such as quicklime. Thus the energy required is greater than that for conventional alcohol distillation. (This step may not be necessary if the alcohol is emulsified in diesel fuel.) If sugar cane is used as the source material, the energy for refining can be obtained by burning the bagasse (the cane residue). In fact, more steam is available from burning the bagasse than is required for the refining operations, and it is possible to take advantage of this energy by establishing an energy balance for the system.

Table 6-4 summarizes data for the most energy-efficient system that could be used in the United States, i.e., sugar cane production in Louisiana.[24] Three cases that differ in the degree of utilization of the energy in the bagasse are covered. For case 1, credit is taken for all the energy obtainable from burning the bagasse to generate steam, case 2 credits the bagasse just with the energy required for operation of the refinery, while case 3 makes no use of the bagasse as fuel. From the energy standpoint the best yield is for the case 1 conditions, but even for this most favorable case the energy equivalent of roughly half of the alcohol produced would have to go back into the agricultural operations.

The capital charge for the land use is high. Using the cane yield of 53 tons per hectare per year and the net energy production of 16.3×10^6 kcal per hectare per year of Table 6-4 and a value of $1000 per acre for the agricultural land gives a capital charge of $\sim \$6/10^6$ Btu for the land. In addition, there would be capital charges for the agricultural equipment, trucks, etc. The capital cost of the refinery is also high; e.g., according to Ref. 22, in 1978 the cost of a refinery for producing 10^5 L per day of ethanol from sugar cane was around $\$10 \times 10^6$. This represents a capital charge of $\sim 5\textcent/L$, or $\sim \$2.00/10^6$ Btu for the gross output. (Both of these figures should be roughly doubled if they are to be put in terms of the net useful output.) Although the cost of the refinery per unit output could be reduced by going to a greater capacity (petroleum refineries commonly have capacities ~ 100 times greater), this is not practicable, because the costs of

TABLE 6-4 Energy Balance of Alcohol Production[24]

Quantity	Sub-total	Total	Net energy balance	Output-input
Agricultural yield, [ton/(ha · year)]		53		
Alcohol production, L				
Per ton		66		
Per hectare per year		3498		
Energy expended [$\times 10^6$ kcal/(ha · year)]				
Agricultural	8.5			
Industrial structure	0.4			
Industrial fuel	10.4	19.3		
Energy produced [$\times 10^6$ kcal/(ha · year)]				
Case 1				
Converting all bagasse to steam	17.2			
Content of alcohol produced	18.4	35.6	+16.3	1.8:1
Case 2				
Converting enough bagasse to meet all industrial requirements	10.4			
Content of alcohol produced	18.4	28.8	+9.5	1.5:1
Case 3				
Burning fossil fuel to meet all industrial requirements	0			
Content of alcohol produced	18.4	18.4	−0.9	0.9:1

hauling sugar cane from the field to the refinery become excessive if a larger cane field area is served.[22] Unfortunately, this limitation also affects the value of the excess steam that can be produced from the bagasse. If employed for generating electricity, as assumed in Refs. 22 and 24, the steam-electric plant will be small (~ 7 MWe), and hence both its operating costs and its efficiency will be much less favorable than would be the case for a large plant.

In the world as a whole, the production of alcohol for fuel has occurred principally in Brazil, where the addition of alcohol to gasoline has been required by law since the 1940s. More favorable growing conditions and a more labor-intensive economy have yielded a higher value than indicated by Table 6-4 for the energy output-input ratio,[23] i.e., ~ 2.4:1 as compared to the 1.8:1 of Table 6-4.

A recent Brazilian study by one of the authors of Ref. 22 indicates that, by burning the bagasse in a high-pressure, high-temperature steam generator and taking a more favorable credit for the steam generated, Brazil may be able to achieve an energy output-input ratio as high as 3.6:1 for sugar cane. The same study indicates that methanol might be produced from wood to give an even more favorable hydrocarbon fuel energy output-input ratio of ~ 10. No details were given in the paper, but this probably assumed that most of the harvesting would be done by manual labor and draft animals. It was stated that the capital cost of the refinery and the gross energy output per hectare should be about the same as for ethanol production from sugar cane. The yield of methanol from the destructive distillation of wood is not high (~ 25 kg/t of wood), but most methanol is synthesized by the reaction $CO + 2H_2 \rightarrow CH_3OH$, using a catalyst and operating at ~ 200 atm and ~ 350°C. (This process is under development for making methanol from coal.) If wood is used as the source material, it was claimed that a methanol yield of 300 kg/t of wood can be obtained.[25] Thus the chemical energy in the hydrocarbon fuel produced would be about half that in the wood supplied, a value close to the upper limit theoretically obtainable.

A significant strategic consideration with respect to small-scale ethanol production in rural areas is that if U.S. farms were equipped to generate alcohol from agricultural wastes, they would not be vulnerable to loss of their fuel supply in the event of a major war, and this would be of enormous strategic value.

It can be argued that costs should be ignored, in which case the question becomes one of the land area required. Assuming the use of sugar cane and the data of Table 6-4 as the point of departure, a net production of 16.3×10^6 kcal per hectare per year in the form of ethanol and prime steam from combustion of the bagasse could be obtained for the most favorable combination of conditions for energy production from biomass in the United States. To meet a U.S. annual energy demand of 80 quads per year would require 12×10^6 km². This is about 10 times the total acreage under cultivation in the United States, or about 100 times the area of Louisiana. It has been suggested that plants might be developed to yield a higher efficiency for photosynthesis, but existing plants represent the culmination of over 3×10^9 years of evolution; hence substantial further improvements seem highly unlikely. Further, this energy would not be from a "renewable" resource: U.S. Department of Agriculture figures for soil erosion rates indicate that 7 kg of topsoil would be lost for each kilogram of alcohol produced. This represents a more serious disadvantage than the erosion associated with the unregulated strip mining of coal, and the world food shortage presents humanitarian problems that preclude large-scale diversion of agricultural land to fuel production.

Solar-Thermal-Energy Conversion Systems

Solar heating systems for domestic hot water were fairly common in Florida in the 1920s, and some of these systems are still in use. The installation of new systems was brought to a halt when natural gas became generally available, because this gave a much lower system cost. The use of solar energy for heating houses has been under investigation at M.I.T. since about 1930

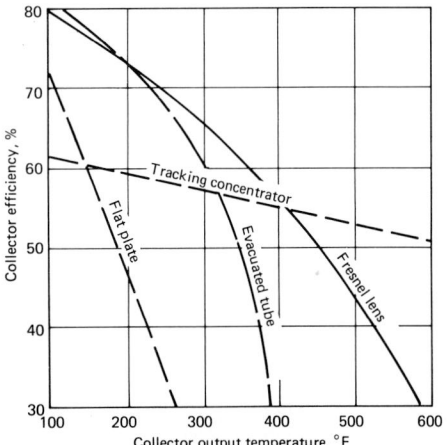

Figure 6-7 Estimated performance of several types of solar collector. Ambient temperature = 21°C (70°F). Solar intensity = 948 W/m² [300 Btu/(h · ft²)], all available as direct.[28] *(Courtesy Barber-Nichols Engineering Co.)*

with funds from a substantial endowment. The M.I.T. work has included the building and testing of a number of single-family houses.[26] Solar heating systems for swimming pools became fairly common in California in the 1970s, extending the swimming season by a month or two in both the spring and the fall. Note that in the domestic hot-water and swimming-pool systems it is easy and inexpensive to provide a large amount of heat capacity relative to the heat loss rate so that diurnal variations in the solar-energy flux do not represent a serious problem. It is for this reason that the most promising application of solar energy is generally considered to be the heating of domestic hot water and buildings. A huge federally funded program of subsidies was under way at the time of writing to encourage these applications of solar energy.

The use of solar energy to generate steam and produce power is also not new; the first power plant of record was built in Egypt in 1912 to pump irrigation water.[27] This plant employed parabolic trough reflectors to concentrate sunlight on boiler tubes. The steam generated was used to drive a 15-hp engine, the largest amount of power to be developed with a solar power plant for the next 65 years.

In the 1950s a substantial program to develop the thermoelectric solar power units for spacecraft was initiated and has been continued up to the time of writing. Much interesting work has been done, but no operational system has been installed in a spacecraft.

It is implicit in the record that production of electric power from a solar-thermal power plant is technically feasible but that the capital costs have been so high that it has not proved to be commercially attractive in spite of the zero cost of solar energy. This section outlines the principal problem areas.

Solar-Energy Collectors

Simple, relatively inexpensive flat-plate collectors are normally employed for domestic hot-water, swimming-pool, and building-heating applications. However, as the temperature at which the energy is collected is increased, the efficiency of the collector falls off rapidly because of increasing losses via thermal radiation and conduction. These losses can be greatly reduced by concentrating the sun's rays on a boiler tube placed at the focus of a parabolic trough reflector and even further by enclosing the boiler tube in an evacuated glass tube to reduce heat losses by thermal convection. Still higher receiver temperatures can be obtained by employing a cavity-type receiver which is well insulated on all sides except for an aperture on the side toward the concentrator.

The efficiencies with which typical collector systems convert the solar energy input into useful heat in the collector are indi-

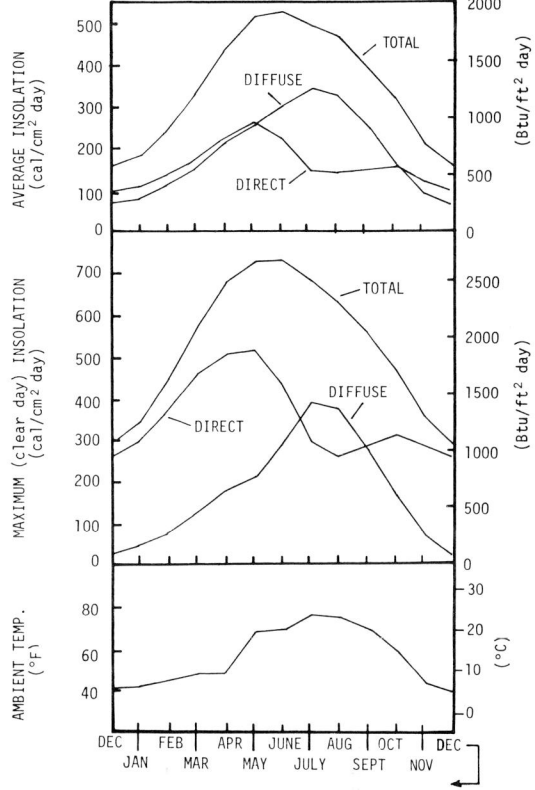

Figure 6-8 Monthly average and maximum insolation incident on a horizontal surface and ambient temperatures based on 16 years of data (1955 to 1971) for the Oak Ridge area (latitude 36.03 degrees).[29]

cated in Fig. 6-7.[28] These data are for operation on a clear day; a substantial amount of haze and/or upper-air turbidity make the comparison more favorable to simple flat plates because a much higher percentage of the incident energy at the first surface is in the form of diffuse rather than direct radiation, and the diffuse radiation is not concentrated by a lens or a parabolic mirror. The average fraction of diffuse radiation varies with the geographical location. Figure 6-8 shows data for the Oak Ridge area in Tennessee.[29] The upper curves show averages on a weekly basis and indicate that over half of the total incident solar radiation is diffuse, and for 4 months of the year the dif-

6–13

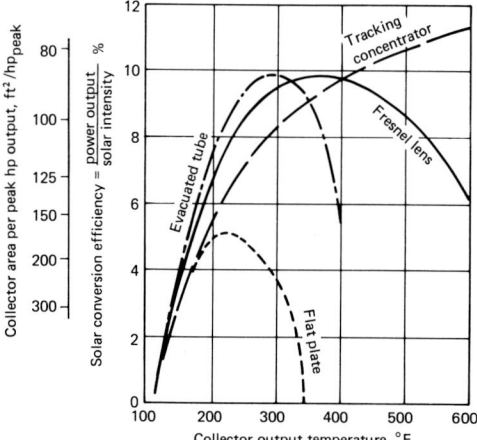

Figure 6-9 Estimated solar conversion system efficiency as a function of collector temperature. Assumptions: (1) total solar intensity = 948 W/m² [300 Btu/(h·ft²)] with 90% direct, (2) maximum cycle temperature = 95% of collector temperature, (3) collector efficiency from Fig. 6-7, (4) Rankine cycle with regeneration, (5) indirect component lost to Fresnel and concentrator.[28] *(Courtesy Barber-Nichols Engineering Co.)*

fuse radiation markedly exceeds the direct radiation. Note, too, that the middle set of curves indicates that even on a nominally clear day the diffuse radiation exceeds the direct radiation for about 3 months of the year. Partly for this reason and partly because of their inherently lower cost, the flat-plate collectors are usually preferred for building heating and domestic hot-water systems. However, the poor thermal efficiency associated with a low collector temperature makes it equally clear that it is best to employ some form of solar-energy concentrator if one wishes to produce electricity via a thermodynamic cycle. Even under the most favorable conditions at midday on a clear summer day, it is difficult to get useful power from a flat-plate collector. This effect is shown in Fig. 6-9, in which the useful power output as a percentage of the solar-energy input is plotted against collector outlet temperature for several types of solar-energy collectors.[28] An additional scale on the ordinate gives the collector area required per peak horsepower output at noon on a clear day.

The nature and magnitude of the losses in a solar-energy collector are illustrated by data for a typical parabolic trough collector designed by the author and shown in Fig. 6-10.[29] To reduce heat losses to the surrounding air, particularly in a strong wind, the unit was covered with a glass plate. This also provided an easily cleaned surface and minimized deterioration in the reflectivity of the parabolic mirror. However, some of the solar energy is reflected from a cover glass surface, and some is absorbed in passing through the glass. In addition, the reflectivity of the mirror surface is less than perfect: commonly about 0.83 for a highly polished, Alzak aluminum surface. Further, depending on the absorptivity of its surface coating, the boiler tube will absorb at best around 90 percent of the incident light energy, the remaining 10 percent being reflected. Even with the glass cover and with thermal insulation lining the box containing the parabolic trough, a substantial amount of energy is lost from the boiler tube by thermal convection and conduction within the insulated enclosure. The cumulative effects of these losses are indicated in Fig. 6-11 for a representative parabolic trough reflector that was not enclosed.[30] Enclosing the parabolic trough and flat plate collectors of Fig. 6-10 reduced the conduction and convection losses sufficiently to much more than offset the losses introduced by the glass cover plate.[29]

The discussion up to this point has been concerned mainly with collector performance when the plane of the reflector is normal to the incident rays of the sun. If this condition is to be maintained, the reflector must be rotated about two axes. Rotation about the polar axis is required to track the sun throughout the course of the day. It is obvious that the elevation of the sun at noon varies with the day of the year, the total variation being 47 degrees through the course of the year. While less obvious, the pointing angle about the horizontal axis also varies with the time of the day except at the equinoxes, and the variation is a maximum at the summer and winter solstices.[29] This effect is shown in Fig. 6-12. A parabolic trough reflector can be mounted on an east-west axis so that it need not track the sun about a polar axis, and its elevation angle can be varied from day to day so that its inlet face will be perpendicular to the sun's rays at noon. However, either a low-concentration factor must be employed or the elevation must be varied throughout the day if a good collector efficiency is to be obtained during the months close to the summer and winter solstices. It was for this reason that vertical fins were placed on the tube of Fig. 6-10; they reduced the concentration factor so that the pointing angle of the collector would not have to be changed throughout the day.

It might be noted that, to keep the cost low, the parabolic mirror of Fig. 6-10 was formed by a simple bowstring tensioning of an initially flat Alzac aluminum sheet to yield a close approximation to a parabolic trough. This gave a concentration factor of approximately 40, an appropriate amount for the mirror aperture width of 47 cm. With the reflector at a fixed elevation angle, variations in pointing error throughout the course of the day near the summer or winter solstice caused the focused band

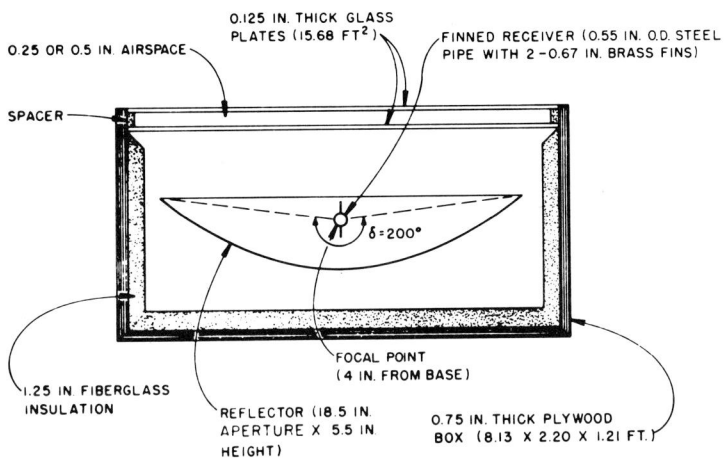

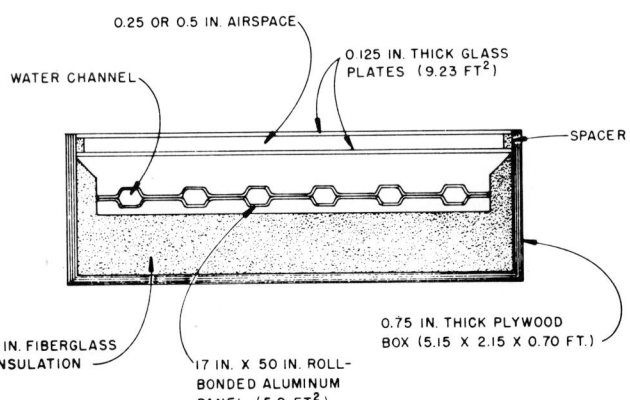

Figure 6-10 Sections through two solar-energy collectors tested.[29]

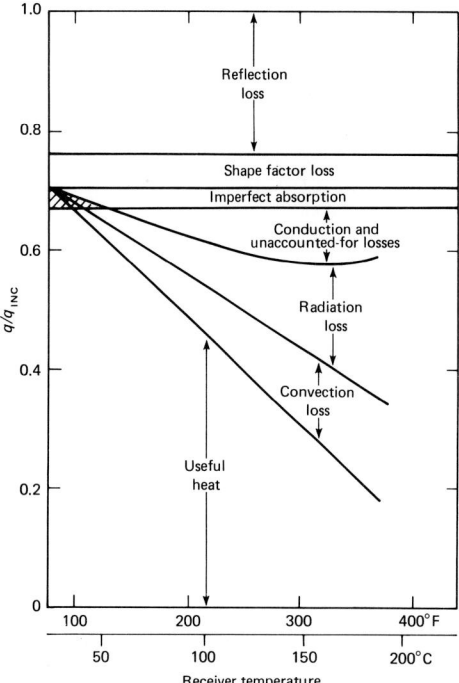

Figure 6-11 Distribution of the incident energy q_{inc} for a 6.19-ft-aperture reflector, 2.375-in-diameter receiver, as a function of receiver surface temperature. (Smoothed data from 15 runs, incident radiation 275 to 325 Btu/min, wind speed 65 to 140 fpm, ambient temperature 70 to 88°F.)[30]

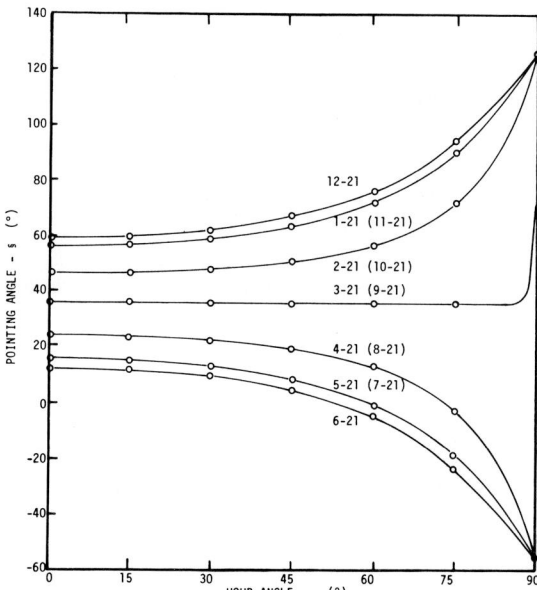

Figure 6-12 Pointing angle ϕ as a function of hour angle ω throughout the year, with $\omega = 0$ corresponding to solar noon. Month-day indicated.[29]

of light to move out onto the fins, whose thermal conduction was high so that the collector tube still functioned. A larger tube could have been employed, but it would have presented a greater area to produce heat losses and would have had a much higher heat capacity—which is undesirable during warm-up conditions early in the day and during transients when clouds pass over, common events throughout the year. The heat capacity is an important consideration, because often, even with the low thermal capacity of the system shown in Fig. 6-10, on partly cloudy days the system would warm up only sufficiently to reach the desired operating temperature before another cloud would come along and the system temperature would drop. In fact, the irregularities in the heat input to the collector caused by occa-sional scattered clouds made it impossible to get a comprehensive set of good heat-balance data in the course of many weeks of testing in the Oak Ridge area. The system would rarely reach thermal equilibrium before another small cloud perturbed it and spoiled the heat balance. It is for this reason that most testing in the United States is carried out either with synthetic energy sources from sun lamps or in the American southwest. But even in the southwest, as indicated by Fig. 6-13, haze, light overcasts, and scattered clouds give a great deal of scatter that poses experimental difficulties.[31] For example, one of the best sets of data available was obtained by a team from Minneapolis which chose a test site near Phoenix, Arizona. Figure 6-14 shows one of their best sets of data as presented in Ref. 32. Note

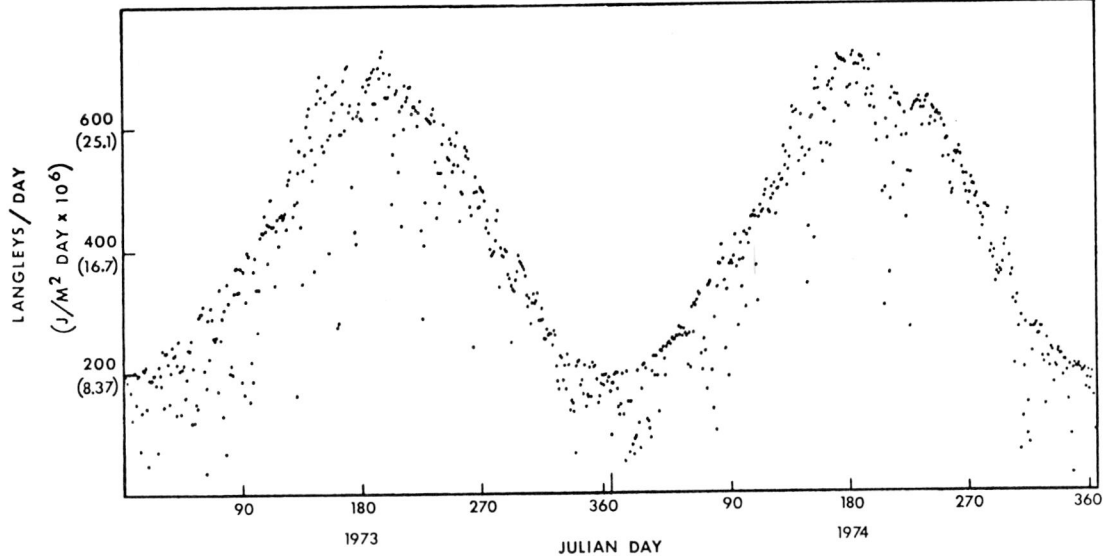

Figure 6-13 Daily total global irradiance for the 2-year period 1973 to 1974 at Las Vegas, Nevada (36 degrees N).[31]

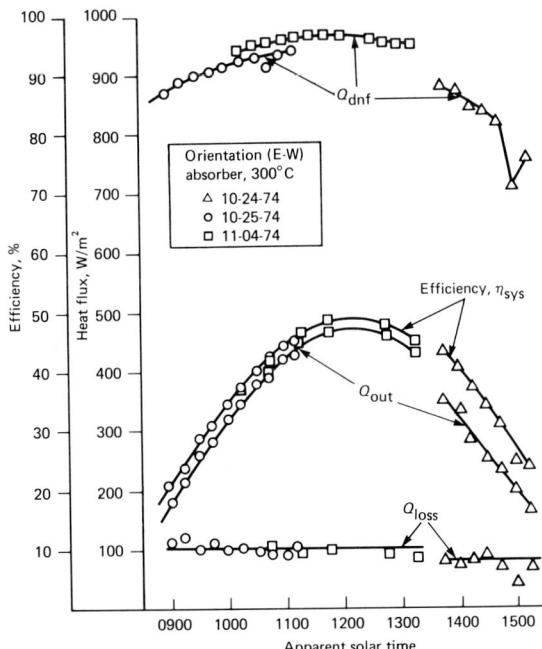

Figure 6-14 Test data from a site near Phoenix, Arizona for a cylindrical parabolic trough reflector with an east-west orientation and a selective coated absorber tube at 300°C. The total direct normal flux is given at the top, the collector efficiency in the middle, and the heat loss at the bottom.[32]

Figure 6-15 Schematic diagram of a solar power tower and heliostat field.

that in preparing Fig. 6-14, they found it necessary to take data obtained on three different days in order to get a complete picture of the collector performance as a function of the time of day. The extreme sensitivity of concentrating solar-energy collectors to variations in superficially clear sky conditions can only be appreciated if one attempts to get good heat-balance data over extended periods.

Solar Power Tower Systems

If one attempts to couple a large number of parabolic trough collectors, such as that of Fig. 6-10, to form a large heat source for a central station power plant, one finds that the extensive system of pipes required to collect the steam, hot water, or other high-temperature heat-transfer fluid leads to a large heat capacity, substantial thermal losses, and high costs. The problems are so serious that a more attractive approach is to employ a large field of two-axis tracking mirrors to concentrate the collected solar energy on a boiler located in a cavity mounted on a tower. A French system employs a field of flat mirrors that direct sunlight to a parabolic mirror that in turn focuses the rays on the boiler.[33] As indicated in Fig. 6-15, a similar system favored in the United States entails focusing the beams from parabolic mirror heliostats directly on the cavity of the boiler.[34] In either case the "power tower" approach has the advantages that the heat losses from the boiler can be kept to a low value—of the order of 5 percent—and the heat capacity is kept to a minimum. This system has the disadvantage that it requires a high degree of accuracy in the two-axis mountings for the mirrors and a fairly complex and expensive tracking system to keep the mirrors focused on the boiler cavity. It also has the disadvantage that abrupt changes in energy input to the boiler tubes associated with a passing cloud induce severe thermal stresses. This effect is shown in Fig. 6-16, which was obtained in tests of the first unit of this type for which test data are available. In this instance the designers had given a great deal of attention to the thermal-stress problem under transient conditions, and test data indicated that by designing for moderate heat fluxes the thermal stresses were in fact kept to acceptable levels. Probably the key question of such an approach is the capital cost of the elegant equipment required for the double-axis tracking system which must maintain a pointing accuracy of less than 1 minute of arc. The mechanism must be sufficiently rugged and rigid to withstand a storm with winds of at least 161 km/h (100 mph). The mirror itself and the frame supporting it must also be strong and rigid to take severe aerodynamic forces without distortion that would spoil the high precision of the optical surfaces required to give the accurate focusing necessary.

A disadvantage of the power tower–heliostat approach is that the individual heliostats must be spaced fairly well apart so that one does not shadow another at any time during the day. In practice this means that the area of the heliostat field must be 6 to 10 times the area of the heliostats[34] (Table 6-5).

Average Solar-Energy Flux in the United States

The distribution of the solar-energy flux for the United States is shown in Figs. 6-17 and 6-18 for both direct radiation to surfaces continuously oriented so as to be roughly normal to the sun's rays (as for the heliostats of Fig. 6-15) and for the total energy flux—both normal and diffuse—to horizontal surfaces (such as simple flat plates for heating hot water). These maps indicate that, as one would expect, the solar-energy flux in December varies by as much as a factor of 3 between the southwest and the northwest, and for most regions by a factor of about 3 between June and December. The solar energy that can be collected and used varies by a much greater factor, because heat losses from the collection system commonly represent about 20 percent of the energy collected at noon and perhaps 50 percent of the energy collected at 9 A.M. or 3 P.M. on a clear summer day (Fig. 6-14). Unfortunately, up to 1980 no experimental data have been obtained to show a yearly summation of the net energy collected by a typical system at a high enough temperature to be useful for a thermodynamic cycle. It is partly because of this problem that efforts to develop solar-thermal power systems are concentrated in the hot, dry, desert areas of the U.S. southwest, where conditions are the most favorable.

Cost of Electricity

The costs of solar-thermal-electric systems are high. Even the enthusiasts cite capital costs several times those for conventional systems, and these costs are based on the peak system output at noon in the summer with no allowances for energy storage. The situation is particularly confusing because the high costs of the experimental systems that have been built (e.g., $300/m² for heliostats in 1979) are dismissed by advocates as not indicative of the cost of actual production systems, which they believe should be much lower. Thus, for purposes of this text it appears that the most meaningful approach to the cost problem is to estimate the minimum cost per pound of equipment that one might hope to get for heliostats on the basis of experience in manufacturing high production items such as trucks, and, from this, estimate the cost of electricity on an annual basis with due allowance for diurnal and seasonal variations in the energy collection and utilization. Accepted values for the cost of the other equipment in the plant, which would be essentially conventional, would be used in the estimate.

In attempting to estimate the cost of heliostats in large-scale production, one finds, on inspection of designs that have been proposed and experimental units that have been built, that a substantial amount of structure is required to provide the necessary rigidity for the mirror. In addition, the mounting for the two-axis tracking system also entails a substantial amount of weight. Examination of several designs indicates that it is unlikely that two-axis tracking heliostats can be built for less than about 22 kg/m² (5 lb/ft²). Extensive experience in manufacturing roughly comparable equipment indicates that with large-scale production the unit cost would be at least $5/kg ($2.30/lb) in 1979 dollars, or about the same cost per pound as a pickup truck. Note that the cost of precision machinery such as

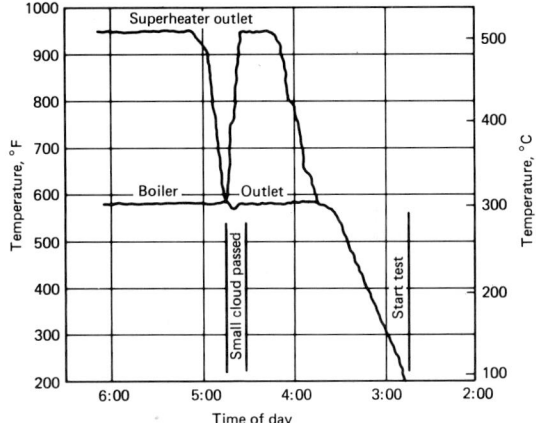

Figure 6-16 Test data from a typical run using a steam generator receiver with the heliostat field of the Centre National de la Recherche Scientifique (CNRS) facility in the Pyrenees. Curves from the strip chart recorder give the boiler drum and superheater outlet temperatures as functions of time following a startup at 2:48 P.M. It took until 4:15 P.M. to reach the design conditions of 8619 kPa/510°C (1250 psig/950°F). A cloud passed at 4:37 P.M.; equilibrium at design conditions was re-established by 5:10 P.M. The temperature rise rate for the boiler in the startup was 278°C (500°F) per hour, and the superheater outlet temperature rise rate after the cloud passed was 26°C (47°F) per minute.[34]

TABLE 6-5 Heliostat and Heliostat Field Areas Required for Typical System Peak Outputs at Noon on a Clear Summer Day
(Based on area data in Ref. 20)

Power, MWe	No. of power towers required	Heliostat area, m²	Heliostat field area, km²
10	1	71,100	0.53
100	5	711,000	5.3
1000	50	7,110,000	53

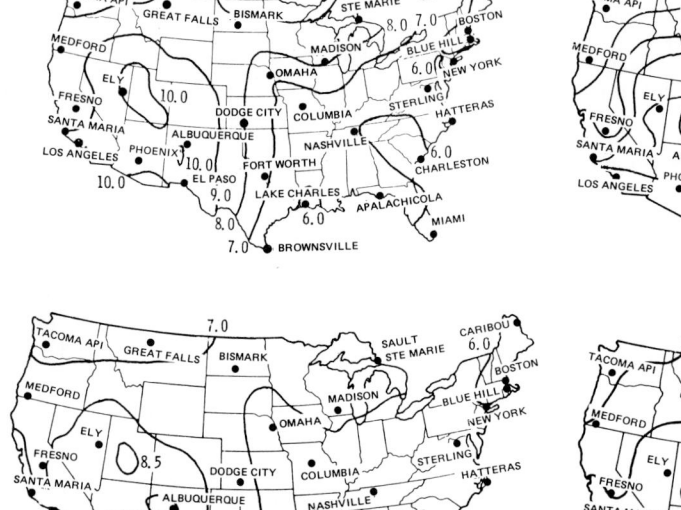

Figure 6-17 Mean daily direct-normal (top) and total-horizontal (bottom) solar radiation for June (kWh/m²) (Ref. 35).

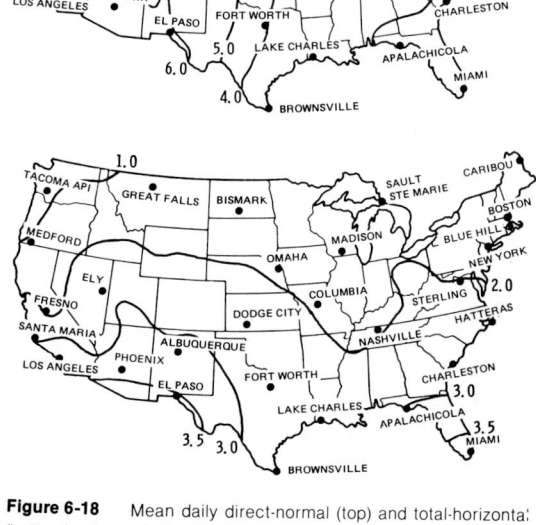

Figure 6-18 Mean daily direct-normal (top) and total-horizontal (bottom) solar radiation for December (kWh/m²) (Ref. 35).

lathes may be appropriate and is very much higher. On the "pickup truck" unit cost basis, a low estimate for the cost of heliostats in large-scale production would be $110/m², including the two-axis mounting and tracking system. In estimating the useful energy collected on an annual basis, one finds that the sun is high enough to give a useful electric output for an average of 6 h per day. In the American southwest the output is further reduced by the weather conditions to about 80 percent of the time, and on the average, even on nominally sunny days, the direct radiation will comprise no more than about 90 percent of the total, the balance being in the form of diffuse radiation which cannot be concentrated on the boiler cavity. Further, the reflectivity of the mirror surface will probably be about 80 percent. Thus the annual total thermal energy input to the boiler, if 0.9 kW/m² is assumed to be the insolation rate to the earth's surface, is given by the following equation:

Thermal energy collected

$$= 0.9 \times 365 \times 6 \times 0.9 \times 0.8 \times 0.8$$
$$= 1135 \text{ kWh/m}^2 \text{ per year}$$

Taking the capital charge for the system as being 16 percent, the cost of this energy would be $17.60, or $4.31/10⁹ J ($4.55/10⁶ Btu). At first glance this appears to be about double the 1979 cost of fuel oil and about four times the cost of coal. However, it does not include allowances for storage costs and energy losses associated with storage operations. If the energy collected were used immediately to generate electricity and the electricity employed in a pumped-hydro storage system, the efficiency of the storage system would be about 70 percent, and hence the effective cost of the thermal energy would be increased by about

40 percent. If the electric energy were transmitted to a load center 1600 km (1000 mi) away, the effective cost would be increased further by about $3.33/10^9 J ($3.50/10^6 Btu). The capital charges for the steam plant utilizing the solar heat collected would be higher than those for a conventional plant by roughly a factor of 5 because of the low utilization factor, and the capital charges for the pumped-hydro system would be much higher than for the usual peaking-power pumped-hydro system because of the much longer drawdown period and the shorter pumping period.

A thermal-energy storage system such as the molten-salt system discussed in the previous chapter may be preferable to a pumped-storage hydro system because, although its costs may be somewhat higher, the capital cost of the steam plant can be much reduced. On the basis of data given in the preceding chapter, and assuming a heat-storage capacity sufficient to carry through one cloudy day, the capital cost of the high-temperature thermal-energy storage system would be equivalent to about $60/m^2 of collector. Allowing for heat losses, this implies about a 65 percent increase in the cost of the solar energy collected.

The cumulative effects of the above factors indicate an overall cost of electricity from a solar-thermal power system to be about 10 times that for electricity from a coal-fired power plant. These costs may seem high, but actually the estimates are probably low because they are based on a number of optimistic assumptions, such as a low cost for the heliostats required by the solar-thermal system.

In view of these unfavorably high costs, it appears that the 1135 kWh per year per square meter estimated above for a heliostat system should be checked by making the estimate on a quite different basis, e.g., the insolation maps of Figs. 6-17 and 6-18. If the solar-energy system is to be self-sufficient, its average output in the winter must be as great as in the summer, because the storage of energy over a period of many months would be prohibitively expensive. Thus the full capacity of the heliostats could not be used to good advantage in the summer because the system output must be based on the insolation rate in December. Using southern Arizona and Fig. 6-18 as the basis, one finds that the average daily energy input to the heliostats would be approximately 6 kWh/m^2. Allowing for an 80 percent mirror efficiency and a 90 percent boiler heat absorption efficiency gives an average net energy collection rate of 4.3 kWh/m^2 per day, or 1580 kWh/m^2 per year. The actual rate would be substantially lower because of heat losses during start-up in the morning and during the passage of intermittent clouds, together with the poor performance early in the morning and late in the afternoon. Thus, the value of 1135 kWh per year per square meter appears to be a reasonable estimate. In fact, neither of these estimates includes allowances for the overall plant availability; even if an unusually high degree of reliability were obtained in the mechanical and electrical equipment so that an availability as high as 80 percent could be obtained, both estimates would have to be reduced by 20 percent.

The high costs of energy storage suggest that it might be best to attempt only to save fuel during the day and provide 100 percent backup in the form of coal-fired plants. If this approach were followed, the capital cost of the solar plant would still have to be written off against a low operating time per year, giving capital charges at least six times, and more probably 10 times, those for a coal-fired plant. The cost of the solar-thermal energy would still be high, and the amount of fossil fuel saved would be only about 20 percent of the usual coal plant consumption. Also, the severe thermal cycling would increase maintenance costs for the coal-fired plant.

The above rough estimates do not include allowances for degradation in performance stemming from reductions in the specular reflectivity of the mirror surface as a consequence of accumulations of dirt or etching by sandstorms—which are fairly common in the southwest. Note that a single severe sandstorm could completely ruin the specular surfaces.

Solar Photovoltaic Systems

Solar cells have been a useful source of electricity since about 1960 and have been finding increasingly widespread use for small amounts of electric energy in remote locations. They have proved particularly well suited for use in spacecraft; in fact, much of the photovoltaic R&D has been funded by the space program. By 1977 the total annual sales amounted to about 750 kW, largely in the form of silicon solar cells. The efficiency of the individual cells is about 12 percent, but the voltage output is low so that, by the time allowances are made for losses in the connections and the area lost to gaps between individual cells, the overall efficiency of the array is commonly only 6 to 8 percent. In 1979 the cost amounted to about $15,000/kW on the basis of the peak output on a clear day with the array normal to the incident light. Fortunately, the cells respond to both direct and diffuse sunlight so that, in general, the cell output is essentially proportional to the sine of the angle of incidence of the direct rays.

Principles of Operations

Light photons can give up their energy to electrons in certain semiconductors such as silicon, and this extra energy may be sufficient to dislodge the electrons from their positions in the crystal lattice.[36] To utilize this reaction in silicon solar cells to produce an electric current, a thin layer on the front face of the crystal lattice of the silicon is "doped" with a trace amount of foreign atoms, such as phosphorus, which take the places in the crystal lattice of some of the silicon atoms. Phosphorus has five valence electrons as compared to four in silicon. Thus there will be a small excess of electrons, which causes electrons energized by photons to migrate with their negative charge toward a positive electrode on the face of the crystal lattice. The balance of the crystal lattice is "doped" with a small percentage of atoms such as boron. Each boron atom has only three valence electrons. Hence the region "doped" with boron will be deficient in electrons, giving electron vacancies, or "electron holes," that have a positive charge and will tend to migrate through the crystal lattice toward the negatively charged opposite face. The electrons activated by photons can then flow from the conductor on the front face of the cell through an external circuit to the conductor on the rear face to fill the "holes" that migrate there, giving a driving voltage of ~1 electronvolt (eV). For good efficiency the electron-rich region on the front face (the negative, or N-type crystal lattice) should be only ~0.5 μm thick so that incident photons dislodge electrons from their lattice locations in the region close to the interface between the two types of crystal lattice. Thus the resulting holes migrate through the positive, or P-type, crystal lattice to a layer of metallic conductor on the rear face of the composite crystal, while the electrons move to a fine grid of metallic conductors on the front face through which sunlight enters the crystal lattice.

Solar-Cell Construction

Up to the time of writing, most of the solar cells have been prepared by growing a silicon crystal from a pool of molten silicon that is very pure except for a small amount of boron to give a P-type crystal lattice. The crystal is then sliced into thin layers with a diamond saw, and one surface of each slice is exposed to a phosphorus atmosphere at high temperature for long enough to give the proper thickness for the N-type layer. Both surfaces are plated with a high-electric-conductivity metal such as silver, and then, by photographic techniques, the silver is etched away from most of the area of the front face to leave a fine grid of conductors (such as that in Fig. 6-19). Thus ~90 percent of the front face is open for light to enter the N-type layer. To protect the silver grid from gradual deterioration by corrosion, the conductor grid on the front face is ordinarily given a protective coating of a metal such as nickel, titanium, gold, or palladium.[37] To reduce the losses caused by light reflection from the surface of the silicon, the front face is given a final coating of antireflection material. The conductor network on the front face is then coupled to the intercell wiring system by soldering[37] (Fig. 6-19).

The above method of fabricating solar cells is inherently complex and expensive. Other materials combinations and fabrication techniques are under development in an effort to reduce costs. However, it is implicit in the high degree of purity required in the materials (≪ 1 ppm of impurities), the complexity of the fabrication processes, and the requirements for tight quality control in all phases of the fabrication that the cost of photovoltaic cells tends to be high.

Basic Factors Affecting the Efficiency of Solar Cells

The efficiency of conversion of light energy into electricity in solar cells is much less than 100 percent because of the cumulative effects of many factors.[36,38] In the first place, ideally the energy of the incident photon should be just that required to dislodge an electron from the crystal lattice so that it becomes mobile. Photons of lower energy will be ineffective, while photons of somewhat higher energy than that required will yield

Figure 6-19 Photograph of a 1-cm-diameter GaAlAs/GaAs concentrator cell after packaging in an aluminum holder/heat sink. The contact ring is connected to the electrode plate with 2-mil gold wire by means of multiple ultrasonic bonds. Visible in the left bottom area are the thermocouple leads used to measure the cell temperature.[37] *(Courtesy Rockwell International.)*

the excess energy as heat, and photons of too high an energy will be completely ineffective. The electron energy range in which the crystal becomes conducting is known as the *conduction band*. As a consequence, only a certain band width of the solar spectrum can be effective in generating electricity in any given photovoltaic material. The resulting *quantum efficiency* as a function of photon energy is shown in Fig. 6-20. The overall efficiencies ideally obtainable from sunlight outside the atmosphere are shown in Fig. 6-21 for some typical semiconductors. Note that AlGaAs is inherently more efficient than silicon and hence is receiving much attention. In practice, small defects in the crystals and other basic deviations from ideality reduce the efficiency by a factor of ~20 percent. Further losses stem from reflection of some of the incident light from the front surface, the obscuring effect of the conductor grid which blocks off ~10 percent of the area of the front face, and cell internal electrical resistance. (These reductions in cell efficiency are not included in Fig. 6-21.) Additional losses are introduced when cells are coupled in arrays, e.g., gaps between cells for the supporting framework, the resistance in intercell conductors, and losses in the power-conditioning equipment including conversion from direct to alternating current.

The efficiency of solar cells depends on operating conditions, one of the most important factors being the current density. Figure 6-22 shows a typical curve relating the voltage output to the current for an AlGaAs cell. A curve for the cell temperature is also shown, the minimum temperature indicating the region for minimum losses and the maximum efficiency. For some types of cell the efficiency can be increased by concentrating the sunlight with a Fresnel lens or a parabolic mirror; Fig. 6-23 shows this effect for the AlGaAs cell of Fig. 6-19. However, if a concentrator is used, it has the disadvantage that it utilizes only the direct rays of the sun, and so, as discussed above for thermal energy systems, about one-third of the total incident sunlight is lost on an annual basis.

If a concentrator is employed, not only can the cost of the cell per unit of electric output be greatly reduced, but also it is possible to split the concentrated beam with a filter into two spectral bands and focus each on a type of cell suited to that band.[39] Figure 6-24 shows the performance of such a filter used with silicon and AlGaAs cells to give an overall efficiency for the filter and cell subsystem of 28.5 percent in terms of the input in the form of the concentrated light beam. This is a most encouraging development, though its cost-effectiveness is reduced because two cells plus a beam-splitting filter are required instead of a single cell. Another approach to the use of a greater fraction of the spectrum is to employ a multiplicity of layers of very thin cells of different types to give multicolor cells—an excellent concept in principle but difficult to apply in practice.

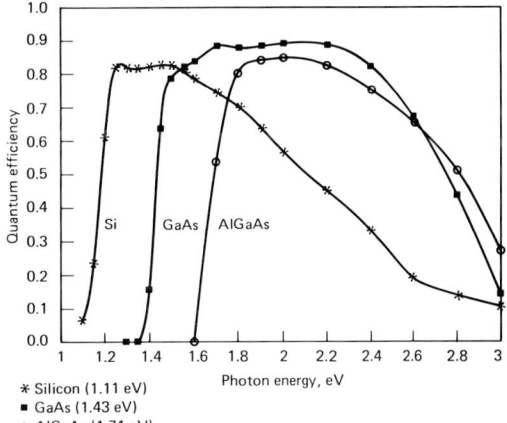

* Silicon (1.11 eV)
■ GaAs (1.43 eV)
o AlGaAs (1.71 eV)

Figure 6-20 Spectral response of Si, GaAs, and AlGaAs solar cells. Quantum efficiency is not corrected for 10 percent contact obscuration. Area = 0.56 cm².[39] *(Courtesy Hewlett Packard.)*

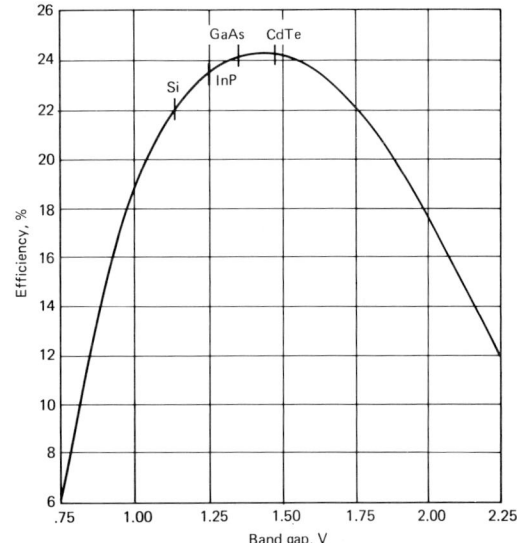

Figure 6-21 The theoretical efficiency of photovoltaic cells as a function of the band gap (conducting region) in electron-volts. The band gaps for four typical semiconductors are indicated on the theoretical curve.[36] *(Courtesy Scientific American.)*

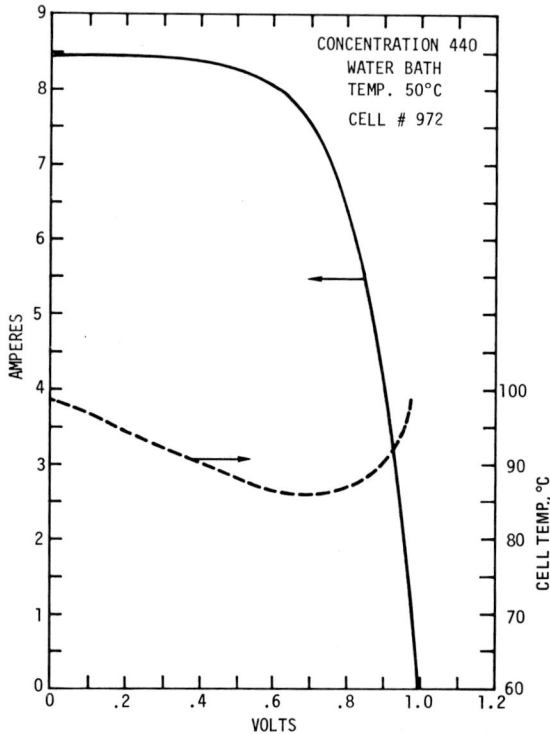

Figure 6-22 The cell temperature variation during *I-V* curve measurement. For this particular plot, the recirculating water temperature and, therefore, the heat-sink temperature, was maintained at 50°C. The cell temperature was measured by a thermocouple placed directly under the center of the cell.[37] *(Courtesy Rockwell International.)*

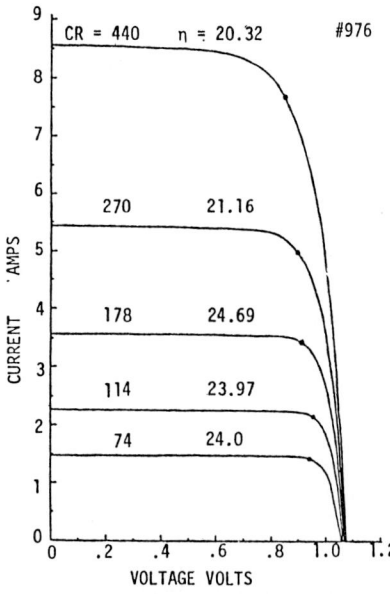

Figure 6-23 *I-V* curves for a concentrator cell No. 976 measured at Table Mountain, California. Each curve is also labeled with the concentration ratio (CR) and the efficiency (η) at the maximum power point (the dot just below the knee of each curve). The cell temperature at the maximum power point was maintained at 50°C.[37] *(Courtesy Rockwell International.)*

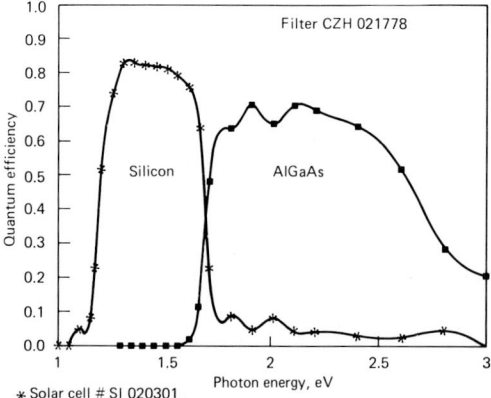

* Solar cell # SI 020301
■ Solar cell # AGA 010478HLR1

Figure 6-24 Spectral response of Si and AlGaAs concentrator cells mounted with a spectral splitting filter. Each cell is 0.56 cm² in area.[39] *(Courtesy Hewlett Packard.)*

Cell	I_{sc}	V_{oc}	P_m	Fill factor	Efficiency, %
AlGaAs (1.61 eV)	1.382	1.26	1.440	0.827	17.4
Si (1.1 eV)	1.711	0.738	0.915	0.725	11.1

Developmental Choices

The development engineers and physicists have a difficult choice. To increase the efficiency of simple flat-plate arrays seems likely to increase the cost of the cells. If concentrators are used, the concentrator cost is likely to be high. Cost estimates by advocates of concentrator systems are commonly low, but where reasonably detailed structural designs that include weight estimates are available, the weights run from 50 to 68 kg/m² (Refs. 40 and 41), or two to three times the 22 kg/m cited earlier as the minimum likely for the simpler heliostats of the solar-thermal system. Hence, even if the solar cells were inexpensive, following the same reasoning as used for heliostats but crediting the photovoltaics with both a 30 percent efficiency and a longer effective day because they are not sensitive to heat losses, one finds the costs of the concentrators to be at least $0.10/kWh in 1980 dollars. Allowance for other charges would probably more than double this amount, giving an overall cost of electricity that would be at least five times the average 1980 charge to consumers—too high to be acceptable to the public except in a few special applications.

Space Applications

Data on solar-cell efficiencies sometimes appear inconsistent because some of the cells are intended for space applications, others for terrestrial applications. The broader solar spectrum in space means that a smaller fraction of the incident light can be utilized by the cell and hence the cell efficiency is roughly 20 percent lower for space applications. Another problem with solar cells in space applications is that they are subject to degradation as a consequence of damage to the crystal lattice caused by protons and electrons in the solar wind or the Van Allen radiation belt. This effect is shown in Fig. 6-25.[42]

In reviewing the more comprehensive record for space solar-cell development to assess the probable rate of future solar-cell developments, one finds that by 1960 after about 5 years of development the efficiency of silicon cells for space applications was as high as 15 percent, and extensive work was under way on other types of cell.[42,43] A 1978 status report listed the best efficiency achieved in the more promising types of solar cell for space applications and gave an efficiency for silicon cells of 15.6 percent.[44] These values compare with a theoretical upper limit of 22 percent. The most encouraging developments have been for gallium-arsenide cells, the efficiency of which for space applications was increased from around 6 percent in 1960 to around 17 percent in 1977. The latter compares with an upper theoretical limit of 25 percent efficiency. These values, of course, are for individual cells and are summarized in Table 6-6. When solar cells are integrated into arrays, additional losses are substantial. As an example, according to 1978 data for a typical array, a flat panel of silicon cells oriented normal to the incident solar radiation was used for the Orbiting Solar Observatory and gave an overall energy conversion efficiency of 7 percent during its first year of operation.[45] In short, the most vital problem is the cell efficiency. The solid-state physics involved is beyond the scope of this text. Hence it will simply be stated that according to expert opinion the highest efficiency ideally obtainable is around 40 percent,[32] and an efficiency of the order of 20 to 30 percent will be required with relatively inexpensive materials to yield an attractive system.[33-40] This is well above values obtained in actual solar-cell arrays. Thus many people believe that the most promising avenue to achieving costs suffi-

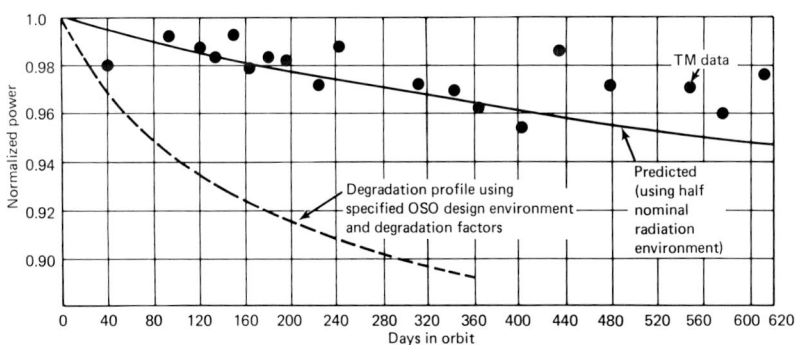

Figure 6-25 Orbiting satellite observatory OSO-8 in-orbit predicted performance and telemetry data versus time in orbit.[45]

TABLE 6-6 Maximum Efficiency in Percent of Incident Solar Energy for the Principal Types of Solar Cell Reported in 1961 and 1978

	Operation in space			Terrestrial operation	
	1961 (Ref. 42)	1978 (Ref. 45)	Theoretical limit	1980	Theoretical limit
Single crystal silicon	15	15.6	22	19	
Gallium arsenide	7	18.5	25	23	
Cadmium sulfide	6	8.5			
Cu_2O	0.5				
Selenium	1				
InPCdS		15			
InP/InSnO		15			
CuInSe		3	17		
Cd/CuInSe		6.2	17		
Two-color cell				28.5	~35
Multicolor cell					~60

ciently low to make solar cells commercially attractive for electric utility applications is to carry out basic research that would yield materials both low in cost and capable of utilizing a large fraction of the energy in the complete solar spectrum. Note that a satisfactory cell must also have a long life. Also, its characteristics in production should be quite uniform so that large numbers of cells can be coupled in series-parallel arrays to give a well-balanced system with a minimum of losses associated with local imbalances in electric current flows. Some notion of the complexity of the problem is given by Figs. 6-22 and 6-23, which indicate the effects of current flow and insolation rate, respectively, on cell output and efficiency.

One of the major advantages of solar cells is that their efficiency is at its best in small arrays. Hence, if the cost were reasonable, solar-cell arrays could be installed on the buildings in which the electric energy is to be used. This arrangement would avoid the losses and expenses associated with transmission systems from a central utility. It would also, however, introduce the requirement for a much larger electric energy storage capacity than would be required if the solar cells were in central stations because there would not be a network of transmission lines to supply power from a multiplicity of sources.

Satellite Power Systems

In the 1960s Peter Glaser proposed that a large array of solar cells be placed in a geosynchronous orbit and that the electric power generated be beamed to the earth in the form of microwave radiation. It seemed doubtful that a beam could be focused sharply enough for the energy to be transmitted efficiently from a satellite 35,000 km (22,000 mi) above the earth, but an experiment carried out at JPL in 1975 showed that 30 kW could be transmitted for a distance of over a mile with an overall efficiency of 82 percent.[46] Analyses have indicated that by scaling up the system to a very large size—at least 5000 MWe—an overall efficiency of around 60 percent might be obtained with a transmitting antenna diameter of about 1 km and a rectenna or receiver array having a diameter of around 10 km. A fairly detailed picture of what such a system might look like is given in Refs. 46 to 49, and data for a typical design are given in Table 6-7. The solar-cell array would be huge, e.g., 21.4 km long by 5.3 km wide and mounted on a triangular girder structure having a girder depth of about half a kilometer. At either end there would be an antenna having a diameter of 1 km, each antenna directing a beam at a receiver on the earth's surface with a microwave frequency of 2.45 GHz. Solar cells would generate 25,500 MW of electricity which would go through a collection system and be converted to beams of microwave energy totaling 17,000 MW. The output of the terrestrial receivers would be 10,000 MW. The energy distribution envisioned for the beams at the earth's surface is indicated by Fig. 6-26.

The problems in developing a system of this sort include roughly doubling the efficiency of solar-cell arrays, cutting the specific weight of the cells by a factor of 4, and increasing their life by a factor of roughly 6. At the same time the cost of the cells would have to be reduced by a factor of roughly 100 through large-scale production. For the system to be viable, all these objectives must be realized. The 1-km-diameter antenna and the

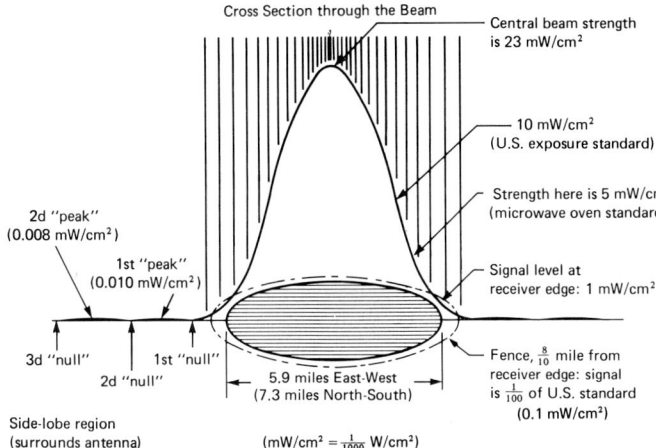

Figure 6-26 Energy distribution across the microwave beam from a satellite power plant in the vicinity of the rectenna at the receiving station.[49] *(Courtesy Boeing Aerospace Co.)*

~10-km-diameter rectenna must be built with a high degree of precision; the position of each conductor in the arrays must be held within ~1 cm. This will be very difficult to do in these large-diameter arrays, particularly with the light structure required for the satellite antenna. Perturbations to be expected include thermal distortion during terrestrial or lunar eclipses, the distorting forces of the gravity gradient to which the satellite will be subjected, and inertial forces associated with keeping the satellite solar array normal to the sun's rays while at the same time keeping the antennae directed toward the earth. There is also a question as to whether the beam can be directed with the required precision so that it lands squarely on the rectenna. For safety reasons the peak energy flux in the center of the beam is to be only 23 mW/cm², which is only a little more than twice the standard U.S. tolerance value for microwave radiation. Thus, if as a consequence of a pointing error the beam should stray from the receiver, it should not represent a serious hazard to the personnel in the surrounding area unless by some fluke it became more sharply focused and yielded more concentrated energy in a local region.

Another basic feasibility question is concerned with the cost of putting the equipment in orbit, assembling it, and maintaining it. Much of the projected cost for such a system involves the space shuttle system required to transport roughly 100,000 tons of material into orbit for each 10,000-MW power plant (i.e., ~10 percent of the weight of the Menkaura pyramid at Gizeh). This will require a reduction in the cost per unit of weight conveyed into orbit by a factor of 10 relative to the cost projected for the first NASA space shuttle system. To obtain this large reduction in cost, it will be necessary to have much larger shuttles

TABLE 6-7 Design Data for a Proposed Space Power System in a Geosynchronous Orbit (35,924 km altitude) (Ref. 49)

Design power output	
Solar cells, MWe	25,500
Energy to antennas, MWe	17,000
Terrestrial receiver, MWe	10,000
Satellite structure	
Length, km	24.6 (14.96 mi)
Width, km	5.3 (3.29 mi)
Depth, km	0.47 (0.292 mi)
Solar-cell array	
Area of array, km²	114.5 (44 mi²)
Solar-cell efficiency, %	17.3
No. of solar cells	21×10^9
Size of solar cells, cm	6.6×7.4 (2.6 × 2.9 in)
Cell thickness, η m	50 (0.002 in)
Cover thickness, η m	75 (0.003 in)
Substrate thickness, η m	50 (0.002 in)
Antennas	
No. required	2
Diameter, km	1.0 (0.6214 mi)
Beam frequency, GHz	2.45
Rectenna	
Number required	2
Shape	Elliptical
Size (each), km	9.5×11.8 (5.9 × 7.3 mi)
Microwave beam power density	
Peak, mW/cm²	23
Value at perimeter fence, mW/cm²	0.01
Total weight in orbit, t	100,000 (220,000,000 lb)

TABLE 6-8 Energy Cost of Materials*

	kWh/kg	Btu/ton × 10^{-6}
Metals		
Aluminum	78.6	244
Chromium	45.7	142
Copper	36.1	112
Lead	9.7	30
Magnesium	127	395
Manganese	16.5	55
Molybdenum	51.5	160
Nickel	140	436
Steel	8.1	25
Titanium	155	482
Silicon	72	224
Zinc	23.2	72
Ceramics		
Cement (Portland)	2.7	8.4
Concrete	0.35	1.1
Firebrick	1.3	4.2
Glass	8	25
Mica	5.8	18
Organics		
Lumber	0.0003	0.0009
Paper	7.1	22
Polyethylene	34	106
Polystyrene	21	64
Polyvinylchloride	15.8	49

*Data from Refs. 50 and 51 and from J. P. Albers et al., *Demand and Supply of Nonfuel Minerals and Materials for the U.S. Energy Industry, 1975-1990—A Preliminary Report,* Geological Survey Professional Paper 1006-A, B, U.S. Government Printing Office, Washington, 1976.

and operate them on a round-the-clock, year-round basis for many years to recuperate the initial investment in the shuttle vehicles. This in turn implies the construction of a complex of space power plant satellites that will provide a very large fraction of the U.S. electric power requirements.

Monetary and Energy Costs of Solar Power Plants

Estimates of the monetary costs of the various solar power plants outlined above generally run from $2000/kW to $20,000/kW of peak output, depending on the degree of optimism of the estimator. Advocates of particular systems generally seem to be able to make a superficially plausible case for a cost of $2000/kW. For terrestrial power plants the effective capital cost would be ~10 times higher because the year-round utilization factor would be low—certainly less than 20 percent—and there would be large additional capital charges. If the plant could operate 24 h per day, as described for satellite plants, a cost of $2000/kW would make such a plant competitive with a fossil fuel or nuclear plant if the cost of fossil fuel were to double or the capital cost of nuclear plants were to double as a consequence of environmental requirements. It may be noted, however, that the huge increases in fuel costs that have followed the Yom Kippur War in 1973 have been accompanied by roughly similar increases in capital charges; thus the relative importance of capital charges and fuel costs probably will not change greatly. One reason for this situation is that the energy costs of producing materials represent a substantial fraction of the total production costs.[50,51]

The energy costs of material, i.e., the amount of energy that must be invested in producing a given quantity of material, is substantial, as indicated by Table 6-8.[50] The energy investment in making silicon solar cells, for example, in the commonly used processes is equal to the total output obtainable from the solar cells during their first 10 years of operating life if they are used without concentrators.[51] This is particularly important if the operating life of the cells proves to be only 10 or 20 years. If an aluminum mirror concentrator is employed, the energy investment in a 1-mm-thick aluminum parabolic reflector for a favorable location in the U.S. southwest is equivalent to the total amount of electric energy that could be produced from the power plant in a period of 3 years. This is one reason why it will be necessary to minimize the quantity of material required by making use of thin-film solar cells if solar cells are to prove a practicable approach to the production of electric energy. For comparison, it should be mentioned that the energy investment in a conventional coal-fired fossil fuel plant is roughly equal to the energy produced by the plant during its first month of operation, and for a nuclear plant the energy investment in materials is about equal to the energy received from the plant during 2 months of operation. It should be noted also that labor involves an energy cost, because people are consumers of energy, and consequently each man-day should properly carry with it an energy charge. For the 1970s in the United States, this appears to have been about 3.5×10^9 J per man-day (3.5×10^6 Btu per man-day), or about 30 gal of fuel oil per man-day.

It is not easy to appraise the degree of difficulty associated with the many developmental problems of the various solar-thermal and photovoltaic energy systems that are under active development, as well as the probable capital cost of the completed power plants if built on a large scale in large numbers. This situation is further confused by the emotional arguments of zealous advocates. The best indication of the prospects is probably given by the results of studies by committees or panels of experts who do not have large personal stakes in the solar-energy program. One such panel organized by the American Physical Society at the request of the White House and the Department of Energy reported early in 1979 that "major scientific and technological advances" will be required before the direct conversion of sunlight into electricity can become a significant source of power.[52] The panel concluded that at best a 20 billion dollar investment in photovoltaic cells might serve to provide 1 percent of the estimated U.S. electric power capacity by the year 2000, and that even this would require some major developments which the panel was not certain could be achieved.

WINDMILLS

As mentioned in the first chapter, windmills have been in use in Europe since at least the eleventh century and were a major factor in initiating the industrial revolution.[53,54] They were introduced into the United States shortly after the initial settlement of New Amsterdam, and they became widely used in this country by the middle of the nineteenth century for pumping water on farms.[55] About 5 million were built, of which around 150,000 were still in use in 1979. The majority had 8-ft-diameter rotors with about 12 blades. In the early 1920s the need for electricity on farms led to the production of wind-powered electric generators, and over a million of these were produced in the 1920s and 1930s. These machines were expensive, however, and rural electrification provided a much less expensive and more dependable source of power.[55] In Europe, steam engines largely supplanted windmills by the early part of this century, and, at the time of writing, even in the Netherlands the very few windmills still being operated were supported with a strong government subsidy.

(It costs about $1000 per year for maintenance on one of the Dutch windmills.) Further, they cannot be allowed to operate unattended, because variations in wind velocity require continual trimming of the sails.

The first effort to build a wind-powered electric generator for utility service in the United States was constructed in Burlington, New Jersey, in 1933.[55] This plant depended on a set of vertical rotors mounted on railcars which operated on a circular railroad track. The system was destroyed in a storm before it could be tested and the concept was abandoned. The next serious effort to produce electric power for utilities was carried out with a 1.25-MW unit put into service by the Central Vermont Public Service Corporation in 1941 and operated through 1945, when the rotor failed.[55,56] No further operation was attempted because the cost of repairs appeared higher than could be justified economically. Note that this machine was designed with the help of the eminent aerodynamicist Theodore von Karman and by professors on the staff at M.I.T.

The abrupt change in the cost of energy in the 1970s led to a number of efforts to produce new systems that would exploit the latest technology to supply economical wind power. Through one of these efforts the Grumman Corporation developed a 20-kW generator with a 7.62-m-diameter (25-ft-diameter) three-bladed rotor. A much more ambitious NASA program has led to the construction of a series of machines ranging from a 100-kW unit at Plumbrook, Ohio, to a 2-MW unit on a mountain (elevation 1347 m, or 4420 ft) near Boone, North Carolina.[57,58] The principal design data for these machines, together with data on older windmills, are summarized in Table 6-9.

TABLE 6-9 Characteristics of Typical Windmills

Type of windmill	Rotor diameter, m	Tower height, m	Wind velocity			Design power output, kW	Cost in 1977 dollars, $/kW
			Min., m/s	Max., m/s	Design, m/s		
Nineteenth-century Dutch drainage mill	18				~8.	~25	
U.S. farm mill for pumping water	2.44	10				~0.5	~2400
U.S. farm mill for generating electricity	6					3	
Grumman Windstream	7.62	40	4	22.3	12.5	20	1093†
Smith-Putman generator, Grandpa's Knob, Vt.	55	34				1250	
NASA MOD-O	38	30	4.5	13.3	8.1	100	5500*
NASA MOD-OA	38	30	3.6	17.8	10	200	
NASA MOD-1	61	42.7	7	20	16	1500	2900
NASA MOD-2	91					2500	
Indian NAL 6-blade windmill	10	12	1.67		2.8	100	11,800

*NASA estimates production quantities will cost ~$1500/kWe.
†Cost is for a kit; does not include assembly and installation.

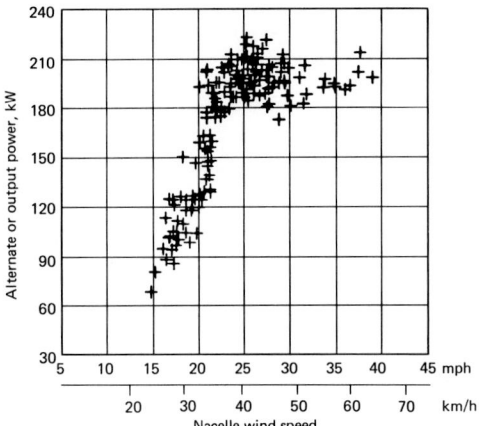

Figure 6-27 Test data for the NASA MOD-0 100-kW wind turbine. Each point represents the average for one revolution of the rotor (1.5 s).[57] *(Courtesy NASA.)*

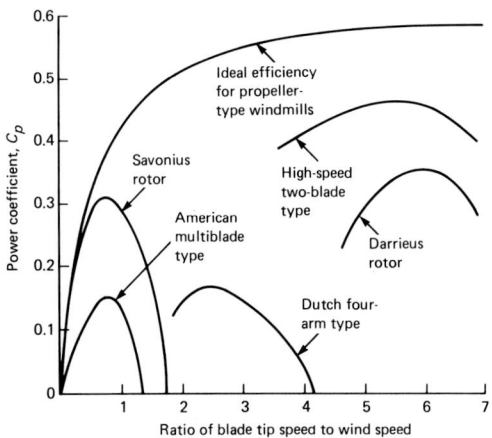

Figure 6-28 Typical performance curves for various types of wind machines.[55]

Characteristics and Problems

Probably the most difficult set of problems posed by windmill operation is the continually fluctuating velocity of the wind. Even if the wind appears to be steady, there are always minor fluctuations that tend to cause the windmill speed to vary. In the big old-fashioned windmills, small changes in wind speed were accommodated by turning the windmill so that the plane of the rotor was inclined away from the wind direction in order to reduce the output. Larger changes were accommodated by reefing the cloth sails. In the new Grumman and NASA machines, variations in wind velocity are accommodated by varying the pitch of the rotor blades. In practice, however, the fluctuations in wind velocity are rapid enough to make holding the rotor speed constant a major problem.[57,58,59] An indication of this is given by the width of the scatter band of points obtained in the test covered by Fig. 6-27. For large rotors these fluctuations are not uniform across the diameter of the disk swept by the rotor, and the resulting fluctuations in the forces on the blades lead to serious vibration problems with the blades commonly vibrating in many different modes. The resulting vibratory stress problems appear to be so severe that many experts consider the maximum practicable rotor diameter to be around 60 m (200 ft),[55] i.e., the size of the MOD-1 NASA machine in Table 6-9. Although blade vibration in that machine has proved to be a problem, the problems appear to be tractable, and a 91-m-diameter (300-ft-diameter) windmill is to be built for NASA in 1981. Even though windmills are designed so that the rotor is stopped if the average wind velocity exceeds a value somewhere between 15 and 25 m/s, severe gusts in storms are likely to produce failures such as that experienced with the Grandpa's Knob rotor in 1945.

A second major problem with windmills is that when the wind speed drops below the value for which the machine was designed to produce its rated output, the power output drops rapidly and normally falls to zero at a wind speed of around 4 m/s (9 mph) as a result of frictional and electrical losses. Thus, under the low-wind-velocity conditions that prevail through much of the summer and fall, the windmill will produce no power at all. If one looks at this quantitatively, the energy ideally available in an airstream with standard sea-level air density is given by $E = \frac{1}{2}\varrho v^3 = 0.0006124\ v^3$, where v is in meters per second and the energy is in kilowatts. Because of the finite number of blades in the rotor and aerodynamic losses, the actual energy that can be removed from the airstream is substantially less and is commonly obtained by multiplying the ideal energy available by the power coefficient. Figure 6-28 gives typical values for the power coefficient for representative types of rotor as a function of the ratio of the rotor tip speed to the wind speed.[55] Thus, if the power coefficient for a given rotor is 0.4, and the design wind speed is 10 m/s, the energy obtainable would be 0.61

kW/m² of rotor disk area. However, as stated before, the useful power output of this rotor would drop to zero for a wind speed of about 4 m/s where the energy in the airstream would be 39 W/m².

Windmills also give problems with icing during winter storms, and large windmills have caused both a high noise level and electromagnetic interference with TV sets. The latter problem can be eased by using blades constructed of fiberglass rather than metal, or by installing cable TV systems for viewers in the area (a step that NASA found necessary for its windmill in Rhode Island). The only solution to the noise problem appears to be the location of windmills remote from inhabited areas. (The noise problem was a major factor when NASA decided to dismantle the MOD-1 windmill near Boone, North Carolina.)

The gusty character of the wind also gives frequency control problems even when the unit is coupled into a utility grid that can tolerate the resulting fluctuations in power output. The problem is particularly difficult in mountainous terrain in which the ridges inherently generate large-scale eddies, evidence of which is easily seen in low clouds rolling over mountain ridges. Both the control and blade vibration problems are aggravated by the boundary-layer effect, which gives a substantial gradient in velocity from the ground surface to the highest level swept by the blade tips.

In attempting to assess the applicability of windmills to electric-utility power generation, it is necessary to consider the local wind velocities available. Figure 6-29 shows the average energy available in the wind as a function of geographical position in the United States.[60] The solid contour lines represent a series of constant values for the energy ideally available, and the dashed lines indicate boundaries for mountainous terrain. Note that most of the metropolitan areas in the United States are in regions in which the energy in the wind during the summer would average less than 100 W/m²—barely enough to turn the rotor of a windmill. This is quite a serious matter because, as indicated in the previous chapter, the cost of storing energy for more than perhaps 6 h becomes very high indeed. Further, the availability of windmill energy is not much better in the autumn months. Thus, as was concluded regarding solar-energy systems in the previous section, the capital charges for the electricity produced by windmills would be much higher than implied by the nominal capital cost per kilowatt of design capacity. Of course, it would be possible to design windmills so that they would function at the lower wind velocities implied by Fig. 6-29. This, however, would reduce the peak output obtainable from a given size of rotor and would actually increase the cost of the electricity that could be produced, because the tower and rotor strength, weight, and cost are determined by gust forces in storms rather than by the wind velocity at which the windmill is designed to function.

A good insight into the cost problem of windmills is given by a proposal for their use in India as sources of power for irrigation.[61] The Indian National Aeronautical Laboratory (NAL) has developed a low-cost unit using cloth sails in an effort to give the 70 percent of Indian farms that do not have electricity a means for irrigating their fields during much or all of the 9-month dry season, thus doubling or tripling the crop output in the many areas where ample groundwater supplies are available. Figure 6-30 gives NAL estimates, which compare the costs of irrigation using various energy sources, including two types of windmill, bullocks, diesels, and electric motors. Two sets of curves are given: one for an area of 1 ha (2.5 acres) and one for 3 ha (7.5 acres). The former is for an average plot, while the latter is for the maximum size that can usually be served from a single well. Two cases are shown for electrically-driven pumps, one including the total capital charge for the electric-power transmission system and the other neglecting that charge if it is written off against other uses for the electricity. It seems highly significant that windmills are barely competitive with other systems even for this favorable application involving irrigation water pumping in areas where labor costs are low and the wind speed during the pumping season would be above the design value of 2.8 m/s for an average of 10 h per day.

WAVE POWER SYSTEMS

It has been recognized for a long time that there is an enormous amount of energy in ocean waves, which in a sense concentrate some of the energy in the wind flowing over the ocean surface. At the time of writing, small-scale efforts to tap this energy were under way in Norway, the United States, Great Britain, and Japan, with the latter two nations giving their projects greater emphasis. Five principal problems are inherent with wave power systems:[56,62,63,64]

1. The energy is diffuse, hence systems are inherently large.

2. The wave forces in storms can be enormous; hence the structure must be very rugged.

3. Depending on wind conditions, waves vary widely in size and wavelength and vary in direction both on short- and long-term scales.

4. The mean water surface level changes with the tide.

5. The energy available varies widely and sometimes is nearly zero.

Many types of mechanisms have appeared promising as a means of converting wave energy into electricity. One employed by the Japanese is a cylindrical buoy whose mass and buoyancy

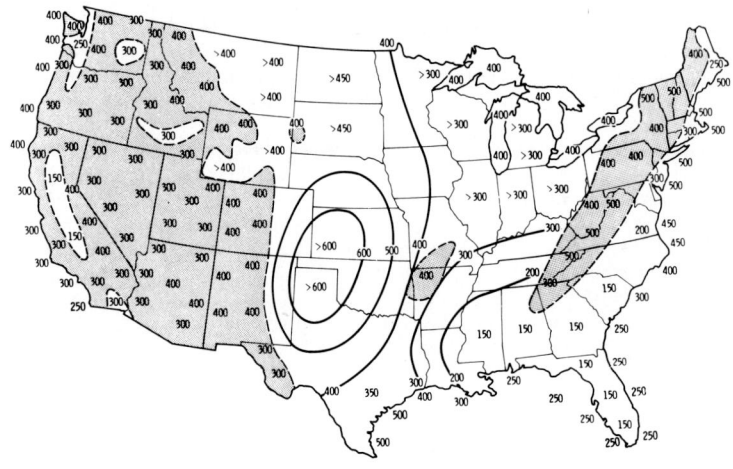

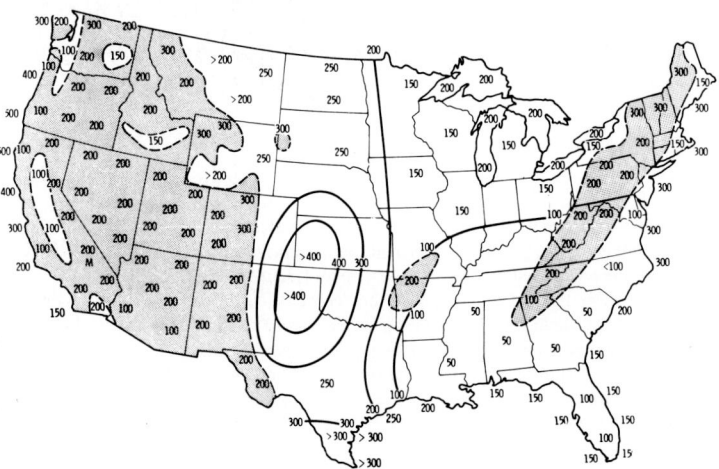

Figure 6-29 Available wind power in watts per square meter of rotor disk area at an elevation of 50 m above the ground. The solid contour lines are for constant values of wind energy while the dashed lines indicate the boundaries of mountainous terrain.[60]

FALL-WIND POWER (WATTS/M^2) AT 50 M

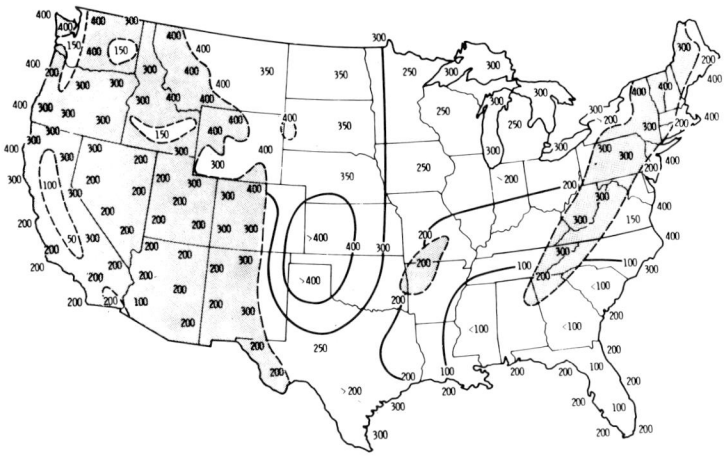

WINTER-WIND POWER (WATTS/M^2) AT 50 M

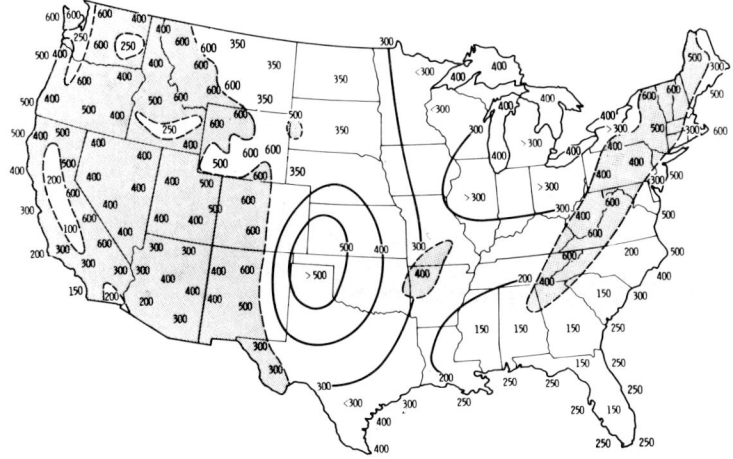

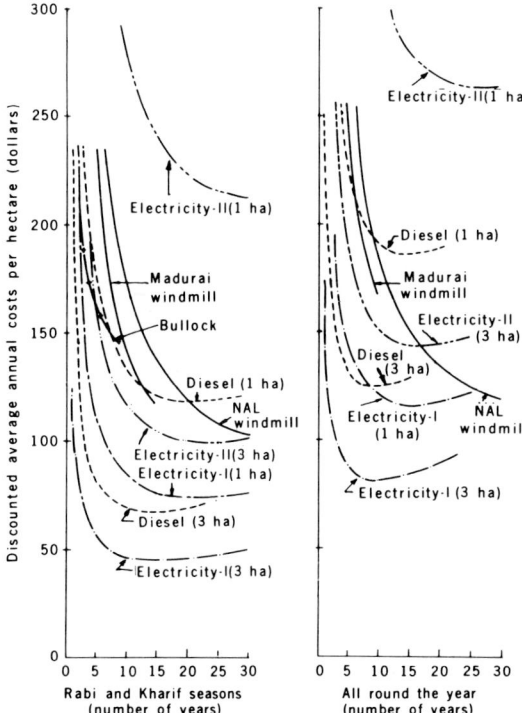

Figure 6-30 Discounted average annual costs in India for irrigating 1 ha (under wheat or equivalent water-consuming crops) from open wells with an average depth of 6.67 m for the water table. ("Electricity-II" includes transmission system costs, "Electricity-I" charges transmission system costs to other uses.) Windmills would prove to be economical if they were designed to operate effectively at low wind speeds and all year round.[61]

can be adjusted so that its natural frequency for bobbing with the waves is the same as the wave frequency, and thus its amplitude is perhaps three times that of the wave height. The vertical motion through the water drives a propeller turbine whose blade angle automatically changes when the buoy reverses direction so that the generator is always driven in the same direction by the turbine. Some of these units have been used to supply power to lights and horns on marker buoys, with a storage battery providing power when the sea is relatively quiet.

Other mechanisms of interest include one in which water acts as a piston to compress air in a cylinder and toroidal floats that move up and down over a central cylinder. In all cases the problem is to obtain a simple, reliable device that will withstand severe storms and yet with its complete electric energy collection system be cost-effective. The meagre cost data available do not look promising.

OCEAN THERMAL ENERGY CONVERSION

The roughly 20°C temperature difference between surface and deep water in tropical regions has been considered a potential source of power for a century. In the 1920s a plant designed to exploit this source was built on the coast of Cuba, but it failed to produce useful power.[56,65] The concept was ardently promoted in the 1970s, and major funding has led to extensive studies. These indicate that the cold water must be brought up from a depth of 400 m or more and that, to keep the pumping power to an acceptable level, the pipe size must be large (to keep the L/D ratio low) and should extend vertically downward from a floating power plant.[65,66] This favors a minimum capacity of ~ 100 MWe in a plant located ~ 130 km (80 mi) off shore.

Size and Costs

The inherently low thermal efficiency of the ideal Carnot cycle (~ 4 percent) means that enormous quantities of water must be handled and the heat exchangers and turbine must be very large. There are two principal approaches. The first involves a flash boiler to obtain steam directly from the warm seawater and accepting the problems of designing for a very large turbine rotor diameter—45 m (150 ft). The second is to use a thermodynamic working fluid such as ammonia or Freon and accept an additional temperature loss in the heat-transfer matrix of a boiler. The large cost of the turbine casing for the former approach has caused many to favor the latter with special surfaces giving improved heat-transfer coefficients.

Some insight into the cost of the heat exchangers required can be gained by drawing a parallel with condensers for conventional steam plants, because these are essentially similar. Table 6-10 gives data for both the TVA Bull Run plant and what appeared to the author to be the most promising OTEC design at the time of writing.[65] Perhaps the most significant points are the amount of surface area and the cooling-water pumping power required per net kilowatt of output. Note from line 10 that the condenser surface area required is almost 200 times as great for the OTEC plant, and note from line 8 that the condenser cooling-water pumping power requirements are 75 times as great. The pumping-power requirements for circulating the warm seawater and for deaerating the condenser are also large—about 60 and 70 percent, respectively, of the pumping power required for the cold seawater. This large parasitic power—equal to about half of the net output of even a very large power plant—is a serious matter, because if these estimates prove to be optimistic and the actual parasitic pumping power proves to be half again as great, the net plant output would be halved and the capital cost doubled. Similarly, the estimated cost of the condenser, $370/kWe, represents about 25 percent of the cost of the plant, and if that cost were to be increased, it would have a major effect on the overall plant cost. Note that the actual cost of the condenser for a conventional nuclear steam plant built in 1978 ran $10/ft^2, or nearly twice the value of $6.10 used in the estimates of Ref. 65. The cost of the condenser would be higher if ammonia or Freon were used as the working fluid because they give lower heat-transfer coefficients than water. This, coupled with the extra temperature loss in the boiler heat-transfer matrix required for an intermediate fluid, led the design team of Ref. 65 to conclude that an "open" cycle with a flash boiler would be a much more promising approach if the problems presented by a 45-m-diameter (150-ft-diameter) turbine rotor could be solved. They evolved a clever and promising rotor design based on helicopter rotor technology. The problems of getting a reasonable cost for the structure of the turbine casing, which must withstand the enormous external pressure loads involved (the turbine must operate in a near vacuum—1 percent of atmospheric pressure), were solved by making the turbine casing of reinforced concrete and an integral element of the plant structure. Figure 6-31 shows a section through the complete plant in which most of the walls are surfaces of revolution.[65] Note that this design is particularly clean from the hydrodynamic standpoint, and the bulk of the structure (virtually all of which is reinforced concrete) serves the dual functions of providing fluid flow passages and hull structure.

Two interesting and difficult problems not always considered in OTEC plant designs but accommodated in the design of Fig. 6-31 stem from wave motion in high seas. One problem with an open cycle is that waves could cause large percentage variations in the pressure in the flash boiler or large fluctuations in the local flow rates unless the plant hull is designed so that its elevation stays constant in relation to the mean sea level and the warm seawater inlets and outlets are sufficiently far below the sea surface to keep the local pressures constant. A second problem is that rolling motions induce severe bending stresses of the joint between the cold seawater inlet pipe and the hull. These problems in the dynamics of the system of Ref. 65 were handled by providing much of the hull flotation force with the toroidal air cushion shown in Fig. 6-31, using a large diameter hull (107 m), and choosing a set of proportions and a mass such that the plant would heave with really large waves in very heavy seas but would

TABLE 6-10   Data Related to Condenser Size and Cost for Typical Coal-Fired and OTEC Plants

Line no.	Parameter	TVA Bull Run plant (Ref. 17 of Chap. 11)	Westinghouse OTEC plant (Ref. 65)
1.	Plant net output, MWe	850	100
2.	Plant gross output, MWe	914	148
3.	Steam flow, kg/s (lb/h)	478.2 (3.787 × 10^6)	1379 (10.9 × 10^6)
4.	Steam flow, m^3/s (ft^3/s)	1533 (53,870)	1692 (5.95 × 10^6)
5.	Steam flow, heat load, MWt (Btu/h)	1043 (3.561 × 10^9)	3296 (11.25 × 10^9)
6.	Condenser cooling water flow, kg/s (gpm)	25,070 (397,500)	419,936 (6.657 × 10^6)
7.	Condenser cooling water pumping power, kWe	2355	21,000
8.	Condenser cooling water pumping power, %	0.28	21.0
9.	Condenser surface area, m^2 (ft^2)	29,730 (320,000)	606,700 (6.5 × 10^6)
10.	Condenser surface area, m^2/kW (ft^2/kW)	0.03498 (0.376)	6.07 (65)
11.	Condenser steam temperature, °C (°F)	30.5 (87)	7.78 (46)
12.	Cooling water inlet temperature, °C (°F)	12.8 (55)	4.44 (40)
13.	Cooling water exit temperature, °C (°F)	22.7 (72.9)	6.25 (43.2)
14.	LMTD, °C (°F)	12.2 (22)	2.33 (4.2)
15.	Heat flux, W/m^2 [Btu/(h · ft^2)]	35,098 (11,128)	5441 (1725)
16.	Overall heat-transfer coefficient, W/(m^2 · °C) [Btu/(h · ft^2 · °F)]	89.1 (506)	72.4 (411)
17.	Cooling water pipe length, m (ft)		973 (3192)

not respond to short-wavelength sea surface motions.

The cost of the plant of Fig. 6-31 was estimated to be about $1500/kWe in 1977 dollars not including provisions for mooring or power transmission to shore, both of which would be substantial. A separate but related study yielded a cost of about $420/kWe in 1976 dollars for a dc transmission line to a plant 130 km (80 mi) off shore. The costs for ac transmission lines were higher.

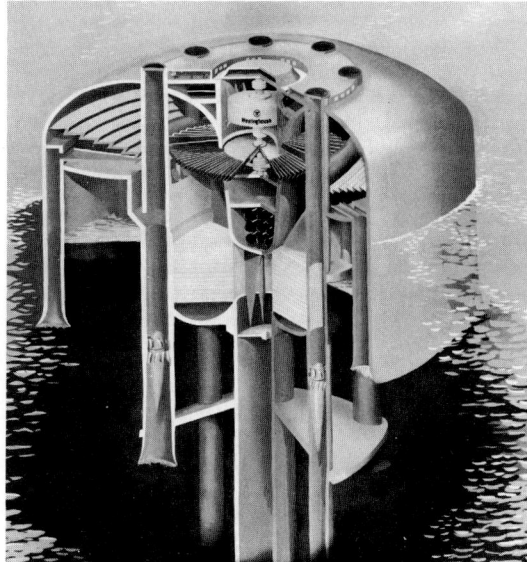

Figure 6-31 Conceptual design for a 100-MWe OTEC power plant having a diameter of 107 m (351 ft).[56] *(Courtesy Westinghouse Electric Corp.)*

Operating Experience

Although several systems have been built, the first to produce a net power output was an experimental OTEC unit built by the state of Hawaii and tested off its coast in 1979.[67] The turbine produced 50 kWe with a 22 °C (37 °F) difference in temperature between the surface water and water drawn up at a rate of 2700 gpm from a depth of 655 m (2150 ft) through a 56-cm-ID (22.1-in-ID) polyethylene pipe. By using a low velocity (0.68 m/s, or 2.2 ft/s) in the cold-water intake pipe, the pumping power required for the system was kept down to 40 kW so that there was a net output of ~10 kWe.[66] Testing was terminated after about six weeks when the cold-water intake pipe was broken off in a storm.

Combining an OTEC Plant with Mariculture

It has long been recognized that the plant nutrients in cold, deep ocean water can greatly increase the productivity of the surface water. In fact, the open ocean (90 percent of the total area) produces only ~0.7 percent of the fish because most of the nutrients in the surface water are extracted by plants and drift down to the ocean floor in the remains of plant or animal life. The waters in the coastal zones are continually supplied with fresh nutrients in the runoff from the adjacent land, and hence support a high level of plant life activity and produce 54 percent of the fish. Only 0.1 percent of the ocean area lies in the upwelling regions, where nutrient-laden water is brought up from the ocean depths, yet these regions produce 44 percent of the fish.[68] The reason for this spectacular difference can be seen in Table 21.11, which shows that the nitrate and phosphorus concentrations in deep seawater are ~150 and 5 times, respectively, their concentrations in surface water at a typical site.

There have been several efforts to produce an upwelling effect. At the urging of the oceanographer C. O. Iselin at Wood's Hole in 1956, the author examined the possibility of using a system of nuclear power plants on the sea floor to produce upwelling but concluded that the cost would be excessive. However, in

TABLE 6-11 Concentrations of Nutrients in Both Surface Water and Water from a Depth of 870 m at St. Croix in the U.S. Virgin Islands[68]

	Nutrients (μg-atoms/L)*				
	$(NO_3 + NO_2) - N$	$NO_2 - N$	$NH_3 - N$	$PO_4 - P$	$SiO_4 - Si$
Surface water (3 km offshore)	0.2	0.2	0.9	0.2	4.9
870-m-deep water	31.3	0.2	0.7	2.1	20.6

*Concentrations given for the key nutrient atoms, and do not include other atoms of O or H in the ion.

the 1970s, O. A. Roels studied the possibility of using a shore-based OTEC plant to supply nutrient-laden water to a mariculture system, and since 1972 he has supported his analyses with a series of experiments carried out at St. Croix in the U.S. Virgin Islands. At that site the ocean is 1000 m deep only 1.6 km offshore. Three polyethylene pipelines 6.9 cm (ID) and 1830 m long have brought ~250 L/min of bottom water into 5-m^3 pools where diatoms from laboratory cultures are grown. The food-laden effluent flows through metered channels to pools where shellfish are raised. The resulting protein production rate has been excellent; 78 percent of the inorganic nitrogen in the deep seawater has been converted to phytoplankton protein-nitrogen, and 22 percent of that was converted to clam-meat-protein-nitrogen. This compares with plant-protein/animal-protein conversion ratios of 31 percent for cows' milk production and 6.5 percent for feedlot beef production.

Coupling an OTEC plant to a mariculture operation would require shore-based plants because in the open sea the cold water from the depths would sink back down with relatively little mixing. For a shore-based plant roughly 50 km^2 would be required for the mariculture tanks per 100 MWe produced. Suitable sites with deep water close to shore would be limited largely to volcanic islands, hence the bulk of the electric power produced would probably go into energy-intensive operations such as the production of aluminum or nitrogen fertilizer. Chlorine probably could not be used to prevent biofouling of heat exchanger surfaces, hence the open cycle of the system of Fig. 6-31 would probably be preferable. These and related disadvantages of coupling OTEC and mariculture systems appear to be much more than offset by the value of the sea food produced when Roels estimates would run about 5 times the value of the electricity. (See Table 6-12.)

TABLE 6-12 Estimate of the Annual Output from a Combined OTEC-Mariculture System Using a Deep-Water Flow Rate of 3.75 m^3/s (Ref. 68)

System	Product	Assumptions	Unit price	Gross sales value
OTEC	1MWe	90% of time on line	4¢/kWh bus bar	$ 315,360
Mariculture	1693 t shellfish	42% meat	$1/lb meat	1,564,200
	Whole wet weight	711 t meat	(= $2.20/kg)	

TABLE 6-13 Distribution of Gross Energy Usage in the U.S. in 1977*

INDUSTRIAL

	Quads	Percent of U.S. energy
Process steam	9.1	12.0
Direct heat	6.3	8.3
Electric drives/lighting	7.2	9.5
Feedstocks	4.7	6.2
Subtotal	27.3	36.0

TRANSPORTATION

	Quads	Percent of U.S. energy
Roads		
Automobiles	11.2	14.8
Buses	0.1	0.1
Trucks	4.9	6.5
Other		
Railroads	0.8	1.0
Air	1.7	2.3
Ships	1.1	1.5
Pipelines	0.8	1.0
Other	0.6	0.8
Subtotal	21.2	28.0

RESIDENTIAL

	Quads	Percent of U.S. energy
Space heating	8.6	11.3
Water heating	2.4	3.2
Large appliances	2.3	3.0
Lighting	1.8	2.4
Air conditioning	1.0	1.3
Subtotal	16.1	21.2

COMMERCIAL

	Quads	Percent of U.S. energy
Space heating	4.7	6.2
Water heating	0.6	0.8
Large appliances	0.8	1.1
Lighting	2.6	3.4
Air conditioning	1.7	2.2
Miscellaneous	0.8	1.1
Subtotal	11.2	14.8
Total for U.S.	75.8	100

*United States Energy Data Book, 1979, General Electric Company, Energy Systems and Technology Division, Sunnyvale Calif.

REFERENCES

1. Burnham, J. B., and D. H. Stewart: *Foreign and Domestic Discussions on Natural Geothermal Power and Potential Use of Plowshare to Stimulate These Natural Systems*, Battelle Northwest Laboratories Report No. BNWL-B-110, July 1971.

2. Lengquist, R., and F. Hirschfeld: "Geothermal Power, the 'Sleeper' in the Energy Race," *Mechanical Engineering*, vol. 98, no. 12, December 1976.

3. Kiewicz, J. Pietrusz: "Optimum Design Conditions for a Power Plant at a Vapor-Dominated Geothermal Resource, P.G. and E.'s Geysers Power Plant Unit No. 16," *Proceedings of the Thirteenth IECEC*, Aug. 20-25, 1978.

4. "Geothermal Energy-Central America," *Mechanical Engineering*, vol. 97, no. 9, September 1975, p. 57.

5. "Furnace Beneath Your Feet," *Mechanical Engineering*, vol. 100, no. 9, September 1978, p. 47.

6. Milora, S. L., and J. W. Tester: *Geothermal Energy as a Source of Electric Power*, The M.I.T. Press, Cambridge, Mass., 1976.

7. White, D. E., and D. L. Williams: *Assessment of Geothermal Resources of the United States—1975*, Geological Survey Circular 726, U.S. Geological Survey, Reston, Va., 1975.

8. Samuels, G.: "Geopressure Energy Resource Evaluation," Oak Ridge National Laboratory Report No. ORNL/PPA-79/2. May 1979.

9. Barker, L. M., et al.: "Geothermal Environmental Effects on Drill Bit Life," *Proceedings of the Eleventh IECEC*, Sept. 12-17, 1976, p. 711.

10. Cristy, G. A., and W. C. McClain: "Examination of High Pressure Water Jets for Use in Rock Excavation," Oak Ridge National Laboratory Report No. ORNL/HUD-1, January 1970.

11. *The Second International Symposium on Jet Cutting Technology*, BHRA Fluid Engineering, Cambridge, England, April 1974.

12. Altseimer, J. H.: "The Subterrene Rock-Melting Concept Applied to the Production of Deep Geothermal Wells," *Proceedings of the Eleventh IECEC*, Sept. 12-17, 1976, p. 717.

13. Dutcher, J. L., and L. H. Moir: "Geothermal Steam Pricing at the Geysers, Lake and Sonoma Counties, Cal.," *Proceedings of the Eleventh IECEC*, Sept. 12-17, 1976, p. 786.

14. Allen, C. A., et al.: "Fluidized Bed Heat Exchangers for Geothermal Application," *Proceedings of the Eleventh IECEC*, Sept. 12-17, 1976, p. 761.

15. Johnson, F. S.: "The Solar Constant," *Journal of Meteorology*, November 1959, p. 431.

16. Spurr, S. H.: "Silviculture," *Scientific American*, vol. 240, no. 2, February 1979, p. 76.

17. "Utilities Put the Sun to Work," *EPRI Journal*, vol. 3, no. 2, March 1978, p. 26.

18. Hager, K. G., and C. A. Berg: "Wood Residue-Fired Gas Turbine," paper presented at the Forest Product Research Society Conference, Atlanta, Ga., November 1976.

19. *An Assessment of Solar Energy as a National Energy Resource*, NSF/NASA Solar Energy Panel, December 1972.

20. Reding, J. T., and B. P. Shepherd: *Energy Consumption: Fuel Utilization and Conservation in Industry*, Dow Chemical, U.S.A., Texas Division, prepared for U.S. Environmental Protection Agency, EPA-650/2-75-032-d, August 1975.

21. Parker, H. W.: "Disadvantages of Fermentation Ethanol as a Motor Fuel," paper presented at the Central Carolinas Section of the AIChE, Mar. 20, 1980.

22. Goldemberg, J.: "A Madeira como Fonte de Carburantes Liquidos, Anais, V Encontro," *Jornal dos Reflorestadores*, ano 1, no. 2, Brazil, April 1979, pp. 13-17.

23. da Silva, J. G., et al.: "Energy Balance for Ethyl Alcohol Production from Crops," *Science*, vol. 201, no. 4359, September 1978, pp. 903-906.

24. Hopkinson, C. S., Jr., and J. W. Day, Jr.: "Net Energy Analysis of Alcohol Production from Sugarcane," *Science*, vol. 207, no. 4428, Jan. 18, 1980, pp. 302-303.

25. Villar Ferrin, A. B.: "Sintese do Metanol Derivado dos Residuos da Madeira, Anais, V Encontro," *Jornal dos Reflorestadores*, ano 1, no. 2, Brazil, April 1979, p. 28.

26. Hottel, H. C.: "Residential Uses of Solar Energy," *Proceedings of the World Symposium on Applied Solar Energy, Phoenix, Ariz.*, 1955.

27. Yellott, J. I.: "Power from Solar Energy," *Trans. ASME*, vol. 79, 1957, pp. 1349–1359.

28. Barber, R. E.: "Current Costs of Solar-Powered Organic Rankine Cycle Engines," *Solar Energy*, vol. 20, no. 1, 1978, pp. 1–6.

29. Tester, J. W., et al.: "Comparative Performance Characteristics of Cylindrical Parabolic and Flat Plate Solar Energy Collectors," ASME paper, 74-WA/Ener-3, November 1974.

30. Lof, G. O. G., et al.: "Energy Balance on a Parabolic Cylinder Solar Collector," *Journal of Engineering for Power, Trans. ASME*, vol. 84, no. 1, January 1962, pp. 24–32.

31. Spight, L. D.: "Solar Insolation Measurements at Las Vegas, Nevada," *Solar Energy*, vol. 20, no. 2, 1978, pp. 197–203.

32. Ramsey, J. W., et al.: "Experimental Evaluation of a Cylindrical Parabolic Solar Collector," *Journal of Heat Transfer, Trans. ASME*, vol. 99, 1977, pp. 163–173.

33. Tracey, T. R., et al.: "1 MW(t) Solar Cavity Steam Generator Solar Test Program," *Proceedings of the Twelfth IECEC*, 1977, pp. 1224–1230.

34. Schweinberg, R. W., and J. N. Reeves: "Solar One Project—A 10-MW Solar Thermal Central Receiver Pilot Plant," *Proceedings of the Fourteenth IECEC*, Aug. 5–10, 1979, pp. 181–182.

35. Kreider, J. F., and F. Kreith: *Solar Energy Handbook*, McGraw-Hill, New York, 1981.

36. Chalmers, B.: "The Photovoltaic Generation of Electricity," *Scientific American*, vol. 235, no. 4, October 1976, pp. 34–43.

37. Sahai, R., et al.: "High Efficiency AlGaAs/GaAs Solar Cell Development," *Proceedings of the Thirteenth IEEE Photovoltaic Specialists Conference*, 1978, pp. 946–952.

38. Queisser, H. J., and W. Shockley: "Some Theoretical Aspects of the Physics of Solar Cells," Energy Conversion for Space Power, vol. 3, *Progress in Astronautics and Rocketry*, Academic Press, New York, 1961, p. 317.

39. Moon, R. L., et al.: "Multigap Solar Cell Requirements and the Performance of AlGaAs and Si Cells in Concentrated Sunlight," *Proceedings of the Thirteenth IEEE Photovoltaic Specialists Conference*, 1978, pp. 859–867.

40. Donovan, R. L., and S. Broadbent: "10 kW Photovoltaic Concentrator Array, Martin-Marietta Aerospace Corp.," Sandia Laboratories Report No. SAND-78-702A, May 1978.

41. "Concentrating Array Production Process Design," Final Report, General Electric Company Report No. 78SDS4266, Sandia Report No. 78-7072, Dec. 15, 1978.

42. Wolf, M.: "Advances in Silicon Solar Cell Development," Energy Conversion for Space Power, vol. 3, *Progress in Astronautics and Rocketry*, Academic Press, New York, 1961, p. 231.

43. Middleton, A. E., et al.: "Evaporated C & S Film Photovoltaic Cells for Solar Energy Conversion," Energy Conversion for Space Power, vol. 3, *Progress in Astronautics and Rocketry*, Academic Press, New York, 1961, p. 275.

44. Donovan, R. L., and S. Broadbent: "Photovoltaic Concentrating Array," *Proceedings of the IECEC*, 1978, p. 1593.

45. Brooks, G. R., et al.: "Orbiting Solar Observatory (050-8) Solar Panel Design and In-Orbit Performance," *Proceedings of the IECEC*, 1978, pp. 105–109.

46. Denman, O. S.: "From Sunlight in Space to 60 Hz on Earth—the Losses along the Way," *Proceedings of the IECEC*, 1978, pp. 178–182.

47. Caputo, R.: "An Initial Comparative Assessment of Orbital and Terrestrial Central Power Systems," Final Report, Jet Propulsion Laboratory Technical Report No. 900-780, March 1977.

48. "Solar Power Satellites," *Proceedings of the IECEC*, 1978, pp. 178–204.

49. Woodcock, G. R., *Solar Power Satellite*, vol. 1, Boeing Aerospace Co. Report No. D180-22876-1, December 1977.

50. Payne, P. R., and D. W. Doyle: "The Fossil Fuel Cost of Solar Heating," *Proceedings of the IECEC*, 1978, p. 1650.

51. Hunt, L. P.: "Total Energy Use in the Production of Silicon Solar Cells from Raw Materials to Finished Product," *Proceedings of the Twelfth IEEE Photovoltaic Specialists Conference—1976*, Nov. 15–18, 1976, pp. 347–352.

52. Robinson, A. L.: "American Physical Society Gives a Long-Term Yes to Electricity from the Sun," *Science*, vol. 203, no. 16, February 1979, p. 629.

53. Freese, S.: *Windmills and Millwrighting*, David and Charles, Ltd., Devon, England, 1971.

54. Spier, P.: *Of Dikes and Windmills*, Doubleday & Company, Inc., Garden City, N.Y., 1969.

55. Hirschfeld, F.: "Wind Power—Pipe Dream or Reality?" *Mechanical Engineering*, vol. 99, no. 9, September 1977, p. 20.

56. "The Earth as a Solar Heat Engine," *EPRI Journal*, March 1978, p. 43.

57. Glasgow, J. C., and A. G. Birchenough: "Design and Operating Experience on the U.S. D.O.E. Experimental MOD-O 100 kW Wind Turbine," *Proceedings of the Thirteenth IECEC*, Aug. 20–25, 1978, pp. 2052–2055.

58. Richards, T.R., and H. E. Neustadter: "DOE/NASA MOD-OA Turbine Performance," *Proceedings of the Thirteenth IECEC*, Aug. 20–25, 1978, pp. 2060–2062.

59. Linscott, B. S., et al.: "Experimental Data and Theoretical Analysis of an Operating 100 kW Wind Turbine," *Proceedings of the Twelfth IECEC*, 1977, p. 1633.

60. Elliott, D. L.: "Synthesis of National Wind Energy Assessments," Pacific Northwest Laboratories Report No. BNWL-2220, WIND-5, July 1977.

61. Tewari, S. K.: "Economics of Wind Energy Use for Irrigation in India," *Science*, vol. 202, no. 4367, Nov. 3, 1978, p. 481.

62. "Will Japan Be First with Solar Wave Power?" *Mechanical Engineering*, vol. 100, no. 1, January 1978, p. 53.

63. Merriam, M. F.: "Wind, Waves, and Tides," *Annual Review Energy*, 1978, p. 29.

64. Swann, M.: "Power from the Sea," *Environment*, vol. 18, no. 4, May 1976, p. 25.

65. "100 MW(e) OTEC Alternate Power Systems," Final Report, Westinghouse Electric Corp., Power Generation Divisions, Nov. 29, 1978.

66. *A Conceptual Feasibility and Cost Study for a 100 MW(e) Sea Solar Power Plant*, United Engineers and Constructors, Inc., 1976.

67. Hartline, B. K.: "Tapping Sun-Warmed Ocean Water for Power," *Science*, vol. 209, no. 4458, Aug. 15, 1980, pp. 794–796.

68. Roels, O. A.: "From the Deep Sea; Food, Energy, and Fresh Water," *Mechanical Engineering*, vol. 102, no. 6, June 1980, pp. 37–43.

Section **7**

Energy from Municipal Wastes

Waste generation from residential and industrial sources continues to grow. With this growth comes greater problems in finding suitable disposal areas—usually landfill. The combination of waste growth, reduced landfill areas, and a reasonably high heating value of refuse—5000 to 6000 Btu/lb—has led designers to consider generating energy from municipal wastes.

Early waste-to-energy facilities have had problems. Some facilities were never completed; others were inoperable. From this experience has emerged a number of techniques that permit designers to plan and build successful plants to convert waste to energy. These techniques are covered in this section.

Two important techniques successfully used to burn waste to generate energy are considered—(1) mass-burning, and (2) processed-fuel. Interestingly, as a reflection of just one of the problems in using municipal wastes to generate energy, both techniques remove large noncombustible items—such as refrigerators and engine blocks—from the waste stream before feeding it to a boiler.

A variety of other problems face the designer. These include topics such as communicating with the community, financing the project, pollution myths, recovering landfill gas, power from sludge, and gas combustion. Each of these is discussed in this section to assist the designer in making the wisest choice of system for waste-to-energy conversion.

From "Energy from Municipal Wastes," *Power*, vol. 129, no. 12, December 1985. Copyright © 1985. Used by permission of McGraw-Hill, Inc. All rights reserved.

Each year in the US, over 300-million tons of refuse are created from residential and commercial sources. To dispose of this waste, historically by landfill, municipalities spend an average of $25/ton, totalling $7.5-billion/yr. If collection and transportation costs are included, the figure approaches $40-billion. This landfilled refuse has an average heating value of between 5000 and 6000 Btu/lb, translating roughly into 15,500 MW of potential capacity which could produce 135-billion kWh of electricity. Thus, it is no wonder that waste disposal has ceased to be a simple matter of landfilling, but has suddenly transformed into a multi-billion-dollar industry called resource recovery. David Sokol, president and chief executive officer of Ogden Martin Systems Inc, Paramus, NJ, estimates that, today, $4-billion annually is expected to be spent on this growing technology in the future.

The fact that resource recovery is rapidly gaining acceptance was evident at the recent meeting, *Energy from Municipal Wastes,* sponsored by POWER magazine and the newsletters *Synfuels* and *Waste to Energy Report,* all McGraw-Hill Inc publications. The nearly 200 attendees let two dozen speakers take them through the process of developing, building, and operating a municipal waste-to-energy facility. They heard that the industry is on the edge of full development with the years of experiment and failure in the past. The technical obstacles have largely been overcome, so that developers are free to concentrate on the equally difficult problems of community participation, financing, and environment. As the speakers explained, these can be overcome as well, as long as a project has been carefully prepared and developed.

Learning from mistakes

Waste-to-energy-facility developers learned many things from those early years of uncompleted and inoperable projects. And these experiences have been applied to the recent, more successful plants in operation today. Charles A Johnson, technical director of the National Solid Wastes Management Assn, Washington, DC, compiled a list of the five significant factors contributing to a successful resource-recovery facility:

1. Waste reduction—*not* energy production—is by far the more important function of a resource-recovery plant. Anyone who thought that garbage simply was a free fuel typically did not have a successful project. What has moved projects ahead, Johnson asserted, has been a community's realization that it is rapidly running out of disposal space.
2. A full-service contractor is the preferable procurement method. Traditionally, municipalities have contracted with multiple vendors. For waste-to-energy projects, they have turned, many for the first time, to a single source responsible for design, construction, operation, and even ownership, of the facility.
3. Tax incentives have helped to keep facility costs in line with conventional disposal options. Tax-exempt financing, the investment tax credit, and accelerated depreciation can reduce disposal costs by up to $20/ton to a project if it is structured to take maximum advantage of the incentives. One or more of these incentives may be taken away by the proposed tax-law changes being debated in Congress.
4. Project financing has forced developers to eliminate all unacceptable risks before a project can get under way. The project's financial underwriters refuse sponsorship until these risks are resolved. Johnson felt that this results in stronger, more successful facilities.
5. Proven waste-to-energy technology has been whittled down to two options: Mass burning and processed fuel. The "high-tech" options have all failed. Mass-burn and processed-fuel technologies are considered fully proven by industry engineers.

A mass-burning facility is just as its name describes: The refuse is burned as received. Any processing the fuel receives involves only hand removal of large noncombustible material (refrigerators, engine blocks, etc) from refuse stream and rough mixing of the remainder equipped with a clamshell bucket. The facility's boiler typically has a moving grate.

In a processed-fuel—also called refuse-derived-fuel (RDF)—system, the solid waste is, obviously, processed before firing. This is a two-stage operation. The first stage, using equipment such as trommels and separators, divides the noncombustibles from the combustible materials. Trommels screen out glass, grit, and aluminum cans; magnets remove metals that can contribute material-handling problems in boiler feeding and ash removal.

The second stage of a typical RDF system takes the remaining waste—now composed largely of combustible material—and reduces it to uniform-sized pieces using a hammermill-type shredder. These pieces, no larger than a few inches in size, are delivered to the boiler for combustion.

The additional equipment required by an RDF plant makes for a more complex operation, but RDF is a more desirable fuel in the view of Don Kaminski, general manager at Heil Engineered Systems Co, Milwaukee, Wis. He listed several reasons:

- Homogenous fuel permits controlled combustion resulting in fewer air pollutants.
- Improved burn-out of fuel produces a more valuable and environmentally acceptable ash.
- Capital and maintenance costs are lower.
- Reliability and availability are greater.
- Less boiler combustion air and smaller fans and electrostatic precipitators mean lower power consumption.

- Material processing can continue while boilers are down.
- Boilers can closely follow steam demand.

There is another option for resource recovery that can be implemented using existing landfills. This involves drawing off the gas produced by the decomposition of organic materials. The gas contains about 500 Btu/ft^3 and can be burned in a conventional boiler, gas turbine, or internal-combustion engine. This method does not, however, address the environmental and space problems associated with landfilling practices today.

Remaining obstacles

Since the technology question has been successfully solved, why aren't waste-to-energy projects burgeoning throughout the US? The answer lies in the unique nature of these facilities. For most municipalities, waste disposal has involved trucking garbage to a landfill and depositing it there. Most of the nation's municipalities have little or no expertise in power generation. And a resource-recovery facility is probably the largest expense by far that a community has ever contemplated, involving financial risks never before encountered. In short, for most communities, a waste-to-energy plant means breaking new ground.

Groundbreaking means answering the questions of a significant number of constituencies. These questions do not lie in the technological area—where few uncertainties remain. They are related to the community, regarding where the plant will be located; to economics, as in how will it be paid for; and to environment, specifically air pollution. Answering these concerns takes time, but as the answers become more generic, the pace of new projects will pick up, adding momentum to the already quickening industry. Here is how the McGraw-Hill conference suggested facing these nontechnical obstacles to successful waste-to-energy projects.

Communicate with the community

Acceptance by the community is the first requirement of any resource-recovery project. And this acceptance will clearly not be forthcoming without fulfillment of two important elements: A clear and real garbage crisis and enthusiastic support and endorsement from the community's political leaders. Certainly, if politicians aren't convinced of rapidly diminishing landfill capacity, then the citizens who elected them won't be either. It is worthwhile considering the credibility of the politicians, as well.

Once the problem has been certified and support lined up, the next requirement is enlisting the backing of the community as a whole, but particularly those citizens who will call the facility neighbor. Regardless of where the plant is located, someone will find it too close for comfort. And all the political support available will not be able to overcome the stone wall of community opposition to a chosen site. As a result, "this aspect of the project must be addressed early and unequivocally resolved well before implementation begins," said Harvey W Gershman, president of Gershman, Brickner & Bratton Inc, Washington, DC. He recommends communicating project plans and expected impacts early and often. Public-relations advisors can develop a series of newsletters, brochures, information, and public presentations to keep the issue before the community. John W Rogers, regional planner at Rogers, Golden & Halpern in Philadelphia suggests enlisting interested citizens for public discussions of the plant, emphasizing the waste-disposal problem that engendered it.

If the public is not kept informed of all project developments, Rogers warned, "all hell breaks loose." The projects become LULUs (locally unwanted land use) and the citizens Nimbys (not in my back yard). Both Rogers and Gershman believe that this kind of reaction is caused largely by a lack of understanding about the project.

While communication is the single most important factor in dealing with the public, projects succeed by using other tactics as well. Siting on land already considered undesirable, such as industrial areas and old landfills, can quiet the opposition. Remember that a resource-recovery plant does not eliminate the need for a landfill. About 30-50% by weight and 5-15% by volume of the waste burned in a mass-burn facility leaves in the form of bottom ash and flyash. Landfilling of large noncombustible waste picked from the refuse stream prior to combustion may also be necessary, and capacity must be available for disposing of raw garbage while the plant is down, either scheduled or unscheduled.

Offering incentives to the citizens most significantly affected by a plant site has also overcome public opposition. These can be reduced tipping fees, a capital fund generated from a portion of the fees, road improvement, and future land use as a park, golf course, or the like. Insurance to cover plant neighbors in case of leaking from landfills also has been available over the last few years. Past successful projects, Rogers noted, have not been shy about dispensing incentives.

Project financing

Once the community has thrown its support behind the project, the next step is finding the money to build it. This can be a lengthy process, as the projects are—more often than not—perceived as capital intensive and highly risky. Neither the municipality nor the project's vendor is willing to commit unconditionally to repay project debt, John Schopfer, vice president at Kidder Peabody & Co, New York, explained, so another source of debt guarantee must be found. Existing projects have found this third source to be the plant's revenues—from tipping fees, energy sales, and, in some cases, resource recovery (such as aluminum cans and other metals).

This financing, called revenue-bond financing, involves commitments from project's various participants. Oversimplified, the commitments are as follows: The municipality agrees to provide waste to the facility and pay for its disposal. The vendor commits to building the facility for a set price and operating it for a fee. A customer for the energy must be found who agrees both to taking and paying for the energy. These are a project's *technical* risks.

The bondholder, yet another party—usually an investment banker—adopts the *investment* risk. In return, he receives tax-exempt interest at rates ranging from 6% to 9.5%.

Some projects are considered more risky, and thus require additional debt security. This has included that backup taxing power of the local or state government, municipal-bond insurance, efficacy insurance for system performance, and additional contributions from the vendor.

Revenue-bond financing can be arranged with either public or private ownership. Private ownership has an added advantage in that the owner can capture the tax benefits of accelerated depreciation and the investment tax credit unavailable to publicly owned projects. A private owner would make an equity contribution towards construction costs that he would reap later when paying taxes. This contribution lowers the project's costs so it requires fewer bonds. Consequently, debt service is reduced.

Waste-to-energy project financing is now in a very critical period, according to Harvey Gershman. The next generation of projects will place heavy demands on capital resources over the next several years. Capital markets may tighten even further if the changes in the tax law proposed by President Reagan are approved by Congress. This proposal threatens to eliminate the tax advantages that resource-recovery projects have enjoyed—specifically it would disallow the use of the investment tax credit. In the face of mounting concern over this, John F Beatty, vice president at Morgan Stanley & Co, New York, tried to calm his audience: "The proposed tax law is not the end of the world." In fact, he said, it may lead to new sources of capital, while reverting project ownership from private contractors back to municipalities.

There are inherent disadvantages in the present tax scheme, Beatty explained. The investment tax credit is cumbersome and inefficient, making investors wait two to three years to earn their credits. There are few investors who will know their liability several years in advance. Accelerated depreciation can create problems for an investor if a project defaults. It can leave him or her tax liable for the remaining debt of the project.

The projects that the proposal is playing havoc with are the ones trying to get started now. No date has been set by Congress for when the proposed law would be effective, although it is likely to be retroactive. As a result, Beatty encouraged those with pending projects to quickly arrange tax-exempt debt in the hope of getting on the right side of the transition date. Don't wait around for the best deal, he warned.

David Sokol saw a much darker picture if the tax law eliminates advantages for private owners. If the market moves back to largely public sponsorship, without the financial support of vendors, Sokol thought, landfilling may once again be more economically viable than resource recovery.

Any project depending on tax benefits for success is much less likely to be completed, Beatty stated. The bottom line for any waste-to-energy project is the tremendous need for safe and efficient waste disposal. Unlike tax benefits, this need will increase over time.

Pollution myths

Of all the uncertainties facing waste-to-energy plants, their environmental impact has proved the most difficult to resolve. While engineers and regulators are satisfied that emission of dangerous toxins remains well within safety parameters, convincing the public of this has been so far mostly unsuccessful.

In 1981, EPA concluded from a study of five waste-to-energy plants that dioxins did not pose a public health hazard. Over the ensuing four years, however, the public concern over the pollutant has mounted. There is a "multi-million-dollar/yr business in research" that perpetuates "the dioxin myth," said David Sussman, manager of the waste-combustion program for EPA's Office of Solid Waste. The issue will never be resolved until that research is cut off. Dioxin emissions pose the same health threat as living in New York City for three days, Sussman added. The agency has not determined an acceptable level of dioxin emissions, because "we don't develop standards for something that is not a hazard."

Sussman explained that there is no way to get rid of garbage—a pollutant—without having an impact wherever it is disposed. EPA believes that the best medium for disposal is not the ground, but the air. While public opposition is centered on dioxins, there are several more toxic pollutants present—notably those from metals, such as arsenic, cadmium, and lead.

These views were echoed by Clinton S Kemp, executive vice president at the Canruf Co, Toronto, Ontario, Canada. He recommended that the waste-to-energy industry concentrate on bolstering the confidence of regulators and the public that dioxins are not a threat in well-designed and -operated facilities. In fact, Kemp explained, a well-run plant can eliminate nearly all air-emission problems.

Kemp described four main sources for air pollution in a waste-to-energy facility:
- Solid waste prior to combustion.
- Solid waste during combustion.
- Incomplete combustion of solid waste.
- Too complete combustion of solid waste.

Preventing the escape of odor from decomposing garbage is simply accomplished by drawing the primary combustion air from the outdoors through the tipping-hall area. This continually flushes the air of odor, as well as truck-exhaust fumes. A moderately reduced pressure in the building prevents air leaks to the outside.

Keeping emissions that result from combustion below accepted standards is a two-part process at any well-run resource-recovery plant. One part of the effort involves keeping the wastes that produce the more toxic emissions from getting into the fuel stream in the first place. Success at this requires a vigilant operations crew that will turn away unfamiliar trucks at the plant gate and carefully examine the waste stream for unusual quantities of tires and sealed containers.

Many wastes that create emissions, however, are acceptable to the fuel stream and those emissions must be controlled—most notably particulate matter. This is collected in conventional post-combustion pollution-control systems such as electrostatic precipitators and fabric filters. Kemp noted that the varying degrees of control among the states calls into doubt what the most efficient levels are. Most states have set limits at 0.03 gr/scf, while California and Connecticut have called for 0.01 gr/scf and New York for 0.02 gr/scf.

It is the third pollution source that Kemp finds responsible for dioxins and similar pollutants. If there is incomplete combustion in the secondary firing zone, research has found, the level of these materials increases significantly. Yet another pollutant is formed if too complete combustion occurs, NO_x.

Conditions for complete combustion seem to be an 8-9% oxygen level in the flue gas, CO below 100 ppm by volume, and flue gas just above the flame zone at 1500F for a second or two. Technologically, Kemp added, the industry knows how to design and operate waste-to-energy plants in this fashion.

Perfect sponsor

Perhaps the single most difficult part of developing a waste-to-energy facility is successfully uniting all of the disparate elements and participants towards the same goal—operation of a resource-recovery plant.

Every project needs one sponsor willing to accept responsibility for development and implementation. Without such a leader, Harvey Gershman stated, inevitable pitfalls and barriers will delay or even terminate the project.

The sponsor can be from the municipality, industry, utility, or vendor, but he or she must have more stellar qualities than a Boy Scout, John Rogers added: authority, credibility, jurisdiction, tenacity, and responsibility.

Everyone involved in the industry must reflect such values, David Sokol concluded, "maintaining the highest level of ethics and quality."

Recovering landfill gas

While recovery and combustion of landfill gas does not—as noted earlier—confront the environmental and space problems of solid-waste disposal, it is still a viable method of producing power from waste. At costs ranging from $0.45 to $5/million Btu, it is considerably cheaper than many energy sources. In raw form, it has a reasonably high heating value—500 Btu/ft^3—and it can be upgraded to pipeline quality. Studies have shown it burns more cleanly than natural gas.

Recovering landfill gas does eliminate some environmental hazards. It will reduce the odors emanating from many landfills resulting from decomposing garbage. Emissions from the landfills often kill vegetation, making land reclamation difficult; gas recovery can reduce these as well. Finally, the danger of explosion and fire—resulting from the buildup of methane gas trapped underground—can be almost eliminated.

Many communities are garnering additional revenue from their landfills by selling the recovered gas (or the right to recover it) to pipeline companies, utilities, or industrial plants. Like those burning solid waste, other communities are becoming power generators themselves.

For all of its benefits, the landfill-gas-recovery industry took a long time to develop, but like solid-waste combustion, it is expected to take off dramatically in the next several years. The first contract to recover gas in the US was signed in 1973; in mid-1985, there were 44 facilities—23 of these are recovery plants and the remaining are electric-generating stations. Within the next year however, over twice as many facilities are expected to be on-line, 95 in all. This new industry estimates that at least 3400 of the nation's 14,000 landfills could be economically adapted for gas recovery.

But don't get the idea that the gas can be merely pumped out of a few holes in a landfill and burned. The rate, quality, and quantity of gas generated can vary by as much as 300% between landfills of similar size. The parameters influencing gas production are dependent on a number of factors, including the age of the landfill, refuse composition and density, pH, the bacterial strain present, weather conditions (particularly rainfall), the temperatures existing in the landfill, and the depth of the wells. This information is necessary to determine the economics of any landfill-gas-recovery project.

Methane in a landfill is generated by anaerobic bacteria in the presence of moisture and heat once the waste is covered to restrict the inward diffusion of O_2. Biodegradation yields a raw gas that contains nominally 45-60% methane and 35-50% CO_2 by volume; the balance is nitrogen, O_2, trace hydrocarbons, and hydrogen sulfide. This last component is usually less than 0.3% by weight of the total. An interesting point: Landfill gas from sites on the East Coast tends to have a higher methane content than it does in California, mostly because of the difference in available moisture. These East Coast facilities will have a shorter life span, however, 5-7 years as compared to 15-18 years.

Collection of landfill gas is similar to recovery operations at natural-gas fields, except that the system is under vacuum up to the gas compressor. After wells are drilled into the landfill, pipe made of polyvinyl chloride, fiberglass-reinforced plastic, or high-density polyethylene, is used to construct the collection system.

Because of the vacuum, a substantial amount of water is generally removed with the gas. This water contains small amounts of suspended matter that has filtered through the decaying material. Local weather conditions—such as heavier-than-normal rainfall or high differential temperature between the gas and the outside air—can aggravate the moisture-intrusion problem. Provisions must be made to minimize the amount of water entering the system with the gas, to separate out the water that does enter, and to clean lines of any water that may accumulate.

Once the gas is collected and compressed, it must be treated to remove the nonmethane hydrocarbons and any remaining moisture. If the gas is to be upgraded to a high-Btu stream (so-called pipeline quality), then the CO_2, and possibly the hydrogen sulfide, must also be removed.

A wide variety of processes are available to remove CO_2. One of the most popular involves contacting the gas with a chemical solvent. In a two-step process, the solvent "strips" nonmethane hydrocarbons and CO_2 from the landfill gas. Once these are separated, the solvents are regenerated and returned to the stripping column.

Physical stripping processes, solid-bed absorption, chemical conversion, and cryogenic processes are also available. The user must evaluate the advantages and disadvantages of each.

Another approach that may be more economical for smaller sites—those generating less than 100,000 ft^3/day—is based on the principle of membrane separation. Here, the gas under pressure diffuses through hollow-fiber membranes encased in a cartridge. CO_2 travels to the hollow inside of the fiber and exits at one end of the cartridge; methane gas exits under high pressure at the other end. The technology has been demonstrated at a landfill site in Alabama.

Moisture removal is critical to prevent corrosion caused by contaminants present in the condensate. At a large landfill site, located at Fresh Kills, Staten Island, NY, moisture separation involves three steps: entrainment separation, aftercooling, and desiccant drying. Keep in mind that this moisture-removal train at Fresh Kills is for a small landfill-gas-to-electric-power demonstration. Most gas recovered here is more extensively treated for pipeline use. About 5-million ft^3 of the upgraded gas is fed into the network of Brooklyn Union Gas Co.

The entrainment separator, located ahead of the gas compressor, is a large-diameter fiberglass tank with a demister pad. Entrained water is separated from the gas stream by changes in velocity and direction of the stream. After the compressor comes a chiller system, consisting of finned tubes. Air is blown across the tubes as the hot compressed gas flows inside them. Following the aftercooler is the desiccant dryer which removes the final traces of moisture.

Gas combustion

Laboratory combustion tests with a natural-gas-fired burner design show that landfill gas exhibits stable flames over a nominal two-to-one turndown range of fuel input, similar to that of natural gas. A noticeable difference is that the landfill-gas flame is shorter because of improved fuel/air mixing. The volumetric flow of landfill gas is about twice that of natural gas because of its lower heating value, and results in higher fuel-injection velocities. Note that this characteristic also keeps CO emissions lower when landfill gas is used in a gas-fired boiler.

Flame and flue-gas temperatures attained when landfill gas is burned with 10% excess air are similar to those for natural gas when it is burned with 20% excess air. In some cases, this may mean reduced NO_x emissions. However, the increase in fuel-gas flow resulting from diluents in the landfill gas may mean an increase in flue-gas pressure drop and possibly require larger ducts or more induced-draft-fan power.

The Central Contra Costa Sanitary District in California has been burning landfill gas in two packaged boilers for over two years now. Both boilers retain their oil/gas-firing capability. Major modifications to the boilers to accommodate the new fuel include: a new burner for each boiler; separate, 4-in. landfill-gas piping trains; a 6-in. supply line to the boiler room; a new pressure-reducing and metering station; and new boiler-stack O_2 analyzers to maintain the correct air-to-fuel ratios during fluctuating gas-supply conditions. The system is designed to handle instantaneous swings in Btu content of up to 5%, a lower heating value of 450 Btu/ft^3, and a normal value of 580 Btu/ft^3. Gas dehydration and other moisture-removal equipment precedes the boiler.

An automatic shutoff valve is linked to an O_2 analyzer in the gas-supply line. If the O_2 in the landfill gas is too high, the valve cuts off supply to the boiler.

Operating performance to date has been positive. The variability in Btu content of the gas was less than expected. This situation allowed operators to abandon the O_2-trim system, which had never lived up to expectations.

Boiler-performance tests showed no degradation in efficiency either. One reason is that the landfill gas contains up to 1% O_2. This reduces excess-air requirements by about 5%, compared to natural-gas firing. No corrosion has been found in the boiler or on the inside of the gas-supply line. However, extensive corrosion occurred on the outside of the supply line leading to the burner and on the burner housing itself, but this was attributed to abnormal conditions that caused condensation of combustion products.

For a gas-turbine-based system, the large amount of CO_2 may be an advantage. Reason: Less air has to be compressed. Gas turbines normally require 300-400% excess air for cooling purposes so that the blades can withstand the high temperature of the gas. The CO_2 enters the turbine already compressed and displaces some of the air-input requirement. Corrosion of internal parts is thought to be a potential problem for gas turbines and engines burning landfill gas. Limited experience with some small installations has demonstrated corrosion-free operation, but it is too early to judge the long term impact on design, operation, and reliability.

Most of the landfills where gas is recovered are located in California. Ironically, air-quality regulations recently enacted may bring the industry there to a sudden halt. The regulations impose strict limits on the levels of methane which can be emitted from a landfill. They were developed in order to limit reactive organic emissions; instead they may limit the amount of gas the recovery facilities can produce.

Not all action on the regulatory front is detrimental. Recently, the state of Illinois recognized that gas recovery is a form of pollution control.

Power from sludge

Sewage sludge is another waste product that can be successfully burned in boilers to generate power—making wastewater treatment at industrial plants and municipalities more energy-efficient. As defined here, sludge is the material left over after process water or sewage has been treated for eventual return to the receiving waters. The material is usually no more than 5% solids after primary and secondary treatment, but the dry solid material has a heating value of between 3000 and 10,000 Btu/lb.

Like other energy-rich waste products, sludge is usually landfilled or land-applied in the US if the acreage is nearby and available, or, if a facility is near the coast, it is disposed of in the ocean. Where these options have not been available, the sludge has traditionally been incinerated, often with high premium-fuel consumption, and with little or no heat recovery.

These practices are changing, especially in heavily populated areas. Demand for better management of sludge wastes is coming about because of high transportation costs, mandates against ocean dumping, solid-waste regulations from EPA, the threat of new regulations regarding groundwater contamination from landfills, the near impossibility of permitting new landfills, land scarcity, and expensive and questionable long-term supplies of premium fuels.

While energy recovery from sludge has broad technical and social acceptability, problems still abound—both engineering and institutional. To illustrate: Many municipalities reverted to landfilling and retired existing incinerators after air-pollution regulations forced them to either buy expensive control equipment or shut down. Today, when a community proposes to build an energy-recovery plant with incineration, people in the area remember the odors and dust that were around with the previous operation. The "Nimby" syndrome often prevails.

On the technological side, sludge combustion can be compared to other waste-burning processes, but its development is slightly behind its refuse cousins. Solid waste has been through its experimental stage, and now has a string of successes as more and more waste-to-energy facilities come on stream. A similar development is in store for sludge combustion, although the pace is contingent on EPA support and whether agricultural uses become more acceptable or economical. The technology and equipment are available, if not the level of experience needed to gain full confidence in implementation.

Before sludge can be burned autogenously—without auxiliary fuel—some level of water removal is necessary. Regardless of the combustion process, sludge burns more easily if some or all of the water is removed. A sludge with an average heating value of 6000 Btu/lb will burn autogenously at between 25% and 30% solids. Conventional gravity settling brings wastewater sludge up to an average of 5% solids. Dewatering beyond this level requires a continuous filtering process. Four pieces of equipment popular in this service are vacuum filters, centrifuges, belt presses, and filter presses.

Any dewatering process usually requires some upstream sludge conditioning for good continuous operation—the extent depends on composition of the sludge and its ability to retain water within the interstices of a sludge conglomerate. Sludge is a colloidal gel composed of particles contained within sheaths that bind a high proportion of the water content. Releasing this water makes filtering easier. Both chemical- and thermal-conditioning methods are available to unlock this water.

Using sludge as a fuel, however, requires water removal beyond what can be accomplished by conditioning and dewatering. Water can be evaporated from dewatered sludge by using either a direct- or an indirect-drying process. An indirect dryer is essentially a sludge/low-pressure-steam heat exchanger, since the sludge never contacts the steam. In direct drying, commonly called flash drying, sludge and hot gases at up to 1200F are intimately mixed and brought down to temperatures between 250F and 500F and the water is flash-vaporized.

One important criterion for choosing between the two processes is the degree of drying desired. Indirect steam dryers achieve solids contents between 60% and 75%. Beyond this level, the exchanger surface area increases sharply because the heat-transfer coefficient from the tube wall to the product decreases rapidly as the feed material loses water. Direct dryers, on the other hand, can bring the solids up to 90% because of the intimate mixing that takes place between the hot gases and the sludge.

If the mechanical complexities and energy requirements of direct and indirect sludge dryers are too imposing, then consider a multiple-effect evaporation process. One process, called Carver-Greenfield after its inventors, makes use of a carrier oil to help transport the difficult-to-handle sludge, improve heat transfer, and protect equipment from corrosion and fouling. It has been used commercially to dry several types of sludges—ranging from pharmaceutical and brewery wastes to municipal sewage—up to a final solids content of 95% and over.

After the sludge has been dewatered and/or dried, the next step is to optimize the combustion process with energy recovery. Many wastewater-treatment facilities have fluidized-bed-combustion (FBC) units or multiple-hearth furnaces on site that were originally intended for sludge disposal only. Therefore, older units tend to have little or no energy-recovery capability and also consume a great deal of auxiliary fuel to sustain combustion. To illustrate: One major supplier of FBC units has supplied energy-recovery equipment—in the form of heat-recovery boilers or economizers—to over one-third of its customers in the US over the last 20 years.

Today, the basic equipment has been redesigned for energy efficiency. Successful heat-recovery techniques are available to eliminate fuel consumption and, in the case of larger installations, to generate excess energy as well.

One of the most attractive ways to make use of sludge is to burn the material together with municipal waste, coal, or other solid fuel. Part of this concept—extracting boiler or incinerator flue gas to dry sludge—has been discussed.

Co-combustion can be accomplished by modifying a multiple-hearth furnace or fluidized-bed combustor to burn carefully prepared refuse-derived fuel with sludge, although only the FBC route has been used commercially in the US. A mass-burning or RDF-designed steam generator can be retrofitted for simultaneous sludge combustion. Experience to date overwhelmingly favors this second approach.

Converting the heating value of sludge to a more useful form is yet another option in the energy-recovery-from-sludge scenario. For example, the age-old anaerobic-digestion process is receiving renewed interest as a method of converting wastewater sludge into a methane-rich gas, nominally 500 Btu/ft^3. Like landfill gas, the fuel is suitable for direct combustion in gas turbines, gas-fired boilers, and gas engines—in many cases, without modifications to the mechanical equipment. Municipal sewage plants have been practicing the technique for years. Today, many companies are adapting the process to the requirements of industrial wastewater-treatment plants.

Section 8

Energy Storage Systems

The objective of energy storage is to counteract the disadvantages that result from fluctuations in demand for electric energy by assuring a steady output from existing power plants. When demand is lower than capacity, energy is stored. When demand is higher than capacity, the stored energy is released.

Storage of energy allows a plant to supply electricity to its load reliably, efficiently, and economically. Peak electrical demands can be met on short notice without overloading the plant.

The need for energy storage was not so acute when generating plants were cheap and the fuel supply plentiful. With increasing supply problems for all fuels, the need to conserve resources, and the increasing unit capital costs ($/kW) and production costs (mills/kWh), suitable methods of energy management—which includes storage—became necessary.

To evaluate any proposed storage method, a designer must know what methods are available, how they compare from an efficiency standpoint, and the relative costs of the methods. This section provides that information for a variety of storage methods.

While energy storage is not considered too often in the usual technical literature, it is a topic that arises often during plant selection. For this reason, the power-plant designer should be familiar with the variety of choices available today.

From *Powerplant Technology,* M.M. El-Wakil. Copyright © 1984. Used by permission of McGraw-Hill, Inc. All rights reserved.

INTRODUCTION

The need for energy storage arises because the demand for electric energy in a utility system is characterized by hourly, daily, and seasonal variations, whereas the supply from that system, in the majority of cases, has a fixed capacity. That capacity must be selected to correspond to the maximum demand plus a reasonable excess to take care of scheduled plant shutdowns for maintenance and unscheduled shutdowns due to abnormal occurrences. The result of this is large, expensive plants that operate much of the time below capacity, thus causing high operating and capital costs.

An example of electric-energy fluctuations in consumption during a typical week in the life of a largely university town is shown in Figs. 8-1 and 8-2 [1]. They demonstrate the differences between daytime and nighttime consumption, between weekdays and weekends, and between summer and winter. More severe fluctuations occur in an industrial or commercial region, where demands drop on weekends. Seasonal fluctuations in all regions also occur. The picture is more clouded, and the need for energy storage is greater, if plants using renewable forms of energy such as solar and wind are used to generate electricity. It is the output of these plants that fluctuates severely because of their input energy intermittency. Their conversion systems are also much more expensive than those of conventional plants.

The objective of energy storage, therefore, is to counteract the disadvantages that result from the fluctuations in demand for electric energy by assuring a steady high output from existing powerplants. When the demand is lower than capacity, energy is stored. When the demand is higher than that capacity, the stored energy is released. The result then is to be able to supply electricity reliably, efficiently and economically,

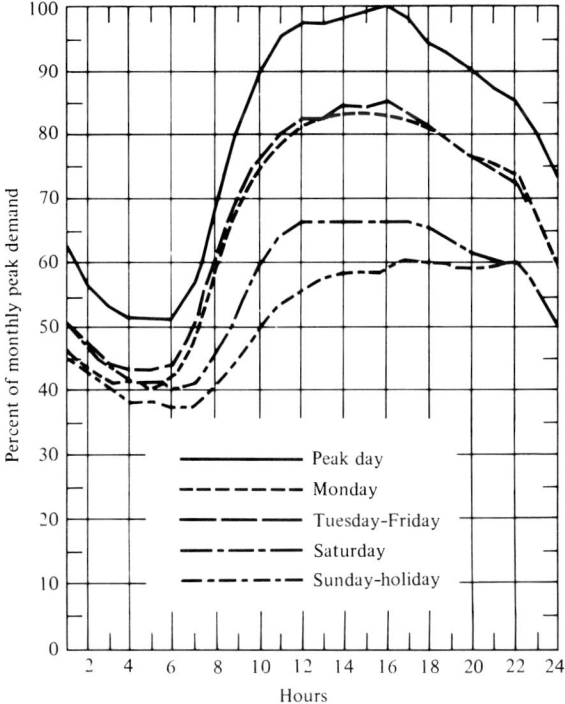

Figure 8-1 Hourly and daily electric-power consumption during summer (average July 1976–1980) in Madison, Wis. [1].

while being able to provide peak electrical demands on short notice during certain times of the day or week.

The need for energy storage was not acute when the generating plants were cheap and the fuel supply plentiful. Indeed, energy storage has, in a sense, been historically done in the form of the latent energy stored by nature in the fuels themselves. The energy density in this "natural" storage is large. Fossil fuels, coal, oil, and gas (at 1000 psia), all contain about a million Btu/ft^3 ($\sim 37 \times 10^6$ kJ/m^3). Natural uranium metal (0.071 percent U^{235}) contains about 3×10^{12} Btu/ft^3 ($\sim 10^{14}$ kJ/m^3). With increasing supply problems of all fuels, the need to conserve resources, and the increasing unit capital costs ($/KW) and production costs (mills*/kWh) of electric-generating plants, suitable methods of *energy management* become necessary. Examples of these are:

1. Supply power peaks by interconnecting power networks that might have different power demands on them.
2. Use newer and more efficient powerplants for base-load generation and use older less efficient plants for peak-power generation.

---

* 1 mill = one thousandth of a dollar = 0.1¢.

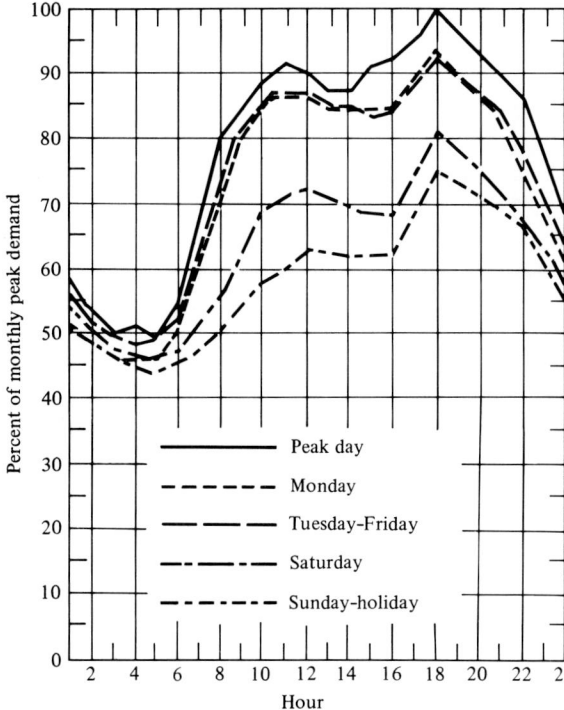

Figure 8-2 Hourly and daily electric-power consumption during winter (average December 1976–1980) in Madison, Wis. [1].

3. Construct smaller, low capital cost, though not so efficient powerplants, as power-peaking units. Examples are special steam plants, small hydroelectric plants, or gas-turbine peaking plants.
4. Add energy-storage systems.

In general, reliability and economy of electrical supply can best be achieved by having a mix of three types of powerplants: a base-load plant, a cycling plant, and a peaking plant.

Base-load plants are used to provide a base electrical load to the grid. Such plants are usually large, efficient, steam-generating, Rankine-cycle type stations powered by fossil or nuclear fuels. They operate continuously except for scheduled maintenance or forced outages. They have a power operating factor (POF) between 60 and 70 percent. This relatively high POF results in a comparatively low unit cost of power (mills/kWh).

Cycling plants, also called *intermediate plants,* usually are older, less efficient steam plants, or new ones specifically designed for cyclic operation, such as combined cycles. They operate primarily during hours of high load demand and have an annual POF between 25 and 50 percent. This rather wide range is primarily the result of seasonal variations, such as those due to periods of high industrial output, air-conditioning loads in the summertime, etc.

Peaking plants are specifically designed to provide relatively inexpensive power during peak-demand periods, such as due to abnormal air-conditioning loads and peak-hour domestic demands in the evenings. They operate at a low annual POF of 5 to 15 percent. They also operate at a low availability factor, i.e., intermittently according to system requirements. Their operation may be for as little as 2 or as much as 12 h/day, for as many as 5 days/week.

The last of the courses of action, energy storage, is the one discussed in this chapter. We are here concerned with large-scale storage suitable for incorporation with utility electrical powerplants. Energy storage would allow the plants to be designed for nearly constant load operation below peak demand, a process called *peak shaving*, which would thus reduce the high capital cost of the initial plants. Energy storage, of course, becomes attractive only when the capital and operating costs of the storage system are more than offset by the reduction in the corresponding costs of the original system.

One drawback to all energy-storage systems is that their energy densities are much lower than those mentioned above for fossil and nuclear fuels.

ENERGY-STORAGE SYSTEMS

There are basically two generic approaches to energy storage in utility systems (Fig. 8-3). These are (1) electrical storage and (2) thermal storage.

Electrical Storage

The primary electric-generating plant is continuously operated in a base-load mode, which results in excess electricity production during the off-peak periods $ab + cd$ (Fig. 8-4). Electrical storage is then used to hold this excess electricity for use during peak demand, period bc. Note that the total energy stored is greater than the total energy supplied because of conversion losses to and from storage. Storage schemes in this category that are in use or under investigation are:

1. Electrical-mechanical energy storage
    a. Potential, pumped hydro
    b. Potential, compressed air
    c. Potential, springs, torsion bars, mass elevation
    d. Kinetic, flywheels
2. Direct electrical energy storage
    a. Batteries
    b. Superconducting coils

Thermal Storage

In thermal storage, all schemes deal with storing energy in a thermal form in a material during periods of low power demand and releasing it back during periods of high demand. The primary electric-generating plant is operated to meet the real-time elec-

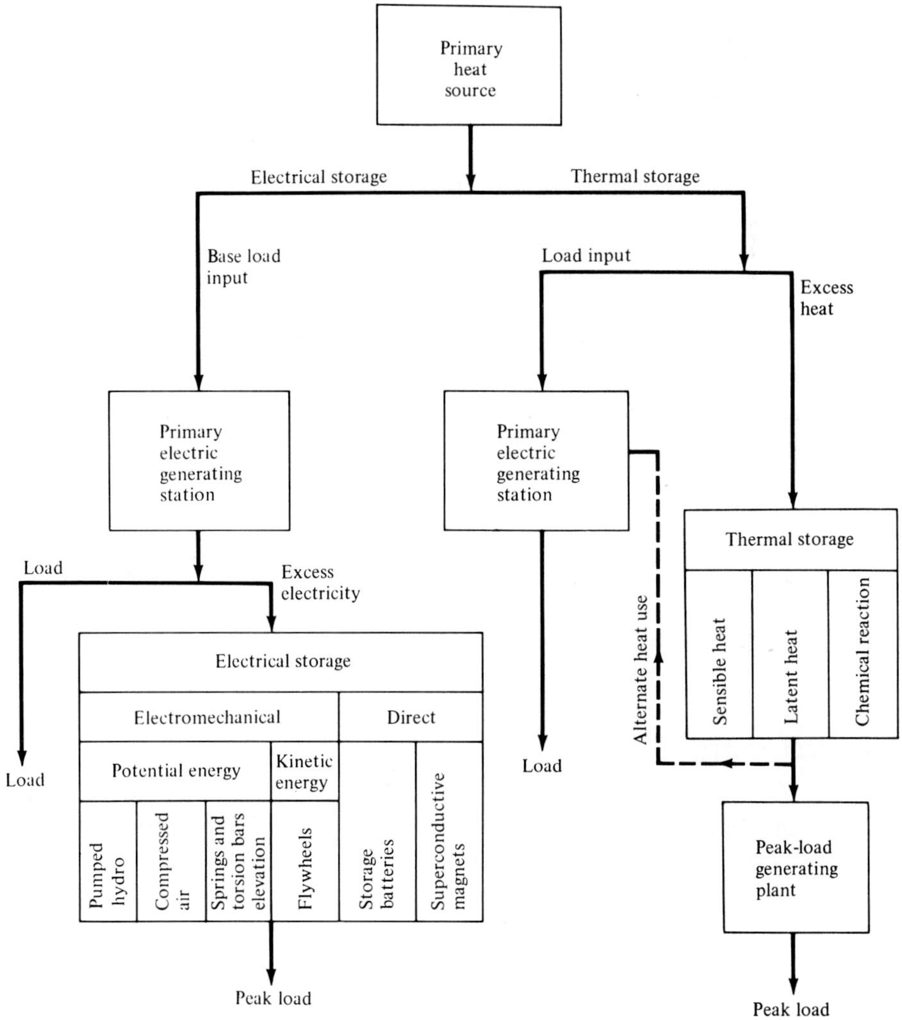

Figure 8-3 Energy storage systems.

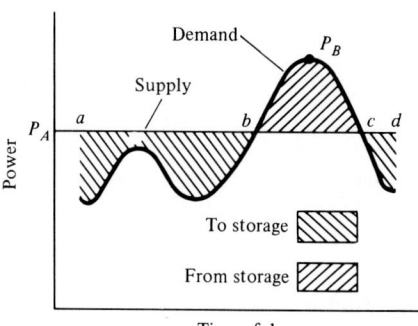

Figure 8-4 Thermal-energy storage with constant thermal input as from fossil or nuclear fuel.

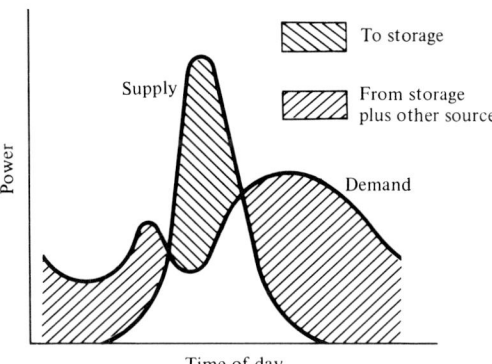

Figure 8-5 Thermal energy storage with hypothetical varying thermal input as from solar incidence.

trical demands during off-peak hours. The available thermal energy input to the plant may be essentially constant, as is that from fossil or nuclear fuel (Fig. 8-4), or varying, as from solar incidence (Fig. 8-5). The excess thermal energy is stored as such and withdrawn to be converted to meet peak electrical demands. Conversion could occur in the primary plant itself or in a separate peaking plant (Fig. 8-4). Again, note that because of storage losses and conversion inefficiencies the total stored energy is greater than that supplied.

Thermal storage schemes include:

1. Sensible heat
2. Latent heat
3. Chemical reaction

Not all the various electrical and thermal energy-storage schemes are suitable for large utility energy storage. Some, like springs, torsion bars, and mass elevation, are very low-capacity systems that are used to power such small devices as watches, clocks, toys, and instruments. They will not be covered in this book. Others, like flywheels and batteries, are in the developmental stages and will probably be suitable for intermediate storage. A few, like pumped hydro, compressed air, and superconductivity are, or will be, suitable for large utility energy storage. These and the various other schemes are covered in the following sections.

PUMPED HYDRO

Pumped hydro, like compressed air, is a potential-energy storage system suitable for large utility energy storage. It is the most developed and used of all storage systems. The principle behind pumped hydro is simple and follows the law of potential energy PE that is, the raising of mass to an elevation, height, or head H. It is given by

$$\text{PE} = \frac{g}{g_c} mH \tag{8-1}$$

where PE = potential energy, ft · lb$_f$ or J

g = gravitational acceleration = 32.2 ft/s² or 9.81 m/s²

g_c = conversion factor = 32.2 lb$_m$ · ft/(lb$_f$ · s²) or 1.0 kg/(N · s²)

m = mass, lb$_m$ or kg

H = head, ft or m

The operating heads on the pump turbine in the pumping mode H_p and in the turbine-generating mode H_T are different and are made up of two components each

$$H_p = H + H_\ell \quad (8\text{-}2)$$

and
$$H_T = H - H_\ell \quad (8\text{-}3)$$

where H is the static head or height and H_ℓ represents the losses during flow conditions (which are different because of the different flow rates).

The pumping and generation powers are given by replacing the mass in Eq. (8-1) with the mass-flow rate $\dot{m}$ lb$_m$/s or kg/s and using the proper head, or with

$$P_p = \frac{g}{g_c} \rho \dot{Q}_p H_p \quad (8\text{-}4)$$

and
$$P_T = \frac{g}{g_c} \rho \dot{Q}_T H_T \quad (8\text{-}5)$$

where P_p and P_T = pumping and turbine powers, respectively, ft · lb$_f$/s or W

ρ = density of water, lb$_m$/ft³ or kg/m³

$\dot{Q}_p$ and $\dot{Q}_T$ = volumetric flow rates in pumping and generation, respectively, ft³/s or m³/s

Equation (8-1) shows that 1000 kg raised 100 m will store 9.81×10^5 J or 0.2725 kWh. Thus large masses must be elevated to sufficiently large heights to store large quantities of energy. Fortunately large masses are available in pumped-hydro systems by the elevation of large quantities of water from a lower to an upper reservoir. One or both of these reservoirs may be artificially excavated or may be a natural river or lake.

Pumped-hydro systems, unfortunately, require a suitable topography that will allow the design and construction or selection of two reservoirs with sufficient capacities, maximum available elevation difference H, and minimum horizontal distance L between them to reduce flow losses. Values of $L/H < 2$ are considered very favorable, although most existing plants average L/H between 4 and 6 with some nearly as high as 10. Good topographies are obviously not available everywhere.

Although high heads are desirable, some topographies do not allow them, and thus pumped-hydro systems are often classified as *above-ground*, which includes the preferred *high head* and *medium head*, and *underground*.

Above-Ground Pumped Hydro

In *high-head* installations, the upper reservoir may have originally been a stream descending a steep slope which has been dammed to form the reservoir. From that reservoir, water is diverted into a horizontal pressure tunnel driven through the rock to the valve house from which the main steel pipeline slopes down to the powerhouse. At the head of the steel pipeline there usually is a surge tank and a valve house. It contains the main sluice valves, which are automatic isolating valves that come into operation in the case of pipeline burst. Automatic air valves, also, may be used. These contain buoyancy floats that fall when sufficient air separates from the water. These floats are attached to a spindle which then opens the valve to vent the air to the atmosphere. Other automatic air valves allow air to enter the pipeline in case the pipeline is drained. They safeguard the pipeline against internal collapse when thus emptied. A surge tank or surge chamber is built near the mouth of the pressure tunnel to relieve the pipes of undue inertia pressure set up in the tunnel when the flow is checked following a reduction of load. Should this pressure exceed a predetermined amount, water merely spills over the lip of the surge tank. The surge tank also provides a reservoir of water that can be drawn upon when the load on the turbine suddenly increases. At every point of deviation of the pipelines, either in the horizontal or the vertical plane, anchorages are constructed with expansion joints provided immediately below. The powerhouse itself is located as close as possible to the lower reservoir into which the tail race discharges.

In *medium-head* installations, a nearly horizontal open canal or conduit may be carried along the side of the valley as far as the powerhouse site. A relatively short pressure pipe, often called a *penstock,* leads to the turbine.

A typical pumped-hydro system of a conventional design is shown in Fig. 8-6.

Underground Pumped Hydro

To overcome the requirement of a suitable topography, *underground pumped hydro* is being considered in this system. The upper reservoir may be at or near ground level. The lower reservoir is placed underground in natural caverns, old mines, or other underground cavities. such a system is shown in Fig. 8-7.

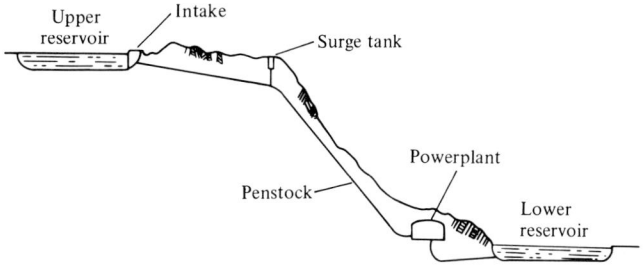

Figure 8-6 Schematic of a conventional above-ground pumped-hydro storage system.

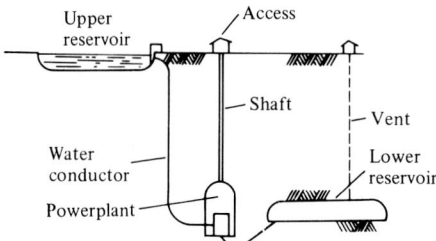

Figure 8-7 Schematic of an underground pumped-hydro storage system.

In all systems, a principal piece of equipment is a reversible pump-turbine or motor-generator set or sets. The excess electric energy supplied by the primary power station during off-peak hours is used to drive it in the motor-pump mode to pump water from the lower to the upper reservoir. During periods of peak demand, the system reverses to the turbine-generator mode to generate the excess electricity needed. (Some old installations use separate conventional pumps and turbines rather than reversible machines.)

The losses in pumped-hydro systems include motor and pump losses and flow losses during upflow; seepage into ground, leakage from pipes and equipment, and evaporation during storage; and turbine and generator losses and flow losses during downflow. The combined efficiency of a pumped-hydro system, called the *turnaround efficiency*, is defined as the total energy output divided by the total energy input during a charge-discharge cycle. In most plants, the turnaround efficiency is in the neighborhood of 65 percent. Pumped-hydro systems are rated according to their power output, usually in megawatts (MW). The maximum power output in the turbine-generator mode is usually greater than the maximum power input in the motor-pump mode, but operation in the latter lasts longer than the former, so the input energy is greater. (Recall that energy = power × time.)

COMPRESSED-AIR STORAGE

Compressed-air energy storage is analogous to pumped-hydro energy storage. Whereas in the latter excess energy generated by a base-loaded plant during periods of low demand is used to increase the potential energy or hydrostatic pressure of water, compressed-air energy storage compresses and stores air in reservoirs, aquifers, or caverns. The stored energy is then released during periods of peak demand by expansion of the air through an air turbine. In general, the turnaround efficiency of compressed-air storage is comparable to that of pumped-hydro storage.

Reservoirs

The underground compressed-air reservoirs are subjected to repeated fluctuations in pressure, humidity, and temperature. The long-range effects of such fluctuations remain to be determined. Usually multiple reservoirs, operating in parallel, serve one storage system. Three types of reservoirs show the most promise [2]: (1) salt caverns, (2) aquifiers, and (3) hard-rock caverns.

Salt caverns These have been used in the past to store petroleum products. Research so far indicates that they are stable under compressed-air storage loadings for the duration of plant life. The major concerns are cavern geometry, size and spacings, long-term creep and creep-rupture of rock salt, and air leakage.

Aquifers These are naturally occurring porous-rock formations. They have been used for natural gas storage for over 50 years but with annual rather than daily cycling. The effects of the different physical properties of air and its oxygen and the high temperatures of storing remains to be evaluated. Among other concerns are cyclic fatigue of the porous rock, air-water interface movement (water is usually present in aquifers), and the generation and transport of fine particulate matter.

Hard-rock caverns Because of their size these require water-compensating surface reservoirs to maintain air pressure and therefore are more costly than the two reservoir types above. However, they are believed to be most stable in the absence of severe temperature fluctuations (50°C). The major concerns here include the effervescence of air in the water shaft (called the champagne effect), hard-rock properties under cyclic conditions, and the residual strength of hard rock after an initial failure.

Adiabatic and Hybrid Systems

When air is compressed for storage, its temperature will rise (since it is a compressible gas) according to the relationship

$$T_2 = T_1 \left(\frac{P_2}{P_1}\right)^{(n-1)/n} \tag{8-6}$$

where T and P are the absolute temperature and pressure and the subscripts 1 and 2 refer to before and after compression, respectively. n is the polytropic exponent for the nonreversible compression process.

The heat of compression may be retained in the compressed air or in another heat-storage medium and then restored to the air before expanding through the turbine. This is called *adiabatic storage* and results in high storage efficiency. Recall that at a given pressure ratio, turbine work is directly proportional to the inlet absolute temperature. Recall also that constant-pressure lines on a temperature-entropy diagram for gases diverge at high temperatures so that isentropic work, equal to the vertical distance between any two constant-pressure lines, increases with temperature. Restoring the heat to the air also prevents the turbine parts from freezing if low-temperature air is allowed to expand through it. If the heat of compression is allowed to dissipate, additional heat could be added by fuel combustion to retain the high storage efficiency, but the results would be extra expense and maintenance problems. This is called a *hybrid system*.

Figure 8-8 shows a simple adiabatic compressed-air energy-storage system. The

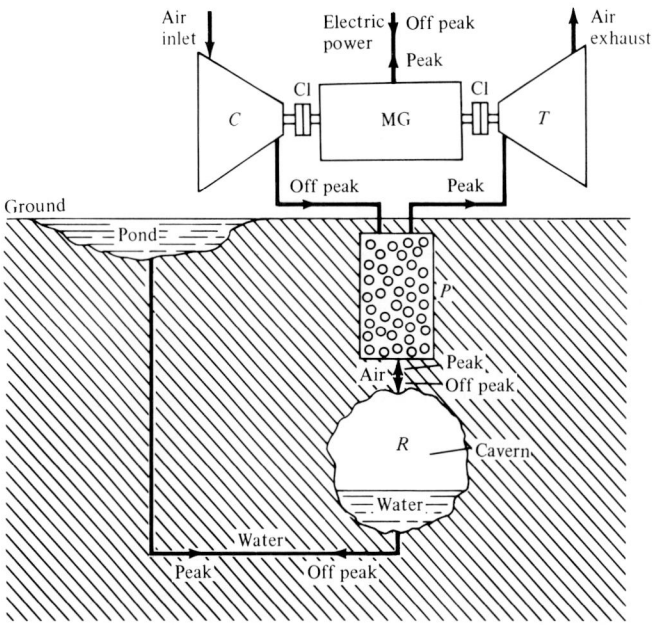

Figure 8-8 Schematic of a simple single-stage adiabatic compressed-air energy storage system with pressure-compensation pond. C = compressor, T = turbine, MG = motor-generator set, P = packed-bed thermal-energy storage, R = air-storage reservoir.

main plant is not shown. During off-peak hours, electric energy from the main plant generator is used by the motor-generator set (MG) operating in the motor mode to drive the compressor (C). The compressed air passes first through a packed bed (P) for sensible thermal-energy storage, then to a constant-pressure underground reservoir (R). The constant pressure is obtained by displacing water to a pressure-compensation pond that has a nearly constant head above the reservoir. During peak hours, air from the reservoir flows through the packed bed picking back sensible heat, then through the air turbine that now drives the motor-generator set in the generator mode. Clutches (Cl) separate the compressor during peak (generation) periods and the turbine during off-peak (storage) periods.

As expected, the air-reservoir (cavern) volume is a strong function of the storage pressure. For a peak unit capacity of 1500 MWh that volume is estimated at nearly 2,000,000 m^3 for 10 bar, or 64,000 m^3 for 100 bar storage pressures. The packed bed thermal-energy storage volume is about a tenth of that of the storage reservoir in most cases. Thus to reduce storage volume and hence cost, operation at high pressure is necessary.

Example 8-1 Calculate the airflow, compressed-air temperature, and storage volume for a 1500-MWh peaking unit charging for 7.5 h. Assume compressor inlet is at 1 bar and 20°C, compressor exit at 100 bar, a compressor polytropic efficiency of 70 percent, a peaking turbine efficiency of 60 percent, and a constant specific heat for air $c_p = 1.05$ kJ/(kg · °C). The air-gas constant $R = 284.75$ kJ/(kg · K).

SOLUTION For a compressor polytropic efficiency of 70 percent and constant specific heat

$$0.7 = \frac{h_{2s} - h_1}{h_2 - h_1} = \frac{T_{2s} - T_1}{T_2 - T_1}$$

where the subscripts 1, 2, and 2s are for compressor inlet, exit, and isentropic exit conditions, respectively.

$$T_{2s} = T_1 \left(\frac{P_2}{P_1}\right)^{(k-1)/k} = (20 + 273)\left(\frac{100}{1}\right)^{(1.4-1)/1.4} = 1092 \text{ K} = 819°C$$

$$T_2 = \frac{819 - 20}{0.7} + 20 = 1162°C$$

(This corresponds to a polytropic exponent $n = 1.5266$.)
 For a turbine output of 1500 MWh

$$\text{Storage capacity} = \frac{1500}{0.60} = 2500 \text{ MWh}$$

$$\text{Mass of air required} = \frac{2500 \times 3.6 \times 10^6}{1.05(1162 - 20)} = 7.5 \times 10^6 \text{ kg}$$

Assuming air is stored in the cavern at 100 bar (10^7 Pa) and 20°C

$$\text{Total volume needed} = \frac{7.5 \times 10^6 \times 284.75(20 + 273)}{10^7} = 62{,}575 \text{ m}^3$$

Average air flow to cavern during 7.5 h of charging = 8343 m³/h

It can be seen that a system such as that described above requires a very large compressor with an inlet airflow at 1 bar of about 834,000 m³/h (~490,000 ft³/m), capable of an exit pressure of about 100 bar (1450 psia), and an exit temperature of more than 1100°C (~2000°F). Such compressors or combination of compressors need to be developed before a large utility compressed-air energy-storage system can be a reality.
 A method of reducing the above temperature is the so-called two-stage compressed-air energy-storage system. It employs two compressor-motor-generator-turbine sets.

The Huntorf Compressed-Air Storage System

The first compressed-air storage system to be built is a 290-MW plant designed by Brown Boveri and built at Huntorf, West Germany [3]. It provides storage for the Nordwestdeutsche Kraftwerke (NWK) utility of Hamburg. The plant (Fig. 8-9) has

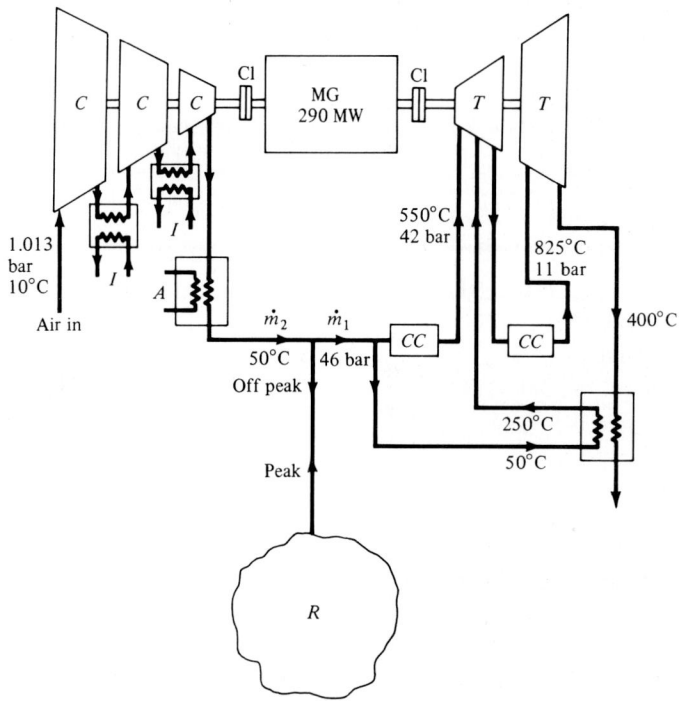

Figure 8-9 The Huntorf hybrid air-storage gas-turbine powerplant. I = intercooler, A = aftercooler, CC = combustion chamber, mass-flow rates: $\dot{m}_2$ = 0.9 to $0.25\dot{m}_1$ [2].

been in operation since 1978. It uses two 150-m-high salt caverns with a total volume of 300,000 m³ created by leaching a salt dome at a depth between 650 and 800 m below ground. The system is composed of a motor-generator set connected by clutches to a three-stage compressor with intercoolers and a two-stage gas turbine with reheat. It is of the hybrid variety that requires heat addition prior to the gas turbine.

In the storing mode, the compressor pumps atmospheric air into the caverns, where it is stored at 50 to 70 bar (725 to 1015 psia). In the generation mode, the stored air, reduced in pressure to 46 bar (667 psia), enters a natural gas–fueled combustion chamber before the high-pressure section of the gas turbine. Reheat is accomplished by a low-pressure natural-gas burner. Storage occurs daily for about 8 h, generation for about 2 h. The compressor and turbine are each sized independently to suit the power requirements during these periods, an advantage over the usual gas-turbine cycle in which the compressor absorbs more than two-thirds of the turbine output.

Huntorf has shown good availability exceeding 98 percent at times and good reliability. The caverns have shown no detectable creep or stability problems.

In the United States, a 220-MW compressed-air energy-storage system is in the planning stages by Soyland Power Cooperative of Decatur, Illinois. It will have hard-rock reservoirs, with water-compensating surface reservoirs. Siting studies have been conducted by Batelle Pacific Northwest Laboratories. Brown Boveri has been awarded the contract for the electrical and mechanical equipment. Soyland hopes to have the plant in operation by 1985, and they estimate a savings due to energy storage of $34 million (in 1981 dollars) over the first 16 years of operation.

ENERGY STORAGE BY FLYWHEELS

Flywheels store off-peak energy as kinetic energy. They have been used extensively to smooth out power pulses from reciprocating engines. They are physically connected to the engine crankshafts and are larger the smaller the number of cylinders per engine. (They, for example, were very large for old single-cylinder steam engines.) They operate by storing some of the energy given by the cylinders and releasing it during periods of no power pulses so that the speed and power delivery of the crankshaft are steady and continuous. More recently interest in flywheel energy storage has been generated by motor vehicle designers. In the so-called hybrid automobile, for example, the flywheel stores some of the energy of the gasoline engine during periods of low vehicle demands and releases it during periods of high demands, such as during acceleration, hill climbing, etc., and thus operates the engine at a more steady and hence more efficient output.*

The use of flywheel energy storage by utilities was tried only a few years ago. In this the flywheel rotor is physically connected to a motor-generator set. In the charging mode, during off-peak periods, the motor adds energy to the flywheel. In the generation mode, during periods of peak demand, the flywheel rotor coasts driving the generator.

The fluctuations in speed caused by torque variations are reduced to a minimum by the use of flywheels. As kinetic energy is proportional to the mass times velocity squared, the changes in velocity from the addition or subtraction of kinetic energy are reduced by the use of a large mass. Conversely the energy stored in a flywheel can be increased by increasing the velocity. The velocity of a flywheel is defined as $2\pi Rn$. The energy stored in a flywheel is equal to the kinetic energy, given by

$$E = \frac{1}{2g_c} m(2\pi Rn)^2 = \frac{2\pi^2}{g_c} mR^2n^2 \qquad (8\text{-}7)$$

where E = energy, ft · lb_f or J

m = mass of flywheel, lb_m or kg

g_c = conversion factor = 32.2 lb_m · ft/(lb_f · s^2) = 1.0 kg/(N · s^2)

* Other uses for flywheels include: a bus electrogyro, which is recharged at bus stops; regenerative braking of subway cars that provides acceleration upon start; electrically powered earth movers to limit peak-power demands; elevator drives including regenerative breaking; and plasma physics experiments to provide very high power peaks.

R = radius of gyration,* ft or m

n = revolutions per second = (r/min)/60

The energy E absorbed (or released) by a flywheel between speeds of rotation n_1 and n_2 is thus given by

$$\Delta E = \frac{2\pi^2}{g_c} mR^2(n_2^2 - n_1^2) \tag{8-8}$$

The ratio of the variation in rotational speed to the mean speed n is called the *coefficient of speed fluctuation* k_s, given by

$$k_s = \frac{n_2 - n_1}{n} = \frac{2(n_2 - n_1)}{n_1 + n_2} \tag{8-9}$$

where

$$n = \frac{n_1 + n_2}{2} \tag{8-10}$$

Combining Eqs. (8-8) through (8-10) results in

$$\Delta E = \frac{4\pi^2}{g_c} k_s mR^2 n^2 \tag{8-11}$$

The value of the coefficient k_s depends upon the desired closeness of speed regulation. For engines, it may vary from 0.005 for fine to .2 for coarse regulation. Thus for a given energy absorption ΔE, m and/or R^2 must be high for close speed regulation.

Another important consideration in flywheel design is the stress level a flywheel rotating at very high speed is subjected to. The *theoretical maximum specific energy* (energy stored per unit mass) is dependent upon the stress-to-density ratio and is given by [4].

$$\left(\frac{E}{m}\right)_{max} = 3.77 \times 10^{-7} k_m \frac{\sigma}{\rho} \tag{8-12}$$

where

E/m = specific energy, kWh/ lb$_m$

k_m = *mass-efficiency factor*, dimensionless

σ = allowable stress, lb$_f$/ft^2

ρ = density, lb$_m$/ft^3

---

* The radius at which the total mass is considered to be concentrated. For a disc of uniform density ρ, uniform thickness t, and outer radius R_o, with r as a variable radius between 0 and R_o:

$$\int_0^{R_o} \frac{1}{2g_c}(2\pi r\, dr\, t\rho)(2\pi rn)^2 = \frac{1}{g_c}(\pi R_o^2 t\rho)(2\pi Rn)^2$$

from which $R = R_o/\sqrt{2} = 0.7071 R_o$. Another use of the radius of gyration is in calculating the moment of inertia $I = R_m^2$.

k_m expresses how well a particular flywheel design utilizes the material strength, being a maximum if the stress is uniform throughout the flywheel. The maximum values of k_m are 1.0 in optimum designs with isotropic materials, where radial and tangential stresses are uniform and equal, and 0.5 in optimum designs with materials where only one stress direction can be utilized, such as fiber-reinforced composites. These two maximum values are obtained only in very slender designs where the energy absorbed per unit volume is minimal. In realistic designs k_m and the specific energy are reduced in order to reduce space, mass, and the cost of the safety shield and gas or vacuum chamber (below). The *volumetric specific energy,* energy per unit volume, is given by

$$\left(\frac{E}{V}\right)_{max} = 3.77 \times 10^{-7} k_v \sigma \tag{8-13}$$

where
E/V = volumetric specific energy, kWh/ft³

k_v = *volume-efficiency ratio,* dimensionless

k_v expresses how well a particular flywheel design utilizes the material strength and fills the cylindrical volume around the flywheel. For a uniform-density material, k_v equals k_m times the fraction of that cylindrical volume occupied by the flywheel.

The principal parameters that determine the suitability of flywheels for energy storage are the two efficiency ratios k_m and k_v as well as the stress and density. The values of k_m and k_v depend upon the type of material (isotropic, uniaxial composite, variable density) as well as flywheel shape (disc, drum, rod). Figure 8-10 shows the relationship between k_m and k_v for high-performance flywheel designs.

The *strength-density ratio* σ/ρ is high for such materials as glass or silica fibers. However, inevitable manufacturing flaws, which also tend to grow because of stress corrosion, cause these high values to be realized only for short periods or at cryogenic temperatures. Cyclic operation expected in energy storage and release also causes fatigue and growth of small flaws and cracks and is expected to be strength-limiting for most materials.

Vibrational frequencies, coupled with high-cycle fatigue, are also expected to be strength-limiting, especially for slender designs such as thin discs or hoops. Thus the suitability of a design for energy storage depends on the design, on the material, and on the extent of manufacturing flaws and the methods for detecting and reinspecting them—in other words, the stringency of the quality control standards.

Materials for energy-storage flywheels must have high strengths, high strength-density ratio, high resistances to cyclic-crack growth, and high strength density-to-cost ratios. Those under consideration include some alloys, such as the so-called maraging steels, and more promising, composites such as fiber-reinforced plastics. One composite that shows particular promise is a 62 volume percent S-glass in epoxy composite. It has a density of 122.7 lb_m/ft^3, an estimated working stress of 21×10^6 lb_f/ft^2 for 10^4 cycles (16×10^6 for 10^5 cycles), a stress-density ratio of 170,000 $(lb_f/ft^2)/(lb_m/ft^3)$ for 10^4 cycles (130,000 for 10^5 cycles), and a cost based on volume production of \$0.80/lb_m. Other attractive composites are graphite-epoxy and kelvar-epoxy [4]. By contrast maraging steel has a density of 500 lb_f/ft^3, a working stress

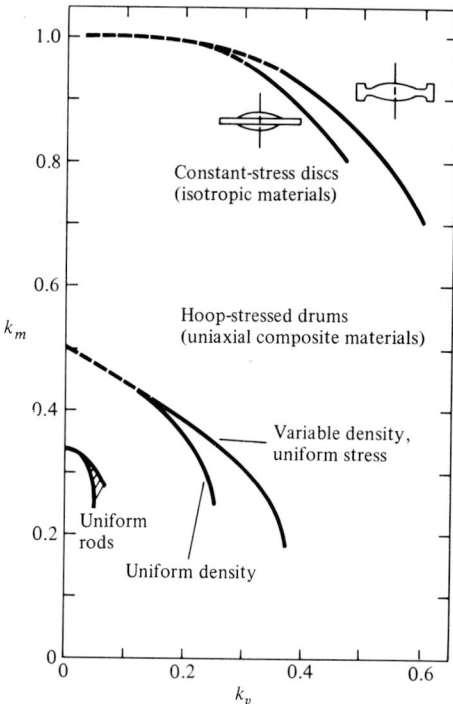

Figure 8-10 Relationship between the mass- and volume-efficiency ratios for high-performance flywheel designs [4].

of $14.5 \times 10^6 \text{lb}_f/\text{ft}^2$ and a stress-density ratio of 29,000, all for 10^4 cycles, and a cost of $3/\text{lb}_m$.

Flywheels for energy storage are systems that include, besides the flywheel itself, a number of subsystems. These are a housing; bearings, with ball bearings believed the most suitable; a vacuum pump to minimize windage losses inside the housing; seals to minimize oil and air leakage into the vacuum chamber; and sometimes a containment ring to protect nearby personnel and equipment from flying fragments in case of flywheel rotor fracture.

Losses in a flywheel energy-storage system include windage, bearing and seal friction, vacuum pump input power, and eddy current (hysteresis) and other inefficiencies in the motor generator (or in transmission systems). In early designs these losses were prohibitively large, and much developmental work still needs to be done to arrive at a technically attractive system.

A NASA conceptual design for a flywheel electric-energy storage system that has nearly zero losses is shown in Fig. 8-11. It is designed for a 10,000-kWh substation. The flywheel rotor, made of a filamentary anisotropic composite material to achieve high strength to mass ratio, has inner and outer diameters and height of 14, 20, and 10 ft (4.27, 6.1, and 3.05 m), respectively. It weighs 735,000 lb_m (333.4 metric tons) and rotates at 1250 r/min. To eliminate rotor-bearing friction, the rotor is magnetically levitated by permanent magnets that are incorporated in the stator with sets of teeth

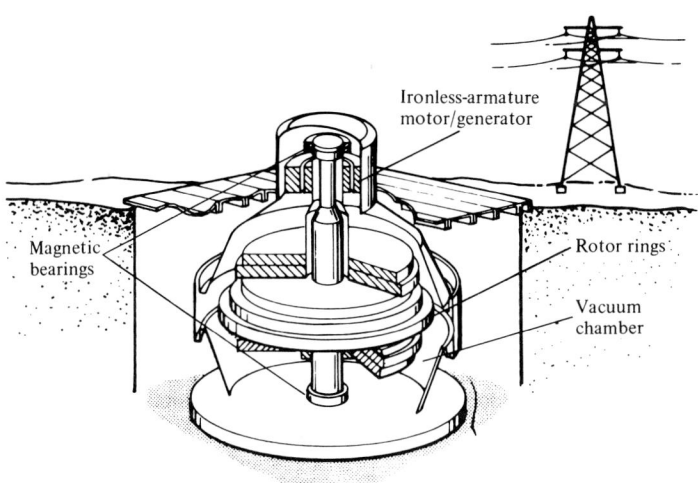

Figure 8-11 A NASA conceptual design of a 10-MWh-capacity flywheel energy storage system.

on both rims facing each other to maintain axial alignment. Alignment in the radial direction is provided by a set of electromagnets in shunt with the permanent magnets. Optical sensors situated along the air gap monitor any radial eccentricity and supply an error signal to the electromagnets to correct the eccentricity.

To eliminate losses from hysteresis and eddy currents, normally present in iron armatures, an ironless motor-generator armature is provided. A contactless electronic commutator is used to control current flow in the motor-generator coils. A permanent magnet in the stator produces a magnetic flux in the gap between rotor and stator. In the motor mode, alternating current is applied to the armature via the electronic commutation that interacts with the flux to produce the necessary force on the rotor. In the generator mode, commutation diodes are connected to the proper phases of the armature coil.

Windage losses are virtually eliminated by operating the flywheel in a sealed evacuated enclosure (vacuum chamber) pumped down to 10^{-3}- to 10^{-5}-torr (1.33×10^{-6} to 1.33×10^{-8} bar) pressure. A safety rotor ring is provided. In addition, the entire system is to be located below ground.

ELECTRICAL BATTERY STORAGE

The familiar *lead-acid battery* used in motor vehicles is a direct-current battery. It contains a number of voltaic cells (six in a 12-V battery, for example) that are connected in series. Each cell contains several lead plates, connected in parallel, made of grids that are filled with a spongy gray lead, Pb, and which form the anode. Alternating with these are plates of similar design but containing lead oxide, PbO_2, which form the cathode. All plates are immersed in a water solution of sulfuric acid H_2SO_4, which acts as an electrolyte. The electrolyte of each cell is housed separately in its compartment.

In the discharge mode, direct current is generated. The lead in the anode oxidizes to Pb^{2+} ions that immediately precipitate on the plates as lead sulfate $PbSO_4$. The lead oxide in the cathode reduces to Pb^{2+} that also precipitates as $PbSO_4$. The electrochemical reactions are

Anode: $\qquad Pb(s) + SO_4^{2-}(aq) \rightarrow PbSO_4(s) + 2e^-$ (8-14a)

Cathode: $\qquad PbO_2(s) + 4H^+(aq) + SO_4^{2-}(aq) + 2e^- \rightarrow PbSO_4(s) + 2H_2O(aq)$
(8-14b)

where (s) and (aq) indicate solid and liquid (aqueous) conditions, respectively.

Thus during discharge all plates are slowly covered by $PbSO_4$, which replaces the lead in the anode and the lead oxide in the cathode, and the concentration of sulfuric acid in the solution slowly decreases. For each molecule of Pb, two molecules of H_2SO_4 ($4H^+$ and $2SO_4^{2-}$) are replaced by two molecules of water ($2H_2O$). This is why the extent of charge of a lead-acid storage battery can be checked by measuring the density of the electrolyte; low density indicates low sulfuric acid concentration and a partially discharged cell.

In the charge mode, the battery can be restored to its original condition by reversing the direction of the current (and electron flow). The reactions represented by Eqs. (8-14) are simply reversed. The charging current must be obtained from a direct-current generator (as in older vehicles) or by an alternator equipped with a rectifier (as in modern vehicles).

The lead-acid battery, then, can be charged and discharged over many cycles and has been widely used to satisfy motor vehicle starter, instruments, lights, and other accessory requirements. Their use as the prime movers of all-electrical vehicles, where they would power drive motors and get charged while idle, as during the night, has run into several main drawbacks. These are unacceptable energy-mass and energy-cost ratios and low cycle life, in which they are almost fully discharged and recharged. These same drawbacks have prevented their use as energy-storage systems for large utility purposes.

The lead-acid battery is therefore limited to the small specialized uses it now enjoys for which the relatively low energy-mass ratio and the high cost of chemicals are not crucial factors.

Research and development has been going on for a number of years to develop advanced storage-battery systems that would have greater energy-mass ratios, lower costs, and greater cycle life. One of these is the *nickel-cadmium battery*, which uses a nickel hydroxide cathode, a cadmium anode, and a potassium-hydroxide-solution electrolyte. This battery is characterized by low mass and is primarily used in portable equipment such as radios and cordless appliances. Another is the *silver-zinc battery* which uses a solution of potassium hydroxide, saturated with zinc hydroxide, as an electrolyte. It has a high energy-mass ratio but also a high cost. It is primarily used in applications in which low mass is more important than cost. It also suffers from low cycle life, with only 30 to 300 charge-discharge cycles reported for some designs.

Battery systems that are potentially more suitable for utility applications, however, use soluble or liquid reactants and operate at temperatures other than atmospheric. The ones with the most promise at present are:

1. *Sodium-sulfur batteries*. These use molten sodium as one electrode, a sulfur and sodium sulfite mixture as the other, and a solid aluminum oxide electrolyte. They have a high energy-mass ratio and operate at temperatures of about 250°C (480°F). This relatively low temperature allows the use of Teflon seals and aluminum or glass containers. These batteries have a long cycle life because of the lack of solid-solid transformations. The low mass makes the sodium-sulfur battery a contender for the battery of the electrical car of the future.
2. *Lithium-chlorine* and *lithium-telluride batteries*. These are less developed than sodium-sulfur batteries but have similar favorable characteristics.
3. *Zinc-chlorine batteries*. Here a zinc chloride solution is pumped through graphite cells on which the zinc is deposited and the chloride is liberated in gaseous form that is drawn away, cooled in a heat exchanger, and stored in a separate tank. Chlorine, being toxic, is stored as a relatively harmless icy slush. In the generating mode, the chilled chlorine is pumped as a solution back to the cells where it reacts with the zinc to produce electricity. The Gulf and Western Company had originally developed this battery for the purpose of utility-type energy storage but has initially adapted it to a motor vehicle. (In one test, it ran a car for 175 mi at 42 mi/h). This battery has the advantage of constant electric output that does not drop off during discharge as is the case with most other batteries, for as long as there is chlorine in the storage tank, the battery system develops essentially constant power. The system is of course more complex than can be tolerated for personal automobiles because of the need for a complex charger, pumping, and heat-exchanger equipment.

The turnaround (charge-discharge) efficiencies of most batteries are good, about 70 to 80 percent, compared with some 60 percent for pumped hydro. However, it is clear that at this time high-energy, low-mass, long-cycle-life batteries of a type that would meet large utility peak-load demands are still not commercial realities.

SUPERCONDUCTING MAGNETIC ENERGY STORAGE

In 1911, the dependence of the electrical resistance of metals on temperature was observed by the Dutch physicist Kamerlingh Onnes. In his investigations, he also discovered that the electrical resistance of mercury dropped suddenly to zero when it was cooled to within a few degrees of absolute zero, a phenomenon which Onnes called *superconductivity*.

Other metals exhibit the same phenomenon. The temperature below which they become superconductive is called the *transition* or the *critical temperature*. All superconducting metals have transition temperatures in the *cryogenic* range*. The phenomenon arbitrarily has found many applications, and cryogenic engineering has become a specialized science as well as an industry all its own. Besides superconductivity, cryogenic applications can be found in medicine (cryogenic surgery, cryogenic ophthalmology, etc.), food preservation, pollution control, and other areas.

In 1970, the main application was the construction of superconducting electromagnets. These were experimentally used for magnetohydrodynamic power generation; bubble chambers to cool electrical generators, motors and transformers; and electric-power transmission and distribution. The latter application promises no-loss transmission. In the 1970s it was determined that it can best be accomplished by the use of high-purity aluminum cables operating at liquid hydrogen temperatures 20 K ($-253°C$, $-423°F$). Commerical success depends upon whether the metal, the gas, and the refrigeration systems can be acquired and operated economically.

Superconducting magnetic energy storage is a concept that initially received attention for pulsed energy storage in which the charge and discharge times have been short. It subsequently became apparent that such a concept is suitable for large-scale energy storage by an electric utility [5]. The concept is based on the principle that energy can be stored in the magnetic field associated with a coil. If the coil is made of a material in a superconducting state, i.e., maintained at a temperature below its critical temperature, then once it is charged, the current will not decay and the magnetic energy can be stored indefinitely. The stored energy can be released back to the network by discharging the coil.

The energy E stored in a coil in which a current I circulates is given by

$$E = \frac{1}{2} LI^2 \tag{8-15}$$

where
E = energy, J (joule = watt × second)

L = inductance, H [henry = (volt × second)/ampere]

I = current, A (ampere)

---

* The term is derived from the Greek *kryo,* which means "icy cold or frost." Cryogenics is said to have been born in 1877 with the liquefaction of small quantities of oxygen at 90 K ($-183°C$, $-298°F$). In 1908, helium was first liquified at 4.2 K, though a commercial method for liquifying helium was not developed until 1947. Goddard used liquid oxygen to propel a rocket in 1926. Germany used it on a large scale to propel V-2 rockets during World War II, an application that brought cryogenics to the attention of the world. It is now widely used in the space program. Modern techniques can produce temperatures to within a minute fraction of a degree of absolute zero. The cryogenic range has been arbitrarily defined as extending from absolute zero, or 0 K ($-273°C$, $-460°F$), to an upper limit of 123 K ($-150°C$, $-238°F$), although most of the research has been done below the boiling point of oxygen, about 90 K. The range is considerably below normal refrigeration temperatures. Besides electrical resistance, other material, physical, and chemical properties such as strength, ductility, and thermal conductivity are also greatly affected when they reach cryogenic temperatures.

The inductance L of a coil is a function of its dimensions, which are characterized, for a coil with conductors of a rectangular cross section (Fig. 8-12) by

$$\xi = \frac{2R}{\sqrt{ab}} \qquad (8\text{-}16)$$

$$\delta = \frac{a}{b} \qquad (8\text{-}17)$$

and

$$V = 2\pi Rab = \frac{8\pi R^3}{\xi^2} \qquad (8\text{-}18)$$

where R = mean radius of coil, m
 a and b = width and depth of conductor, m
 V = volume of conductor in one coil turn, m³

and L is given by

$$L = f(\xi, \delta) R N^2 \qquad (8\text{-}19)$$

where $f(\xi, \delta)$ = form function having the units (V · s)/(A · m)
 N = number of turns of coil

The energy stored in a coil, Eq. (8-15), can now be written as

$$E = \frac{1}{2} f(\xi, \delta) R N^2 I^2 \qquad (8\text{-}20)$$

Using a current density j given by

$$j = \frac{NI}{ab} \qquad (8\text{-}21)$$

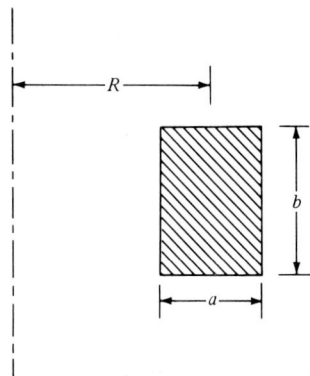

Figure 8-12 Dimensions of a cylindrical coil with rectangular conductors.

gives

$$E = \frac{1}{4}\pi^{-5/3}f(\xi,\delta)\xi^{-2/3}V^{5/3}j^2 \qquad (8\text{-}22)$$

A coil that gives the maximum value of inductance to volume L/V is called a *Brooks coil* [6]. It has the dimensions

$$a = b \quad \text{and} \quad R = \frac{3}{2}b \qquad (8\text{-}23)$$

so that for a Brooks coil, $\delta = 1$, $\xi = 3$, and the energy stored in it E_B is given by

$$E_B = 3.028 \times 10^{-8} V^{5/3} j^2 \qquad (8\text{-}24)$$

For a cylindrical coil, other than Brooks, the energy E is given as a fraction E_B, Eq. (8-24), by a factor F less than 1.0

$$F = \frac{E}{E_B} \qquad (8\text{-}25)$$

F is a function of ξ and δ of the particular coil and is given in Fig. 8-13.

An important parameter is the volume of material per unit energy stored. This is given by

$$\frac{V}{E} = \frac{V}{FE_B} = \frac{0.33 \times 10^8}{FV^{2/3}j^2} \qquad (8\text{-}26)$$

where for a Brooks coil $F = 1.0$. Equation (8-26) demonstrates the economy of scale of coils as the cost of the coil is proportional to its volume and thus the cost per

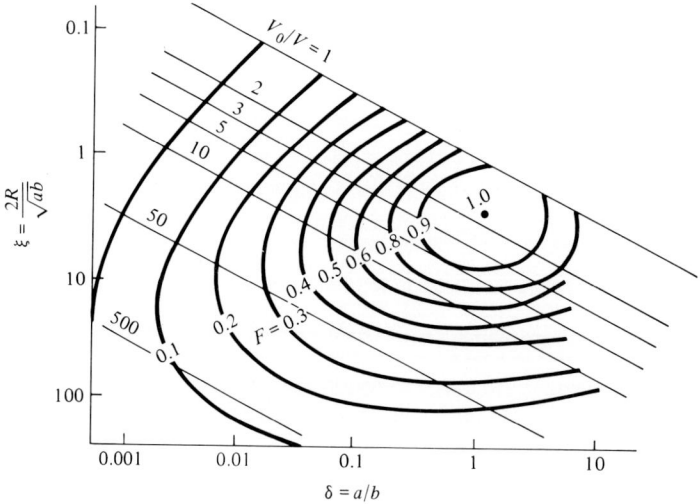

Figure 8-13 Lines of constant factor F relating energy of a cylindrical coil to that of a Brooks coil.

unit energy stored is inversely proportional to its volume to the power 2/3 and to the current density squared. The current density is, however, limited by stability considerations to values between 50×10^6 and 100×10^6 A/m².

The main mechanical design problem in magnetic energy storage arises because of the need of a very large structural mass to contain the magnetic field energy. This causes a large radial outward force from the solenoid. The mass is proportional to the material density and the stored energy, and is inversely proportional to the stresses. Such mass, if made of stainless steel, would amount to about 160 kg/kWh and result in unacceptable costs [7]. This consideration led to the selection of bedrock as the structural material with an excavated circular tunnel that would bear the radial outward force and transmit it to the surrounding bedrock.

Figure 8-14 shows an artist's sketch of a section of a proposed 5500-MWh magnetic storage system in which the superconductors are cooled in dewar by superfluid helium (He-II)* at 1 atm pressure and 1.8 K. The total structure is a low-aspect ratio cylinder that has an inner diameter of 1.57 km, a height of 15.7 m, and a thickness of 5 m. The solenoid has 112 turns and carries a 765,000-A current. The conductors are made of aluminum plus NbTi. The struts are made of fiberglass epoxy. The ripples in the conductor and dewar walls are designed to counteract excessive motion on cool down and to reduce magnetic tensile loads. The system has a total radial force of 3.1×10^{10} N, a total axial force of 3.1×10^{11} N, and an average radial pressure of about 4 bar.

The energy transfer between the three-phase ac current from the grid to the dc magnet is accomplished by an ac-dc power converter and inductor using a Graetz thyristor bridge circuitry. The turnaround efficiency (charge-discharge) of this circuitry is said to be greater than 95 percent.

Much of the activity of magnetic energy storage for utility use has taken place in the USA and Japan. A 1-MWh demonstration unit was planned to be built in Japan for the International Science Exposition to be held in Tsukuba Science and Education City in 1985. It would have 12 coils carrying 5000 A, an inductance of 288 H, and major and minor radii of 6.9 and 3.5 m. This plan has now been dropped in favor of a larger 10-MWh unit planned for later construction.

THERMAL SENSIBLE ENERGY STORAGE

Sensible energy storage is the first of the thermal-energy storage systems to be discussed. In general, thermal-energy storage systems can operate at many desired temperature levels depending upon use and choice of system and material, ranging from refrigeration temperatures to 1250°C. They have found wide use in many industrial applications, such as in the manufacture of cement, iron and steel, glass, aluminum,

---

* Helium becomes a liquid at 4.2 K and is called helium I. The isotope He⁴ (99.99987 percent of all helium) behaves unusually below 2.17 K, the so-called lambda point. It becomes a superfluid, called helium II, that has a thermal conductivity three or four orders of magnitude better than copper at low temperature and surface heat-transfer characteristics about an order of magnitude better than He-I.

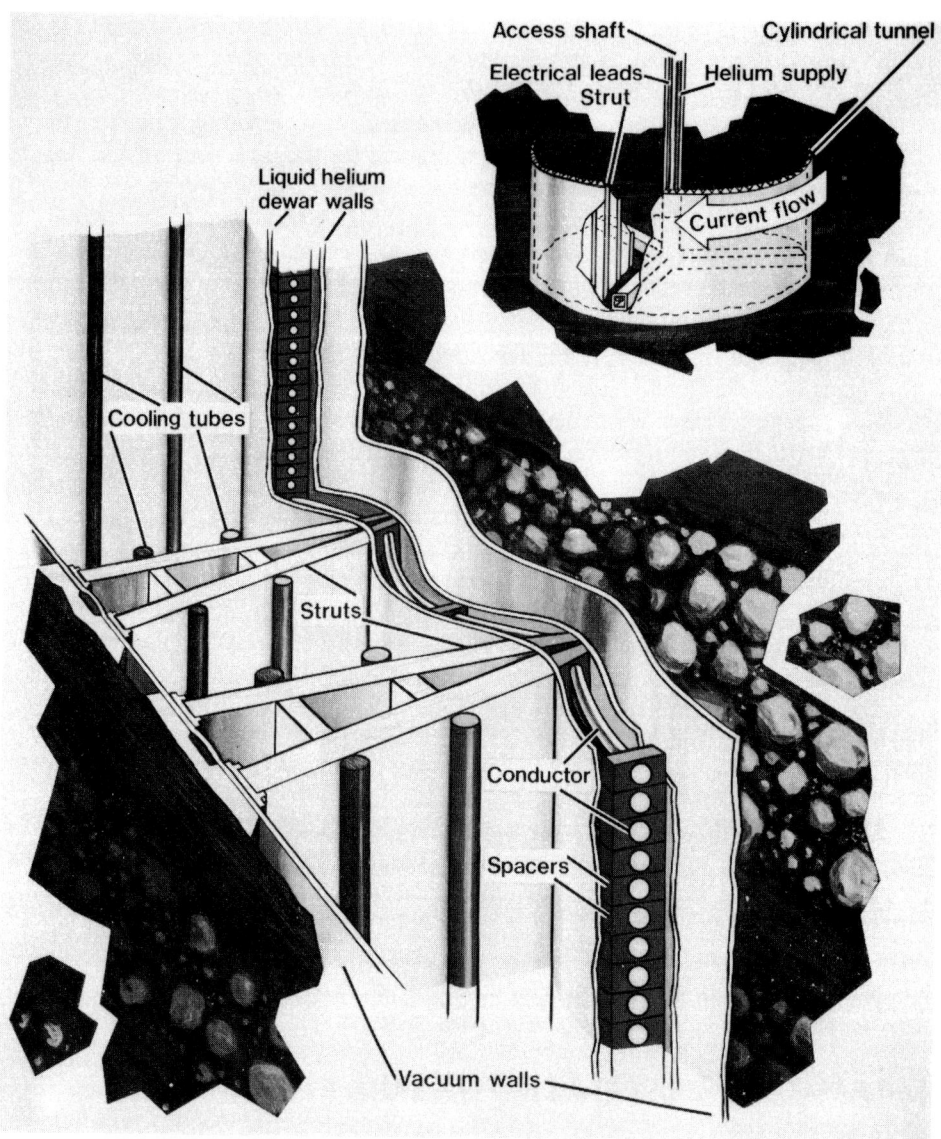

Figure 8-14 Artist's conception of a section of a proposed 5500-MWh magnetic energy storage unit. (*Courtesy of the University of Wisconsin Applied Superconductivity Center.*)

paper, plastics and rubber, and in food processing. Here we will deal with powerplant-related thermal-energy storage, in particular the Rankine-type steam-generator-turbine-condenser cycles.

Sensible energy storage is accomplished by raising the temperature of a material, such as water, an organic liquid, or a solid. The storage density, J/m^3 or Btu/ft^3, is equal to the product of the temperature difference, the specific heat, and the density of the material chosen. This system is simple in concept but has the disadvantages of variable temperature operation and relatively low storage density. Depending upon the coefficient of thermal expansion of the material, large volume changes may be encountered. Sensible energy storage could employ one of the following devices:

1. Pressurized-water storage
2. Organic liquid storage
3. Packed solid beds
4. Fluidized solid beds

Figure 8-15 shows an example of pressurized-water sensible energy storage [8] system in a powerplant in which the primary heat source is either a nuclear reactor or a fossil-fueled furnace. The base-loaded portion of the plant is capable of supplying more steam than needed during periods of low demand. The excess steam is bled at high pressure via turbine extraction (as in feedwater heating) during these periods of low demand. This extracted steam is fed to steel accumulators and mixed with water, thus producing saturated pressurized water. The accumulators are later discharged

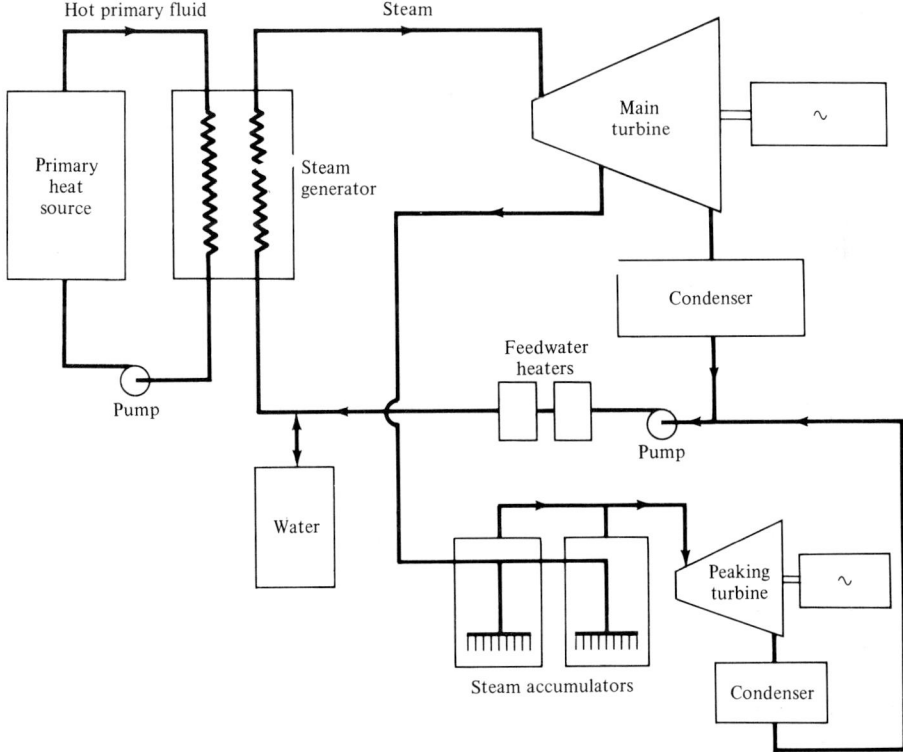

Figure 8-15 Schematic flow diagram of a powerplant with a pressurized-water sensible energy storage system.

through a small peaking turbine during periods of high demand. Discharge continues until a low specified pressure is reached in the accumulators. It has been seen that this results in low and varying steam temperature entering the peaking turbine. Typical values of accumulator high and low pressures are 20 bar, corresponding to a saturation temperature of about 212°C (414°F), and 2 bar, corresponding to a saturation temperature of about 120°C (248°F). Note that while this system involves steam condensing in water during accumulator charge and reevaporating during discharge, the storage medium is the pressurized water in the accumulators and operates over a relatively wide temperature range. Hence it is an example of sensible rather than latent energy storage. The system is sometimes referred to by the misnomer "steam storage."

The storage density of the thermal energy utilized in the peaking turbine per unit volume of the higher-pressure saturated water is given by

$$\text{Storage density} = \frac{1}{v_{f,1}}(h_{f,1} - h_{f,2}) \tag{8-27}$$

where v_f and h_f are the specific volume and enthalpy of saturated water, respectively, and the subscripts 1 and 2 refer to the stored and emptied pressures, respectively. Using the steam tables (SI units), App. A, the storage density for 20 and 2 bars is:

$$\text{Storage density} = \frac{1}{0.0011766}(908.5 - 504.8) = 343{,}107 \text{ kJ/m}^3$$

$$= 95.3 \text{ kWh/m}^3 = 2.7 \text{ kWh/ft}^3$$

Over the range of temperature cited these correspond to about 1 kWh/(m^3 · °C), 0.016 kWh/(ft^3 · °F), and 55.6 Btu/(ft^3 · °F). The electric-energy density obtained by the peaking turbine-generator depends upon two efficiencies. The first is *thermal turnaround efficiency* η_{ta}, and the second is the *peaking turbine-generator efficiency* η_p. The former is a complex function of the losses associated with sensible heat transfer to and from the steel walls, structural members of the accumulators, and interconnecting pipework, and the time-dependent convective heat losses to the environment.

The portion representing the ratio of energy stored in the accumulator structure to that stored in the contained water at a given pressure can be shown to be obtained by noting that the thickness t of a cylinder of diameter D is given by $DP/2\sigma$, and hence the ratio for a cylinder height L would be represented by

$$\frac{\pi D L t \rho_s c_s}{(\pi D^2/4) L \rho_f c_f} = 2\left(\frac{P}{\sigma}\right)\frac{\rho_s c_s}{\rho_f c_f} \tag{8-28}$$

where P is the pressure, σ the wall stress, ρ the density, and c the specific heat. The subscripts s and f denote solid and liquid, respectively. The volumetric heat capacities, given by the product ρc, are roughly equal. The ratio P/f is of the order of 0.03 for steel. Thus the contribution of losses by sensible heat transfer to the walls is rather minimal and may be ignored.

The convective heat losses from the water to the environment are therefore the major contributor to the thermal turnaround efficiency. They vary with time and depend upon the water temperature and the overall heat-transfer coefficient U between the water and the outside environment. The time constant τ of the system is given by the product of heat capacity and heat-flow resistance. Thus

$$\tau = \left(\frac{\pi D^2}{4} \rho_f c_f L\right)\left(\frac{1}{\pi D L U}\right)$$

$$= \frac{D \rho_f c_f}{4U} \tag{8-29}$$

If the temperature of the liquid at fully charged conditions is T_1, its time-dependent temperature of $T(\theta)$ decreases with time θ following fully charged conditions at (time zero) due to heat losses alone (no energy withdrawal) (Fig. 8-16). Assuming a lumped capacity system, common in transient heat-transfer calculations where the external heat-flow conductance is low compared with the internal heat-flow conductance (low Biot number), $T(\theta)$ is given by the familiar equation

$$\frac{T(\theta) - T_1}{T_\infty - T_1} = 1 - e^{-\theta/\tau} \tag{8-30}$$

where T_∞ is the environmental temperature. Assuming a storage period of θ_s at which the water temperature decreases to T_s

$$\frac{T_s - T_1}{T_\infty - T_1} = 1 - e^{-\theta_s/\tau} \tag{8-31}$$

The thermal turnaround efficiency is given by the energy left in storage at T_s after heat losses divided by the original energy stored. Thus

$$\eta_{ta} = \frac{h_s - h_2}{h_1 - h_2} \tag{8-32}$$

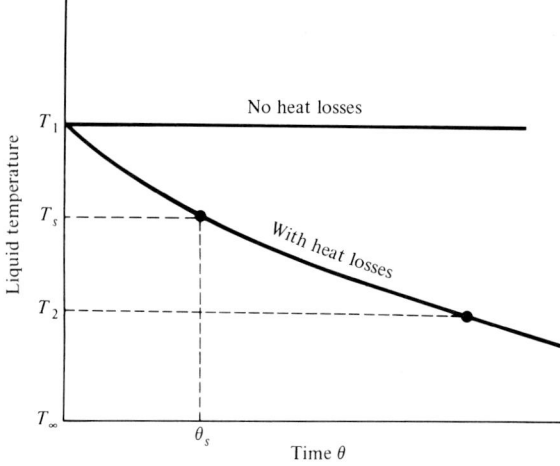

Figure 8-16 Stored liquid temperature-time change in sensible energy storage. $T_1 =$ fully charged; $T_2 =$ end of energy withdrawal; $T_\infty =$ ambient.

where h_1, h_2, and h_s are the compressed liquid enthalpies at T_1, T_2, and T_s, respectively. Assuming, for simplicity, equal specific heats over the temperature ranges in question

$$\eta_{ta} = \frac{T_s - T_2}{T_1 - T_2} \qquad (8\text{-}33)$$

Combining Eqs. (8-31) and (8-33)

$$\eta_{ta} = 1 - \frac{T_1 - T_\infty}{T_1 - T_2}(1 - e^{-\theta_s/\tau}) \qquad (8\text{-}34)$$

It can be seen that this efficiency is a strong function of the ratio θ_s/τ. θ_s is several hours in a daily storage system. The only important variable in τ is U, Eq. (8-29), which depends heavily on accumulator insulations, design, and location. Good insulation and design result in low values of U and therefore high values of τ.

Accumulators may be constructed above ground or underground. Underground accumulators are more costly but have higher insulation and generally higher T_∞ and therefore higher values of η_{ta}. Accumulators in general pose safety problems. A pipe or vessel rupture can release large amounts of energy. The choice of above-ground or underground accumulators, therefore, should be based on the careful evaluation of efficiency, cost, operational problems, and safety. Natural underground caverns may be used if such are available near plant sites or if plants can be located near them, and if they can be properly prepared for high pressure hot water storage. They may result in much lower costs.

The efficiency of the peaking turbogenerator is probably low because of variable inlet conditions and the use of low-temperature saturated steam, small size, insufficiency or absence of feedwater heating, and other factors. An efficiency of 20 to 25 percent is probably reasonable, compared with 33 to 40 percent for the base-load plant.

Example 8-2 A base-loaded 1000-MW powerplant is designed with a sensible thermal-energy storage of the type shown in Fig. 8-15. The thermal energy stored is called upon to produce 4000 MWh daily. The accumulators are 4 m in diameter each and are well insulated so that $U = 5$ kJ/(h · m² · K). The storage time is 15 h. Maximum and minimum storage pressures are 20 and 2 bar. Calculate:

1. The thermal turnaround efficiency.
2. The total accumulator volume.
3. The storage cost in $/kWh if the accumulators cost $300/m³.
4. The minimum released energy if an accumulator vessel or pipe ruptured at 20 bar.

The ambient temperature is 20°C. Take the specific heat of water as 4.35 kJ/(kg · K). Assume a peaking plant efficiency of 25 percent.

SOLUTION   Following are the water-saturation conditions from the steam tables.

At 20 bar: $T_1 = 212.37°C$ $h_{f,1} = 908.5$ kJ/kg $v_{f,1} = 0.0011766$ m³/kg

At 2 bar: $T_2 = 120.23°C$ $h_{f,2} = 504.8$ kJ/kg $v_{f,2} = 0.0010605$ m³/kg

At ambient conditions: $T_\infty = 20°C$ $h_{f,\infty} = 293$ kJ/kg

Take an average density of the water as

$$\frac{1}{2}\left(\frac{1}{v_{f,1}} + \frac{1}{v_{f,2}}\right) = 896.43 \text{ m}^3/\text{kg}$$

then

$$\tau = \frac{D\rho_f c_f}{4U} = \frac{4 \times 896.43 \times 4.38}{4 \times 5} = 785.3 \text{ h}$$

1. Therefore

$$\eta_{ta} = 1 - \frac{212.37 - 20}{212.37 - 120.23}(1 - e^{-15/785.3})$$

$$= 1 - 2.0878(1 - 0.981) = 0.96$$

For a peaking plant efficiency of 25 percent, the thermal storage needed is given by

$$\frac{4000}{0.25 \times 0.96} = 16{,}667 \text{ MWh}$$

$$16{,}667 \times 3.6 \times 10^6 = 6 \times 10^{10} \text{ kJ}$$

The mass of water needed to be flashed to steam is given by

$$\frac{6 \times 10^{10}}{h_{f,1} - h_{f,2}} = \frac{6 \times 10^{10}}{908.5 - 504.8} = 1.486 \times 10^8 \text{ kg}$$

2. The water volume at 20 bar is

$$1.486 \times 10^8 \times v_{f,1} = 1.486 \times 10^8 \times 0.0011766$$

$$= 174{,}840 \text{ m}^3$$

The volume of the accumulator will be larger to accommodate a steam blanket on top and the water left at end of discharge.

3. The minimum cost of accumulators would therefore be

$$\$300 \times 174{,}840{,}000 \approx \$52.5 \times 10^6$$

4. The energy contained that would be released if a rupture occurred is equal to the mass of the flashed water times the difference between the enthalpy of water at 20 bar at ambient conditions (20°C). The minimum corresponds to the above mass and equals

$$1.486 \times 10^8 \times 908.5 - 293 = 9.146 \times 10^{10} \text{ kJ}$$

LATENT HEAT ENERGY STORAGE

In this system energy is stored in the form of the latent heat caused by phase change, either by melting a solid or vaporizing a liquid. Energy release is accomplished by reversing the process, i.e., solidifying the liquid or condensing the vapor. The storage density here is equal to the product of the latent heat of fusion (or vaporization) times the density of the storage material. It is greater than that in sensible heat storage because the latent heats are much larger than the specific heats of the single phases of the materials. The system has the additional advantage of operating at essentially constant temperature with low volume changes during phase changes. It also has the advantage of a wide choice of materials with different fusion and evaporation temperatures, which allows a choice of operating temperatures and the ability to generate high-temperature steam for the peaking unit. Some sensible heat storage may be added to latent heat storage by further raising the temperature of the resulting molten solid or vapor.

Latent heat energy storage is not, at this writing, considered a simple, operationally reliable solution to the problem of energy storage in electric-generating powerplants. It is, however, included here as a potential solution along with the problems that must be overcome if such a solution is to become a viable one. Although little work has been done on the application of latent heat energy storage to large powerplants, much work has been done on its use for residential and solar heating applications using fused salts that are available for high- and low-temperature operating ranges [9].

Storage materials must possess, in addition to proper transition temperatures and high latent heat, many other necessary physical and chemical properties. Some of these are good thermal conductivity, containability, stability (considering cyclic operation), nontoxicity, and low cost. No material meets all these requirements but some fluoride salts meet some of them. One of the salts considered most suitable for latent heat storage is the 70% NaF–30% FeF_2 eutectic salt, which has a fusion temperature of about 680°C (2256°F) and potentially possesses the highest storage energy density of any thermal-energy storage material, about 1500 MJ/m^3 (~40,000 Btu/ft^3). $ZnCl_2$ is another with a fusion temperature of about 370°C (~700°F) and a potential storage energy density of about 400 MJ/m^3 (~11,000 Btu/ft^3).

Other materials being suggested are silicon, germanium, and sulfides of germanium. These have high heats of fusion, and like water, they expand upon freezing, so that they tend to float upon solidification, which has advantages in heat transfer. Silicon and germanium, however, have fusion temperatures that are too high for powerplant operation and are very reactive. Germanium sulfides have usable fusion temperatures but tend to solidify to a glassy consistency rather than crystallizing, thus posing an undesirable heat-transfer barrier.

In addition to finding a suitable medium, studies have to evaluate the extents of corrosion, erosion, plant start-up and shutdown, etc. Corrosion problems require that the system be free of oxygen and water vapor, which enter the system because of salt volume changes during heat addition and withdrawal, thus posing interesting engineering problems. It can be seen that there are many design and developmental problems to be overcome before a reliable, economical latent heat energy storage system can be incorporated into an electric-generating powerplant.

A latent heat energy storage conceptual design using the 70% NaF–30% FeF$_2$ eutectic salt as the storage medium has been proposed by Bundy [10]. The study had the purpose of identifying the technical difficulties and obtaining preliminary cost estimates. It envisaged a high-temperature gas-cooled reactor (HTGR) as the heat source (Fig. 8-17). The helium coolant operates at 48 bar (700 psia), and the various temperatures are indicated on the figure. The storage-system capacity is 7200 MWh, whereas the charge-discharge rates are 600 MW. The peak electric-generating capacity of the plant is 200 MW for 12 h.

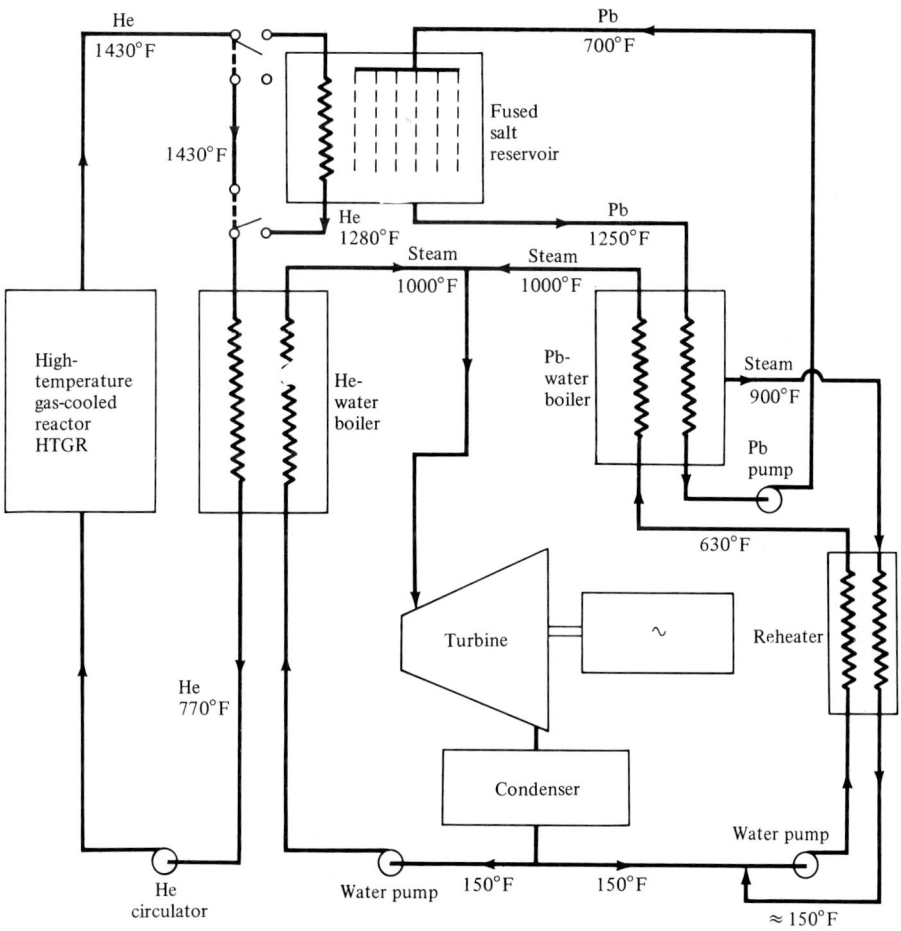

Figure 8-17 A schematic flow diagram of a high-temperature gas-cooled reactor (HTGR) powerplant using a latent heat energy storage system [10].

For the storage capacity of 7200 MWh, 42,000 tons of the eutectic are to be used. In order to give the necessary heat-absorption and heat-release rates, the eutectic is not allowed to freeze completely but instead operates as a slurry. (Recall the undesirability of glassy freezing as with the germanium sulfides.) The total latent heat of fusion is not utilized, and a salt mass of about 80,000 tons is needed. This requires a huge containment vessel about 36 m (120 ft) in diameter and 30 m (100 ft) high.

Another feature of the design is the addition of a secondary heat-transfer loop using molten lead as a heat carrier. Shell-and-tube heat exchangers between helium and the salt slurry, and steam and the salt slurry, have been considered. However, the build-up of solids on the tubes during heat withdrawal would seriously impede the heat-transfer rate and, therefore, the effectiveness of the heat exchanger. This necessitated the addition of the above-mentioned molten-lead loop. During heat withdrawal, therefore, lead is heated from 380°C (700°F) to 675°C (1250°F) in the slurry tank and used in a lead-steam boiler to produce steam for peak-load operation. Lead heating is accomplished by having globules of lead "rain" on top of the slurry, thus sufficiently stirring the reservoir. The globules distort as they fall and thus are expected to "shed" any thin skins of solidified salt. The design lead flow is 2 m^3/s (70 ft^3/s). After passing through the lead-steam boiler, the lead is pumped back to the top of the reservoir against more than 500 psi static pressure via an 800-kW pump.

In addition it is important that the temperature of feedwater entering the lead boiler not be lower than the melting point of lead (325°C, 620°F). This necessitated the addition of a feedwater preheater that uses wet steam from the same boiler. The plant operation, therefore, has the following modes:

1. During base-load operation, helium from the reactor at 775°C (1430°F) is short-circuited directly to the helium-water boiler.
2. During periods of low demand, helium is shunted to the fused salt reservoir, thus storing heat in the slurry at the fusion temperature of 680°C (1260°F) and leaving at 690°C (1280°F) to the helium-water boiler.
3. During periods of peak demand lead is circulated to the reservoir, leaving at 675°C (1250°F) to the lead-water boiler. In all cases steam with the proper flow rate is generated at 540°C (1000°F) and admitted to the same turbine-condenser system. Condensate at 65°C (150°F) is fed back to the helium-water boiler and, during peak demand, also to the lead-water boiler via the preheater.

A good turnaround efficiency for the storage system, more than 90 percent, is expected because the percentage heat losses per cycle are expected to be quite small once the reservoir and other components reach thermal equilibrium.

An economic evaluation, based on 1974 dollars, shows salt costing $2.15/lb$_m$ in large lots, with the price down to $0.1 to $0.5/lb$_m$ if large scale production is attained. This is said to result in a unit energy storage cost of $3 to $13/kWh, a unit molten-lead loop cost of $33/kW, and a unit peak powerplant equipment cost of $90 to $110/kW.

CHEMICAL-REACTION STORAGE

In this mode of energy storage the heat of reaction of reversible chemical reactions is used to store thermal energy during endothermic reactions and to release it during exothermic reactions. Like latent energy storage, this form also offers large energy storage densities and thus has been considered an attractive alternative for some time. (Besides reversible chemical reactions, this category also includes the solution and dissolution of a solid in a liquid and a gas in a solid.) Initial interest focused on low-temperature energy storage for residential heating and cooling. More recent interest contemplates its use to store high-temperature thermal energy suitable for power-generation cycles. In pioneering work by Schulten et al. [11], the following reaction was suggested for long-distance transmission of gas-cooled nuclear-reactor thermal energy

$$CO + 3H_2 \rightleftarrows CH_4 + H_2O \qquad (8\text{-}35)$$

This and other reversible reactions are listed in Table 8-1 with their operating ranges and heats of reaction.

Chemical-energy storage is now explained with the help of Eq. (8-35). Heat is stored by absorbing it in the endothermic direction of the reaction, from right to left. The enthalpy of formation [13] of $CH_4 + H_2O$ (liquid) at 25°C is given by $(-74.9) + (-286) = 360.9$ kJ/(g · mol). The enthalpy of formation of $CO + 3H_2$ at 25°C is $(-110.6) + 0 = -110.6$ kJ/(g · mol). Thus, moving from right to left with both reactants and products maintained at 25°C results in a net energy transfer of $(-110.6) - (-360.9) = +250.3$ kJ/(g · mol). The positive sign indicates energy added to the reaction, i.e., energy is absorbed and the reaction is endothermic. The reverse reaction, from left to right, results in -250.3 kJ/(g · mol), i.e., energy is

Table 8-1 Reversible chemical reactions under consideration for energy storage*

Reaction	Temperature range, K	Heat of reaction at 298 K	
		Btu/(lb · mol)	kJ/(g · mol)
$CO + 3H_2 \rightleftarrows CH_4 + H_2O$	700–1200	107,640+	250.3†
$2CO + 2H_2 \rightleftarrows CH_4 + CO_2$	700–1200	106,380	247.4‡
$C_6H_6 + 3H_2 \rightleftarrows C_6H_{12}$	500–750	89,100	207.2
$C_7H_8 + 3H_2 \rightleftarrows C_7H_{14}$	450–700	91,800	213.5
$C_{10}H_8 + 5H_2 \rightleftarrows C_{10}H_{18}$	450–700	135,000	314.0
$C_2H_4 + HCl \rightleftarrows C_2H_5Cl$	420–770	24,120	56.1
$CO + Cl_2 \rightleftarrows COCl_2$		48,420	112.6

*From Ref. [12].
† Higher heating value, including heat of condensation of H_2O.
‡ Heat of reaction per g · mol of CH_4.

released and the reaction is exothermic. The signs have been ignored in Table 8-1. Although the exothermic reaction is common, it will not occur at low temperatures or in the absence of a catalyst or an "igniter," which leads to potentially long storage times. The endothermic reaction is called *reformation** and the exothermic reaction is called *methanation*.†

A schematic of a powerplant with a chemical storage system using the reaction in Eq. (8-35) is shown in Fig. 8-18. During periods of low demand, some heat from the primary heat source is diverted to the reformer (endothermic reactor) to convert the products $CH_4 + H_2O$ to the reactants $CO + 3H_2$, which are stored in a vessel at high pressure, probably 70 bar, but at ambient temperature. During periods of high demand, these reactants are fed to the methanator (exothermic reactor) where heat is generated to run a peak turbine (or generate more steam for the main turbine). In the methanator the reactants are converted to the products $CH_4 + H_2O$, which are stored in a separate vessel for later use in the reformer during periods of low demand.

A thermal turnaround efficiency of this system is estimated (but yet unproven) at 85 to 90 percent [14]. The losses are mainly heat losses to storage vessels and piping and pumping losses of the gases.

The two storage vessels and the two reactors will all have to operate at different pressures. Storage pressures need to be high to minimize vessel size and cost, and the reformer has to operate at low pressures to maximize the rate of the endothermic reaction $CH_4 + H_2O \rightarrow CO + H_2$. There will, therefore, be a large pressure differential between the tube side, which carries the primary heat source fluid (e.g. helium at about 40 bar from a high-temperature gas-cooled reactor), and the gases on the shell side. A similar pressure differential occurs between the steam loop and the gases in the methanator. Such pressure differences require careful design. A complicated scheme to minimize these difficulties envisages the use of compressor-expander sets with a compressor between the reformer and the reactants storage tank and an expander (turbine) between the latter and the methanator. A similar compressor-expander set would be put between the methanator, the products storage vessel, and the reformer, with equal complication. The products expander may supply some of the work required by the reformer compressor because it operates at about the same off-peak time during the cycle. Similarly the reactants expander may supply some of the work required by the methanator compressor during peak hours. Noting the various compressor-expander inefficiencies and the fact that the number of moles of reactants and products are not the same, there will be a net work available during peak operation that may be reabsorbed during off-peak operation. This imbalance in net work of the compressor-expander sets may be remedied by electrical or mechanical storage, and thus we end up with two (or three) forms of energy storage in the same plant.

Some of the problems to be solved before such a storage system can become a

---

* In general, *reformation* is a process in which low-grade or low-molecular-weight hydrocarbon is catalytically reformed to a higher-grade or higher-molecular-weight hydrocarbon. The term also applies to the endothermic reforming of methane (the process under consideration above) for the production of hydrogen by the reaction of methane and steam in the presence of nickel catalysts.

† *Methanation* is the production of methane from a mixture of carbon monoxide and hydrogen.

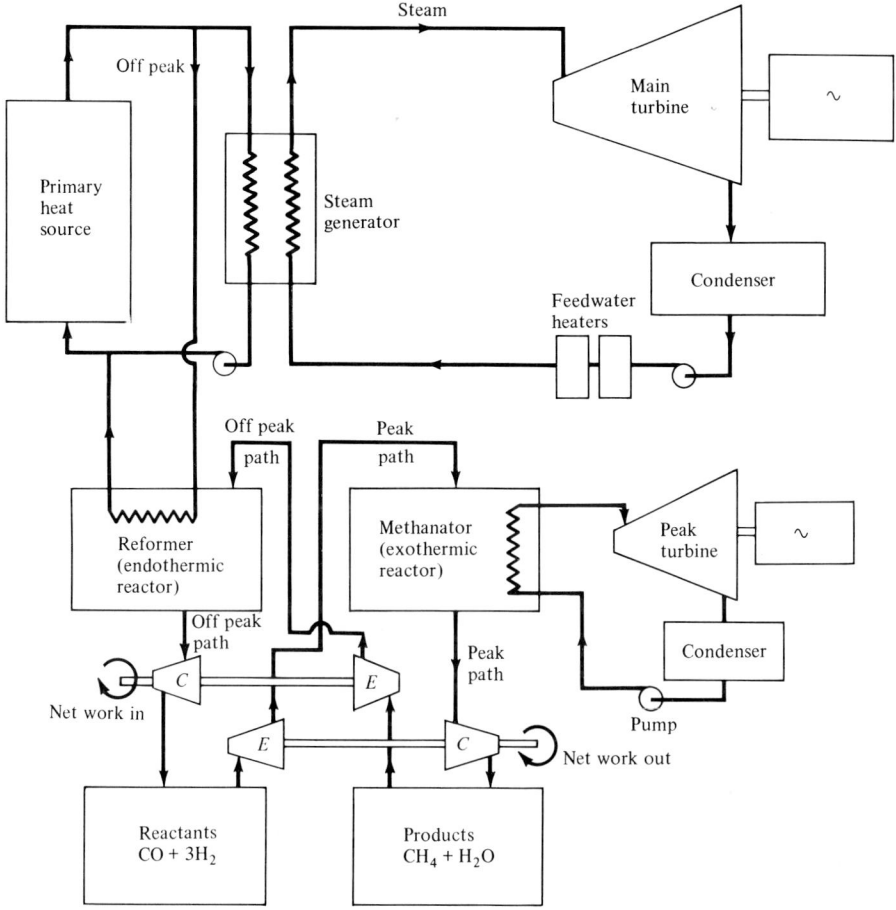

Figure 8-18 Schematic flow diagram of a powerplant with a chemical storage system using the reaction $CO + 3H_2 \rightleftharpoons CH_4 + H_2O$.

serious contender in powerplant use are safety and developmental problems. The major safety problems are those associated with the storage of large volumes of high-pressure flammable and poisonous gases. Developmental problems include the operation of methanators, which in the past were limited to relatively low temperatures to generate steam at high enough temperatures for efficient turbine operation. This entails the availability and use of a suitable catalyst that can operate at high temperatures with a reasonably long lifetime. Other significant problems involve the optimization of the entire cycle to reduce irreversibilities and increase overall efficiency and proper heat-exchanger design in the reformer and methanator. One idea that has been proposed involves the extraction of electrical work from the chemical reaction directly by electrochemical means.

The size of the storage vessels, even with operation near a pressure of 70 bar, which is compatible with current technology for gas storage, is still very large, of the order of hundreds of thousands of cubic meters. The vessels may either be steel tanks or, to reduce costs, underground caverns if suitable sites are available. Caverns, however, would pose problems of contamination of the gases by impurity gases from underground rocks. Such impurities may cause corrosion in the system and result in poisoning of the catalysts used in the reactors. Careful consideration must also be given to the problem of the diffusion of light gases, such as H_2, through rocks and fissures.

A preliminary, and possibly overoptimistic, economic estimate of the above system, based on cavern storage at $0.5 to $2/ft^3, a turnaround efficiency of 85 to 90 percent, and 70 percent conversion of original methane shows storage plant costs of $35 to $60/kW and a unit peak powerplant equipment cost of $90 to $110/kW, all based on 1975 dollars [14].

References

1. Data supplied by Madison Gas and Electric Co., Madison, Wis., 1982.
2. CAES—It's More than Hot Air, Staff Report, *Mech. Eng.*, vol. 104, no. 1, pp. 20-23, January 1982.
3. Hoffeins, H., N. Romeyke, and D. Hebel: Commissioning the First Air Storage Gas Turbine Set, *Brown Boveri Rev.*, vol. 67, pp. 465-473, August 1980.
4. Fullman, R. L.: "Energy Storage by Flywheels," General Electric Company Report No. 75CRD051, April 1975.
5. Boom, R. W. and H. A. Peterson: Superconductive Energy Storage for Power Systems, "Proceedings of the 1972 Intermag Conference," Kyoto, Japan, April 1972.
6. Grover, F. W.: Formulas for Mutual and Self Inductance, *Bull. U.S. Bur. Stand.*, vol. 8, no. 1, 1912.
7. Boom, R. W., and R. F. Bischke: Inductor-Converter Superconductive Magnetic Energy Storage for Electric Utility Usage, *Phys. Technol.*, vol. 13, pp. 18-27, 1982.
8. Gilli, P. V., and G. Beckmann: "The Nuclear Storage Plant—An Economic Method of Peak Power Generation," Paper No. 4.1.10, Ninth World Energy Conference, Detroit, 1974.
9. Gawron, V., and J. Schröder: "Proceedings of the Fourth European Symposium on Fluorine Chemistry," Ljublijana, Yugoslavia, 1972.
10. Bundy, F. G.: "Power Generating Plant with Nuclear Reactor-Heat Storage System Combination," U.S. Patent No. 3,848,416.
11. Schulten R., C. B. Van der Decken, K. Kugeler, and H. Barnert: Chemical Latent Heat for Transport of Nuclear Energy over Long Distances, "Proceedings of the British Nuclear Energy Society International Conference on The High Temperature Reactor and Process Applications," 1974.
12. Hanneman, R. E., H. B. Vakil, and R. H. Wentorf, Jr.: Closed Loop Chemical Systems for Energy Transmission, Conversion and Storage, "Proceedings of the Ninth Intersociety Energy Conversion Engineering Conference," 1974.
13. Van Wylen, G. J., and R. E. Sonntag: "Fundamentals of Classical Thermodynamics," 2d ed., John Wiley & Son, Inc., New York, 1973, chap. 12.
14. Golibersuch, D. G., F. P. Bundy, P. G. Kosky, and H. B. Vakil: "Thermal Energy Storage for Utility Applications," General Electric Company Report No. 75CRD256, December 1975.

Section **9**

Environmental Aspects of Power Generation

Growing concern for the environment—and the effect of power generation on it—has led to increased interest in power-plant pollutants and their control. Designers must now give greater attention to what a power plant emits than ever before.

Acid rain and acid snow seem to be in the headlines almost every day of the week. Their control is argued over by local and national governments—often with less than friendly exchanges. Yet their control is often—at least to some extent—in the hands of the power-plant designer. Today the newest offender is acid fog.

Controlling atmospheric pollution usually comes down to two major considerations—namely design and money. Since design always involves money (in the choice of equipment), pollution control can be said to be primarily a matter of money. Why? If there is enough money, almost any pollutant can be reduced to an acceptable level.

This section discusses a variety of the known and proven ways to control air pollution caused by power plants. Likewise, thermal pollution—wherein heat is rejected to the environment—is also considered. Most thermal power plants add more heat to the environment than they produce in energy output.

Finally, nuclear power plants pose radiation, thermal, and waste-disposal considerations unique to themselves. These are discussed in detail in this section.

From *Powerplant Technology*, M.M. El-Wakil. Copyright © 1984. Used by permission of McGraw-Hill, Inc. All rights reserved.

INTRODUCTION

It should be recognized at the outset that there is virtually nothing people can do to improve their life standards, or even to maintain these standards in the face of growing populations, not to mention bringing up the standards of the billions of people living in substandard conditions around the world, that does not have adverse side effects, particularly on the environment. We can no longer expect the return of the pristine environment that existed in the early history of humankind. This is true whether people are producing increasing quantities of food and clothing, which have become energy-intensive; building dams and other irrigation schemes; or building powerplants to produce the electricity so necessary for industry and domestic uses and the maintenance of that life standard. The important question is whether or not the benefits of these necessary systems outweigh their adverse effects. Some sentiments have been expressed to the effect that the adverse effects of power generation can be arrested solely by conservation. Conservation means more efficient use of our resources and more efficient electric-energy production and use and is commendable. However, conservation alone can only reduce the rate of increase of environmental degradation. It is thus at least of equal importance to study the causes and effects of the adverse effects on the environment and to minimize them as much as possible.

Another factor to recognize is that nature itself contributes large quantities of contaminants to the environment, mainly due to the natural processes of plant and animal decay and natural background radioactivity.

At this point one needs to distinguish between the often used terms contaminant and pollutant. *Contaminants* are those materials, radiations, or thermal effects that are added to environment beyond what nature itself puts into it. In the 1960s it was estimated that, globally, nature puts into the environment some 10 times the amount of contaminants that people put into it. The contribution of nature is, however, diffuse and thus largely harmless, whereas the contribution by human beings is more localized and concentrated. It follows that *pollutants* are contaminants in concentrations high enough to adversely affect something that people value, such as their environment and health.

Table 9-1 presents a summary of chemical air pollutants from all sources in the United States in 1966. It is to be noted that motor vehicles are the largest contributors, followed by industry and powerplants, and then by small contributions from space heating and refuse. Not included in Table 9-1 is thermal pollution, the warming up of bodies of water and the atmosphere from the above sources as well as nuclear powerplants.

The case of radioactive pollution from nuclear powerplants is even less troublesome than the above sources when compared with the contributions of nature. Table 9-2 presents a summary of average annual exposures per person from all sources. Note that nature is the largest contributor, followed by medical irradiation, whereas releases attributed to the nuclear industry as a whole are relatively miniscule.

Even in abnormal occurrences the picture presented in Table 9-2 holds true. This can be illustrated by two such occurrences, one artificially made, the March 1979 accident at Three Mile Island (TMI-2) powerplant, the other naturally occurring, the May 1980 volcanic eruption of Mount St. Helens in the state of Washington, USA.

The radioactive release from Mount St. Helens was far more significant than that from TMI-2. It is reasonable to assume that volcanic eruptions and other such natural occurrences have released and will continue to release large amounts of radioactivity into the atmosphere.

Powerplants are, therefore, not the sole or largest contributors to environmental problems. They are, however, a growing concern, as their numbers and sizes will continue to increase in the decades ahead. The powerplant pollutants of most concern are:

Table 9-1 Total United States chemical air pollution (1966)*

Source	Pollutants	
	tons/year	%
Motor vehicles	86	60.6
Industry	23	16.8
Fossil powerplants	20	14.1
Space heating	8	5.6
Refuse disposal	5	3.5

*From Ref. [1].

Table 9-2 Average radioactive exposures per person*

Source	Exposure mrem/yr	%
Natural background	100.0	67.60
Medical irradiation	45.0	30.70
Fallout	0.9	0.60
Miscellaneous	0.7	0.50
Occupational	0.7	0.45
Nuclear industry	0.2	0.15

*From Ref. [2].

1. From fossil powerplants:
   a. sulfur oxide
   b. nitrogen oxides
   c. carbon oxides
   d. particulate matter
   e. thermal pollution
2. From nuclear powerplants:
   a. radioactivity release
   b. radioactive wastes
   c. thermal pollution

In addition, pollutants such as lead and hydrocarbons are contributed primarily by motor vehicles and will not be covered here.

This chapter is devoted to the discussion of these various aspects of electric-power generation and methods to minimize or alleviate their effects. The first part of the chapter deals with the effects of fossil-fuel powerplants, the latter part with those of nuclear powerplants.

CONSTITUENTS OF THE ATMOSPHERE

The effects of powerplant pollutants on the environment are primarily to the air and water and, to a lesser extent, the land.

The total mass of the earth's atmosphere is estimated at 5.7×10^{15} tons of various major and trace constituents. Table 9-3 shows approximate masses in tons of the major constituents, N_2 and O_2, as well as other constituents that are of main concern in the earth's atmosphere. The table lists the global (total atmosphere) tonnage as well as the tonnage in a portion of the atmosphere roughly lying between 30 to 60°N latitude and below 20,000 ft altitude. This portion of the atmosphere is less than 20 percent of the total, but it corresponds to the area of much of the world's industry and is therefore the receptacle for much of humankind's additions of contaminants to the

Table 9-3 Major and selected constituents of the atmosphere, tons*

Component	Tonnage		Lifetime
	Global	Between 30–60° N latitude	
N_2	4.25×10^{15}	0.55×10^{15}	Indefinite
O_2	1.30×10^{15}	0.17×10^5	Indefinite
CO_2	2.80×10^{12}	—	Years
CH_4	5.00×10^9	—	Years
CO	6.00×10^8	—	Years
NO_x	9.00×10^6	2.00×10^6	Days
SO_2	2.50×10^6	5.00×10^5	Days
Particulate matter	1.55×10^8	3.00×10^7	Days to years

*Data from Ref. [1].

global atmosphere. In addition, the rate of atmospheric mixing between the northern and southern hemisphere is slow, so that the region mentioned above may be considered isolated except for extended periods of time.

The atmosphere is recognized as a self-cleansing environment in which contaminants are continually added and removed. Thus, the amount of any one contaminant at any given time is affected by the mass rates of addition and removal of that contaminant. The concentration of that pollutant is affected by these rates as well as the size of the atmosphere itself. The addition of contaminants is due to both natural causes and artificial causes that are of fairly recent origin. The last column in Table 9-3 represents the average lifetime of each constituent. This is the average residence time of the constituent between addition and removal. The figures in this and the tables to follow in this chapter are only ball park figures because of the doubtful accuracy of measurements, so they should be viewed with caution. They are also bound to vary with time. The relative magnitudes are, however, informative.

We will now discuss the most important contaminants, their effects, and methods of coping with them.

OXIDES OF SULFUR

Sulfur in the atmosphere exists essentially in three forms: sulfur dioxide, SO_2; hydrogen sulfide, H_2S; and various sulfates. H_2S comes primarily from natural sources. The sulfates come from sea spray and from the oxidation of SO_2.

Sulfur dioxide, SO_2, which primarily comes from artificial causes, is therefore our primary concern, even though it is believed to be responsible for something less than 25 percent of all sulfur in the atmosphere. It is produced in the combustion of coal and oil, mainly in powerplants, but also in steel mills, smelters, and similar

Table 9-4 SO$_2$ flowchart in the 30° to 60° N latitude*

Additions	Natural (volcanoes)	Powerplants	Industry	Space heating	Solid waste	Motor vehicles
tons/year	Negligible	5×10^7	1×10^7	0.6×10^7	0.3×10^7	0.2×10^7
Steady state		5×10^5 tons		0.2 ppb, average		
Removal			Precipitation	Gravitation		

*Data from Ref. [1].

industries, and space heaters (all also produce some H$_2$S), with coal responsible for about 70 percent of the total.

Although the mass of SO$_2$ in the 30 to 60°N latitude region of the atmosphere, which is about 5×10^5 tons (Table 9-3), represents an average of about 0.2 ppb (parts per billion) in that region, the yearly mass added, and removed, is much higher, being of the order of 10^8 tons (Table 9-4). Local SO$_2$ concentrations vary widely and are usually measured in ppm (parts per million). Urban industrial areas often reach 3.2 ppm, and peaks of 11 ppm have been recorded. SO$_2$ concentrations below 0.6 ppm produce no ill effects in human beings. Most people, however, become cognizant of sulfur at about 5 ppm and become irritated at about 10 ppm. A 1-h exposure to 10 ppm can cause breathing problems and mucus removal. The effects are further complicated at high air temperatures and humidities, in the presence of aerosols, etc., and they become much more serious when SO$_2$ is combined with particulates and enters the digestive system.

OXIDES OF NITROGEN

The oxides of nitrogen come in different forms: nitric oxide, NO; nitrous oxide, N$_2$O; nitrogen dioxide, NO$_2$; nitrogen trioxide, NO$_3$; nitric anhydride, N$_2$O$_5$; and nitrous anhydride, N$_2$O$_3$. Of these only NO and NO$_2$ are the significant artificially made oxides; they are commonly referred to as NO$_x$, pronounced "nox."

Nitric oxide, NO, is formed in the combustion of all fossil fuels. The rate of formation is strongly dependent upon the combustion temperature, being significant only at high temperature. Formation is also dependent upon the oxygen concentration present during combustion and the time allowed for the combustion process. The primary contributors to NO are the motor vehicles, in which combustion does occur at high temperatures, with lesser contributions from the combustion of coal and oil in powerplants, which occurs at lower temperatures. Concentrations of NO in the exhaust of gasoline automobiles depends upon the fuel-air ratio used and vary between about 3000 ppm near stoichiometric conditions to about 6000 ppm for 80 percent lean mixtures.

In the atmosphere NO rapidly oxidizes to NO_2, a process greatly accelerated by photochemical effects (the presence of sunlight) and by the presence of organic material in the air. NO_2 has a more adverse health effect on human beings than NO. It has an affinity for hemoglobin, which carries oxygen to body tissues. It thus deprives them of oxygen. It also forms acid in the lungs and hence is much more toxic than CO (below) for the same concentrations. It also reduces atmospheric visibility. People begin to recognize the existence of NO_2 by its odor when it reaches concentrations of 0.4 ppm and higher. A continuous exposure to 0.06 to 0.1 ppm of NO_2, however, can cause respiratory illness. A few minutes exposure to 150 to 200 ppm causes obliterations of the bronchiols (the smallest divisions of a bronchial tube), and a few minutes exposure to 500 ppm causes acute edema (swelling from the effusion of a watery liquid into cellular tissue).

The atmosphere in the 30 to 60°N latitude region contains 2×10^6 tons of NO_x (Table 9-3), which corresponds to an average of 1 ppb. The additions, and removals, per year are 2×10^8 tons from natural causes, mainly biological, and 5×10^7 tons from artificial causes (Table 9-5). It should be noted here that the two NO_x compounds represent only a small fraction of the total nitrogen compounds entering and leaving the atmosphere. For example ammonia, NH_3, enters and leaves the atmosphere at a rate some 10 times that of NO_x.

While the average in the 30 to 60°N latitude region is about 1 ppb, local urban concentrations of NO_x can be much higher, ranging between 0.02 and 0.9 ppm. Temporary but severely high concentrations up to 3.9 ppm have been recorded in Los Angeles, which has a rate of addition 100 times the world average. (It is instructive to note that tobacco smoke can contain about 250 ppm.)

Although contributions of NO_x by artificial causes on a global scale are far less than those by natural events, local human contributions could be far larger and could

Table 9-5 NO_x flowchart in the 30 to 60°N latitude*

Additions	Natural biological reactions	Motor vehicles	Industry	Powerplants	Space heating	Solid waste
tons/year	20×10^7	2×10^7	2×10^7	0.8×10^7	0.2×10^7	0.2×10^7
Steady state	2×10^6 tons, 1.0 ppb, average					
Removal	As nitrates by precipitation, vegetation					

*Data from Ref. [1].

thus give rise to severe health and environmental problems. The effects of NO_x on bodies of water and land will also be covered later.

OXIDES OF CARBON

Carbon monoxide, CO, methane, CH_4 and carbon dioxide, CO_2, are the most widely produced of all contaminants, with CO_2 being the largest of all (Table 9-3).

Carbon monoxide is caused in part by natural causes such as marsh gas, coal mines, vegetation, lightning, and forest fires. However, their contribution to production of CO is small compared with that from human-generated causes, of which more than 90 percent is produced by motor vehicles (Table 9-6). Powerplants produce less than 1 percent. The removal of CO is almost entirely due to natural events whose exact nature is not known with certainty.

The CO additions to the entire earth's atmosphere total about 230 million tons. Local additions are often serious. For example, in the Los Angeles basin alone, some 4 million tons are added annually. The basin average CO concentration increased from about 7 ppm to about 11 ppm between 1957 and 1963 compared with the global 0.1 ppm. Emission-control devices are causing a slow decrease in these concentrations. People who smoke voluntarily expose themselves momentarily to concentrations said to reach 42,000 ppm.

The health hazard of CO, like NO, stems from its depriving body tissues of oxygen because of its affinity for hemoglobin, which carries oxygen to body tissues. A CO concentration of 100 ppm causes headache, 500 ppm causes collapse, and 1000 ppm is fatal.

Carbon dioxide, CO_2, is more abundant and, unlike CO, is largely contributed by powerplants and hence of more concern to us here. Natural causes, such as the decay of organic matter, however, contribute much greater amounts of CO_2 than artificial causes (Table 9-7). Although the CO_2 added to the atmosphere by these

Table 9-6 Carbon monoxide flow chart, entire atmosphere*

Additions	Natural events	Motor vehicles	Industry	Space heating	Powerplants	Solid waste
tons/year	0.2×10^8	2.0×10^8	0.04×10^8	0.04×10^8	0.02×10^8	0.02×10^8
	↓	↓	↓			
Steady state	6×10^8 tons, 0.1 ppm, average					
	↓	↓	↓			
Removal	Oxidation to CO_2 in atmosphere, dissolution in water, consumption by vegetation, other unknowns					

*Data from Ref. [1].

Table 9-7 Carbon dioxide flowchart, entire atmosphere*

	Natural causes			Artificial causes	
Additions	Organic decay	Volcanoes	Respiration	Combustion	Industry
tons/year	2×10^{11}	1×10^{10}	1×10^{9}	2×10^{10}	2×10^{6}
	↓	↓		↓	
Steady state	3×10^{12} tons, 325 ppm average				
	↓	↓		↓	
Removal	Photosynthesis (2×10^{11} tons/year), ocean absorption, rock weathering				

*Data from Ref. [1].

causes is considered a contaminant, it is not normally considered a pollutant because it is essential to plant and, therefore, human life. The main removal mechanism of CO_2 is *photosynthesis*. This is the process that occurs in green plants in the presence of light by which water, CO_2, and minerals are converted back to oxygen and various organic compounds.

Despite the necessity mentioned above of CO_2 in the atmosphere, there is increasing concern that the growing combustion of fossil fuels for all types of energy will cause the CO_2 concentration in the entire atmosphere to continue to increase with serious effects on the earth's climate as a result of the greenhouse effect.

THE GREENHOUSE EFFECT

The concentrations of CO_2 in the atmosphere have slowly but continually increased since 1880 with the evergrowing worldwide combustion of fossil fuels. Since the 1950s, the concentrations have risen at an increasing rate (Fig. 9-1). The addition of CO_2 in the latter part of the decade of the 1970s is believed by some to have reached a level that is beyond the capacity of plant life and the oceans to completely remove it, and about half of the CO_2 added is retained in the atmosphere.

The existence of CO_2 in the atmosphere causes a *greenhouse effect*. The atmosphere, analogous to the glass panes of a greenhouse, transmits the radiation from the sun. Because the surface of the sun is at about 6000 K, most of that radiation is in the shortwave and visible portion of the spectrum, and only a small portion of this radiation is absorbed or scattered back to space by the atmosphere. The transmitted radiation is largely absorbed by the surface of the earth, which is warmed by it.

Part of the resulting heat of the earth is transmitted by various modes of heat transfer (conduction, convection, evaporation, and condensation, which has a negative contribution) to the atmosphere and is ultimately reradiated away from the surface. Becaue the surface temperatures are low, this reradiation is mostly in the infrared portion of the spectrum. CO_2 has no emission and absorption bands in the short and

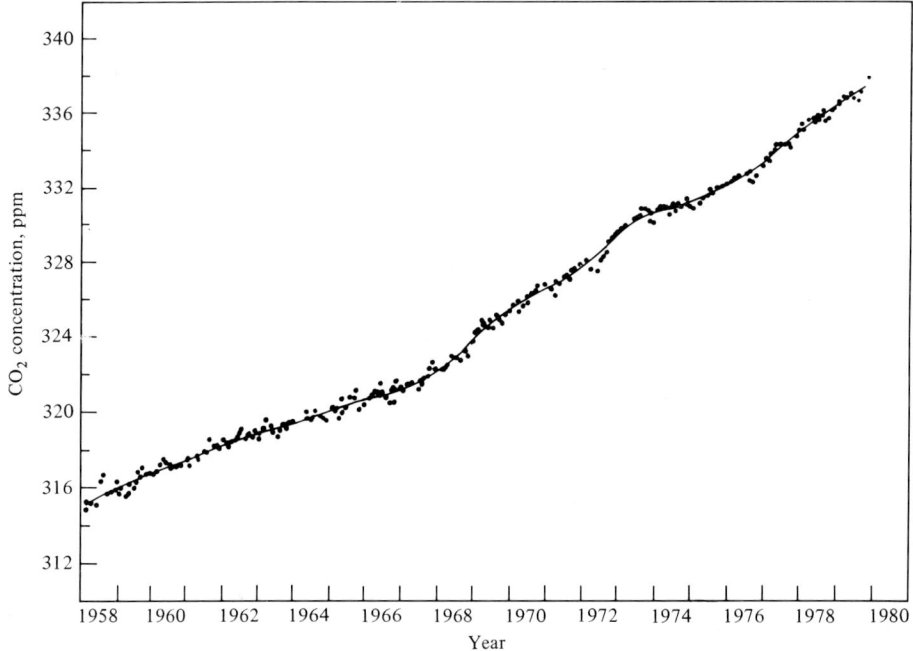

Figure 9-1 Average CO_2 concentrations in the atmosphere up to 1980 with seasonal effects removed. Seasonal fluctuations about average: slightly lower in spring due to absorption by new plants; slightly higher in fall due to release from decaying plants. *(Based on work by C. D. Keeling et al. from Ref. 3.)*

visible wavelengths of the spectrum but does have several infrared bands between the wavelengths 2.36 to 3.02 μm, 4.01 to 4.08 μm, and 12.5 to 16.5 μm. Water vapor, likewise, has only infrared emission and absorption bands in the wavelengths 2.24 to 3.27 μm, 4.8 to 8.5 μm, and 12 to 25 μm. When both CO_2 and H_2O are present, the absorptions are, to a good approximation, additive.

The presence of CO_2 and H_2O in the atmosphere results in absorption of a large portion of the longwave infrared radiations from the surface of the earth and partial radiation of them back to it. In essence, therefore, the atmosphere is not completely transparent to the reradiated energy and, like the panes of a greenhouse, traps much of the energy of the sun. Whereas an equilibrium of sorts existed over the centuries, the growing levels of concentration of CO_2 are expected to cause the earth and its lower atmosphere to warm up to higher temperatures than would otherwise be the case. Indeed, measurements have shown a small warming up already (Fig. 9-2).

It is estimated that if all the fossil-fuel reserves are burned, the current CO_2 concentration in the atmosphere would increase 5- to 10-fold. It is feared, however, that long before this ultimate situation happens, climatic changes might occur that would result in disasterous consequences, such as melting of the polar ice caps, which would result in the raising of sea level and flooding of many coastal areas of the world. Some evidence that this process has already started has been cited.

Somewhat counterbalancing the increasing effect of CO_2 on the atmosphere is the

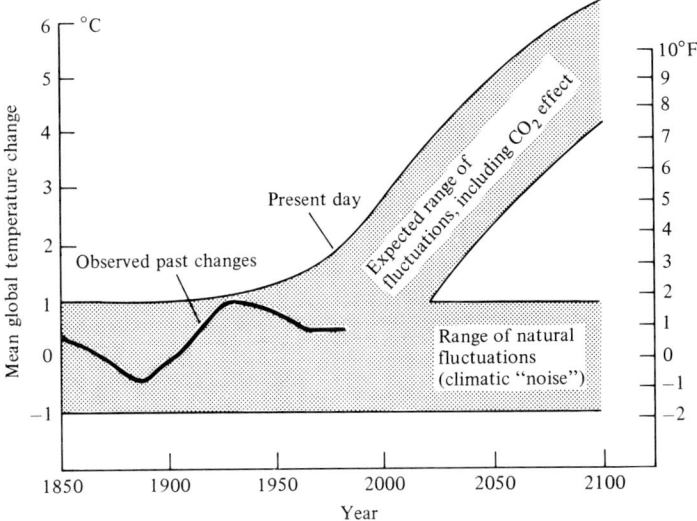

Figure 9-2 Changes in mean global temperature with and without projected CO_2 effects. *(Based on work by J. M. Mitchell, from Ref. 3.)*

increasing content of particulate matter. Particulate matter helps reflect some of the incident radiation from the sun back to space and thus prevents it from reaching the surface of the earth.

ACID PRECIPITATION

Acid precipitation is the return to earth of the oxides of sulfur and nitrogen in acid form. They take several forms: acid rain and acid snow, dry deposition, and acid fog.

Acid Rain and Acid Snow

Acid fallout in the form of *acid rain* or *acid snow* is one of the more serious environmental hazards of increased concentrations of sulfur and nitrogen oxides in the atmosphere. Pure water has a neutral pH value of 7.0.* The average pH of normal rainfall is a slightly acidic 5.6 because of the formation of mild carbonic acid H_2CO_3 when pure rainwater combines with natural and human-generated carbon dioxide in the atmosphere; it is considered harmless. However, biologists usually consider rainfall with pH values lower than 5.6 to be potentially harmful to plant, animal, and human life. In many parts of the world, however, rainfall has grown increasingly more acidic in recent years. pH values as low as 2.4, about the same acidity as vinegar, have been recorded in England and in widely separated regions of the world.

* *Acidity* and *alkalinity* are measured on a pH scale of 0 to 14. pH = 7.0 denotes neutrality. $0 \leq pH < 7$ represents acidity. $7 < pH \leq 14$ represents alkalinity. The strongest acids have pH values below 1.0.

Acid rain and acid snow are caused by sulfur dioxide, SO_2, hydrogen sulfide, H_2S, and the oxides of nitrogen, NO_x, in the atmosphere. Over a period of time after these gases are emitted, usually hours or days, they are carried along by wind currents and combine with water molecules in the water vapor of the atmosphere to form tiny drops, mainly of nitric acid, HNO_3, and sulfuric acid, H_2SO_4. These aerosollike acids are then returned to the surface of the earth when they encounter rain- or snow-producing clouds at various distances from their place of origin, often as far as hundreds or thousands of kilometers away.

In general SO_2 contributes about 60 percent of the acidity of such precipitation, whereas NO_x contributes about 35 percent. More is known about the relatively simple mechanism of formation of acid rain and snow from SO_2 than the more complex mechanism due to NO_x.

Acid precipitation was first observed in Scandinavia in the 1950s. Excessive acidity is a more recent phenomenon. It has been observed on the North American continent in such places as the northeastern parts of the United States and eastern Canada (Fig. 9-3), where the toxic fumes have crossed the international border in both directions, although their source is often blamed on powerplants in the Ohio valley that export them on prevailing winds. Other affected areas in the United States are the northwestern region, the southern Appalacians, and portions of Florida. In Europe, industrialized countries such as Britain, Germany, and France produce millions of tons of these gases that affect many other countries, some as far away as Scandinavia. Acid rain has also been found in parts of Asia and other world spots.

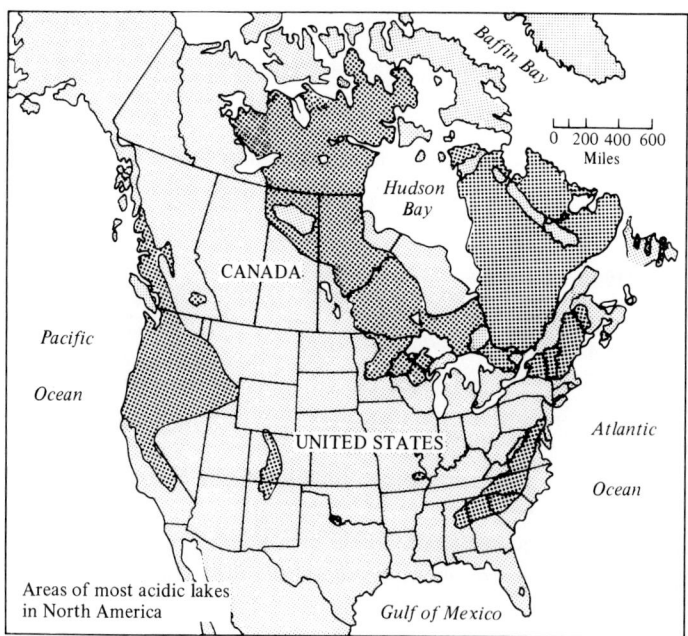

Figure 9-3 Areas of most acidic lakes in North America [4].

This unwelcome traffic across national and international boundaries has prompted multigovernmental conferences to discuss the causes, sources, and extent of acid rain, the outlook for the future, and if possible, to find corrective measures. Treaties between nations defining responsibilities and outlining joint actions may have to be drawn. Expressions of concern were voiced at a 1979 international conference in Toronto by Canadian officials who stated that acid rain is their "most serious environmental problem"; by U.S. officials who announced that "too much damage has already been done by acid rain—too many trout lakes and salmon streams have already been rendered lethal to fish, and valuable wilderness areas are beginning to show signs of acidification"; and by others.

The effect of acid on streams and lakes has resulted in the lowering of pH values of some of them to values betwen 2.8 and 5.0. This acidification causes the decimation or malformation of fish, particularly trout and salmon. It has been estimated that 50,000 lakes in the Adirondack Mountains and Ontario, Canada, have become acidified to the extent that fish population has declined or ceased to exist. High levels of acidity are also said to cause the dissolving of toxic metals such as cadmium, lead, and aluminum out of minerals in the bottom sediments. These toxic metals, like the acids, can destroy fish and also contaminate drinking water.

The effect of acid rain on soil can cause the leaching of essential plant nutrients from the soil and reduce nitrogen fixation by microorganisms, which would make the soil less fertile. It can also dissolve aluminum and cadmium out of soil minerals, as mentioned above, and thus allow them to enter roots and kill trees. A recent (1982) study on the effects of acidity on forests at the University of Vermont showed that on Camel's Hump, a 4100-ft peak in the Green Mountains, half the red spruce trees have died. In West Germany, acid rain is blamed for the death of about 3700 acres of evergreen trees in Bavaria.

Dry Deposition

A particularly lethal form of acidity is called *dry deposition*. This occurs when sulfate particles fall on tree leaves before mixing with water. This dry fallout is said to account for about half the total acid fallout. When these particles mix with surface or rain water, they become concentrated sulfuric acid.

Attempts to alleviate the problem so far (1982) have met with little success. The most serious of these attempts involves dispersal and detoxification of the emissions that cause the acid rain in the first place. Other plans involve the breeding of new plant or animal species that would be acid-resistant. Probably the most effective method is the use of smokestack scrubbers that remove much of the sulfur from powerplants and industrial furnaces, though they have no effect on nitrogen oxides. They, however, are used only in a few large installations and are expensive to install and operate. Ironically, the dispersal of the emissions by the use of tall smokestacks, thought to be the answer at one time (the solution of pollution is dilution), is effective only in removing particulates and exhaust locally. It has actually encouraged the formation of acid rain by more efficiently sending the emissions to those air currents that transport them to rain clouds, often far away from their point of origin.

Detoxification of lakes by the addition of lime has proven counterproductive because lime has been found to combine with heavy metals in the bottom sediment and to free them, actually increasing the toxicity of water.

The idea of breeding of acid-resistant fish is a proposition that is not taken very seriously by most zoologists.

Acid Fog

A recently noted major acid pollutant is *acid fog*. Its origin is the same as acid rain or snow, i.e., sulfuric and nitric oxides from powerplants and, to a lesser extent, motor vehicles. It forms by the mixing of these pollutants with water vapor near the ground. The acid vapors then begin to condense around very tiny particles of fog or smog, pick up more water vapor from the humid air, and turn into acid fog. When the water in the fog burns off (evaporates) due to the sun or other causes, drops of nearly pure sulfuric acid are left behind. It is these drops that make acid fog so acidic. In Los Angeles and Bakersfield in southern California, the mists have a pH of 3.0 compared with 4 or 4.5 for acid rain. Acid fog 100 times as acidic as acid rain has been detected. Cases have been reported where people had trouble breathing when it was foggy. The problems of fog are now believed by some to be more serious than those of smog in these areas.

Many researchers consider the effects of acid precipitation, especially the changing of soil chemistry, to be irreversible and fear its long-range effects. Monitoring programs of air, soil, and water are being instituted to ascertain these long-range effects. However, the uncertainty about the real extent of the problem adds to the prevailing disquiet regarding it.

PARTICULATE MATTER

Particulate matter in the atmosphere is composed of smoke, dust, and other solids made of a wide variety of organics and metals. Samples taken in urban areas of the United States between 1960 and 1965 showed average and maximum particulate concentrations in the atmosphere of about 100 and 1250 $\mu g/m^3$, respectively. The Los Angeles basin, by now shown to be a particularly bad case, is estimated to put into the atmosphere some 170,000 tons/year of particulate matter, including that due to SO_2, about 30 times the mass of the atmosphere over that basin. During the London atmospheric pollution crisis of 1962 the particulate concentration rose to 2000 $\mu g/m^3$. There is evidence that air-circulation patterns over industrialized areas cause self-contained dust domes that aggravate matters over these areas (Fig. 9-4). Nonurban areas in the United States averaged 28 $\mu g/m^3$.

The total particulate matter in the 30 to 60°N latitude region of the atmosphere is about 3×10^7 tons, averaging about 30 $\mu g/m^3$ (Table 9-8). Both natural and human activities seem equally to blame for the presence of particulate matter in the atmosphere. Natural causes include natural dust caused by wind, storms, volcanoes and natural fires, metoritic dust, and fog. Fog, although not strictly a pollutant, con-

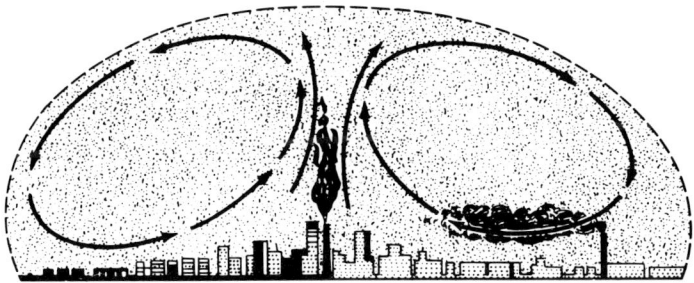

Figure 9-4 Air-circulation pattern over a large industrial urban area, creating a self-contained dust dome.

tributes to undesirable climatic conditions, especially when combined with smoke (smog). Of the human causes, dust and ash that emanate from industrial processes, fossil powerplants, and other combustion processes are the largest contributors, of which powerplants contribute about one-third. Sulfur compounds are larger contributors to particulate matter. This comes about by the SO_2 in the atmosphere oxidizing to sulfur trioxide, SO_3, which forms H_2SO_4 (sulfuric acid) mist, which in turn reacts with other materials in the air to form, among other things, ammonium and calcium sulfates.

The effects of particulate matter in the atmosphere are many and varied. Besides the obvious effects of decreasing visibility and increasing soiling and corrosion, and the already mentioned effects on climatic conditions, there is a health hazard that is a complex function of concentration and particle size. The size distribution is given by the usual log mean normal distribution curve (below). Numerically, most particles have a diameter below 2 μm, with a numerical average about 1.27 μm. The larger particles, however, although fewer in number, represent a greater mass fraction of the

Table 9-8 Flowchart of particulates in the 30 to 60° N latitude*

Additions	Natural causes			Artificial causes				
	Natural	Meteors	Fog	Industrial	Space heating	Motor vehicles	Solid waste	Sulphates
Tons/year	6×10^7	1×10^3	3×10^7	3×10^7	2×10^6	2×10^6	2×10^6	6×10^7
	↓			↓		↓		
Steady state	3×10^7 tons, 30 μg/m³ average							
			↓		↓		↓	
Removal	Gravitation, impaction, drop nucleation, washing							

*From Ref. [1].

total. It is estimated that an individual breathes about 1 mg of particulate matter per day during times of heavy pollution, with the larger particles depositing in the mucous lining and the smaller ones in the deeper parts of the lungs. Particulate matter in the atmosphere is intrinsically toxic; it absorbs toxic substances and obstructs respiratory passages. An annual mean 100 to 200 $\mu g/m^3$ results in respiratory illnesses, whereas 300 to 600 $\mu g/m^3$ causes a large increase in the number of bronchitic patients.

Particulate-Matter Distribution

A *Gaussian* or *normal distribution* of a statistical quantity, such as the diameter of particles in a large sample, is one in which the numbers of particles of given diameters fall on a bell-shaped curve, symmetrical about a mean diameter, when plotted against the diameter on a linear scale. It is given by

$$N(d) = \frac{1}{\sqrt{2\pi}S} e^{-\frac{1}{2}\left(\frac{d - d_n}{S}\right)^2} \tag{9-1}$$

where
$N(d)$ = number of particles with diameters between d and $d + dd$

d = particle diameter

d_n = mean particle diameter

S = standard deviation of d, having the same dimensions as d and given by

$$S = \left[\frac{\Sigma\left(d_i - \frac{\Sigma d_i}{N}\right)^2}{N - 1}\right]^{0.5} \tag{9-2}$$

where
N = total number of particles

Most particle samples from emissions, however, have size distributions that follow a *log normal distribution*. This has the same symmetrical bell-shaped curve as the Gaussian distribution except that the variable is the *logarithm* of the diameter rather than the diameter itself. It is thus given by

$$N(\omega) = \frac{1}{\sqrt{2\pi}S_\omega} e^{-\frac{1}{2}\left(\frac{\omega - \omega_n}{S}\right)^2} \tag{9-3}$$

where
$\omega = \ln d$

ω_n = mean value of ω or $\ln d$

$N(\omega)$ = number of particles per unit interval betweeen $\ln d$ and $[\ln d + d(\ln d)]$

S_ω = standard deviation of ω

Figure 9-5a is a plot of the fraction $N(\omega)N$ showing the usual Gaussian curve but with a logarithmic abscissa scale for d. Figure 9-5b shows a log normal plot of the fraction $N(\omega)/N$ with a linear abscissa scale for d.

The notations on Fig. 9-5b have the following meanings.

Count modal diameter The diameter at which the greatest number of particles occurs, represented by the maximum point on the distribution curve

Count or number median diameter The diameter for which 50 percent of all particles or $N/2$ are larger and 50 percent or $N/2$ are smaller by count

Count mean diameter The arithmetic mean diameter of all particles present, e.g., the sum of diameters of all particles present divided by the total numbers of particles N

Area median diameter The diameter for which the surface area of all particles larger than it constitutes 50 percent of the total surface area

Area mean diameter The diameter of a particle that has a surface area equal to the arithmetic mean of the surface areas of all particles

Mass median diameter The diameter for which the mass of all particles with diameters larger than it constitutes 50 percent of the total mass

Mass mean diameter The diameter of a particle that has a mass equal to the arithmetic mean of the masses of all particles

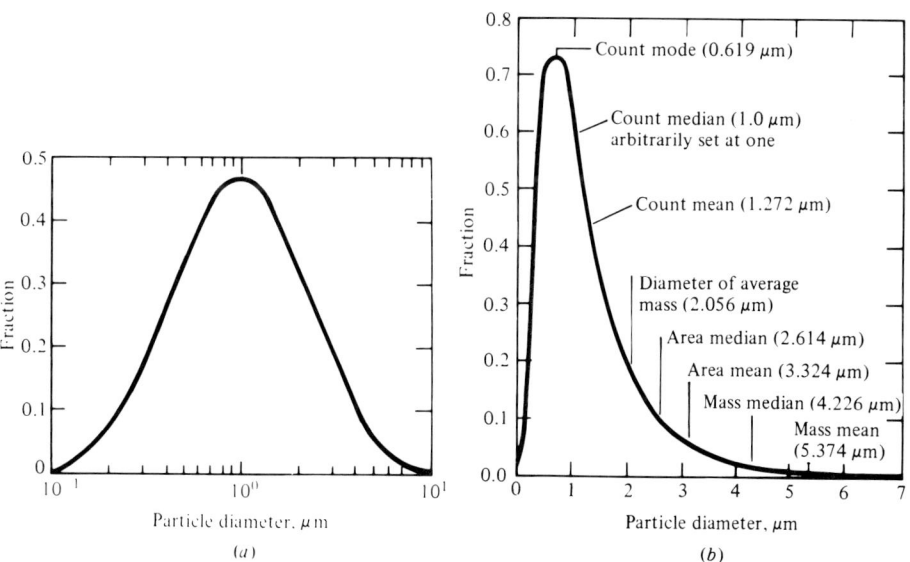

Figure 9-5 Log normal distribution of particle diameters plotted (a) on a semilog scale and (b) on a linear scale.

Devices that are used to clean up gas streams from the particulate matter they carry usually do better with larger particle diameters. They are characterized by (1) a fractional, or grade, efficiency and (2) an overall collection efficiency.

The *fractional, or grade, efficiency* is a collection efficiency, fraction, or percent of mass removed, for a given particle diameter. It is rather poor for the smallest sizes, increases rapidly with size, and becomes nearly 100 percent for the largest sizes. Figure 9-6 shows typical fractional efficiency curves for collection equipment. The *overall collection efficiency*, on the other hand, is the fraction or percent of mass removed of all particles or above a specified minimum diameter. It is to be noted that although liquid drops in suspension are usually spherical, solid particles are not. Thus the exact curve depends on what type of "diameter" representations are selected, e.g., the diameter of a spherical particle having the same surface area as the nonspherical particle, or the same volume, the Stoke's diameter, and others [5].

One must then be careful in evaluating the effectiveness of collection devices, which are usually characterized by the overall collection efficiency often above a given minimum diameter. Because efficiency is a mass ratio, and given the nature of the size distribution, a good collection efficiency might well mean a good mass removal but a poor numeric removal, as a very large number of small particles escape the collection device.

The majority of particulate matter in the atmosphere usually ranges in size (diameter) from 0.1 to 10 μm, although some are as small as 0.001 μm and as large as 500 μm. Below 0.1 μm, the particles behave like molecules and attain random motions characteristic of collisions with gas molecules. Between 1 and 20 μm the particles tend to be carried along with the air in which they are borne. Above 20 μm the particles

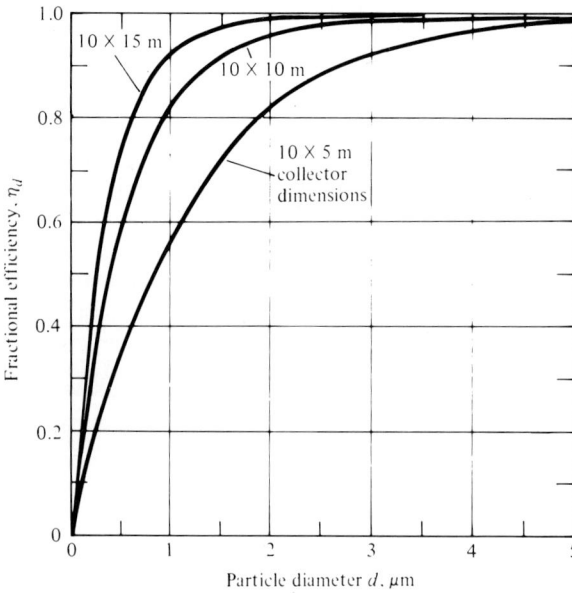

Figure 9-6 Typical fractional efficiency curve for collection equipment. Data plotted using the Deutsch equation for the data of Example 9-1.

attain settling rates that rapidly increase with size (0.3 cm/s for 10 μm, 30 cm/s for 100 μm), and they are, therefore, airborne for relatively short periods of time. The following sizes are of interest:

Particulate matter	Size, μm	Particulate matter	Size, μm
Natural dust	Above 1	Tobacco smoke	0.01–1
Liquid mist	Below 10	Oil smoke	0.03–1
Liquid sprays	Above 10	Coal dust	1–100
Smog	Below 2	Fly ash	1–200
Natural rain	500–10,000	Pulverized coal	3–500

Systems that help solve air quality-control problems associated with fossil-fuel steam-generator exhausts primarily deal with particulate collection and SO_2 emission cleanup. There are four basic types of systems:

1. Flue-gas desulfurization (FGD):
   a. Wet-type scrubbers
   b. Dry-type scrubbers
2. Particulate collection:
   a. Electrostatic precipitators
   b. Fabric filters

FLUE-GAS DESULFURIZATION (FGD) SYSTEMS

Gas desulfurization can be accomplished by wet, dry, or alkali scrubbing. These methods are covered in this section.

The Wet Flue-Gas Desulfurization System

The wet FGD system, also called a *wet scrubber,* is commonly based on low-cost lime-limestone* in the form of an aqueous slurry. This slurry, brought into intimate contact with the flue gas by various technique, absorbs the SO_2 in it.

The wet scrubbing process was originally developed in the 1930s by Imperial Chemical Industries (ICI) in England. In the modern version of the process, the flue gas is scrubbed with a slurry that contains lime (CaO) and limestone ($CaCO_3$) as well as the salts calcium sulfite ($CaSO_3 \cdot 2H_2O$) and calcium sulfate (in hydrate form, or natural gypsum, $CaSO_4 \cdot 2H_2O$). The SO_2 in the flue gas reacts with the slurry to form additional sulfite and sulfate salts, which are recycled with the addition of fresh lime or limestone. The chemical reactions are not known with certainty but are thought to be

---

* Lime is calcium oxide, CaO, a white caustic solid, also known as burnt lime, quicklime, and caustic lime. Limestone is a rock composed almost entirely of calcium carbonate, $CaCO_3$, from which building stones and lime are made. When crystallized by heat and pressure, it becomes marble.

$$
\left.\begin{aligned}
CaO + H_2O &\rightarrow Ca(OH)_2 \\
Ca(OH)_2 + CO_2 &\rightarrow CaCO_3 + H_2O \\
CaCO_3 + CO_2 + H_2O &\rightarrow Ca(HCO_3)_2 \\
Ca(HCO_3)_2 + SO_2 + H_2O &\rightarrow CaSO_3 \cdot 2H_2O \downarrow + 2CO_2 \\
CaSO_3 \cdot 2H_2O + \tfrac{1}{2}O_2 &\rightarrow CaSO_4 \cdot 2H_2O \downarrow
\end{aligned}\right\} \quad (9\text{-}4)
$$

One technique employs a *spray tower* downstream of the particulate-removal system (electrostatic precipitator or fabric filter) (Fig. 9-7). The flue gas is drawn into the spray tower by the main steam-generator induced-draft fan where it flows in countercurrent fashion to the limestone-slurry spray. A mist eliminator at the upper exit of the tower removes any spray droplets entrained by the gas. The gas may have to be slightly reheated before it enters the stack to improve atmospheric dispersion.

The sprayed limestone slurry collects in the bottom of the tower and is recirculated back to the spray nozzles by a pump. A system of feed and bleed charges a fresh slurry, under pH control, and discharges an equivalent amount from the circulating slurry. The fresh slurry is prepared by mixing the lime-limestone with water in a "slaker-grinder" and stirred in a slurry tank. The bled slurry is sent to a dewatering system, which is in the form of thickeners and filters or centrifuges, where water is removed from the calcium-sulfur salts. The reclaimed water is used to help make fresh slurry.

The wet scrubber has the advantages of high SO_2 removal efficiencies, good reliability, and low flue-gas energy requirements. In addition, it is capable of removing from the flue gases residual particulates that might have escaped the particulate-removal system.

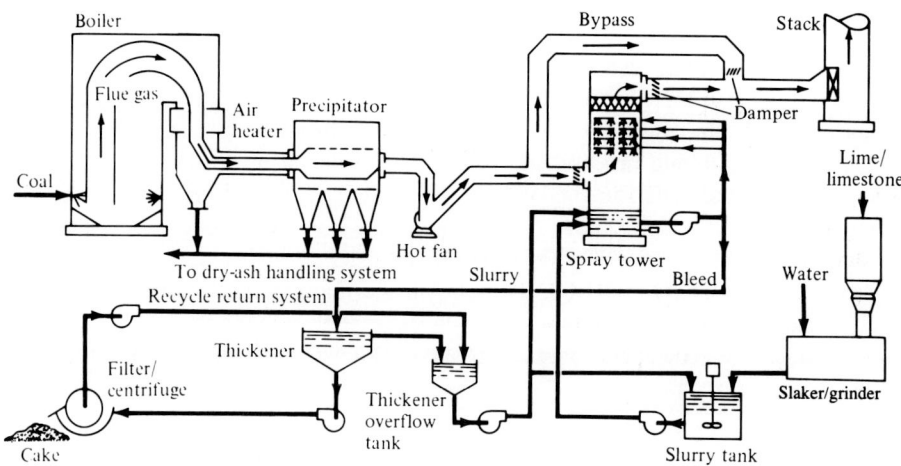

Figure 9-7 Schematic of a wet flue-gas desulfurization (FGD) system with a spray tower (*Electric Forum.*)

A main disadvantage is the buildup of scale in the spray tower and the possibility of plugging. The prevention of such scale is essential to the reliable operation of the tower. Scaling occurs because both calcium sulfite and calcium sulfate have low water solubility, normally around 30 percent, and can therefore form supersaturated water solutions. A minimum liquid-to-gas ratio must therefore be used, its value depending upon the SO_2 content of the flue gas and the expected extent of sulfite oxidation. Precipitation occurs at a finite rate, which necessitates holding the SO_2–absorbing liquor in a delay tank after each pass. An insufficient delay time increases supersaturation and promotes scaling. Another technique for controlling scale is the use of *seed crystals*. These are calcium sulfite and sulfate precipitate crystals, in a supersaturated solution, that are maintained in the SO_2–absorbing liquor. They provide sites around which preferential precipitation takes place and enhance the precipitation rate.

Other disadvantages of the wet scrubber are the reheating of the flue gas, a larger gas pressure drop requiring higher fan power requirements than the dry FGD system (below), and typically higher capital and operating costs.

The waste material from wet scrubbers is a water-logged sludge that poses difficult and costly disposal problems.

The Dry Flue-Gas Desulfurization System

Like the wet scrubber, above, the dry FGD system, also called a *dry scrubber,* utilizes an aqueous slurry of lime, CaO, to capture flue gas SO_2 by forming calcium sulfites and sulfates in spray absorbers (Fig. 9-8). The slurry in this case, however, is atomized, usually by a centrifugal atomizer, into a fine spray that promotes the chemical absorption of SO_2 and, because of the small spray particle size, is quickly dried by the hot flue gases themselves to a particulate suspension that is carried along with the desulfurized gas stream. The reaction particulates as well as those carried by the flue gases (fly ash) are then removed, mainly by a fabric filter, before the gas is drawn by the induced-draft fan to the stack.

A major component of this system is the slurry-generating system. A "slaker" meters lime and water into an agitated tank to prepare a slaked lime slurry which, in turn, is diluted by additional water and processed to remove inert impurities called grits, which are disposed of. The lime slurry is pumped to the spray absorber with the flow controlled by the amount of SO_2 in the flue gas.

Particulates both coming in with the flue gas and generated in the FGD are collected from the absorber and fabric-filter hoppers and sent to a recycling silo for disposal or for recycling of a portion of it with the slurry (depending upon the extent of original utilization of the reactant in the absorber). The recycled slurry is enriched by an alkaline material, such as CaO, MgO, K_2O, or Na_2O.

The main advantages of the dry system are the dry, powdery nature of the waste material, which poses fewer and less costly disposal problems than the wet waste from the wet FGD system (thought these problems are still large), and the mechanical simplicity of the system.

The main disadvantage is that the efficiency of SO_2 removal is lower than that of the wet scrubber. 1979 NSPS (New Source Performance Standards) regulations, which

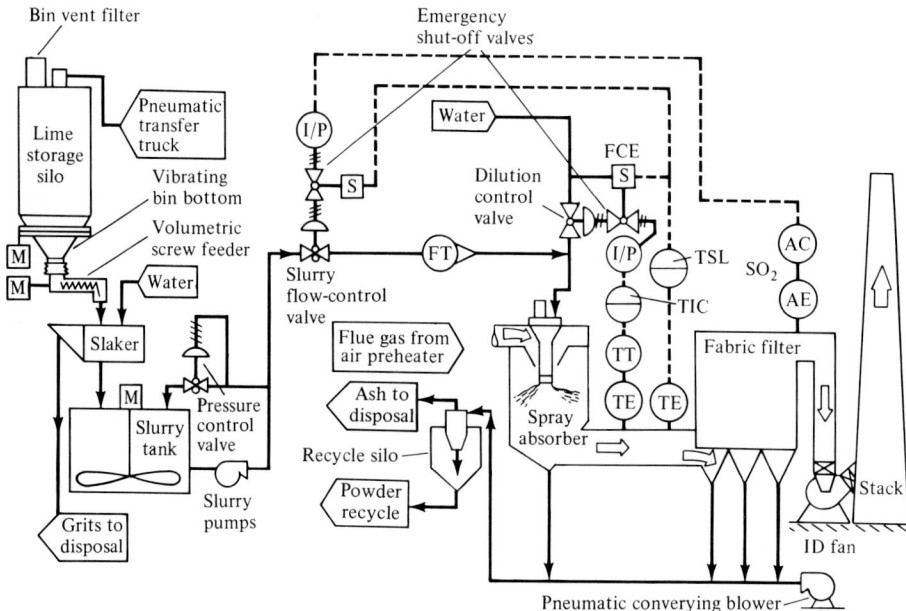

Figure 9-8 Schematic of a dry flue-gas desulfurization (FGD) system (*Electric Forum.*)

specify only 70 percent SO_2 removal in new plants, have encouraged the development of the dry system, however. Other disadvantages are the need for careful design optimization of the spray absorber and the slaker, and the strong dependence of collection efficiency on absorber outlet temperature, which necessitates operating as close as is safe to the saturation temperature that corresponds to the partial pressure of the water vapor in the gas in order to avoid condensation (below the corresponding dew point). This poses problems with filter-bag performance.

Single Alkali Scrubbing

Clear water solutions of either sodium (usually in the form of sodium hydroxide, NaOH, or sodium sulfite, Na_2SO_3) or ammonia (NH_3) are excellent absorbers of SO_2. The advantage of alkali scrubbing is that it avoids the scaling and plugging problems of slurry scrubbing by using alkaline earth. Ammonia scrubbing has the advantage that the scrubber product, ammonium sulfate, can be sold as a fertilizer, but the disadvantage that the process produces troublesome fumes.

A well-developed sodium scrubber is the *Wellman-Lord SO_2 recovery process,* which has found use in powerplants, refineries, sulfuric acid plants, and other industrial installations in the USA and Japan. The process utilizes a water solution of sodium sulfite (Na_2SO_3) for scrubbing and generates a concentrated SO_2 (about 90%), in effect removing the SO_2 gas from other flue gases.

The flue gas from fossil powerplants (or nonferrous smelters) is first pretreated by cooling and removal of particulate matter, such as by electrostatic precipitators, prior to being sent to the absorber (Fig. 9-9). In the absorber the water solution of sodium sulfite absorbs the SO_2 in the pretreated flue gas to produce sodium bisulfite $NaHSO_3$ according to

$$SO_2 + Na_2SO_3 + H_2O \rightarrow 2NaHSO_3 \qquad (9\text{-}5)$$

The desulfurized gas is reheated before going to the stack in order to improve atmospheric dispersion.

The sodium bisulfite is sent to a forced-circulation evaporator-crystallizer via a surge tank. The evaporator-crystallizer is the heart of the system. The surge tank allows steady flow rates into it despite gas flow and concentration fluctuations. Through the application of low-pressure steam, such as from a turbine exhaust, the sulfite is regenerated in the form of a slurry according to

$$2NaHSO_3 \rightarrow Na_2SO_3 \downarrow + SO_2 \uparrow + H_2O \qquad (9\text{-}6)$$

The H_2O is separated from the SO_2 in a condenser and recycled to a dissolving tank where the sulfite slurry is redissolved and sent back to the absorber via a solution surge tank that has the same function as the one mentioned above.

A small amount of the circulating solution oxidizes to nonregenerable sodium sulfate crystals that must be disposed of. This necessitates purging a small stream of solution and adding fresh sulfate.

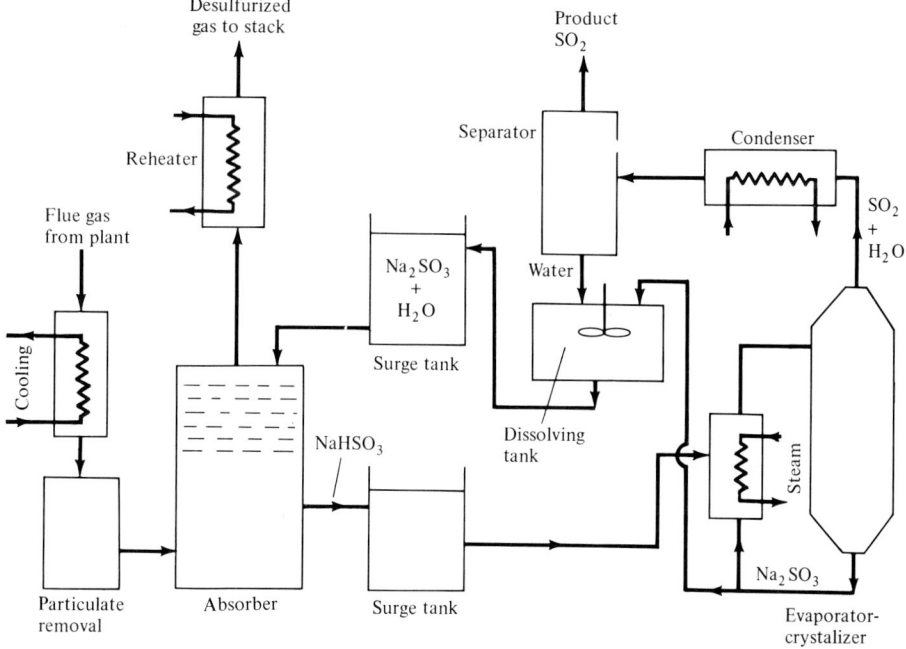

Figure 9-9 Schematic of the Wellman-Lord SO_2 recovery process.

The product SO_2 may be utilized to produce liquid SO_2 or sulfuric acid, on site or in a satellite plant, or to produce elemental sulfur. A well-known process for doing this is called the *Claus process,* which is based on the addition of H_2S according to

$$SO_2 + 2H_2S \rightarrow 3S + 2H_2O \tag{9-7}$$

NO Removal

A process for the removal of NO, also by the addition of H_2S, is proposed. It is given by

$$NO + H_2S \rightarrow S + \frac{1}{2}N_2 + H_2O \tag{9-8}$$

The combined removal of SO_2 and NO is under study. In both reactions, the H_2S must be completely consumed as it is a pollutant itself.

In 1977 the system was estimated to add an additional $120/kW, or some 12 to 15 percent to the base capital cost of a powerplant. It was said operating costs would increase by about $60/MBtu.

Most scrubbers in use by 1981 have been of the wet type. There is not sufficient experience with the dry type to establish which of the two may be selected by utilities in the future. Presently all scrubber systems are large and occupy a sizable area of a powerplant, have capital costs that run in the tens of millions of dollars for 500- to 1000-MW plants, and consume a sizable fraction of the gross electrical output of these plants. They also require a lot of maintenance, which results in the doubling of operation and maintenance personnel and causes, consequently, larger operation and maintenance costs. In addition, they generate huge amounts of waste that has to be disposed of. There are two types of disposal of FGD wastes: *wet disposal,* called *ponding,* and *dry disposal* in landfills, which are getting scarce. In general utilities are not always eager to build these disposal systems. Nevertheless, some 19,000 MW of FGD and sludge disposal systems were in operation, and 26,000 MW were under construction or planned, in 1981. The Electric Power Research Institute (EPRI) has published the *FGD Sludge Disposal Manual* (CS-1515 under RP1685-1), which incorporates the latest waste-disposal technology and regulations and describes how to design an environmentally acceptable waste-disposal system and the options available for processing and disposal of the wastes.

ELECTROSTATIC PRECIPITATORS

The principal components of electrostatic precipitators are *two sets of electrodes*. The first is composed of rows of electrically grounded vertical parallel plates, called the *collection electrodes,* between which the gas to be cleaned flows. The second set of electrodes are wires, called the *discharge electrodes,* that are centrally located between each pair of parallel plates (Fig. 9-10). The wires carry a unidirectional, negatively charged, high-voltage (between 20 and 100 kV, but typically 40 to 50 kV) current

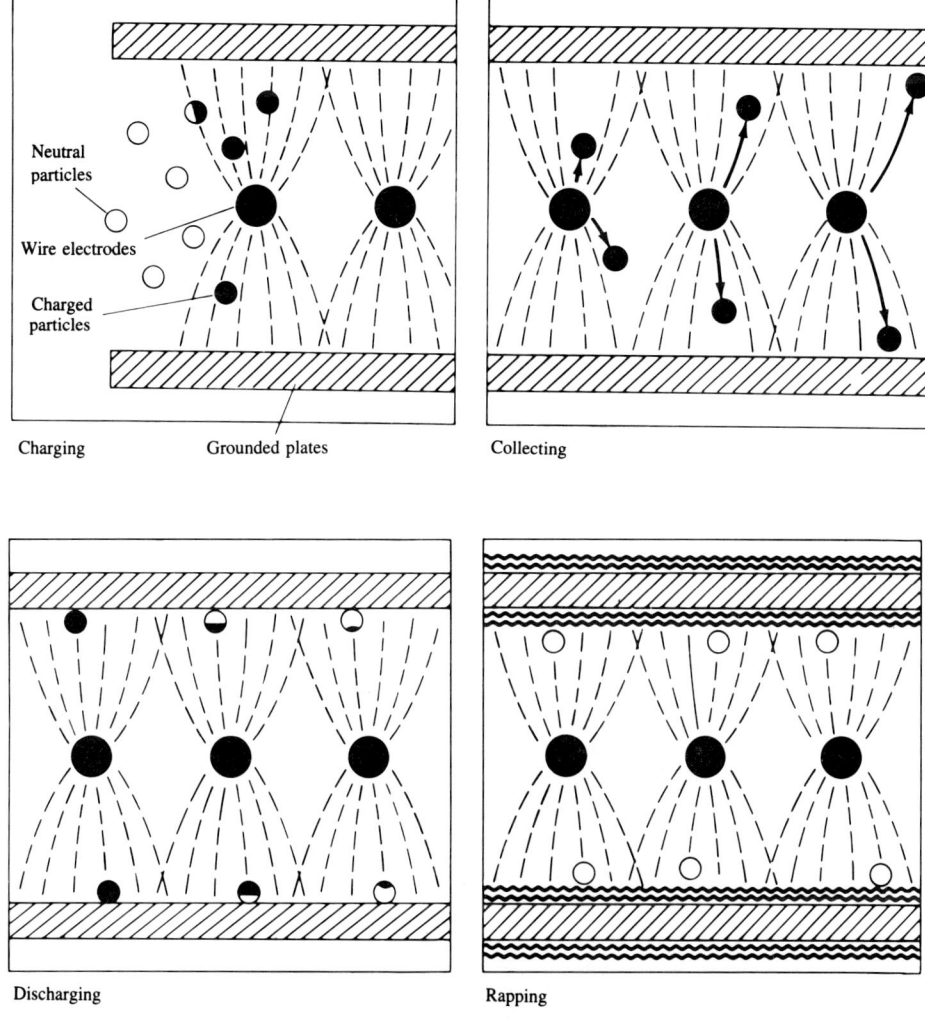

Figure 9-10 Vertical electrodes and grounded plates in an electrostatic precipitator showing four basic operations.

from an external source. The applied high voltage generates a unidirectional, non-uniform electric field whose magnitude is greatest near the discharge electrodes. When that voltage is high enough, a blue luminous glow, called a *corona,* is produced around them. The corona is an indication of the generation of negatively charged gas ions that travel from the wires to the grounded collection electrodes as a result of the strong electric field between them. This process occurs because the electrical forces in the corona accelerate the free electrons present in the gas so that they in turn ionize the gas molecules, thus forming additional electrons and positive gas ions. The new electrons themselves create more free electrons and ions, which results in a chain reaction or avalanche of free electrons and ions.

The positive ions migrate to the negatively charged wire electrodes. The electrons follow the electric field toward the grounded electrodes, but their velocity decreases as they move away from the corona region around the wire electrodes toward the grounded plates. (A sparkover may occur if the applied voltage and the electric-field strength are high enough so that the avalanche formation of positive ions extends across the entire interelectrode space.)

The negative ions that migrate along the electric-field lines collide with the particulate matter in the gas and charge them with a negative potential. (Random thermal motion creates additional collisions that result in what is called diffusion charging.) This continues until the particles have acquired sufficient charge and migrate to the grounded electrodes due to a force that is proportional to the product of the charge and the electric-field strength. The migration velocity is dependent upon these, as well as on the particle dielectric constant, and on size, being higher the larger the particles.

A theoretical model, based on the assumption of particle Stokes' flow, which is applicable for a Reynolds number below 1, and on the absence of turbulent diffusion and other restrictions, results in the following relationship for the migration velocity of a particle of diameter d

$$V_m = \frac{2.95 \times 10^{-12} p (E/s)^2 d}{\mu_g} \tag{9-9}$$

where
- V_m = migration velocity m/s
- p = a function of the particle dielectric constant that varies betweeen 1.50 and 2.40 for many types of dust, with an average of 2.0
- E = applied voltage, V
- s = distance between charging and collecting electrodes, m
- d = particle diameter, m
- μ_g = gas viscosity, kg/(m · s)

Equation (9-9) shows that the migration velocity is directly proportional to the particle diameter and the square of the field strength and inversely proportional to the gas viscosity and, hence, is sensitive to changes in gas temperature because the viscosity of gases increases with temperature. Actual migration velocities may deviate considerably from those predicted by Eq. (9-9), and design values are usually based on empirical observations.

When the particles collect on the grounded plates, they are supposed to lose their charge to ground. The electrical resistivity of the particles, however, causes only partial discharging, and the retained charge contributes to forces holding the particles to the plates. Resistivities that are too high or too low pose problems. High restivity causes retention of most of the charge, which increases the forces holding the particles to the plates and makes removal more difficult.* The maximum particle resistivity is known to occur around 300°F. This can then be corrected by operating at high gas temperatures, for instance by installing a "hot side" precipitator, upstream of the steam-generator air preheater where the gas temperature is usually in excess of 600°F. Low resistivity, found in particles with high carbon content such as fly ash from spreader stokers, causes them to lose their charge quickly to ground, and thus the forces that hold them to the plates, and to become reentrained by the flue gas. This can be corrected by increasing the holding forces by greater power input and decreasing reentrainment by designing low gas velocities, low electrode heights, and aspect ratios.

Particle Removal

When dust builds up on the plates, it deposits in a layer of increasing thickness with possible reentry into the gas stream unless it is periodically removed. This removal is done by *rapping* the plates to cause shock vibrations that shake the dust into hoppers at the bottom of the precipitator. Properly controlled and timed removal is critical for precipitator performance. The intensity and frequency of rapping of individual collection plates varies from the inlet to the outlet of the precipitator to suit the changing characteristics and collection rates of the dust. It is also customary to rap sequentially so that only a fraction of the accumulated dust is disturbed at any one time. A rapper device is used. It contains a vertical plunger that strikes the collection electrode support system to deliver the necessary shock wave and then returns back to its top position to prepare for the next strike. In one design the downward impact motion is effected by an electromagnetic coil and the return by a spring. In another the plunger is heavier, the impact is by gravity, and the return is by an electromagnet.

The length of the precipitator passage in the direction of gas flow that is necessary to remove a given particle size is obtained by seeing to it that the time required for the particle to migrate to the collection electrode has to be less than the time it takes it to pass through the precipitator at the same velocity as the gas. Neglecting the charging time, the required length would be given by

$$\frac{s}{V_m} \leq \frac{L}{V_g}$$

or

$$L \geq s \frac{V_g}{V_m} \qquad (9\text{-}10)$$

where

L = length of passage, m or ft

V_g = gas velocity, m/s or ft/s

---

* High resistivity may also cause the electric field in the dust layer to accelerate electrons sufficiently to produce positive ions, thus reducing the voltage at which sparkover occurs. It may also generate enough positive ions even when voltage is not high enough for sparkover, the result being a *back corona* that neutralizes the unidirectional field. These effects reduce or completely disrupt precipitator performance.

Collection Efficiencies

An electrostatic precipitator, and all other particle-collection devices, has an *efficiency*, already mentioned earlier. The *overall collection efficiency* η_o is given by

$$\eta_o = \frac{\text{mass or concentration of all particles retained by collector}}{\text{mass or concentration of all particles entering collector}} \quad (9\text{-}11)$$

Although the relationships in Eqs. (9-9) and (9-10) are based on highly idealized models, the trends they predict are applicable to the real case. Thus, because the migration velocity V_m is greater for large particles, the length of passage necessary to remove them is smaller. In other words, it is easier to collect larger particles than smaller particles. There is, therefore, a *fractional collection efficiency* η_d (also called the *grade efficiency*) that is given by

$$\eta_d = \frac{\text{mass or concentration of particles of a given size retained by collector}}{\text{mass or concentration of particles of a same size entering collector}} \quad (9\text{-}12)$$

From Eq. (9-12) it can be expected that η_d increases with size of diameter d. η_d is also expected to increase with electrode area and decrease with the flue-gas volume flow rate. Several relationships have been proposed for η_d. A well known one is the *Deutsche expression*, which is derived from physical or probability considerations [6, 7]

$$\eta_d = 1 - e^{-(AV_m/\dot{Q})} \quad (9\text{-}13)$$

where
A = area of collector electrodes, m² or ft²

V_m = migration velocity, m/s or ft/s

$\dot{Q}$ = flue-gas volume flow rate for each plate, m³/s or ft³/s

Example 9-1 Plot the Deutsch fractional efficiency for an electrostatic precipitator that has plates with height × width of 10 × 10 m, and a spacing of 25 cm. The applied voltage is 50,000 V. The mean flue-gas temperature and velocity between plates are 300°C and 1.5 m/s, respectively. Repeat for plate dimensions of 10 × 15 m and 10 × 5 m.

SOLUTION Refer to Eq. (9-9) and use $p = 2.0$, $s = 0.25/2 = 0.125$ m, and $\mu_g = 2.93 \times 10^{-5}$ kg/(m · s).

$$V_m = \frac{2.95 \times 10^{-12} \times 2 \times (50{,}000/0.125)^2 d}{2.93 \times 10^{-5}} = 3.225 \times 10^4 d$$

Refer to Eq. (9-13) and use

$$\dot{Q} = V_g \times \text{cross-sectional area of flow per plate side}$$
$$= 1.5 \times 10 \times 0.25 \times \frac{1}{2} = 1.875 \text{ m}^3/\text{s}$$

$$A = 10 \times 10 = 100 \text{ m}^2$$

Therefore

$$\eta_d = 1 - e^{-\left(\frac{100 \times 3.255 \times 10^4}{1.875}\right)d} = 1 - e^{-1.72 \times 10^6 d}$$

Similarly, for 10×15 m and 10×5 m plates

$$\eta_d = 1 - e^{-2.58 \times 10^6 d} \quad \text{and} \quad 1 - e^{-0.86 \times 10^6 d}$$

respectively.

η_d is plotted in Fig. 9-6 as a function of d in micrometers (μm). It can be seen that the plates with the larger path length are more efficient, that η_d is essentially 100 percent for particle sizes beyond 3 μm, and that the efficiency drops drastically for small particles. This is not in accordance with practical experience, from which measurements in many precipitators show a turnaround in fractional efficiency for very fine particles in the submicron range, with minimum efficiencies occurring in the 0.1 to 0.5 μm range and efficiencies going back to 90 to 95 percent below 0.1 μm.

The Deutsch equation is now recommended for use in estimating the operation of a precipitator at off-design conditions. In this case a value V_{mo} is obtained from experimental studies under different operating conditions and used in lieu of V_m, resulting in an overall collection efficiency η_o. V_{mo} is known as the *effective migration velocity*. Thus

$$\eta_o = 1 - e^{-(AV_{mo}/\dot{Q})} \qquad (9\text{-}14)$$

Equation (9-14) may now be used to estimate the effect of *changes* in volume flow rate $\dot{Q}$ (representing changes in plant load) on η_o, or, for the same V_{mo}, the effect of changes in precipitator design A.

Attempts to arrive at equations more representative of modern high-efficiency precipitator designs [8] are being attempted. Two current ones are

$$\eta_o = 1 - e^{-(AV_{mo}/\dot{Q})^x} \qquad (9\text{-}15)$$

where x is a variable, fitting most data at a value of about 0.5, and

$$\eta_o = 1 - \left(1 + \frac{AV_{mo}}{n\dot{Q}}\right)^{-n} \qquad (9\text{-}16)$$

where n is a variable, fitting most data in a range between 3 and 5 (but can vary between 2 to 8). Putting $n = \infty$ reverts Eq. (9-16) back to the Deutsch equation. In both equations above, V_{mo} is the effective migration or drift velocity. For utility fly ash its value can vary between 4 and 20 cm/s.

Figure 9-11 shows an electrostatic precipitator used to clean flue gases from an electric-generating powerplant.

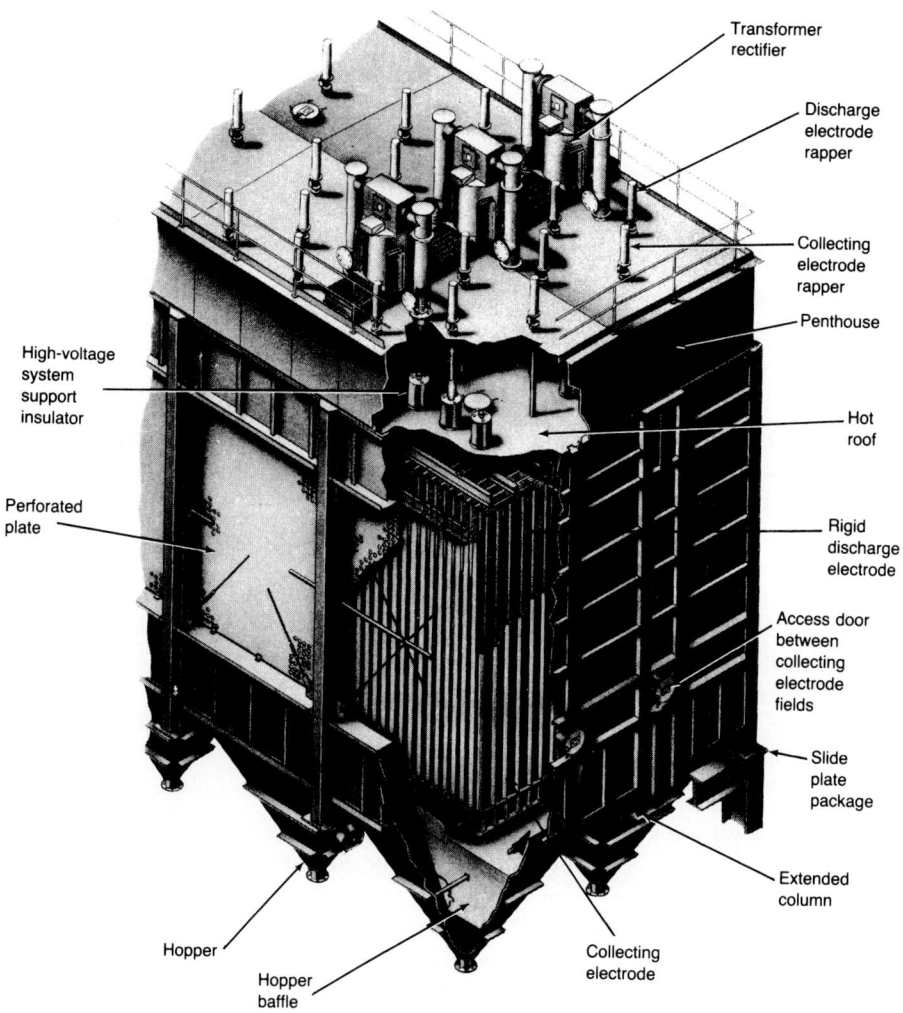

Figure 9-11 An electrostatic precipitator. *(Courtesy Research-Cottrell, Somerville, New Jersey.)*

FABRIC FILTERS AND BAGHOUSES

Fabric filters are used in powerplants to remove dust particles from a gas stream on a principle similar to that of a household vacuum cleaner except, of course, that the size of fabric filters is far greater. They are made of porous material that retains particulate matter as the carrier gas passes through the voids.

A fabric-filter element is usually made in the form of a long, hollow cylindrical tube that provides a large surface per unit of gas volumetric flow rate. The inverse of this parameter, called the *air-to-cloth,* or *filtering, ratio,* is equal to the superficial gas velocity; i.e., it is based on the surface area rather than the void flow area within the fabric. It ranges typically between 0.5 to 4.0 cm/s.

A fabric-filter system usually contains a large number of vertical cylindrical fabric-filter elements arranged in parallel rows. Such a system is called a *baghouse* (Fig. 9-12). A powerplant baghouse might contain several thousand such cylinders, each ranging in diameter from 5 to 14 in and in height up to 40 ft. The exact number is determined by their size, the required total capacity, plus an additional number to allow for shutdown of portions of the baghouse for cleaning with the plant on load. This points to one disadvantage of baghouses: their size is large in comparison with other types of particulate-removal systems.

In general, the elements have an open bottom and closed top. They rest on a tube sheet above a dirty-air plenum. The sheet distributes the gas evenly to the bags, allowing it to enter the elements at bottom, deposit its particulate matter on the inside of the

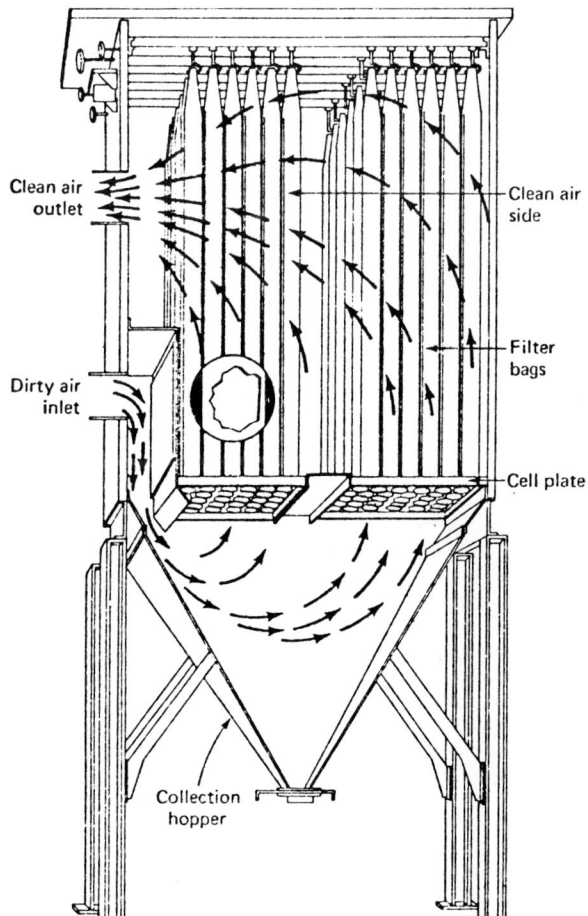

Figure 9-12 Typical baghouse with mechanical shakers. *(Courtesy Wheelabrator Frye, Inc., Mishawaka, Ind.)*

tubes, and pass laterally through the fabric and exit to an outlet manifold where it is drawn out by the plant induced-draft fan.

Collection hoppers, below the tube sheets, are receptacles for collecting the removed particulate matter. This matter usually collects on the inside of the elements in the form of a dust cake that must be made to fall down to the hopper by some means. Such means include mechanical shaking, pulse jets, or reverse airflow.

Mechanical shaking is accomplished by oscillating or vibrating rods attached to the top of the filter elements (Fig. 9-13). The oscillations ripple the bag surfaces, breaking the dust cake and causing it to discharge into the hopper below.

In *pulse-jet* cleaning, the bags are hung from the top, normal flue-gas flow is from the outside in, and dust collects on the outside of the tubes. For cleaning, short pulses of compressed clean air are forced down into the bags through venturis at the top. The pulse fractures and dislodges the dust, which falls to the hopper. The bags in this system are usually shorter than other types (less than 15 ft) to allow for proper cleaning of the tube bottoms.

Most coal-fired powerplant baghouses utilize the *reverse-air method* of cleaning. In reverse-air baghouses, an auxiliary fan forces clean air through a flapper valve into the clean-air plenum of the portion of the baghouse that is shut off for cleaning. This air passes through the filter elements in a reverse direction, from the outside in, resulting in a "backwash" action that collapses the bag and fractures the dust cake. When the bag is brought back on line, it reinflates and dislodges the fractured dust cake, which falls into the hopper. Some designs do away with the auxiliary fan and rely upon the natural collapse and reinflation of the bags due to the loss of vacuum when the clean-air plenum is shut off from the induced-draft fan, and then by returning back on line.

Cleaning may be periodic, which requires shutdown of portions of the baghouse, as explained for the reverse-air system, or continuous, i.e., done on load. Continuous

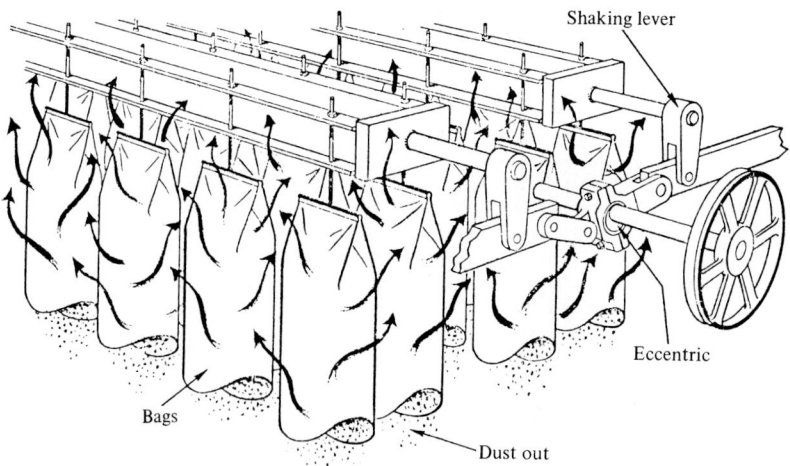

Figure 9-13 Shaker mechanism for the tubular filters of a baghouse. *(Courtesy Western Precipitation Division, Joy Manufacturing Company, Los Angeles, Calif.)*

cleaning is possible with the pulse-jet system because the duration of the pulse is very short.

The bag filter material can fail as a result of high temperature, burning, caking, erosion, chemical attack, and aging. A variety of fabrics have been used, depending upon the type of particulates and flue-gas composition, humidity, and temperature. They include wool, cotton, nylon, glass fiber, polyesters (such as dacron), and aromatic polyamides [9]. Wool and cotton, on one end of the scale, can operate only with low gas temperatures, up to about 175 or 200°F. Glass fibers, at the other end, tolerate high gas temperatures, up to 500 to 550°F. Research is continuing to find fibers that can withstand temperatures beyond 550°F to avoid precooling of the flue gases before they enter the baghouse. A baghouse is usually placed after the air preheater, which has a flue-gas outlet temperature of 300°F or more. High gas temperatures also mean high volume flow rates through the elements. Gas cooling below its dew point, which depends upon the water vapor mole fraction, results in condensation, which is not permissible with fabric filters.

Chemical attack of the fabric is particularly severe with high-sulfur coals. With the shift by many utilities to low-sulfur, high-ash coals, fabric filters have become increasingly attractive in comparison with other particulate-removal systems, such as electrostatic precipitators. Their performance is also less dependent on fuel and flue-gas characteristics, and they have fewer critical design parameters.

Figure 9-14 shows a baghouse of the type used to clean flue gases from fossil-fuel powerplants.

THERMAL POLLUTION

All thermal powerplants (fossil, nuclear, solar) reject low-availability or low-temperature heat to the environment. A comparison of the amounts of heat rejected by powerplants of different efficiencies should be based on their output, not the heat added. Thus

$$\eta = \frac{W}{Q_A} = \frac{W}{W + Q_R} = \frac{1}{1 + (Q_R/W)} \qquad (9\text{-}17a)$$

or

$$\frac{Q_R}{W} = \frac{1}{\eta} - 1 \qquad (9\text{-}17b)$$

where
η = plant net thermal efficiency

W = plant net output

Q_A = heat added

Q_R = heat rejected

For plants of 1000-MW output, the heat added, and the heat rejected, in MW, are given as a function of efficiency in Table 9-9. Thus an increase in thermal efficiency of 10 percent, from 30 to 40, results in a reduction in heat rejected of 35.7 percent,

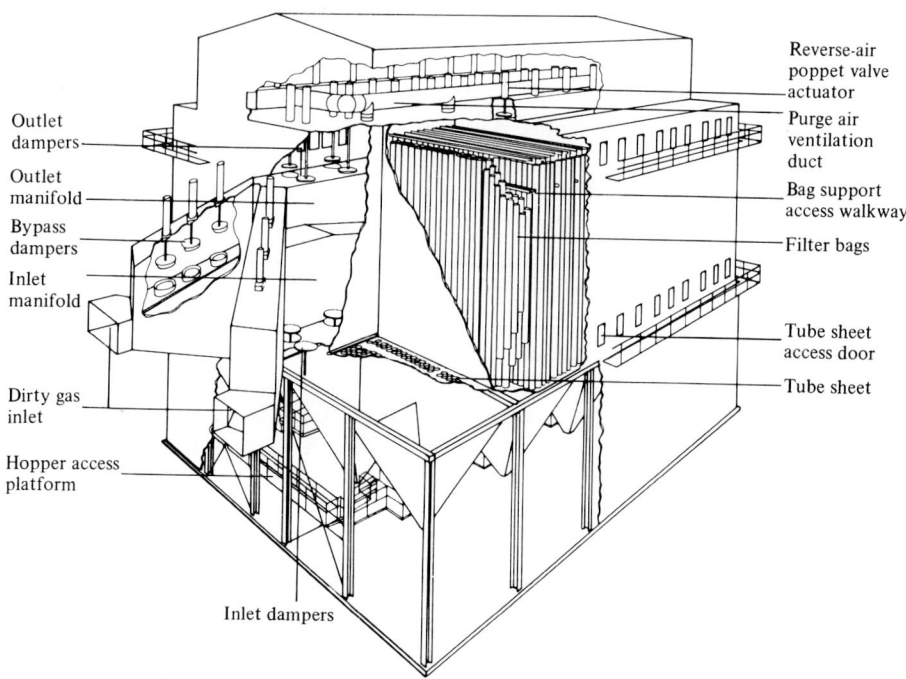

Figure 9-14 A baghouse. *(Courtesy Western Precipitation Division, Joy Manufacturing Company, Los Angeles, Calif.)*

from 2333 to 1500 MW. It can be seen that efficiency has a pronounced effect on heat rejected, greater than the change in its own value.

Because most modern thermal powerplants have efficiencies below 50 percent, actually in the 30 to 40 percent efficiency range, the amount of heat rejected is greater than the plant output. These very large amounts of heat are added to and affect the environment, and because they are mostly added to bodies of water they affect the aquatic ecological system.

The heat rejected in Rankine-cycle type powerplants is usually done via the cooling water of the condenser. The condenser is part of a once-through circulating-water system (the most common) or a closed system. In the once-through system large

Table 9-9 Heat added and rejected, MW, by plants producing 1000 MW, as a function of plant thermal efficiency

Thermal efficiency, %	10	20	25	30	33.3	40	50	60
Plant output W, MW	1,000	1000	1000	1000	1000	1000	1000	1000
Heat added Q_A, MW	10,000	5000	4000	3333	3000	2500	2000	1667
Heat rejected Q_R, MW	9,000	4000	3000	2333	2000	1500	1000	667

amounts of relatively cool water are taken from the environment, passed through the condenser, and discharged at a higher temperature back to the environment. In the closed system only enough water is drawn from the environment to replace the condenser cooling water that is lost, usually by evaporation in a cooling tower or a cooling pond.

Once-Through Systems

In the case of the once-through system, two effects are of concern. The first is the effect of heat on the large volumes of water as it goes through the condenser cooling system. The second is the effect of discharging the same but now warm water on the lake or stream it originated from. A 1000-MW powerplant, depending upon its thermal efficiency, needs between 250×10^6 to 400×10^6 lb_m/h of condenser cooling water. This corresponds to about 65,000 to 100,000 ft^3/min, or about 30 to 50 m^3/s.

Water in a lake or a stream is a habitat for numerous species of animal and plant life. A good intake system must therefore be designed to screen out these organisms. Nevertheless, simple screens still let through small organisms such as fish eggs, larvae, and plankton. And some larger fish, unable to escape because of the speed and volume of intake water flow, impinge and accumulate on the screens along with other debris from the lake or stream, thus requiring frequent cleaning of the intake screens. A method conceived to minimize impingement and entrainment is the use of a *porous dike* upstream of the intake screen. Built of suitably sized rocks, it slows the speed of water at intake so that small fish are not easily drawn toward the intake screen.

The organisms that do filter through the screens pass through pumps, pipes, hot condenser tubes, and other cooling-water system components before being discharged back to the lake or stream. In this path, they are subjected to buffeting, pressure, and heat, resulting in a high mortality rate. The ecosystem is said to be "stressed." Some ecological studies, however, indicate that ecosystems, like populations, have a natural resilience to compensate for adverse impacts on small parts of the whole.

From an engineering point of view, these organisms, dead or alive, clog filters, pipes, and pumps. Furthermore, bacterial slime and algae grow rapidly on the inside surfaces of the tubes, a condition known as *bifouling*. This is usually accompanied by *scaling*, which is caused by corrosion and the deposition of suspended solids in the water on the same surfaces. Bifouling and scaling reduce the bore of the condenser tubes, reduce heat transfer, and thus reduce condenser effectiveness, which in turn raises the back pressure on the turbine and thus reduces plant efficiency. In addition, these conditions increase pressure drops and pumping power in the circulating-water system, further reducing plant net output.

Bifouling and scaling can be combatted by frequent applications of chemicals such as chlorine. But chlorine and its compounds are discharged with the cooling water and are potentially harmful to the ecosystem.* In severe cases, condenser tubes are cleaned by mechanical means that require costly plant shutdowns.

---

* The Ocean Water Act of 1977, which mandates that electric utilities change their cooling systems to the "best available technology," requires them to limit the use of chlorine.

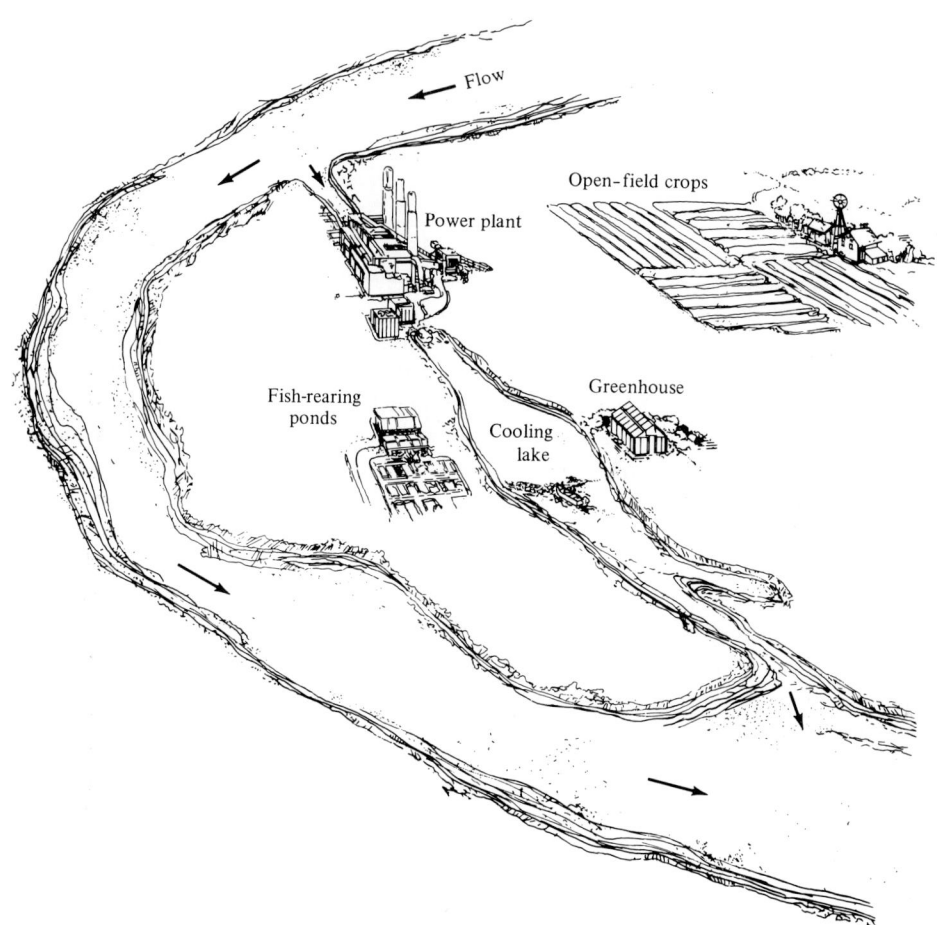

Figure 9-15 Warm cooling water from a powerplant used in fish hatcheries, greenhouses, and open agricultural fields before passing through a cooling lake on its way back to the main river. (*Reprinted from the EPRI Journal.*)

The warm water discharged in a once-through cooling system is some 20 to 25°F (11 to 14°C) above intake. The effect of this on the ecosystem depends upon the plant site, whether the plant is situated on a river, lake, estuary, or ocean, as well as on the workings of the aquatic world it is having an impact upon. Studies of flora and fauna,* necessary to understand the problem completely, are not simple and are time-consuming. Examples are fish spawning and migration patterns, especially in estuaries and sheltered coastal areas where environmental disturbances can cause great damage, the effect of small concentrations of chlorine on various species of aquatic life, etc.

---

* Plants and animals. In classic mythology: Flora is the goddess of flowers; Fauna is a Roman goddess, the sister of Faunus, the god of nature and patron of farming and animals.

Warm-water discharges in large volumes were once suspected of causing major ecological damage. It has, however, been recently demonstated that with good planning and management, warm water can be beneficial. It can for example be used to boost fish production in hatcheries (Fig. 9-15). (Commercial operations are already under way; an example is the Long Island Oyster Farm, operated in conjunction with the Long Island Lighting Company plant in Northport, New York.) It can also be used in agriculture, where it has been found to increase materially the production in greenhouses and open fields in cold climates. (An example is an agricultural facility operated in conjunction with the Northern States Power Company plant in Sherburne County, Minnesota.) Long-range economic questions, however, arise as to the relative gains from increased agricultural production versus costs of constructing the facilities, pumping the water, etc.

Closed Systems

Closed systems take in only sufficient water to make up for losses by evaporation from cooling towers, ponds, etc. They use corrosion-inhibiting additives with the cooling water. The systems are periodically flushed out, which releases these additives to the body of water on which the plant is situated, thus resulting in water pollution.

In general it is believed that fish mortality caused by powerplants is insignificant when compared with their natural mortality rate. It is also a fact that aquatic life is resilient and compensates naturally for the decrease in its population by producing more young or by increasing the survival rate of the normal number of young population, although this compensation is not always automatic and requires further study. Finally, an assessment should take into consideration the relative benefits of supplying the electric energy needed for the welfare of humankind and the loss, regrettable as it is, of an insignificant quantity of aquatic life, especially when the plant sites are chosen with care.

NATURAL AND ARTIFICIAL RADIOACTIVITY

Just like gaseous and particulate matter, which are caused by both natural and artificial events, radioactivity or radiation* in the environment is also caused by both natural and human-generated events. Of all environmental radioactivity, the portion contributed by nuclear powerplants in normal operation is miniscule. The history of nuclear-powerplant operation strongly suggests that even with abnormal operation (accidents), that portion is still insignificant. This suggests that the crisis atmosphere periodically generated by opposition to nuclear powerplants on this issue is largely unjustified.

Natural radioactivity has *always* been present. It comes from the sun and outer-

---

* The term *radiation* is a broad term that includes light (electromagnetic) and radio waves. However, it is often used to mean *radioactive* or *ionizing radiation,* which can present a health hazard. It will be used in the latter sense in this section.

space; it exists in numerous earth materials and in food, water, and air. It is generated by natural eruptions and volcanoes. It even exists within our own bodies; the human body contains about 0.35 percent (by mass) potassium, of which 0.0118 percent is radioactive potassium 40. The total of these is often called *natural background radiation*. Its level varies widely from location to location around the earth but averages about two-thirds of all radiation present.

It is instructive here to compare a natural occurrence with a human-made abnormal one. These are the eruptions from Mount St. Helens in the state of Washington, USA, that first erupted in May 1980, and the March 1979 accident at Three Mile Island nuclear powerplant unit No. 2 (TMI-2). The President's commission to investigate TMI-2, the Kemeny Commission, estimated in its report [10] that a total of about 2.5 million curies (Ci) of noble gases, mostly xenon, were released over the course of the accident. On the other hand, an extensive study by Battelle Pacific Northwest Laboratories of the ash from Mount St. Helens estimates that up to 3 million Ci of radon gas were released in the eruption of one day, May 18, alone.

Both xenon and radon are noble gases that are chemically and biologically inert. The difference, however, is that radon decays to a series of radioactive daughter elements that are chemically and biologically active. Thus the radioactive release from Mount St. Helens due to the gases alone is much more significant than that from TMI-2. In addition, the ash from the eruption at Mount St. Helens that spread over vast areas of the state of Washington and beyond was found to contain significant radioactivity from isotopes such as radium 226, potassium 40, thorium 232, polonium 210, and lead 214. Newly fallen ash was also found to contain rather high concentrations of the short-lived radon daughters lead 214, bismuth 214, and polonium 214. Note also that Mount St. Helens has erupted with varying intensity several times since May 1980.

Of the artificial sources of radioactivity we are subjected to, medical irradiations (x-rays, etc.) account for the largest portion, about 30 percent of the total. Occupational exposures, fallout from nuclear weapons tests, and miscellaneous sources (such as high-altitude flying) account for less than 1 percent each. Releases from the nuclear-power industry, including fuel manufacturing and reprocessing and powerplant emissions, account for some 0.15 percent of the total (Table 9-2).

Both natural and artificially generated radiations may be classified as particles and electromagnetic radiation. The particles include beta (β) or $_{-1}e^0$ (electrons in a free state); alpha (α) or $_2He^4$ (helium nuclei); as well as positrons $_{+1}e^0$ (positively charged β); neutrons, n; protons, p or $_1H^1$; tritons, $_1H^3$; and fission products. The electromagnetic radiation includes gamma (γ) rays, x-rays, and Bremsstrahlung. Examples of radioactive reactions (or disintegrations) that cause these radiations are:

$$\text{Beta:} \quad _{19}K^{40} \xrightarrow{1.28 \times 10^9 \text{ years}} {_{20}C^{40}} + {_{-1}e^0} \quad (9\text{-}18)$$

$$_6C^{14} \xrightarrow{5730 \text{ years}} {_7N^{14}} + {_{-1}e^0} \quad (9\text{-}19)$$

$$\text{Alpha:} \quad _{94}Pu^{239} \xrightarrow{24,000 \text{ years}} {_{92}U^{235}} + {_2He^4} \quad (9\text{-}20)$$

$$_{88}Ra^{226} \xrightarrow{1600 \text{ years}} {_{86}Rn^{222}} + {_2He^4} \tag{9-21}$$

$$\text{Positron: } _{15}P^{30} \xrightarrow{2.5 \text{ min}} {_{14}Si^{30}} + {_{+1}e^0} \tag{9-22}$$

$$\text{Neutron: } _{54}Xe^{137} \xrightarrow{3.9 \text{ min}} {_{54}Xe^{136}} + {_0n^1} \tag{9-23}$$

The original isotope is often called the *parent*. The first isotope on the right-hand side is called the *daughter*. The emitted particle is the radiation. (Gamma decay by itself does not alter the isotope, only its energy level, however γ rays often accompany other radiations.) The daughter may be stable or radioactive; if radioactive, it in turn decays, resulting in a radioactive chain, sometimes of considerable length. The half-life of each parent, shown above the arrows, is defined as the time during which one-half of the original parent nuclei decays and one-half is left. One-quarter is left after two half lives, one-eighth after three, etc.

The Curie

A radiation dose means that a person has been exposed to and has absorbed some radiaton energy. Radiation has both quality and quantity. The quantitative physical unit of radioactivity is the *curie*, Ci (and its fractions millicurie, mCi; picocurie, pCi, 10^{-12} Ci; etc.). 1 Ci = 3.70×10^{10} disintegrations per second. The number of curies emitted by a radioisotope depends upon both its mass and its half-life.

Example 9-2 Compute the activity in disintegrations per second of 1 g of radium 226. Ra^{226} has an atomic mass of 225.0245. It decays into radon gas with the emission of α particles, with a half-life $\theta_{1/2}$ = 1600 year = 5.049×10^{10} s.

SOLUTION Initial number of atoms of Ra^{226} in 1 g is

$$N_o = \frac{\text{Avogardro's number}}{\text{atomic mass}}$$

$$= \frac{6.0225 \times 10^{23}}{226.0245} = 2.6645 \times 10^{21}$$

$$\text{Decay constant } \lambda = \frac{0.6931}{\theta_{1/2}} = \frac{0.6931}{5.049 \times 10^{10}} = 1.3727 \times 10^{-11} \text{ s}^{-1}$$

$$\text{Activity} = \lambda N_o = 1.3727 \times 10^{-11}$$
$$\times 2.6645 \times 10^{21} = 3.6476 \times 10^{10} \text{ dis/s}$$

This activity is small compared with the initial number of atoms. The activity of radium may thus be considered constant, a true phenomenon for any radioactive species with a long half-life. Early measurements of radioactivity indicated that 1 g of radium had an activity of 3.70×10^{10} dis/s instead of the more accurate value given above. 3.7×10^{10} was, and still is, adopted as the numerical value for the curie. A curie is also used to indicate the quantity of any isotope having 1 Ci of radioactivity.

The Rad and the Gray

While the curie is a physical quantitative unit that indicates the number of radioactive events or disintegrations, it must be recognized that different radioisotopes emit different radiations with different energies. A more significant unit from the point of view of biological effect is the *rad*. The rad is a unit of radiation *energy* absorbed. 1 rad = 0.01 J/kg (4.3×10^{-6} Btu/lb$_m$), 1 millirad = mrad = 0.00001 J/kg. Some higher-than-average natural background levels, measured in mrad/year, emanate from monazite sand in Egypt (220 to 475), some beaches in Rio de Janeiro, Brazil (550 to 1250), the city of Kerala, India (800 to 8000), granite areas in Sri Lanka (3000 to 7000), and others. The rad is now being replaced by another unit in the SI system of units, called the *gray* (symbol Gy). 1 Gy = 1 J/kg = 100 rad.

The Rem and the Sievert

The biological effect of radiation on human beings must further take into account not only the total energy absorbed but also the relative biological effects of different types of ionizing radiation on people, such as the number of cells damaged by this radiation. Thus the rad is multiplied by a factor called the *relative biological effectiveness* (RBE), also called the *quality factor Q*, to obtain a more meaningful unit called the *rem*, for *roentgen equivalent man* (a millirem, *mrem*, is one-thousandth of a rem). Thus

$$1 \text{ rem} = 1 \text{ rad} \times \text{RBE} \tag{9-24}$$

some RBE values are 0.6 for 4-MeV gamma rays, 1.4 for 1-MeV electrons, 4 to 5 for thermal neutrons, and 2 to 10 for 1-MeV neutrons. These values are relative, based on an RBE for 200 kVp x-rays equal to 1. In the SI system of units, Gy replaces the rad, and therefore a new unit called the *sievert* (symbol Sv) replaces the rem. 1 Sv = 100 rem.

In addition, the rate at which the radiation is received is important. An analogy here may be made with drinking liquids. Some drinks are more harmful to health than others. A small quantity of a potentially harmful drink may not be harmful. A large quantity, say taken in the course of one evening, is. However, that same quantity would not be harmful if taken over a long period of time, say weeks. Thus the rate, mrem/h or mrem/year, is also important. Table 9-10 lists some examples of doses and dose rates we are exposed to in our normal lives.

Radiation effects are a complex subject. A complete treatment is beyond the scope of this text. They are, however, covered in the next sections where appropriate. For a fuller treatment, the reader is referred to specialized books on radiation and on health physics [11-16].

Table 9-10 Some common doses and dose rates of radioactivity

Source	Dose, mrem	Dose rate, mrem/year
1 diagnostic x-ray	20	
1 transatlantic flight	2	
Cosmic rays		45
Soil		15
Food, water, and air		25
Brick house		50–100
Concrete house		70–100
Wooden house		30–50
TV set		1–10
Living in the vicinity of a nuclear powerplant		1
Average background:		
New York		100
London		100
Paris		120
Denver		125
Kerala, India		400

NUCLEAR POWER AND THE ENVIRONMENT

In the United States, and doubtless in almost all countries constructing nuclear powerplants, federal licensing proceedings for each plant require the inclusion of detailed environmental statements to be issued as public documents. In the United States, these should be in accordance with the National Environmental Policy Act of 1969 (NEPA). Such statements must assess not only the impact upon the environment that is associated with the construction and the operation of the powerplant, but also the effect of the transportation of radioactive materials to and from that plant.

Besides thermal pollution which it shares with almost all types of powerplants, nuclear power's effects on the environment stem mainly from (1) the nuclear fuel cycle, (2) low-level dose radiations from nuclear-powerplant effluents, and (3) low- and high-level dose radiations from wastes.

The Fuel Cycle

Most nuclear powerplants in operation or under construction in the world today are using, and will continue to use for the near future, ordinary (light) water cooled and moderated reactors: the pressurized-water reactor (PWR) and the boiling-water reactor (BWR). A small number use the heavy water cooled and moderated reactor (PHWR). The expectations are that the fast-breeder reactor powerplant and perhaps an improved version of the gas-cooled reactor powerplant will come on line in increasing numbers in the twenty-first century. Almost all current water reactors use slightly enriched uranium dioxide, UO_2, fuel. The fuel has to go through a cycle that includes prereactor preparation, called the *front end,* in-reactor use, and postreactor management, called the *back end.* A typical fuel cycle may or may not incorporate fuel reprocessing in the back end (Fig. 9-16). The different processes are briefly explained below.

1. *Mining* of the uranium ore.
2. *Milling* and *refining* of the ore to produce uranium concentrates, U_3O_8.
3. *Processing* to produce of uranium hexafluoride, UF_6, from the uranium concentrates. This provides feed for isotopic (U^{235}) enrichment.
4. Isotopic *enrichment* of uranium hexafluoride to reach reactor enrichment requirements. This is done invariably now by the gaseous diffusion process.
5. *Fabrication* of the reactor fuel elements. This includes conversion of uranium hexafluoride to uranium dioxide UO_2, pelletizing, encapsulating in rods, and assembling the fuel rods into subassemblies.
6. *Power generation* in the reactor, resulting in *irradiated* or *spent* fuel.
7. *Short-term storage* of the spent fuel.
8. *Reprocessing* of the irradiated fuel and *conversion* of the residual uranium to uranium hexafluoride, UF_6 (for recycling through the gaseous diffusion plant for reenrichment) and/or extraction of Pu^{239} (converted from U^{238}) for recycling to the fuel-fabrication plant. Reprocessing can reuse up to 96 percent of the original material in the irradiated fuel with 4 percent actually becoming waste.
9. *Waste management,* which includes long-term storage of high-level wastes.

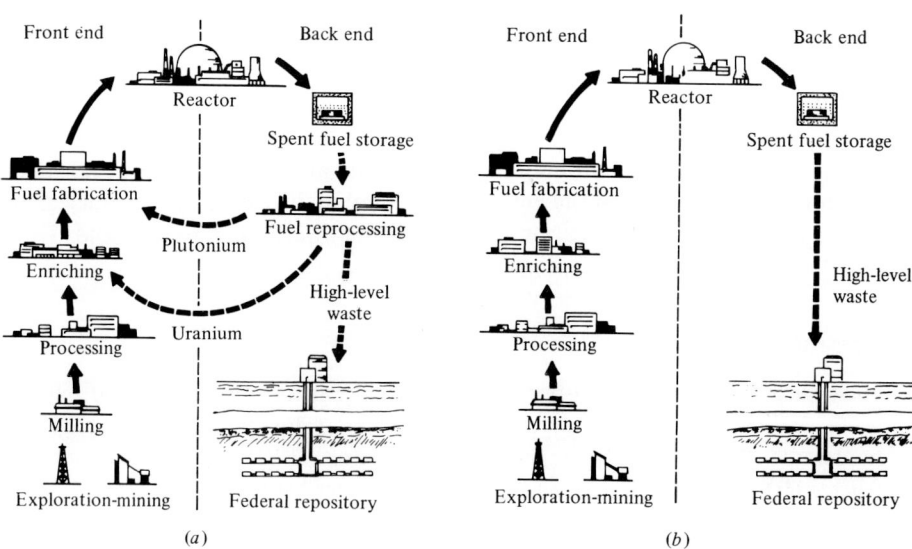

Figure 9-16 A typical nuclear fuel cycle (*a*) with reprocessing and (*b*) without reprocessing [18].

Step 8, reprocessing, may be bypassed, which results in disposal of both reusable fuel and wastes. This is the current (1982) U.S. Department of Energy process for dealing with irradiated fuel. The fuel assemblies are stored for at least 10 years and then buried. This is the so-called *throw-away* fuel cycle (Fig. 9-16b).

Wastes

The wastes associated with nuclear power can be summarized as:

1. *Gaseous effluents.* Under normal operation, these are released slowly from the powerplants into the biosphere and become diluted and dispersed harmlessly.
2. *Uranium mine and mill tailings. Tailings* are residues from uranium mining and milling operations. They contain low concentrations of naturally occurring radioactive materials. They are generated in large volumes and are stored at the mine or mill sites.
3. *Low-level wastes (LLW).* These are classified as wastes that contain less than 10 nCi (nanocuries) per gram of transuranium contaminants and that have low but potentially hazardous concentrations of radioactive materials. They are generated in almost all activities (power generation, medical, industrial, etc.) that involve radioactive materials, require little or no shielding, and are usually disposed of in liquid form by shallow land burial (Fig. 9-17).
4. *High-level wastes (HLW).* These are generated in the reprocessing of spent fuel. They contain essentially all the fission products and most of the transuranium elements not separated during reprocessing. Such wastes are to be disposed of carefully.
5. *Spent fuel.* This is unreprocessed spent fuel that is removed from the reactor core after reaching its end-of-life core service. It is usually removed intact in its fuel-element structural form and then stored for 3 to 4 months under water on the plant

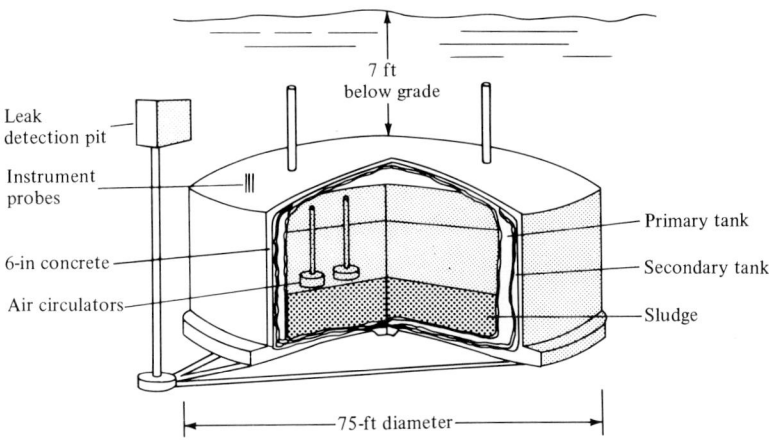

Figure 9-17 A typical low-level liquid-waste storage tank with double-walled containment.

site to give time for the most intense radioactive isotopes (which are the ones with shortest half-lives) to decay before shipment for reprocessing or disposal. Lack of a reprocessing capacity or a disposal policy has resulted in longer on-site storage, however. If the spent fuel is to be disposed of in a throw-away system (without reprocessing), it is treated as high-level waste.

RADIATIONS FROM NUCLEAR-POWERPLANT EFFLUENTS

Radiations from nuclear-powerplant effluents are low-dose-level types of radiations. The effluents are mainly gases and liquids. Environmental concerns about nuclear powerplants are prompted mainly by the effects of these radiations on the populations living near the plants. Sources of effluents vary with the type of reactor.

In both pressurized-water reactors (PWR) and boiling-water reactors (BWR), two important sources of effluents are (1) the condenser steam-jet air ejectors and (2) the turbine gland-seal system. The ejector uses high-pressure steam in a series of nozzles to create a vacuum, higher than that in the condenser, and thus draws air and other noncondensable gases from it. The mixture of steam and gases is collected, the steam portion condenses, and the gases are vented to the atmosphere. In the gland seal, high-pressure steam is used to seal the turbine bearings by passing through a labyrinth from the outside in so that no turbine steam leaks out and, in the case of low-pressure turbines, no air leaks in. The escaping gland-seal steam is also collected and removed. In the BWR, the effluents come directly from the primary system. In the PWR, they come from the secondary system, so there is less likelihood of radioactive material being exhausted from a PWR than a BWR from these sources.

The primary-coolant radioactivity comes about mainly from fuel fission products that find their way into the coolant through the few small cracks that inevitably develop in the very thin cladding of some fuel elements. Such activity is readily detectable. However, to avoid frequent costly shutdowns and repairs, the system is designed to operate as long as the number of affected fuel elements does not exceed a tolerable limit, usually 0.25 to 1 percent of the total. Also, some particulate matter finds its way into the coolant as a result of corrosion and wear (erosion) of the materials of the primary system components. These become radioactive in the rich neutron environment of the reactor core. Corrosion occurs because the radiolytic decomposition of the water passing through the core results in free O_2 and free H and OH radicals as well as some H_2O_2. These lower the pH of the coolant and promote corrosion. Finally, radioactivity in the primary coolant may be caused by so-called *tramp uranium*. This is uranium or uranium dioxide dust that clings to the outside of the fuel elements and is insufficiently cleaned off during fabrication. It will, of course, undergo fission, and its fission products readily enter the coolant. The problem of tramp uranium is being minimized by improved processing and quality control.

The primary coolant is cycled through a *demineralizer system* that removes the contaminants by an ion-exchange process. Corrosion is inhibited by the slow addition of small amounts of lithium hydroxide, LiOH, to control acidity, and by the addition of hydrazine, NH_2NH_2, during shutdown to remove the radiolytic oxygen.

Of the above sources, the neutron activation of the corrosion and wear products represents the major activity in the primary coolant. The materials irradiated include Zircaloy, Inconel, stainless steel, carbon steel, and other steel and copper alloys that may be rich in nickel, chromium, and cobalt. The principal "crud" activity in PWRs is due to Co^{58}, which results from the neutron irradiation of Ni^{58} with the release of a proton (hydrogen nucleus). In BWRs it is due to Co^{60}, which results from the irradiation of Co^{59}. Other activities in both reactor types are due to Fe^{59}, Cr^{51}, Mn^{54}, Zr^{95}. Zircaloy 4, used as cladding in most water reactors, contributes very little to the crud because the oxide film adheres well to the zircaloy.

Another form of activation in the primary coolant in water reactors is the result of the irradiation of the primary coolant itself. The main radioactivity is due to the neutron capture of oxygen in the water. (Neutrons captured by hydrogen convert it to nonradioactive deuterium.) The most important of the oxygen reactions, because of the high energy γ emitted, is the $O^{16}(n,p)N^{16}$ reaction. It has a microscopic cross section of 1.4×10^{-5} barn and is given by

$$_8O^{16} + {_0}n^1 \rightarrow {_7}N^{16} + {_1}H^1 \tag{9-25}$$

N^{16} is a radioactive β and γ emitter, reverting to O^{16}, with a half-life of 7.2 s. The β rays are mainly of 3.8, 4.3, and 0.5-MeV energy. Gamma rays are mainly of 6.13- and 7.10-MeV energy. Another reaction of somewhat less importance is caused by the neutron capture by O^{18}, which is present to the extent of 0.024 percent of all oxygen. The reaction is $O^{18}(n,\gamma)O^{19}$. O^{19} is also a β and γ emitter of 29-s half-life that converts to stable F^{19}. A third reaction of some importance is due to O^{17}, present to the extent of 0.037 percent of all oxygen. The reaction, $O^{17}(n,p)N^{17}$, results in N^{17}, also a β and γ emitter of 4.16-s half-life, which reverts back to O^{17}. Although the cross section for this reaction is greater than the first one by a factor of about 10^3, its effect is reduced by the small concentration of O^{17} in water. These and other resultant isotopes that yield only a few weak radiations are shown in Table 9-11.

Impurities other than corrosion and erosion products also exist in the primary water. The main one is argon 40, which enters into solution in water from the atmosphere, where it exists to the extent of 1 percent by volume. It undergoes the

Table 9-11 Principal activation of water in reactors

Isotope	Half-life	Concentrations	BWR release, μCi/s
N^{16}	7.20 s	1.0×10^2	1.7×10^8
O^{19}	26.80 s	8.0×10^{-1}	1.4×10^6
N^{17}	4.16 s	1.6×10^{-2}	2.6×10^4
N^{13}	9.9 min	6.5×10^{-3}	1.2×10^4
F^{18}	109.8 min	4.0×10^{-3}	7.2×10^3

Ar$^{40}(n,\gamma)$Ar^{41} reaction. Ar^{41} is a β and γ emitter with a 1.83-h half-life that decays to stable potassium 41. With limited water exposure to the atmosphere and hold decay tanks used for effluents, the effect of this reaction is minimal.

Although all these activities can be released with effluents from water-reactor powerplants, the BWR is unique in that the primary coolant is also the working fluid and hence the gaseous isotopes, in particular, are more likely to be released to the atmosphere. The fourth column in Table 9-11 indicates the possible release rates in a 1000-MW BWR.

A problem that is more characteristic of PWRs than BWRs, on the other hand, is that of *tritium*. Tritium, a gaseous emitter of 12.25-year half-life, is primarily the result of using chemical shim with boric acid in the primary water in PWRs (Sec. 10-5). A little less than 20 percent of all boron is B^{10}. Upon neutron capture it undergoes the reaction B$^{10}(n,T)2\alpha$ given by

$$_5B^{10} + {_0}n^1 \rightarrow {_1}H^3 + 2{_2}He^4. \tag{9-26}$$

by which a tritium atom, $_1H^3$ or T, and two α particles (helium nuclei) are released. This reaction has a neutron-energy threshold of about 1.5 MeV, below which it will not take place. The tritium, like ordinary hydrogen, becomes a diatomic gas, $_1H_2^3$ or T_2, or combines with hydrogen to become HT.

Other tritium-producing reactions are due to lithium and deuterium. Lithium enters the primary coolant through the use of LiOH in the demineralizer or ion exchanger, which is used to help maintain the pH value of the primary water at an alkalinic 9.5. Lithium 6 constitutes about 7.4 percent of all lithium. Upon neutron capture, it undergoes the reaction Li(n,α)T, given by

$$_3Li^6 + {_0}n^1 \rightarrow {_1}H^3 + {_2}He^4 \tag{9-27}$$

This reaction occurs at all neutron energies. Again the tritium becomes T_2 or HT gas. Tritium production from this source can be eliminated by substituting KOH or NH$_4$OH for LiOH in the demineralizer. It may be noted here that BWRs use no chemical shim and no lithium compounds, and, hence, the tritium in their liquid effluents is less than 10 percent of that in PWRs.

Deuterium dioxide or heavy water, H$_2^2$O or D$_2$O, is present to the extent of 0.015 percent of all water. Upon neutron capture one or both deuterium atoms in the heavy-water molecule undergoes the reaction D(n,γ)T, given by

$$_1H^2 + {_0}n^1 \rightarrow {_1}H^3 + \gamma \tag{9-28}$$

resulting in DTO or T_2O. This reaction contributes insignificantly to the production of tritium when compared with the preceding two. Tritium is also generated as a result of the use of boron-bearing control rods, which usually contain boron in the form of B$_4$C. (Reactors using Ag-In-Cd control material produce less tritium.) Tritium is also generated in *ternary fission*, i.e., fission that results in three, instead of two, fission fragments, one of which is tritium. A small fraction of this tritium, estimated at 1 percent, escapes the fuel through the Zircaloy cladding and enters the coolant. (Stainless steel cladding allows a much higher percentage of tritium to pass through.) Most of this tritium, therefore, is released in fuel-reprocessing plants rather than at the reactor site.

When tritium, generated as T_2 or HT gas, enters ordinary water H_2O, one T atom replaces one H atom in the water resulting in HTO and releasing hydrogen gas H_2 according to

$$\left.\begin{array}{r} T_2 + H_2O \rightarrow HTO + HT \\ \text{and } HT + H_2O \rightarrow HTO + H_2 \end{array}\right\} \quad (9\text{-}29)$$

and

HTO is physically and chemically similar to H_2O and becomes an integral part of it. Some of this "tritiated water" eventually finds its way into the environment.

In *heavy-water reactors,* as it is to be expected, the nearly 100 percent D_2O coolant accounts for particularly large tritium production, according to Eq. (9-28). As with light-water reactors, heavy-water reactors also suffer from N^{16} production, Eq. (9-25).

In *high-temperature gas-cooled reactors* that are helium-cooled (the HTGR), tritium is generated by the neutron activation of He^3, which is present to the extent of 0.00013 percent of all helium. It is also generated by ternary fission to a greater degree than in water reactors. Tritium, however, can be removed from helium by absorption on a titanium sponge, which requires regeneration, normally every few months.

In *liquid-metal-cooled fast reactors,* the primary coolant, sodium, becomes intensely radioactive during plant operation. All naturally occuring sodium is made up of the isotope Na^{23}. When subjected to neutrons, it undergoes the reaction Na^{23} $(n,\gamma)Na^{24}$. Na^{24} is a radioisotope of about 15-h half-life that decays into stable Mg^{24} while emitting γ radiations, mainly of 2.76 and 1.38 MeV. The primary system, however, is separated from the steam cycle by a sodium intermediate system that does not become radioactive. The primary system operates at low pressures, below those of the intermediate and steam systems, so there is a very low likelihood of the radioactive sodium's leaking into the atmosphere and none of its leaking into the intermediate and steam systems.

One advantage of sodium is its ability to combine with or retain other elements. Some fast-reactor fuel elements are designed so that their gaseous fission products, such as iodine, xenon, and krypton, are vented into the coolant. Others may release fission products upon cladding failure. Sodium tends to retain some of these products, forming sodium iodide or holding cesium in solution, so that they do not escape to the gas blanket (usually argon) that separates sodium from air. Xe and Kr, on the other hand, are not retained by Na and would escape into the gas blanket.

A discussion of the biological effects of low-level radiation on human beings is a large subject and is beyond the scope of this text. It is, however, believed to be slight when compared with other hazards, such as toxic wastes [17].

HIGH-LEVEL WASTES

We will be concerned here with the high-level wastes that result from the spent fuel. High-level wastes, because of the intense exothermic activity, generate too much heat for early passive burial and have to be cooled, usually by air circulation or other means, possibly for decades, before they can be permanently stored.

The level of activity is not as long as generally perceived. Although some fission products have extremely long half-lives, it should be recalled that the intensity of radiation is not a sole function of the half-life but is also a function of the energy generated in the reaction. Figure 9-18 [18, 19] shows the relative energy generated by some of the more important fission products versus time. The level of activity is so high the first 10 years that storage can probably be best accomplished if the wastes

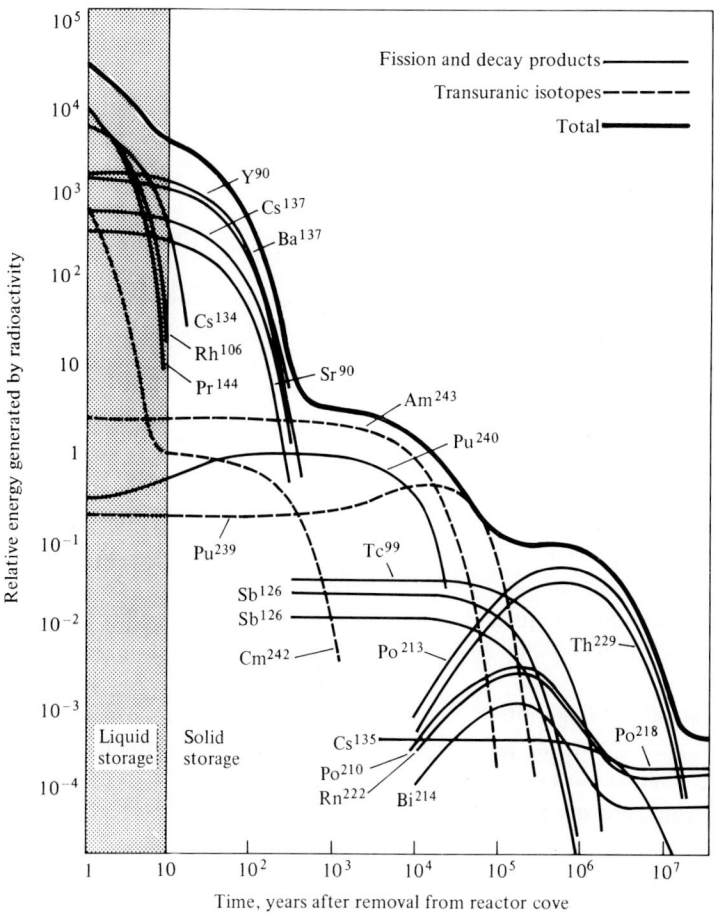

Figure 9-18 Relative energy generated by the more important fission products in high-level waste from spent nuclear fuel [18].

were in liquid form. After 10 years, the level drops by nearly an order of magnitude, thus permitting storage in solid form. It is interesting to note another and sharper drop in energy generated between 100 and 1000 years.

Figure 9-19 shows the relative "hazard" generated by the high-level waste and that generated by the original uranium ore as extracted from the ground. It shows that the high-level waste hazard drops below that of the ore it originally came from after about 800 years. The argument, based only on the half-lives of the isotopes, that stored nuclear waste would be a hazard to humanity for tens of thousands of years is, therefore, an inaccurate one.

Biological Effects of High-Level Wastes

Although all possible efforts are made to isolate high-level wastes from the biological system, it is instructive to study their effects. The principal effect is the destruction of body cells in the vicinity of the irradiated region. The effects are classified as somatic or genetic.

Somatic effects are limited to the exposed individuals. They are a direct result of the doses received by cells and manifest themselves in some form of malignancy. There is much information on high-dose irradiation effects on small animals that suggests that the frequency of effects within their population is proportional to the dose. There is, on the other hand, very little information on human beings. The only groups that have received high doses are the survivors of the atomic bombing of Hiroshima and Nagasaki, those receiving therapeutic radiation cancer treatment, and some occupationally exposed workers, such as those who work in underground uranium mines.

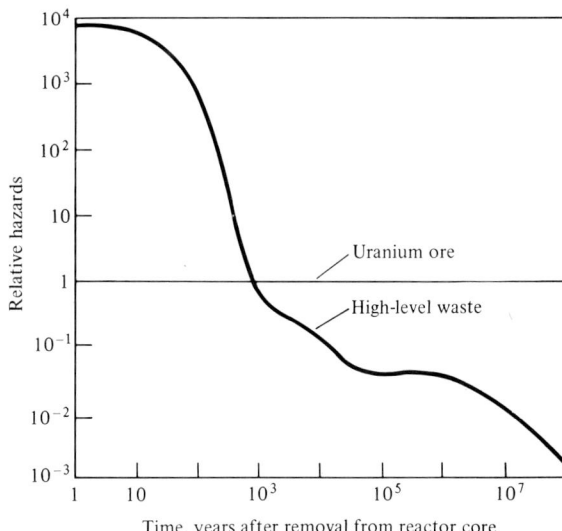

Figure 9-19 Ration of hazards of high-level waste to those of original uranium ore versus time. *(From Ref. 20.)*

High-dose levels of radiation can result in leukemia, which occurs most frequently within a few years after doses of 50 to 500 rad. By 25 years, its frequency drops to levels that are normally encountered in the absence of radiation. Lung cancers have been observed in Hiroshima survivors who received 30 to 100 rad of gamma radiation. Estimates were that 10 cancer cases at 250 rad to 40 cases at 30 rad develop per rad for each million people during the first 25 years after exposure. Other information is available for the incidence of breast cancer (6 to 20 cases), thyroid cancer (40 cases), and other types of cancer (40 cases), all per rad for each million in the first 25 years after exposure to levels between 60 and 400 rad. Uranium miners exposed to at least a few hundred rads of alpha radiation from radon gas and its radioactive daughters show a high incidence of lung cancers. On the other hand, current evidence indicates few, if any, somatic effects can be detected at doses below 10 rad.

Genetic effects can be transmitted to the descendants of exposed individuals and thus affect unexposed generations. They are radiation-induced changes in the genetic materials of sex cells. They manifest themselves in different ways: (1) gene* mutations (changes) that result in changes in the functions of individual genes, (2) chromosome† aberrations due to breakage and reorganization of chromosomes, and (3) changes in the number of chromosomes. Such changes can result in offspring abnormalities, ranging from mild to lethal. Unfortunately (or rather fortunately), useful and adequate human data on the effect of high doses of radiation are not available and estimates of high doses to human reproductive cells are based on research on mice and other species. Such estimates yield ranges as wide as 30 to 1500 mutations per million babies per rad of acute exposures to males, and about half of that to females. Based on the response of mouse ovaries, it is expected that the effect on females may approach nil if conception occurs a sufficient time after exposure. Further, it is believed that dominant gene mutations are induced in the first-generation offspring of an irradiated population but that gene mutations appear at a higher frequency and last many more generations than chromosome mutations. Of the known human cases, however, about 75,000 children born to parents irradiated at Hiroshima and Nagasaki, and examined periodically since 1945, show no increased frequency of congenital malformations, stillbirths, or growth and development abnormalities.

---

* A *gene* is the carrier of information in the nucleus of a living cell that determines the physical characteristics of living things, such as the shape of the eyes in a human or the color of a flower or a plant. Genes are inherited, which is the reason why offspring resemble their parents. Genes were once thought to possess a physical structure. They are now thought of as the function or operational unit by which heredity is transmitted from parent to offspring.

† A chromosome is a threadlike microscopic part which exists in the nucleus of the cell and which carries hereditary information in the form of genes. The chromosome contains deoxyribonucleic acid, DNA, and ribonucleic acid, RNA, attached to a protein core. The arrangement of the components of the DNA molecules determines the genetic information. Each body or somatic cell contains a certain number of chromosomes, called diploid. Reproductive cells contain exactly half that number, called haploid. Fertilization occurs when two such reproductive cells combine to reproduce what is called a zygote with a diploid set of chromosomes.

Fuel Reprocessing

As indicated previously, spent fuel can either be stored directly or reprocessed. There are advantages and disadvantages to both schemes. The throw-away system (the former) avoids the costs and hazards associated with a reprocessing plant. The latter utilizes the unused uranium, converted plutonium, and other radioisotopes for use in a wide variety of services, such as isotopic generators, medicine, agriculture, and industry. There are also indications that the energy generated by the high-level waste produced after reprocessing is lower than that generated by the intact spent fuel by ratios of about 0.83 after 10 years, 0.38 after 100 years, and 0.06 after 1000 years [18].

Reprocessing of spent fuel is done by dissolving it, usually in nitric acid, then removing the converted plutonium and unspent uranium by solvent extraction. The remaining solution contains more than 99.9 percent of the nonvolatile fission products, plus some constituents of the cladding of the fuel elements, traces of plutonium, uranium, and others, and most of the resulting transuranium elements.

This remaining solution constitutes the high-level wastes. It is usually concentrated by evaporation and then stored as an aqueous nitric acid solution, usually in high-integrity stainless steel tanks. Permanent storage in liquid form, however, requires continual supervision and tank replacement over an indefinite period of time.

The experience with short-term tank storage has generally been good, though some leakage has been encountered, particularly in tanks used for storage of military wastes. It is now generally believed that, for the long term, a storage system based on solidification of the wastes would be more acceptable.

The conversion of the liquid wastes to a solid form has been studied since the 1960s, including large-scale pilot operations. Advanced processes are currently being developed. The aim is to convert the wastes to a solid product that is not liable to leakage, requires less supervision, and is more suited to final disposal. This solid product should maintain its mechanical strength. It should have good thermal conductivity, as it will continue to generate heat from radioactive decay for a long time, though at a decreasing rate. Ideally it should have a low leach rate.

Glasses and ceramics are now considered to be the most suitable forms for this final disposal. About 30 different processes have been developed over the past 20 years. The basic processes are shown in Fig. 9-20. The simplest one involves evaporation and denitration (or calcination) to form a granular or solid calcine. This is considered an interim product, since it does not meet all the above requirements; it is treated further by being mixed with additives and is then melted to form glasses or ceramics.

A second process involves mixing the additives with the original waste solution, then evaporating, denitrating, and melting this mixture to form the glasses or ceramics.

A third process uses an adsorption process and treatment at high temperature to produce the ceramics.

In recent years, attempts at improving the properties of the above glass or ceramic final products have been made. These essentially involve the further formation of either the calcine or glass into metal matrixes or coated particles to improve their leach resistance.

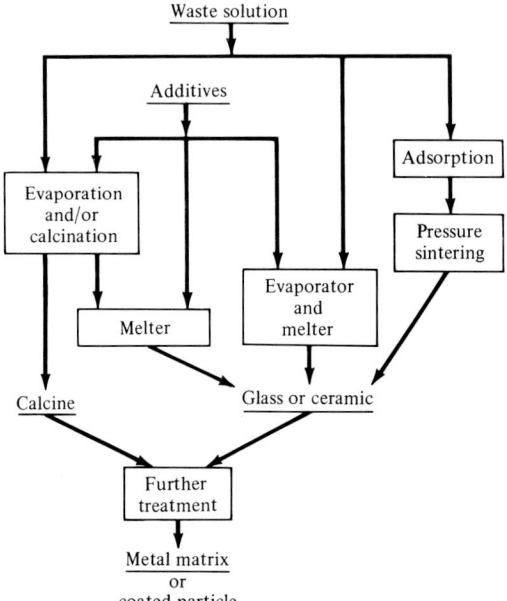

Figure 9-20 Basic high-level waste solidification processes.

Historically, the first attempt at converting wastes into glass was made at Chalk River, Canada, in 1960. This was closely followed by another developed at Harwell, England, and later by several attempts in the United States. In the United States a new plant was under construction in 1980. It is based on a 1963 process, used in a processing plant in Idaho, by which the waste is solidified by spraying it into a fluidized bed and then calcined at 400 to 500°C, thus producing a granular product that is then stored in air-cooled bins in underground vaults. Other promising plants are to be found in several countries, such as France, India, Germany, and the USSR.

Most solidification plants produce off-gases of steam and oxides of nitrogen that usually contain some fine particulate carryover and volatile radionuclides. These gases must be treated. All processes involve high temperatures and, of course, high levels of radioactivity. This combination imposes severe demands on plant design and operation. It is expected, however, that workable, commercially attractive solutions will emerge in the late 1980s. This is not late, as the problem of high-level waste is not expected to be of major effect for a few decades.

Disposal

The final disposition of the wastes, with or without the above treatments, is also of major concern. Many countries are undertaking activities involving underground disposal in deep geological formations. These activities include the investigation of suitable sites and suitable methods of storage in these sites. The main objectives, of course, are the protection of present and future populations from potential hazards. The suitable sites must be free of flowing groundwater, but the storage vessels must demonstrate reliability even in flowing-water conditions.

The disposal of low- and intermediate-level wastes (such as those used in research and medicine) has been done at relatively shallow depths in many countries. For example, since 1967 such wastes in solid form have been packaged in concrete or steel drums and buried in the old Asse Salt Mine in Germany.

If the spent fuel is to be disposed of in a throw-away process, it is buried intact.

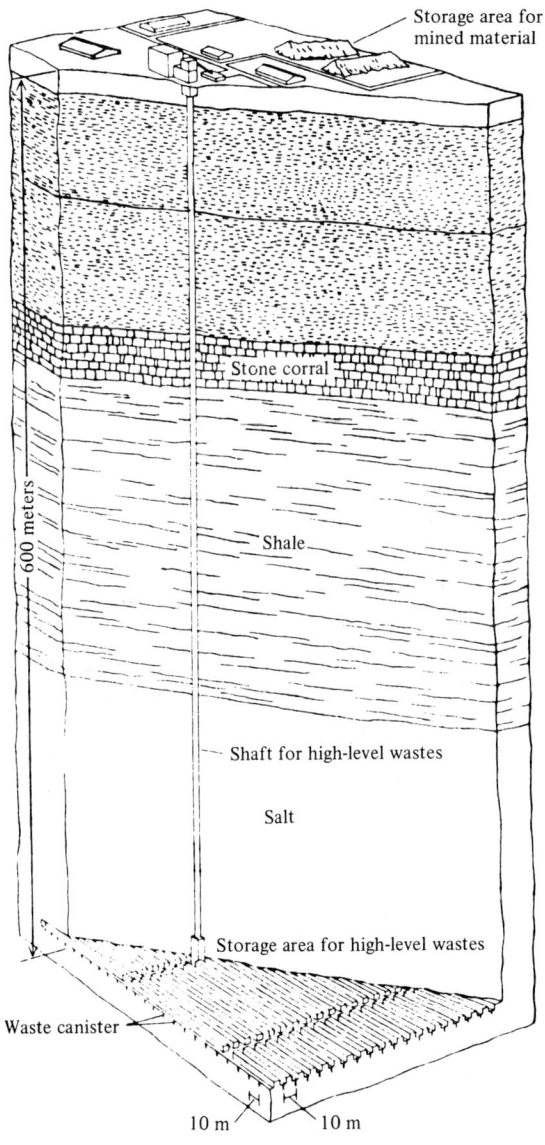

Figure 9-21 A conceptual depository of high-level waste in rock salt formations [18].

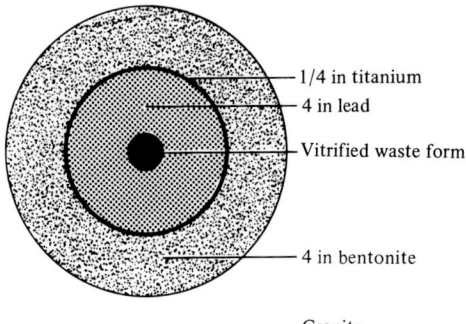

Figure 9-22 Cross section of a proposed canister for vitrified high-level waste.

If it is reprocessed, then it is buried either as a liquid or in solid form. There are several strategies for burial, most of them in deep salt or granite formations below ground or below the sea bed. In the United States, as of this writing, no final decision on high-level waste disposal has been made by the government.

A National Waste Terminal Storage Program was initiated in 1976. It includes the identification of suitable rock formations, in situ tests, and plans aimed at the establishment of up to six federal repositories. A Waste Isolation Pilot Plant (WIPP) in New Mexico is now being studied for the receipt of military wastes by the mid-1980s. Other federal repositories are planned for commercial-powerplant wastes. This also includes the disposal of spent reactor core fuel elements, should a decision be made not to treat them but to operate on a throw-away fuel cycle basis. The search includes sites in salt formations, shale, crystalline rock (volcanic basalt), the injection of liquid waste into isolated porous strata, and the injection into fractures induced in impermeable formations. Repositories are to be licensed and regulated by the U.S. Nuclear Regulatory Commission (NRC).

In France, which has the largest per capita nuclear program, plans are being drawn for long-range storage under the Alps. Other European countries have other diverse and innovative plans. The problem, although technically soluable, is laden with political, emotional, physical, and economic problems. Figure 9-21 shows a conceptual depository for the storage of high-level waste in rock salt formations for thousands of years. In the figure, the solidified waste is placed in canisters that are stored in holes drilled in rock salt with a spacing of about 10 m to allow for the efficient dissipation of energy without exceeding permissible temperature limits of either the canisters or the salt. It is estimated that each canister (producing about 10 kW at the time of storage) will require about 100 m^2 of salt for cooling. This means that about 0.5 km^2 would be needed to store all the high-level waste produced annually in the United States in 1980, if all electrical production were from nuclear powerplants. Figure 9-22 shows the cross section of a canister of Swedish design for disposal in granite. It shows vitrified waste surrounded by 4 in of lead, 0.25 in of titanium, 4 in of bentonite (an absorptive and colloidal clay mineral), and finally, granite.

References

1. Myers, P. S.: "Automobile Emissions—A Study in Environmental Benefits versus Technological Costs," Society of Automotive Engineers paper No. 700182, 1969.
2. Taylor, F. E., and G. A. M. Webb: "Radiation Exposure of the U.K. Population," National Radiological Protection Board Publication No. NRPB-R77, 1978.
3. Council on Environmental Control: "Global Energy Futures and the Carbon Dioxide Problem," U.S. Government Printing Office, January 1981.
4. Cowling, E. B., and J. N. Galloway: Effects of Precipitation on Aquatic and Terrestrial Ecosystems; A Proposed Precipitation Chemistry Network, *J. Air Pollut. Control Assoc.*, vol. 28, no. 3, pp. 228-235, March, 1978.
5. Wark, K., and C. F. Warner: "Air Pollution, its Origin and Control," 2d ed., Harper & Row, Publishers, Inc., New York, 1981.
6. White, H. J.: "Industrial Electrostatic Precipitation," Addison-Wesley Publishing Company, Inc., Reading, Mass., 1973.
7. White, H. J.: Modern Electrical Precipitation, *Ind. Eng. Chem.*, vol. 47, no. 2, pp. 932-939, 1955.
8. McCain, J. D., J. P. Gooch, and W. B. Smith: Results of Field Measurements of Industrial Particulate Sources and Electrostatic Precipitator Performance, *J. Air Pollut. Control Assoc.*, vol. 25, no. 2, pp. 117-121, 1975.
9. Fabric Filter Systems Study, "Handbook of Fabric Filter Technology," vol. 1, PB200-648, APTD-0690, National Technical Information Service, December 1970.
10. The President's Commission on the Accident at Three-Mile Island, J. G. Kemeny, Chairman, Washington, D.C., October 1979.
11. Johns, H. E., and J. R. Cunningham: "The Physics of Radiology," Charles C Thomas, Publisher, Springfield, Ill., 1969.
12. Attix, F. H., and W. C. Roesch (eds.): "Radiation Dosimetry," vol. 1, Academic Press, Inc., New York, 1968.
13. Blatz, H.: "Radiological Health," McGraw-Hill Book Company, New York, 1964.
14. Gloyna, E. F., and H. O. Ledbetter: "Principles of Radiological Health," Marcel Dekker, New York, 1969.
15. Duhamel, A. M. F. (ed.): "Health Physics," vol. 2, pt. 1, Progress in Nuclear Energy Series XII, Pergamon Press, Oxford, 1969.
16. Cember, H.: "Introduction to Health Physics," Pergamon Press, London, 1969.
17. Upton, A. C.: The Biological Effects of Low-Level Ionizing Radiation, *Sci. Am.*, vol. 246, no. 2, pp. 41-49, February 1982.
18. Gilbertson, J.: The Fraudulent Nuclear Waste Controversy, *Fusion*, vol. 2, no. 4, pp. 34-36, January 1979.
19. Cohen, B. L.: The Disposal of Radioactive Wastes from Fission Reactors, *Sci. Am.*, vol. 236, pp. 21-31, June 1977.
20. Sokol, J. P., and M. H. Cooper: Radioactive Waste in Perspective, *Nucl. Eng. Dig.*, Westinghouse Nuclear Energy Systems, pp. 15-21, 1979.

Section **10**

Economic Operation of Power Systems

Any power system is useless unless it operates. And when it operates it should do so as economically as possible. If such a level of operation is not achieved, the designer has failed to produce the best plant possible.

This section considers the variables that enter economic operation of power systems, namely fixed costs, variable costs, thermal plant efficiencies, incremental rates, economic loading of generating units, computers for economic loading, effects of varying fuel costs, transmission losses, and other topics.

Using the data here, the designer can aim at designing a plant that will produce and transmit power to meet the system load at minimum production cost with proper consideration of the effect on system capacity. This objective is achieved when all generation is operated at equal incremental cost with consideration for transmission losses.

Other forms of generation—nuclear, geothermal, solar, wind, and hydro are also considered. These data are useful for any type of end-use power plant being designed today.

From *Power System Operation*, 2d ed., by Robert Herschel Miller. Copyright © 1983. Used by permission of McGraw-Hill, Inc. All rights reserved.

Successful operation of power systems requires attention to safety of personnel and equipment and the provision of service to utility customers without interruption and at the lowest feasible cost. The problem of providing low-cost electric energy is affected by such items as efficiencies of power-generating equipment, cost of installation, and fuel costs for thermal-electric plants. Factors involved in the cost of producing energy can be divided into two categories: those that are fixed and those that are variable.

FIXED COSTS

Fixed costs include capital investment, interest charges on borrowed money, labor, taxes, and other expenses that continue irrespective of the load on the power system. Persons responsible for the operation of a power system have little control over these costs.

VARIABLE COSTS

Variable costs are those costs which are affected by the loading of generating units of different fuel or water rates, control of losses caused by reactive flows, the combination of hydro and thermal generation to meet daily load requirements, and purchase or sale of power. These costs are materially controlled by power system operators. This section will discuss factors in power system operation that can be controlled and methods used to ensure that power generated to carry the power system load is always produced in such a way that minimum costs will result. The savings that can be achieved by proper operation of power resources can be very significant; they may amount to several thousand dollars a day on large power systems.

Many power systems have several alternative sources for electric energy, such as conventional steam-electric plants, nuclear plants, hydro, geothermal, gas turbine, and outside sources from which power may be purchased. Solar, wind power, and fuel cells are also alternative sources of electric power. Considerable work is being done to make the energy cost of such sources competitive with that of more conventional sources of energy. Normally the electrical capacities of the units of such sources are much less than the capacities available from conventional thermal and large hydro units. The continuing problem is to determine at all times the combination of sources, and loads on these sources, which will result in minimum overall production cost.

Fuel supplies for thermal plants can be natural gas, oil, nuclear sources, or coal, with varying costs for each. Furthermore, the load on a power system is continually changing. For this reason the economic supply problem must be reviewed frequently, and, if necessary, load allocations on the various power sources readjusted so that deviations from the most economic operation will be held to a minimum.

Water supplies for hydro generation can have different values from time to time, and the use of hydro power must be integrated into the system power supply so that the lowest overall costs result.

Pumped storage is a special type of hydro power in which water is pumped to an upper reservoir during "off peak" hours when thermal generation costs are at a minimum. The pumped water is released during "peak" hours to generate hydro energy, and thus replace thermal generation when fuel costs would be high. It should be pointed out that pumped-storage generation requires more energy for pumping than is recovered during the generation cycle. However, the value of the power generated during the peak load periods will normally more than offset the cost of the thermally generated power used for pumping.

Power exchanges between interconnected systems can also be used to advantage in minimizing fuel costs when there are significant differences in generation costs of interconnected power systems.

Because the characteristics of the various types of power supplies differ, each type will be considered individually. An effort will then be made to develop the basis for the procedures used to attain overall economy when the different types of power sources are used simultaneously. Consideration will be given first to conventional thermal plants.

THERMAL POWER PLANT EFFICIENCIES

It is a well-established physical principle that as the differences between the temperature and pressure inputs and outputs of a heat-operated device, such as a steam turbine, are increased, more mechanical power output will be developed for the same amount of heat energy input. This is the basic reason for the ever-increasing pressures and temperatures in modern steam-electric generating units. The overall efficiency of thermal units is determined by measuring the heat input and the electric energy output and expressing the results as a ratio at various loads. Curves called input-output curves can be drawn from the results of such tests, with energy input expressed in Btu (British thermal units), equivalent barrels of oil, or Mcf of gas per hour, and the output as kilowatts or megawatts. A characteristic of these curves is that fuel input is increased as electrical output is increased, but not necessarily linearly. Such a curve is shown in Fig. 10-1.

Curves of this type are developed for each generating unit involved. From them it can readily be seen that efficient units will develop a given amount of power with less fuel input than will be needed by units of

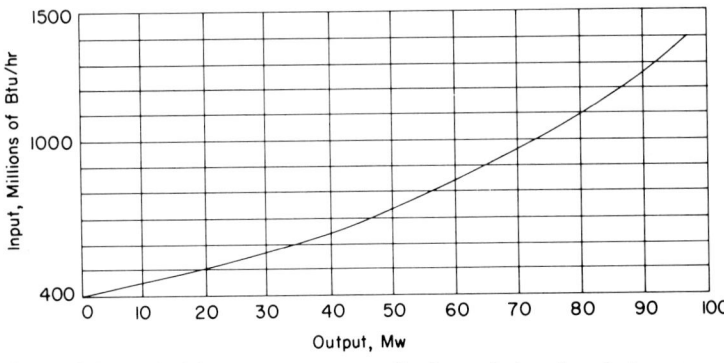

figure 10-1 Typical input-output curve of a thermal-electric unit. Input can be expressed in Btu, equivalent barrels of oil, or Mcf of gas. (An equivalent barrel is 6,250,000 Btu, and an Mcf of gas may vary depending on the source but is usually approximately 1,000,000 Btu.)

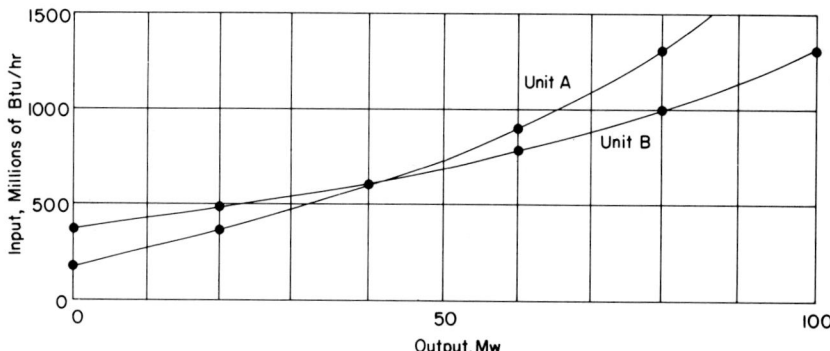

figure 10-2 Input-output curves for two typical thermal units. Even though the no-load fuel for unit *B* is greater than that of unit *A*, at loads above 40 MW the heat input for unit *B* is less than that for unit *A*.

lower efficiency. The first and obvious conclusion might be to load the efficient units before loading the less efficient units. This would, of course, be a better solution than loading the low-efficiency units first, but the desired solution is to load the available units so as to develop the required power at the least possible cost. Techniques to solve this problem have been developed and are in general application throughout the industry.

Consider two 100-MW thermal units *A* and *B* with input-output curves as shown in Fig. 10-2.

INCREMENTAL RATES

It can be shown mathematically that minimum fuel input for any given total load of the two machines will occur when they are operated at equal incremental heat rates. Because fuel has a cost, such as cents per Btu, dollars per equivalent barrel, or cents per Mcf, the above statement can be modified to say that *the minimum cost will occur when the incremental costs are equal.*

The term "incremental" merely means a small increase. Of course, the smaller the increment (increase), the more precise the determination of incremental change. An incremental rate is defined as the slope of a curve from one point to another. Examples of the determination of incremental rates are given in Fig. 10-3.

An inspection of Fig. 10-3 gives a clue to an easy method of determining incremental rates. If it is assumed that the curve is made up of straight-line segments between the numbered points, then we have already calculated the incremental rates (slopes) for points *B* and *D*. The slope at

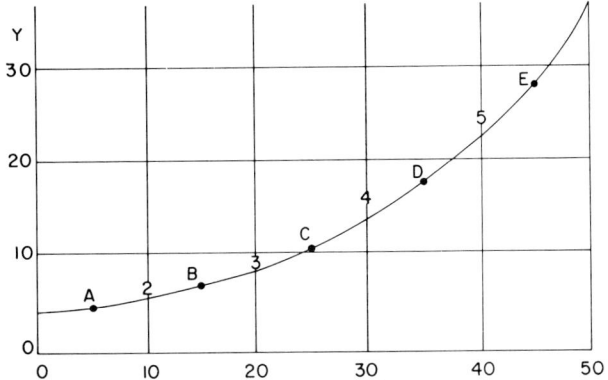

figure 10-3 Curve showing determination of incremental changes. For a small distance along a curve it can be considered to be a straight line. The increments on the x axis in this case are from 10 to 20 between points 2 and 3 and from 30 to 40 between points 4 and 5. When following the curve from 10 to 20 on the x axis, it goes from 6 to 8 on the y axis. The slope of this portion of the curve is a ratio of the differences, or $(8 - 6) \div (20 - 10)$, which equals $2/10 = 0.2$. When following the curve from 30 to 40 on the x axis, it goes from 14 to 22 on the y axis. The slope then is $(22 - 14) \div (40 - 30)$, which equals $8/10 = 0.8$. This method, using smaller and smaller increments until they approach zero, is the basis for differential calculus. For practical purposes in determining incremental costs, little error is introduced by using reasonably small increments other than those approaching zero.

A would be $(6 - 4) \div (10 - 0)$ or $2/10 = 0.2$. At point C the slope would be $(13 - 8) \div (30 - 20)$ or $5/10 = 0.5$.

At point E the slope would be $(36 - 22) \div (50 - 40) = 14/10 = 1.4$. An incremental-rate curve is developed by plotting the points determined by the above calculations. This is shown in Fig. 10-4.

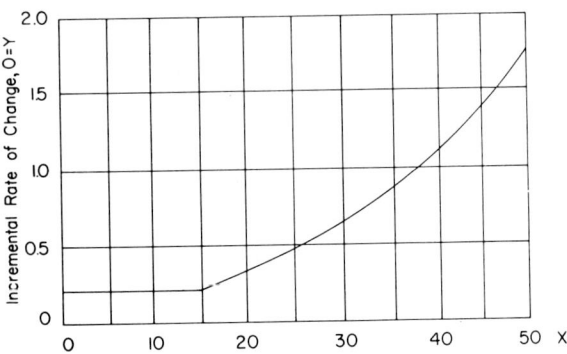

figure 10-4 Curve showing the incremental change of Y as X increases for the curve shown in Fig. 10-3.

TABLE 10-1 Procedure for Determining Load Allocation for Two Units Whose Curves Are Shown in Fig. 10-2*

(1) Load, MW	(2) Million Btu/h	(3) Dollars/h (2) × $3.50	(4) Charge for load increment, dollars/h	(5) Incremental rate, dollars/MWh (4) ÷ (1)
Unit A				
0	200	700		
20	300	1050	350	17.50
40	450	1575	525	26.30
60	650	2275	700	35.00
80	950	3320	1045	52.20
100	1500	5250	1930	96.50
Unit B				
0	250	875		
20	350	1225	350	17.50
40	450	1575	350	17.50
60	600	2100	525	26.20
80	800	2800	700	35.00
100	1050	3675	875	43.80

*The fuel costs shown in this tabulation have been increased by a factor of 10 to more realistically reflect present costs, which have drastically increased since the first edition was prepared.

ECONOMIC LOADING OF GENERATING UNITS

When the basis for determining incremental-rate curves has been developed, this method can be used to determine how to operate electric generating units for minimum production cost. A procedure for determining load allocation for minimum fuel cost between the two units whose input-output curves are shown in Fig. 10-2 is shown in Table 10-1. Since we are primarily interested in cost, the fuel rates will be converted to dollars per hour for various loads. The incremental rate in dollars per megawatt-hour for various loads is determined. A fuel price of $3.50 per million Btu is assumed.[1]

The incremental-rate information developed in Table 10-1 can be plotted as curves for both machines, Figs. 10-5 and 10-6. If 1-h periods are considered, the vertical scale will be incremental cost in dollars per megawatthour, with megawatts as the horizontal scale, as shown in Fig. 10-6.

---

[1] Fuel costs have increased by a factor of at least 10 since the early 1970s. This has significantly increased the cost of electric power for all consumers.

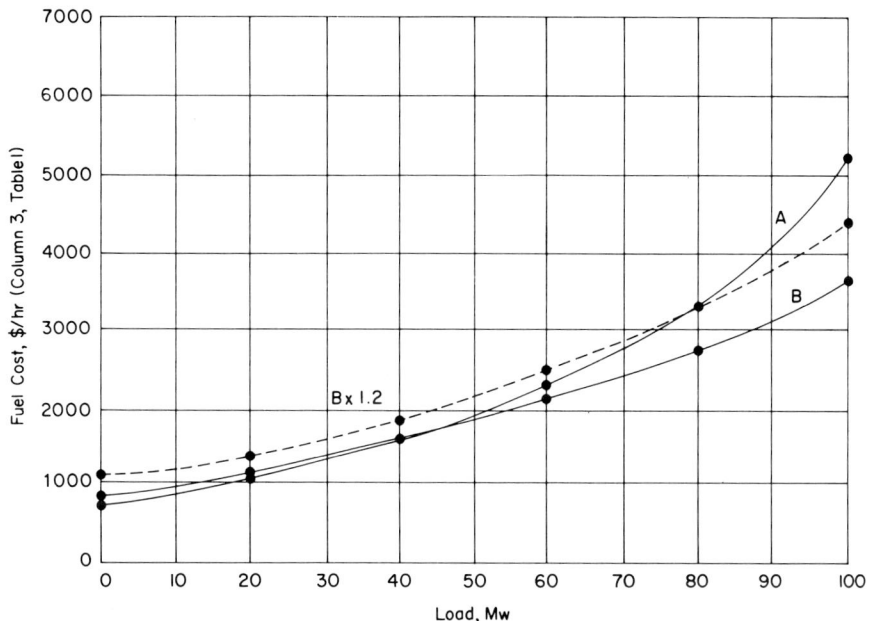

figure 10-5 Input-output curves showing fuel dollars per hour versus MW load.

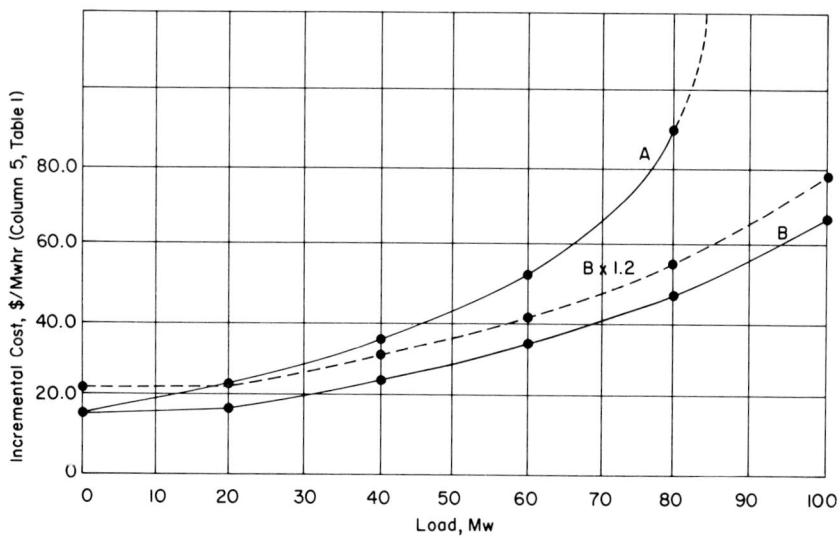

figure 10-6 Incremental cost curves for units *A* and *B*.

From the information shown in Figs. 10-5 and 10-6 it should be possible to determine the proper division of load between the two machines to result in minimum fuel cost. Assume a total load of 100 MW to be carried by the two units. Various combinations of loading can be made, but the objective is to carry the load with minimum cost. The tabulation shown in Table 10-2 was developed by taking points on the incremental fuel cost curves of Fig. 10-6 to match various load conditions and fuel cost per hour from the input-output curves shown in Fig. 10-5, where fuel cost in dollars per hour is plotted against loads in megawatts.

From Table 10-2 it can be seen that the minimum fuel cost occurs when unit A is loaded to 40 MW and unit B to 60 MW with incremental costs of $30 MWh for each machine. If desired, tabulations similar to that in Table 10-2 can be set up for other fuel costs and the minimum cost (fuel rates) determined as further proof of the principle of loading machines for equal incremental costs.

It is obvious that when many units are involved, a manual solution of the economic loading problem is impractical, because many load changes would be needed while solving the problem for only one situation. Various devices have been developed to help solve the economic loading problem rapidly where many generating units are involved.

Probably the simplest device for allocating load on an incremental basis is the economic loading slide rule. These devices were used to a considerable extent prior to the advent of digital computers with economic loading programs.

The slide rules made use of sliding elements showing unit loading on

TABLE 10-2

Unit A			Unit B			
Load, MW	Incr. fuel cost, dollars/MWh	Fuel cost, dollars/h	Load, MW	Incr. fuel cost, dollars/MWh	Fuel cost, dollars/h	Total fuel cost, dollars/h
0	15.50	730.00	100	48.50	3650.00	4380.00
10	17.50	830.00	90	43.80	3220.00	4050.00
20	21.00	1010.00	80	39.00	2800.00	3810.00
30	26.30	1250.00	70	35.00	2450.00	3700.00
40	30.00	1575.00	60	30.00	2100.00	3670.00
50	35.00	1870.00	50	26.50	1825.00	3695.00
60	43.00	2250.00	40	21.00	1575.00	3820.00
70	52.50	2750.00	30	17.50	1400.00	4150.00
80	66.00	3320.00	20	17.50	1228.00	4540.00
90	96.50	4100.00	10	17.50	1050.00	5150.00
100	—	5200.00	0	17.50	900.00	6100.00

logarithmic scales, and a straightedge that could be adjusted in position. The sliding elements could be set to the unit fuel cost, and by moving the straightedge to the appropriate position, unit loadings could be determined for minimum overall fuel cost.

COMPUTERS FOR ECONOMIC LOADING

Digital computers are almost universally used for economic loading of generating units. Although there may still be some analog load-frequency-control (LFC) systems in use, they are rapidly being supplanted by digital systems.

The particular advantage of computers is that they can continuously monitor the system loading conditions, determine the most economical allocation of generation between units, and send control impulses to load the units to the desired values. Computer control, properly applied, can approach an almost exact allocation of unit loadings for minimum fuel cost. In applying computers to online economic loading problems, the unit input-output curves and incremental-fuel-rate curves are stored in the computer, which goes through a process similar to that followed in Table 10-2 to calculate the desired machine loadings. Because of the tremendous speed at which computations can be made in a digital computer, it can solve economic loading problems in very short time intervals and simultaneously carry out other system-control functions.

Automatic generation control (AGC) is commonly included in supervisory control and data acquisition system (SCADA) installations.

EFFECTS OF VARYING FUEL COSTS

Before leaving the problem of incremental loading of thermal plants, the matter of varying fuel costs should be mentioned. The shapes of the input-output and incremental-fuel-rate curves are not changed by different fuels or by changes in the cost of the same fuel. Consequently, if the incremental curves are plotted with incremental cost as the vertical scale, the ratio of the cost of the fuel being burned to the cost of the fuel for which the curves were drawn can be used as a multiplying factor. This factor is employed to correct for fuel-cost changes for any or all of the units. By this means it is possible to solve the economic loading problem under all conditions of fuel cost.

There is a further complication in accounting for losses due to transmission of power from generation to load. This will be discussed in more detail later. It will suffice for the moment to state that transmission losses can be, and are, evaluated and their effect used as a multiplier on the incremental fuel cost of each unit to provide a means of obtaining actual overall economy, including transmission losses.

NUCLEAR GENERATION

The above discussion has been confined to the loading of conventional fossil-fueled thermal plants. Nuclear plants present a special problem. Owing to various factors it is usually desirable to operate nuclear units at base load and at full output at all times. Consequently, it is not necessary to consider them as an incremental source to the system. However, in supplying a portion of the energy that otherwise would have to be supplied by conventional thermal units, nuclear units cause these other thermal units to operate at a reduced incremental cost.

GEOTHERMAL GENERATION

As a result of escalating fuel costs in recent years, there has been greatly increased interest in the use of geothermal energy for electric power generation. This source can be used only where there are sources of natural steam or hot water that can be economically developed. The largest such development is the Geysers of the Pacific Gas and Electric Company in California.

Like nuclear installations, the units at geothermal plants are normally operated as base-load units and not on an incremental basis.

SOLAR AND WIND POWER GENERATION

There has been progress in the development of solar and wind power in recent years. These developments are also due, to a considerable extent, to the increasing costs of fossil fuels.

At present such developments are of relatively small capacity, and because of the variable nature of sun and wind, neither source is adaptable to incremental loading techniques. Power from these sources is developed when sun and wind conditions are favorable, and to the extent available it will reduce the use of normal fossil fuels.

COORDINATION OF HYDRO AND THERMAL GENERATION

The operation of hydro units in a system in which both hydro and thermal generation are used presents an extension of the economic loading problem. There are many conditions connected with hydro operations, such as uncontrolled flows and required releases of water for irrigation or flood control, which take away from the system operator some of the alternatives that might be available if the water could be used entirely as desired for the benefit of power production. However, if a value can be placed on water, usually in dollars per acre-foot, hydro units can be operated incrementally along with thermal units for overall economic operation of the system.

Of course, the value of water changes from time to time, being lowered during periods of high flow, during and immediately following storms, and increased during periods when flows are low or when reservoirs are being drafted at controlled rates of flow. Since each acre-foot of water through a hydro plant will develop a definite amount of energy, depending on the head of the plant, water is equivalent to fuel such as gas or oil for power-producing purposes.

Procedures for integrating the operation of hydro and thermal generation on a system for minimum cost of generation have been developed and are in use. This procedure is called hydrothermal coordination.

Basically, in a hydrothermal-coordination program, input-output curves for each hydro unit are developed, showing acre-feet per hour plotted against load in megawatts. From the input-output curves, incremental-water-rate curves showing the incremental water rate in acre-feet per megawatthour plotted against load in megawatts can be developed by exactly the same method used for thermal plants.

An arbitrary price in dollars per acre-foot is placed on the water for each plant. If it is desired to use more water, the price is reduced, and if less water is to be used, the water price is increased. By proper selection of water prices, exactly the desired amount of water will be used during any desired time period. The hydro plants then will follow incremental loading requirements of the system and help achieve the desired result of overall minimum fuel cost.

The water value in hydrothermal-coordination programs is usually denoted by the Greek letter gamma (γ) to distinguish it from the thermal unit and system fuel cost, which is designated by the Greek letter lambda (λ).

The proper integration of hydro and thermal generation for minimum overall cost is quite complex and can be solved optimally only by a digital computer. Even with a computer, the number of calculations used to determine the most economic operation can be so great that considerable computer time is required to obtain a correct solution to the problem.

TRANSMISSION LOSSES

The preceding discussion has centered on determining the loads to be placed on thermal and hydro units in order to obtain equal incremental fuel cost for minimum overall cost of generation. The problem is only partially solved, however, until transmission losses are considered.

It was mentioned previously that if transmission losses could be evaluated, their effect could be used as a multiplier on fuel cost (or water value for hydro) to compensate for the energy lost in transmission and to arrive at a true economic loading of the system.

In the sections on energy transfer and var flows, it was pointed out that all transmission lines have resistance, determined by the conductor material, conductor size, and length of the line. It was also pointed out that the transmission loss in watts was the product of the line current squared times the resistance of the line ($I^2 R$).

In the simplest possible system, a generating unit connected by a single transmission line to a load, the determination of transmission loss is quite simple. Figure 10-7 illustrates this case.

The generator must, of course, produce enough energy to supply the load plus the transmission losses—in the above case, the load plus 100 kW. The power required to supply the losses will move the generation to a higher point on the incremental cost curve, resulting in an increase in the cost of each kilowatthour of energy.

When two or more generating units are connected to a load via separate transmission lines, the correct allocation of load between the units will result when the incremental costs, including the costs of supplying the energy for transmission losses, are equal.

Here again the problem rapidly compounds in complexity as the number of generators, lines, loads, and tie points is increased. Manual methods of calculating loss factors become impractical, and it is necessary to resort to analog or digital computing devices to determine the effects of transmission losses on a power system.

No effort will be made here to develop the mathematical solution of the transmission-loss problem. For the purposes of this discussion, it should suffice to state that a coordination equation has been developed to determine what is called the "penalty factor." Penalty factor is equal

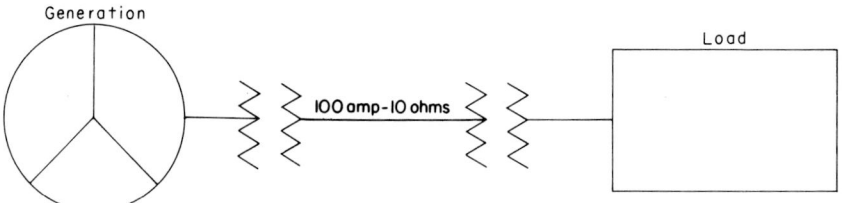

figure 10-7 Simple transmission system of a single generator and load connected by a line carrying 100 A through 10 Ω. Loss is equal to $(100)^2 \times 10 = 100{,}000$ W or 100 kW.

to $1/(1 - \text{loss factor})$, and it can be seen that as the loss factor increases, the penalty factor will increase.

In order to determine penalty factors, it is necessary to develop a mathematical model of the system. After this has been done, an analog penalty-factor computer or a digital computer can be used to determine penalty factors for any load condition for each generating station or tie-line source to the system load center. When penalty-factor calculations are made "offline," they are manually applied to incremental slide-rule slides for each unit or to penalty-factor setters on analog dispatch-control units. By this means the incremental-cost curves are adjusted upward or downward as required by the penalty factor so that the generating units are loaded on a strictly competitive basis for minimum cost, including transmission losses. These methods have become relatively obsolete because of the wide acceptance and application of digital computers for power system control.

When digital computers are used for system control, penalty-factor calculations are made at frequent time intervals, and generation-control impulses are produced, including current penalty factors, so that the system generation is consistently maintained with the most economic allocation between generating units.

It has been shown previously that minimum fuel input occurs when generating units are operated at equal incremental costs. To demonstrate the effect of transmission penalty factors on load division between gen-

TABLE 10-3

	Unit A			Unit B		
Load, MW	Incr. fuel cost, dollars/MWh	Fuel cost, dollars/h	Load, MW	Incr. fuel cost, dollars/MWh	Fuel cost, dollars/h	Total fuel cost, dollars/h
0	15.50	730	100	60.00	4410	5140
10	17.50	830	90	52.50	3800	4630
20	21.00	1050	80	47.50	3360	4410
30	26.30	1270	70	42.50	2830	4100
40	30.00	1575	60	36.50	2520	4095
47	33.00	1780	53	33.00	2220	4000
50	35.00	1870	50	31.50	2180	4050
60	43.00	2250	40	26.00	1890	4140
70	52.20	3320	30	21.00	1650	4970
80	66.00	4100	20	21.00	1470	5570
90	96.50	5200	10	21.00	1230	6430
100	—	—	0	21.00	1050	—

erating units, an example will be worked out using the two machines previously considered, but with a penalty factor of 1.2 applied to unit B and a penalty factor of 1.0 applied to unit A.

Under these conditions the values shown on the input-output and incremental-cost curves of unit B will be multiplied by 1.2 and replotted. This has been done, and the curves for unit B operating with the assumed penalty factor are shown as the dashed curves on Figs. 10-5 and 10-6. The effect is to raise both the input-output and incremental-cost curves. If the penalty factor had been less than 1, it would indicate that system losses would be reduced by adding load to unit B, and the curves would move downward.

The comparative tabulation under the new operating conditions is shown in Table 10-3. This table shows that the minimum fuel cost occurs with 47 MW on unit A and 53 MW on unit B, with an equal incremental fuel cost of $33/MWh.

ECONOMIC INTERCHANGE OF POWER

Another problem that is encountered by a system operator is to determine when it is economical to buy power from or sell power to other systems. Whenever power is purchased and received into a system, the power that must be produced to carry the system load is reduced by the amount of power received from the other system. Conversely, whenever power is sold, power production must equal the system load plus the amount of the sale.

The preceding discussion has demonstrated that when the power output of generating units is increased, the unit incremental cost and also the system incremental cost (lambda) increase. Conversely, when power is received from another system, as unit loading is decreased the system lambda decreases.

When power is sold, the additional (incremental) production cost must be determined in order to be able to quote a price to the prospective purchaser of the power.

When power is purchased, production costs will be reduced, and this saving has a value that must be determined. The value of saving in a purchase transaction is called the "decremental value."

Definitions of these two terms are as follows:

1. Incremental cost is the additional cost incurred to generate an added amount of power.
2. Decremental value is the cost saved by not generating an amount of power.

The units usually used are cents per kilowatthour or dollars per megawatthour.

The method used to determine the incremental cost of a sale transaction is to take the average of the existing system incremental cost and the new incremental cost and to quote this average figure to the prospective purchaser. As an example, assume that the existing cost is $0.03/kWh. If a sale of 100 MW is contemplated, the cost with the new system load condition would be $0.035/kWh. The average incremental cost would then be ($0.03 + 0.035)/2 = $0.0325/kWh.

Exactly the reverse process is used when a power purchase is considered. Assume that the existing cost is $0.03 kWh, and it is desired to purchase 100 MW of power. This amount of received power would reduce the system cost to $0.025/kWh. The decremental value (average saving) would be ($0.03 + 0.025)/2 = 0.0275/kWh.

In considering transactions involving the purchase or sale of power, as in determining how generating units should be loaded for maximum economy, the effect of transmission losses must be considered. As has been pointed out, to determine properly how generating units should be loaded, the unit incremental cost is multiplied by the penalty factor to calculate the worth of the power at the system load center.

When power is being received from another system via a tie line, it is handled exactly as though it were coming from a generating unit at the tie point. The price at the tie point is multiplied by the penalty factor to determine the worth of the purchased power at the load center as compared with that from generating units in the system.

When a power sale is being evaluated, the reverse is true. In this case power is being transmitted from the load center to the tie point with the purchasing system, and it is desired to determine the worth of the power at the tie point. In order to make this determination, the value of the power (system incremental cost) at the load center is divided by the penalty factor.

Examples of both situations will be shown. First, assume a system incremental cost of $0.03/kWh at the load center. A purchase of 100 MW is being considered at a quoted price of $0.026/kWh. The penalty factor from the tie point to the load center has been determined to be 1.15.

To evaluate properly the economics of the proposed purchase, it will be necessary to determine both the cost of the purchased power at the load center and the decremental value of the purchase to the system. The cost at the load center would be $0.026 \times 1.15 = $0.0299/kWh. If the system generation is reduced by 100 MW due to the purchase and the system cost is reduced to $0.027, the decremental value would be

($0.03 + 0.027)/2 = $0.0285/kWh. In this case there would be no saving in purchasing the power because the cost of the purchased power is greater than the decremental value of the saving.

Another situation might be developed in which a system with an incremental cost of $0.03/kWh at existing load was asked to supply 100 MW to another system with an incremental cost of $0.042/kWh at its existing load. Assume that the selling system's incremental cost went to $0.035/kWh with the additional load, and that the penalty factor to the tie point is 1.02 at 100-MW delivery. The quoted price would be

$$\frac{\text{Original cost + new cost}}{2} \times \frac{1}{\text{penalty factor}}$$

or

$$\frac{0.03 + 0.035}{2} \times \frac{1}{1.02} = 0.0318/\text{kWh}$$

The purchasing system would determine its decremental value as follows. Assume that if its generation is reduced by 100 MW, its system cost will be reduced to $0.038/kWh and that the penalty factor from the tie point to load center will be 1.05. The decremental value will be ($0.042 + 0.038)/2 × 1.05 = $0.042/kWh. The difference between the buyer's decremental value and the seller's incremental cost will be $0.042 − 0.0318 = 0.0102/kWh.

In purchase and sale transactions of the type discussed above, it is customary to split the savings between the buying and selling systems. In other words, the average of the sum of the buyer's decremental value and the seller's incremental cost would be determined as in the case just outlined.

$$\text{Buyer's decremental value} = \$0.042/\text{kWh}$$
$$\text{Seller's incremental cost} = \$0.0318/\text{kWh}$$
$$\text{Average} = \frac{0.042 + 0.0318}{2} = \$0.0369/\text{kWh}$$

The purchasing system would pay $0.0369/kWh and would save the difference between what it would have cost to generate the power and the cost of the purchased power. In this case $0.042 − 0.0369 = $0.0051/kWh, which represents, at 100-MW delivery, a saving of $510 per hour. The seller would benefit by the same amount.

Obviously when there is a significant difference between costs on systems, it is mutually desirable to enter into transactions of the type just discussed. These are usually termed "economy energy" transactions. Most contractual arrangements between systems for economy energy have a minimum difference, such as $0.005/kWh, before such transactions are permitted. This is to protect against inaccuracies in estimating load that may fluctuate during a transaction period, which is usually for a fixed period such as an hour. Furthermore, the determination of incremental costs and decremental values may not be precise, and unless a significant difference exists, losses rather than savings may result.

In some cases, instead of averaging the seller's incremental cost and the buyer's decremental value, power sales are made by multiplying the seller's incremental cost by a fixed percentage, such as 15 percent. For example, a sale might be effected at seller's incremental cost of \$0.03/kWh × 115 percent, expressed as \$0.03 × 1.15 = 0.00345/kWh. This method of calculating energy cost is used when the buyer may not know the decremental value. Contracts sometimes permit transactions by either method.

SUMMARY

This section might be summarized by restating that the objective in power system operation is to produce and transmit power to meet the system load at minimum production cost with proper consideration of the effect on system security. This objective is achieved when all generation is operated at equal incremental cost with consideration for transmission losses.

Hydro plants can be integrated into the operation by putting a value on the water used through the hydro plants and then loading them incrementally in competition with the thermal plants.

Material savings in fuel cost can be achieved by careful adherence to economic (incremental) loading of generating units.

Purchase and sale of energy can also be used to minimize power production costs. In considering such transactions, the cost of the purchased power must be compared against the saving that will result by not producing the amount of power involved in the purchase.

Section **11**

Power System Reliability Factors

Every power system must be as reliable as possible, in view of its allowable cost. Balancing these two factors—i.e. reliability and cost—can be a major challenge for any power-plant designer.

A commonly used design criterion is to provide facilities and capacity to withstand one foreseeable contingency, such as the loss of one power line, one transformer, or other credible occurrence. Usually, system design does not provide for second or greater contingencies because of excessive cost and the reduced probability of two events occurring simultaneously.

This section considers factors affecting power-system reliability, spinning reserve, transmission and station capability, effects of temperature on equipment, matching generation with load, and aspects of reliability. Using the data given here, any designer can work towards a more dependable generating system that delivers the desired power within the design budget.

From *Power System Operation,* 2d ed., by Robert Herschel Miller. Copyright © 1983. Used by permission of McGraw-Hill, Inc. All rights reserved.

INTRODUCTION

Providing reliable service from a power system is one of the most important responsibilities of power system operators. This factor receives a great deal of attention in the design and construction of power system equipment and transmission and distribution lines.

Generation and substation equipment is carefully engineered to give many years of reliable service and has design provisions to withstand transient overvoltages due to lightning or switching surges. Equipment is also designed to withstand the mechanical and electrical stresses that may result when it is subjected to high fault currents.

System design provides for sufficient capacity in lines and station equipment so that equipment failure will not ordinarily result in customer load being interrupted in the event of loss of a line, transformer bank, circuit-breaker bushing, or similar trouble.

A commonly used design criterion is to provide facilities and capacity to withstand one foreseeable contingency, such as the loss of one line, one transformer, or other credible occurrence. Usually system design does not provide for second or greater contingencies because of the excessive cost and the reduced probability of two events occurring simultaneously.

After a power system is designed and constructed, it is the responsibility of the system operators to operate the system so that the design limits are not exceeded, to be alert to conditions that may exist that could affect reliability, and to be ready to take action to prevent hazardous situations from developing. Following trouble, when service is lost or equipment is unavailable, the system operator should proceed to restore the system to as near normal operation as possible so that its reliability is maintained at the highest possible level.

FACTORS AFFECTING POWER SYSTEM RELIABILITY

Some of the major factors that affect power system reliability are:

1. Reserve capacity
2. Adequate transmission and station capability
3. The ability to match load with generation
4. Prompt disconnecting of faulted lines or equipment and restoration of facilities
5. The ability to restart generation equipment
6. The ability to operate equipment such as power circuit breakers without dependence on power system energy
7. The ability to provide alternative arrangements of lines or station equipment to restore unfaulted equipment to service promptly
8. Adequate and reliable interconnections with other systems
9. Reliable indication of system conditions and communications with key generation and transmission stations

The above list of items is by no means complete, but it covers major categories that must be reviewed in order to ensure reliability of power system operation. Some items listed are determined by design and are not within the control of system operators. The following discussion will attempt to point out factors over which system operators can exert control in order that maximum reliability can be achieved with the available facilities.

SPINNING RESERVE

Generating capacity that is on the line and that is in excess of the load on the system is called spinning reserve. Adequate spinning reserve is probably the primary security factor in power system operation.

The amount of spinning reserve that a system desires to carry is a policy decision based on risks and economics. Once the spinning-reserve policy has been established, it is the duty of system operators to attempt to meet the criterion each day so that the system will not be jeopardized by inadequate reserve. Because of the no-load fuel costs incurred by excessive reserve, system operators also should see that excessive reserve is not carried.

The amount of spinning reserve can be expressed as a percentage of the daily peak load or be based on the risks of loss of generating capacity that actually exist on the system. The determination of spinning reserve as a percentage of the daily peak load leaves much to be desired, since it may not take into consideration the actual risks that exist on a system. Furthermore, particularly with thermal generating equipment, units usually require several hours from the time that they are ordered to be placed in service before they are actually available. Consequently, it is necessary to make estimates of load, which may be somewhat in error. If load is underestimated, the percentage of spinning reserve at peak time may actually be less than that required by the criteria. Interconnection agreements sometimes provide penalties for inadequate spinning reserve.

Probably a more realistic method of specifying spinning-reserve criteria is to base them on risks, along with allowances for forecast error and regulating requirements. Elements of risk include the load on the most heavily loaded unit or the amount of power being imported into the system if there are interconnecting tie lines. In addition to the risk due to unit or tie-line load, allowance is made for forecast error and for regulation error. These factors are usually 2 to 3 percent each. In some cases another factor is added, an arbitrarily determined amount which takes into consideration abnormal system arrangements or other conditions which might result in a higher-than-normal risk.

An example of calculating required spinning reserve on the basis just outlined is illustrated in Fig. 11-1 for system *B*.

Another factor connected with spinning reserve that should be under the continuous scrutiny of the system operator is its location and makeup. In the event of loss of a large, heavily loaded unit, frequency will drop. The amount of frequency sag will depend on the proportion of total available generation lost, including that on interconnected systems. It is desirable to restore frequency to normal as soon as possible, and in the event of interconnected operation to return tie lines to normal schedule as soon as possible so that instability or overloads will not occur, inadvertent energy accumulations will not become excessive, and the effect of the trouble on the interconnected area will be minimized.

Generating units have limits on the rate at which they can respond

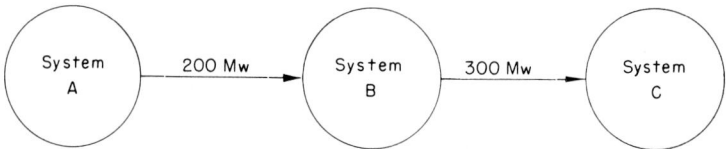

figure 11-1(a) Assume a system peak load of 4000 MW with the most heavily loaded unit 350 MW, a 2 percent forecast error and a 3 percent regulation error, and no abnormal system arrangement.

Most heavily loaded unit	= 350 MW
Forecast error	= 80 MW (0.02 × 4000)
Regulation error	= 120 MW (0.03 × 4000)
Contingency factor	= 0
Required spinning reserve	= 550 MW

figure 11-1(b) Assume the same system and largest unit loads as in A and an arbitrary contingency factor due to abnormal system arrangement of 100 MW.

Net imported power	= 500 MW
Forecast error	= 80 MW
Regulation error	= 120 MW
Contingency factor	= 100 MW
Total spinning reserve	= 800 MW

when a load pickup is required. With hydro units, rates of pickup are usually limited by the rate at which water can be accelerated in the penstocks, and with thermal units, after the initial energy stored in the boilers is used, by the rate at which steam can be produced to sustain a load pickup.

It is possible to determine the percentage of unloaded capacity that can be picked up by the various generating units in time categories such as 5 s, 10 s, 30 s, 1 min, 5 min, etc. With such a determination of spinning reserve, it is possible to predict reasonably well how a system will respond to trouble resulting in a frequency drop.

With large-scale interconnections, the loss of even a large, heavily loaded generating unit will not produce a significant frequency drop. In such cases the power angle of the system losing generation retards, and there is an immediate flow of power from other interconnected systems into the system with deficient generation. Since there is little or no frequency drop, the deviation of tie lines from normal schedules is the only indication of abnormal conditions. Tie-line-load frequency-control signals or telephonic orders to plants with reserve capacity are necessary to restore tie-line schedules to normal. In such cases spinning-reserve response is somewhat slower than when a significant frequency drop occurs.

A very important factor in maintaining proper spinning reserve is to have the reserve distributed over several units throughout the system. If most or all of the reserve is on one large unit, the total response is limited to the rate at which that one unit can pick up load. When reserve is divided among several units, each can provide its share in restoring system conditions to normal, and the possibility of instability, tie-line tripping, or line and equipment overloads is reduced.

To summarize, adequate spinning reserve is one of the major factors in maintaining power system security. The amount of spinning reserve to be carried is based on an evaluation of risk and is a management policy decision. After the policy has been decided, it is the responsibility of the system operators to ensure that the policy is met and that allocation of reserve among available units is such that proper response to loss of generation or tie lines will be achieved.

TRANSMISSION AND STATION CAPABILITY

The power-handling capabilities of transmission lines and substation equipment are design factors, and are not under the control of system operators. However, after lines and equipment are installed and in service, the system operator is in a position to see that capabilities are not exceeded in normal operation. By frequently monitoring load and voltage conditions at various locations in a system, the operator can be kept aware of conditions and can adjust generation or alter arrangements to prevent overload conditions from occurring.

System operators should be familiar with normal and overload ratings of facilities under their jurisdiction. Some equipment, particularly transformers, can be operated at greater than nameplate rating for limited periods of time without damage. Generating equipment ratings are established by the manufacturer and confirmed by operational tests following installation.

EFFECTS OF TEMPERATURE ON EQUIPMENT

The limiting loading factor on all electrical equipment is temperature rise.

Generating and substation equipment maximum operating temperatures are specified by manufacturers and by information provided by system engineering and operating organizations. If the ambient air temperatures are low, it is usually possible to load equipment to higher loads than with high ambients. With modern thermal generating equipment, it is possible to gain appreciable capacity temporarily at the expense of efficiency by cutting out feedwater heaters. In emergency conditions the additional capacity available by this means may prevent overloads on other equipment or emergency load shedding. Increasing pressure in thermal unit boilers, within limits, can also be used to temporarily increase capacity.

POWER-FACTOR CONSIDERATIONS

Another factor that should be under the constant surveillance of the system operator is the power factor of generating equipment. If a unit is supplying a relatively large var output, its total rating may be exceeded even though the MW load is below rating. When leading var are being supplied, there is an increased possibility of heating the end laminations of generator armatures. Temperature-sensing devices such as resistance thermal devices (RTD) or thermocouples are usually provided to monitor such conditions.

TRANSMISSION-LINE RATINGS

Transmission-line ratings are determined by conductor type, size, and line length. For short lines, only the type and size of conductor are of importance because the electrical phase shift of the line under heavily loaded conditions is not sufficient to cause a stability problem. On such lines the thermal rating of the conductor is the limiting factor.

Usually lines where thermal capacity limits capability are given summer and winter ratings. The summer ratings are somewhat lower than the winter ratings because of higher summer ambient temperatures.

On long lines, the rating may be determined by stability limitations rather than by conductor thermal capability. On such lines the stability limits are reached before the conductor current reaches its maximum limit.

A system operator who is familiar with line and station capabilities can take necessary action during both normal and trouble conditions to ensure that capabilities are not exceeded or, if necessary, to modify switching arrangements or shed load to assist in providing maximum service reliability.

MATCHING GENERATION WITH LOAD

When a power system is operating at normal frequency, with tie lines to other systems carrying scheduled loads, generation and load are matched. Any increase or decrease in load must be followed by a corresponding change in generation to match the new load condition.

The system operator is provided with various indicating devices, including system frequency, telemetered tie-line flows, and area control error, so that the operator can be kept continuously aware of these factors. Matching generation and load is the basic responsibility of the system operator. Devices such as load-frequency-control and automatic-dispatch equipment are provided to help match generation and load. It is only after this is accomplished that economic loading of generation is effected.

To make sure that there is always sufficient generation capability to carry any expected load, with provision for the loss of a major generating unit, spinning-reserve capacity is carried, as previously discussed. However, if serious troubles occur, such as the loss of all tie lines or a total generating station from bus trouble, the remaining generation may not be sufficient to carry the load on the system.

When generation is inadequate, system frequency will decline. It is of extreme importance that frequency be prevented from continuing to decline. In thermal plants particularly, auxiliary devices such as boiler feed pumps, draft fans, etc., must operate at or near normal speed and voltage to function properly. A frequency drop of more than a few hertz (5 or 6) may cause a loss of the plant auxiliaries and result in complete loss of the plant, further reducing the available generation and increasing the possibility of a total system collapse.

If frequency decline continues, it is necessary to match load with the available generation. This can be done manually by dropping customer load promptly in sufficient quantity to arrest the frequency decay and start to restore frequency to normal.

Manual load shedding leaves much to be desired, since there usually is very little time for the system operator to assess the situation and take the proper corrective action. As a consequence, it is common practice to install underfrequency relays to disconnect load automatically in amounts sufficient to match the remaining load with available generation. In developing load-shedding programs, it is customary to drop load in increments with declining frequency. For example, if the credible incident could result in a loss of 30 percent of the generation, it might be decided to drop 35 percent of the load in stages, with 5 percent of load being dropped at 59 Hz and additional amounts dropped if frequency continues to decline, until the total 35 percent is dropped at some preselected frequency, such as 58 Hz.

Load-shedding programs vary among systems, but all are planned so that the maximum load is dropped before frequency is reduced to such an extent that plant auxiliaries are lost and a total system collapse occurs.

Another factor that can be used to minimize the possibility of system collapse with declining frequency is to open tie lines at some predetermined frequency if power is being exported from the system. If power is being imported, opening the ties will worsen the situation. Tie-line opening in time of trouble should usually be a last resort and should be instituted only if power is flowing out of the system with continuing decline of frequency. Interconnection agreements and operating orders issued by various systems usually cover this procedure, and it is also covered in North American Power Systems Interconnection Committee (NAPSIC) Guide 9. In most cases, interconnections will serve to stabilize and limit frequency decay.

A more sophisticated procedure by which the rate of frequency decay is used to determine the amount of load to be shed can be applied. Relays that are sensitive to rate of frequency decay have been developed, and these relays can supply frequency information to a central control computer, which can analyze frequency variations and send out control impulses to shed load when conditions require it.

As frequency recovers, load is reconnected in amounts that match the available generation. In some cases this is done automatically by relay, and in others manually. In any event, it is much more desirable to interrupt a portion of a system's load for a short period of time than to permit frequency decline until the system collapses.

Matching generation with load also is necessary in the event that a large block of load is lost, or when tie lines which are exporting power relay. In such circumstances, load will be less than generation, and frequency will rise. The amount of frequency rise must be limited because it is accompanied by voltage rise and may damage customer equipment if it is not limited. Overfrequency is less hazardous to a power system than low frequency. As frequency rises, the additional torque that is required from the prime movers tends to limit frequency rise. However, the total rise in frequency should be limited to some predetermined level. This is usually done manually by tripping generation if governor and automatic control action is not adequate.

The important principle to be remembered is that in power system operation, generation and load must always be matched. Any mismatch results in a variation of frequency from normal or results in deviations of tie lines from schedule. Inadequate generation results in reduced frequency and overgeneration in overfrequency.

It is important to prevent serious frequency decline so that plant auxiliaries will not be lost, causing further mismatch of generation and load, which can result in a total system collapse. Present practice is increasingly trending toward application of underfrequency relays to shed load during periods of serious frequency decline. Upon restoration of frequency, load can be automatically or manually restored.

In interconnected operation the tie lines will ordinarily help maintain frequency. However, if power is being exported at a time when frequency is declining, conditions can be improved by opening the ties, but this should be done only as a last resort, as it will make an adjacent area that is already deficient more deficient.

For overfrequency operation resulting from loss of load, it may be necessary to drop generation to restore the match between generation and load.

The maintenance of the match between generation and load is a primary and continuing responsibility of the system operator.

DISCONNECTING FAULTED LINES OR EQUIPMENT AND RESTORATION OF FACILITIES

One important measure in maintaining power system security is the rapid disconnection of lines or equipment that are in trouble. Because rapid action is necessary, automatic devices usually are relied upon instead of manual operations.

The design and setting of protective relay systems is a function of system engineering and operating staffs and is not ordinarily a matter over which system operators have any control. However, system operators should be aware of the protective devices at important locations in the power system and their expected performance.

Knowledge of the type of protective devices in use and the portions of lines and equipment protected is of value in determining the nature and extent of trouble after a relay operation. Such information should provide the operator with clues on how to proceed in restoring the system to normal, or as nearly so as possible, in minimum time.

Many types of troubles, such as insulator flashovers on transmission lines, often are only momentary. The protection system design therefore frequently provides for automatic reclosure following such incidents. On the other hand, transformer bank differential or generator-elevated neutral relay operations usually indicate more serious troubles.

As a guide to system operators, procedures are normally provided which outline the steps to be taken after different types of relay operations or after unsuccessful reclosure tests. System operators should be thoroughly familiar with such system policies in order to restore the system to as nearly normal as possible with minimum delay.

In the event that automatic reclosure is unsuccessful, or after relay action has occurred that indicates equipment trouble, the equipment should be disconnected and cleared for repairs. Alternative line or station bus arrangements should be effected in order to restore interrupted load or generating equipment to service, so that normal loads will be served and generation margins restored.

RESTARTING GENERATION EQUIPMENT

After a generating unit relays or is lost to the system because of line or station trouble, it should be returned to service promptly in order to restore generation margin if there is no machine damage.

Normally, restarting a generating unit does not present particular problems, other than those of the routine procedures that must be followed in putting it into service. However, if there is no power available at the generating station, from the system, or from local startup or house units, there may be a long delay in getting the unit back into service.

DESIGN FACTORS AFFECTING RELIABILITY

A great deal of effort is expended in the design of generating stations to make them as reliable as possible and ensure maximum availability of the equipment for service. Some of the means used to enhance reliability are spare exciters, startup transformer banks, and battery or pneumatically operated control devices with capability of several operations from stored energy. In the event of a total system or area collapse, however, it may not be possible to restart a plant without sufficient electric power to supply auxiliaries.

In some thermal plants small "house" units are provided which will separate from the system on serious frequency drops and supply station auxiliaries, such as feed pumps, draft fans, lubricating oil pumps, and other devices essential to plant operation, at normal frequency and voltage. Such plants are normally capable of restarting with a minimum of difficulty after a complete separation or system collapse.

In order to increase system reliability, many installations of diesel or gas turbine generators of sufficient capacity to provide starting power are being made. Such startup units are capable of starting rapidly with only a battery or compressed air source, which is supplied as a part of the installation.

Hydro plants are ordinarily less complex than thermal plants, and if bearing oil supplies and control power are available, they can be started and placed in service in very short periods of time.

The features of plants on a system are the result of design considerations that are not within the province of the system operator. However, the operator should be aware of the capabilities for startup after a total shutdown as well as the normal operating capabilities of the various plants for which he or she is responsible. Some of the factors of which the system operator should be informed are:

1. Availability of startup power, and whether by house unit, diesel, gas turbine, or other source

2. Transmission sources to the stations in the event that it is necessary to bring power back to the station for starting

3. Sources within the system that can be used for starting other plants

4. Switching procedures to route startup power to stations that require outside or system power for startup

Most of the above information is usually available in written emergency procedures, which system operators should understand thoroughly so that facilities can be restored to service and normal operations resumed as soon as possible in the event of major system shutdowns.

The foregoing is by no means a complete discussion of all the problems of restarting generating equipment or of emergency procedures. These will vary for each installation and system, and must be covered by detailed information and procedures developed by the engineering and operating departments of each system. However, the subject is one of paramount importance in making it possible to provide maximum service under all conditions.

OPERATING EQUIPMENT WITHOUT NORMAL ENERGY SOURCES

During power system emergencies it may be necessary to operate equipment such as power circuit breakers or motor-operated air switches when normal sources of energy for operating such devices are not available.

Power circuit breakers, whether oil, air, or gas types, are provided with mechanisms to open or close the breakers as desired. Solenoids, pneumatic devices, or operators using energy stored in compressed springs are used for this purpose. In order to make these devices independent of the system power, station batteries of capacity sufficient to provide energy for several open and close operations are ordinarily provided.

It is possible that station battery sources may be lost because of battery cable failure or other cause, and it may be critically important to operate circuit breakers during the period of battery failure. Usually emergency means of operating such devices can be developed. Some spring-operated breaker mechanisms can be hand cranked to compress the operating spring so that the breaker can be closed.

With pneumatically operated breakers, when there is no compressed air available, it is possible to use a nitrogen bottle, temporarily connected to the air system, to operate the breakers.

Motor-operated air switches can usually be manually operated, either directly or by winding a spring operator.

The important factor is that even in the event of loss of normal sources of energy for operation of devices in a power system, methods can frequently be devised to permit operation during emergencies. System operators who are aware of possible emergency procedures can direct restoration of service with much less delay than if they were to wait for repairs to be made to damaged equipment or for system conditions to return to normal.

ALTERNATIVE ARRANGEMENTS

Normal arrangements of transmission and distribution lines and the configuration of lines to station buses permit proper division of load, ensuring minimum risk in the event of bus or transformer bank failure and proper relay action.

During emergencies, system operators are frequently required to devise alternative arrangements of line and station equipment in order to restore service with a minimum of delay. Common procedures are to parallel lines over an auxiliary bus or to make use of bus parallel breakers to replace a breaker that is damaged or out of service for work.

In some cases, in the event of a permanent fault on a line, the line can be sectionalized or jumpers opened at dead-end structures to restore at least a portion of the service until repairs can be made and the system returned to normal.

It is not possible to cover all the alternatives in this brief discussion, but system operators can review potential troubles and develop procedures to be used in such cases. It is common practice on most power systems to prepare standard switching procedures for various contingencies. However, it is never possible to foresee all possible contingencies, and thus intimate knowledge of the system will help the system operator to devise emergency arrangements when necessity arises and no prepared procedure is available.

INTERCONNECTIONS WITH OTHER SYSTEMS

Interconnection with other systems can be of significant assistance to a power system during times of trouble. In the event of loss of a large block of generation, energy will flow from surrounding systems to the system with deficient generation. The ability to provide mutual standby is one of the major incentives for power system interconnection.

In most cases, with moderate loss of generation, interconnection of systems will reduce the amount of frequency drop because the percentage of total interconnected capacity lost is less than if a system operates isolated from other systems.

Serious disturbances can result in tie-line overloads, and in some cases instability, resulting in tie-line tripping, which may lead to cascading relay operation and area shutdowns more extensive than would have resulted with systems operating independently. Some of the measures previously discussed can minimize the possibility of such major pool outages.

Maintenance of adequate spinning reserve with capability of rapid response, load-matching-relay (underfrequency) installations, and proper tie-line-relay settings will ordinarily prevent disturbances from developing into area-wide shutdowns.

System operators are in a position to monitor tie-line flows and other conditions on their systems, and by proper surveillance and action in emergencies can prevent or minimize the effects of troubles. If trouble conditions persist, accompanied by declining frequency, it is desirable to open tie lines if power is outgoing and save at least a portion of the area. As previously mentioned, this procedure is well outlined in NAP-SIC Guide 9. By maintaining a portion of the area in service rather than permitting a total collapse, normal operation can be restored much more rapidly than would be possible following a total area collapse.

Interconnection agreements usually specify minimum spinning-reserve requirements and outline mutual standby conditions. Knowledge of system capabilities, tie-line ratings, spinning reserve available, and other operating features is important so that the system operator can take appropriate action during emergency conditions.

INDICATION OF SYSTEM CONDITIONS AND COMMUNICATION

Because of the nature of their work, system operators must rely on communication and signaling devices to keep them aware of conditions on their power systems. Many of the key locations, large generating plants, points of interconnection, and major switching stations are many miles from the system operator.

In order to provide the system operator with information on the status of the system, key information is remotely indicated in dispatching offices. Control channels and telephone circuits are provided for automatic or supervisory control of equipment and for telephone contact with operators at various stations throughout the system and between control centers of interconnected systems.

The reliability of telemeter, control, and voice channels is of major importance to system reliability. Various means of providing communication facilities are used. Channels are sometimes leased from common-carrier telephone companies, power-line carrier circuits are established on power-transmission circuits, and privately owned microwave systems are installed by the utilities.

In order to ensure reliability of communications, it is common practice to provide more than one communication path to major generation, switching, and tie points. Such alternative channels are usually routed over diverse paths so that the probability of concurrent failure is minimized.

The power supply to terminals and repeater stations of microwave and power-line carrier installations is made independent of system power, or an auxiliary power source is provided for service during a power system interruption.

When adequate care is used in the design and installation of communication and indication facilities, the system operator has reasonable assurance of being able to maintain contact with key locations at all times and will be able to take the necessary actions to obtain the best possible reliability from the system.

SUMMARY

From the preceding discussion it should be apparent that the highest importance is attached to power system reliability. Those responsible for the design and operation of power systems devote a great deal of thought to this subject and invest substantial sums to ensure reliability, including the reliability of communications. Modern SCADA systems frequently include security monitoring, state estimation, and online load-flow programs to provide system operators with indications of potential problems affecting the reliability of their systems. When provided with the facilities outlined in this section, it should be possible for system operators to perform their functions properly in maintaining safe and reliable operation of their systems.

Bibliography

Kirchmeyer, Leon: *Economic Operation of Interconnected Power Systems,* John Wiley and Sons, Inc., New York, 1959.

Anderson, P. M., and A. A. Fouad: *Power System Control and Stability,* Iowa State University Press, Ames, 1977.

Zaborsky, J., and Joseph Rittenhouse: *Electric Power System Transmission,* Rensselaer Book Store, Troy, N.Y., 1969.

Langsdorf, Alexander S.: *Theory of Alternating Current Machinery,* 2d ed., McGraw-Hill Book Company, New York, 1955.

REA Bulletin 66-10: "Supervisory Control and Energy Management Systems," July 1979.

Westinghouse Electric Corporation: "Applied Protective Relaying," 1979.

Fink, D. G., and H. W. Beaty, eds., *Standard Handbook for Electrical Engineers,* 11th ed., McGraw-Hill Book Company, New York, 1978.

North American Power Systems Interconnection Committee (NAPSIC): "Operating Manual."

Index

Above-ground pumped-hydro energy storage, **8**-9
Absorber generator, **4**-64
Absorption cycle, **4**-6
Accelerated cost recovery system (ACRS), **4**-91
AFBC (*see* Atmospheric fluidized-bed combustor)
AFUDC (allowance for funds used during construction), **3**-11
AGC (automatic generation control), **10**-11
Air conditioning, **4**-6, **4**-63
Air pollution, **4**-4, **4**-15
   legislation on, **2**-6, **2**-34
Air preheat, **2**-32
Air preheater, **2**-18
Allowance for funds used during construction (AFUDC), **3**-11
Alternate arrangements of transmission lines, **11**-14
Alternate fuels, **4**-32
Alternative sources for electrical energy, **10**-4
American Society of Mechanical Engineers (ASME), **3**-16
Ammonia, **4**-20
Aquifer, **6**-4, **8**-9
Atmosphere, constituents of, **9**-4
   oxides of carbon, **9**-8, **9**-10
   oxides of nitrogen, **9**-6
   oxides of sulfur, **9**-5
   particulate matter, **9**-14
Atmospheric fluidized-bed combustor (AFBC):
   construction of, **2**-36
   corrosion in, **2**-38
   heat transfer in, **2**-39
Automatic generation control (AGC), **10**-11
Auxiliary power requirements, **2**-30, **6**-35

Back-pressure steam turbines, **4**-10
Baghouse, **2**-6, **9**-30
Base-load power plants, **8**-4
Batteries, electrical, **8**-19 to **8**-21
Bearings, **2**-2
Best available control technology (BACT), **4**-30

Biofouling, **6**-37
Biomass fuels, **6**-9, **6**-11
Bituminous coal, **3**-17
Boiler feedwater, treatment of, **2**-5
Boiler header drum, **2**-4
Boilers, **2**-4 to **2**-6, **4**-33, **4**-49
   construction of, **2**-4, **2**-17
   forced-circulation, **2**-17
   history of, **2**-5
   once-through, **2**-4
   waterwalls, **2**-4
Boiling flow stability, **2**-39
Boiling heat transfer, **2**-39
   burn-out heat flux, **2**-39
   nucleate, **2**-39
   vapor blanket, **2**-39
Boiling heat-transfer coefficient, **2**-39
Bonded debt, **2**-3
Bottoming cycles, **4**-18, **4**-33 to **4**-35
   turbines for, **4**-23
   working fluids for, **4**-20
Breeder reactors, **2**-25
Brooks coil, **8**-24
Bulding and district heating and cooling, **4**-64, **4**-65
Bull Run power plant, **2**-3, **2**-8
Burners, **4**-53

Calcium sulfate, **2**-32
Calcium sulfite, **2**-32
Canister, nuclear waste, **9**-54
Capital costs, **3**-2, **3**-12, **3**-19
   of biomass systems, **6**-11
   effects of energy costs, **3**-21
   of fossil fuel plants, **3**-4
   of goethermal systems, **6**-7
   of hydroelectric units, **5**-14
   of nuclear plants, **3**-4
   of OTEC systems, **6**-34
   of solar systems, **6**-19, **6**-25
   of tidal power, **5**-18
   of windmills, **6**-31
Carbon dioxide:
   in the atmosphere, **9**-8, **9**-10
   and the greenhouse effect, **9**-9
Cash flow, **4**-83
Caverns, **8**-11
Centrifugal chillers, **4**-63

Chemical reaction energy storage, **8**-35
Coal conversion, **2**-7
Coal metering and feed systems, **2**-36, **2**-40
Coefficient of speed fluctuation of flywheels, **8**-16
Cogeneration for urban areas, **4**-3
   with air conditioning, **4**-6
   with desalinaton, **4**-7
   with distillation of sewage, **4**-7
   general energy requirements, **4**-4
   matching electrical and thermal loads, **4**-10, **4**-18
   plant siting, **4**-12
   ratio of heat to electical output, **4**-10, **4**-17
   typical systems, **4**-15
Collection efficiency, **9**-18, **9**-28
Collector, solar, **6**-13
Combined power cycle, **4**-33, **4**-47
Combustion turbine, **4**-36, **4**-41
   cross-section drawings, **4**-37, **4**-41
   industrial, temperatures, materials, and coatings for, **4**-44
   modular, **4**-43
Combustion turbine generator, **4**-36, **4**-45, **4**-68
   offshore platform applications, **4**-42
   power station, **4**-45
Compressed-air energy storage, **8**-10
   adiabatic, **8**-11
   at Huntorf plant, **8**-13
   at Soyland plant, **8**-14
Computers for economic loading, **10**-11
Condensers, **4**-21, **4**-64
Conservation, **4**-3, **4**-29
Construction work in progress (CWIP), **3**-20, **4**-31
Containment shell, **2**-22, **2**-26
Contingency, **3**-12
Control, **4**-78
   of fluidized-bed combustors, **2**-36, **2**-39
   of hydraulic turbines, **5**-10

I-1

Control considerations, load range in, **4**-18, **5**-10
Control rods, reactor, **2**-24, **2**-29
Control room, **4**-67
Conventional power plants, **4**-33
Cooling towers, **2**-41, **4**-75
   costs of, **3**-10
Coordinaton of hydro and thermal generation, **10**-12
Corona in electrostatic precipitators, **9**-25
Corrosion of boilers, **2**-5
Cost escalation, **3**-11, **3**-20
Cost estimating, **3**-16, **3**-20
Costs, **3**-2 to **3**-23
   administrative, **3**-3
   capital (*see* Capital costs)
   of components, **3**-20
      buildings, **3**-4
      coal handling, **3**-7
      condensers, **6**-35
      flue-gas cleanup, **2**-35
      furnace, **3**-4
      heat exchangers, **3**-16
      reactors, **3**-4
      steam generators, **3**-4
      tubing, **3**-16
      turbines, **3**-4, **3**-8
      welds, **3**-17
   contingency, **3**-12
   depreciation, **3**-3
   design, **3**-6, **3**-15
   direct, **3**-12
   effects of construction time on, **3**-10
      extended workweek, **3**-14
      fuel used, **3**-3
      licensing, **3**-11, **3**-14
      type of ownership, **3**-2
   of electricity, **3**-9, **3**-19
   environment and safety, **3**-12
   escalation, **3**-11, **3**-20
   extended workweek, **3**-14
   fixed, **10**-3
   fuel, **3**-3, **3**-17, **4**-82
   incremental, **10**-6, **10**-8
   interest during construction, **3**-11
   licensing, **3**-11
   maintenance, **3**-7, **3**-17
   materials, **3**-19
   operation, **3**-7, **3**-17

Costs (*cont.*)
   overhaul, **3**-2
   of power plants, **3**-2
      coal-fired, **3**-4, **3**-12
      gas-fired, **3**-5, **3**-7
      nuclear, **3**-4, **3**-13
      oil-fired, **3**-5
   production, **3**-2
   safety and environment, **3**-12
   scale, effects of, **3**-5
   taxes, **3**-3, **3**-19
   transmission and distribution, **3**-2
   variable, **10**-4
Creep buckling, **6**-7
Cross-section drawings, combustion turbine, **4**-37, **4**-41
Curie, **9**-39
Current density in solar cells, **6**-21
Curves, input-output, **10**-5, **10**-8, **10**-13
CWIP (construction work in progress), **3**-20, **4**-31
Cycling power plants, **8**-4

Dams, hydroelectric, **5**-2 to **5**-9
Deaerator, **2**-20
Decay heat from spent fuel, **9**-48
Decremental value, **10**-16 to **10**-19
Deforestation, **6**-9
Delivery time, **4**-32, **4**-87, **4**-91
Depository, nuclear waste, **9**-53
Depreciation, **3**-3, **4**-32, **4**-87, **4**-91
Depreciation rates, **4**-91
Desalination, **4**-7
Design, plant, **4**-70
Design costs, **3**-6, **3**-15
Design criteria, **4**-26
Design documentation, **4**-31
Design factors affecting reliability, **11**-12
Desulfurization systems, flue-gas:
   dry process, **9**-21
   single-alkali process, **9**-22
   wet process, **9**-19
   (*See also* Flue-gas desulfurization)
Deutsch equation, **9**-28
Diesel cogeneration facility, **4**-33, **4**-68
Diesel engine, **4**-73
Diesel generator sets, **4**-75

Disconnecting faulted lines or equipment, **11**-11
Disposal:
   of desulfurization waste, **9**-24
   of nuclear waste, **9**-52 to **9**-54
Distillation:
   of seawater, **4**-7
   of sewage, **4**-7, **7**-3
District heating systems, **4**-3, **4**-13
Division of load for minimum fuel cost, **10**-8 to **10**-11
Dowtherm working fluid, **4**-20
Dry cooling tower, **2**-42

Ebullient system, **4**-78
Economic comparisons, **4**-81
   and cash flow, **4**-83
Economic effects of energy, **3**-21
Economic feasibility, **4**-80
Economic interchange of power, **10**-4, **10**-16
Economic loading of generating units, **10**-8 to **10**-11
Economic loading slide rule, **10**-10
Economic operation of power systems, **10**-3 to **10**-19
Economic Recovery Tax Act, **4**-91
Economies of scale, **3**-5, **4**-15
Efficiency, **4**-76
   thermal, **4**-33, **4**-75
Efficient energy sources, **4**-29
Efficient heat recovery, **4**-76
Electric utility, **2**-2, **3**-2, **4**-27
Electrical batteries, **8**-19 to **8**-21
Electrical energy storage, **8**-5
   battery, **8**-19 to **8**-21
   compressed-air, **8**-10, **8**-13
   flywheel, **8**-15, **8**-17
   pumped-hydro, **8**-7, **8**-9
   superconductive, **8**-21
   Wisconsin, **8**-25
Electricity revenue, **4**-86
Electrostatic precipitators, **2**-6, **2**-33, **9**-24 to **9**-30
Energy conservation, **4**-3, **4**-29
Energy consumption:
   for heating, **4**-4
   by industry, **4**-6
   for production of materials, **6**-28
Energy investment tax credit, **4**-92
Energy management systems, **4**-72, **8**-3

Energy price, **4**-86
Energy storage, **8**-2
 electrical, **8**-5
  battery, **8**-19
  compressed-air, **8**-10, **8**-13
  flywheel, **8**-15 to **8**-18
  pumped-hydro, **8**-7, **8**-9
  superconductive, **8**-21
  Wisconsin, **8**-25
 thermal, **8**-7
  chemical, **8**-35
  latent, **8**-32
  sensible, **8**-26
 turnaround efficiency for, **8**-10, **8**-29, **8**-34, **8**-35, **8**-38
Energy utilization, **4**-5, **4**-11, **4**-19
Engine generator sets, **4**-74, **4**-80
Engine heat recovery system, **4**-74
Engines, reciprocating, **2**-2, **4**-73
Enhanced heat-transfer surfaces, **4**-20
Environmental assessment, **4**-27
Environmental considerations, **4**-30
Environmental effects, **4**-15
Environmental Protection Agency (EPA), **2**-6, **2**-34
Evaporator, **4**-63
Exemptions, **4**-32
Expansion ratio, **2**-2
Expansion valve, **4**-64
Explosions, **4**-9
Extended workweek, **3**-14

Fans:
 forced-draft, **2**-19
 induced-draft, **2**-30
FBC (*see* Fluidized-bed combustor)
Federal depreciation allowance, **4**-87, **4**-91
Federal Energy Regulatory Commission (FERC), **3**-5
Federal Power Commission (FPC), **3**-5
Feedwater pumps, **2**-3
Feedwater quality, **4**-29
Feedwater treatment, **2**-5
FERC (Federal Energy Regulatory Commission), **3**-5
Fertilizer, **3**-19
Filters, **2**-6

Fission reactors, **2**-22
 fuel elements, **2**-24, **2**-29
 liquid-metal-cooled fast-breeder (LMFBR), **2**-26
 pressurized-water (PWR) (*see* Pressurized-water reactor)
Flame quenching, **2**-39
Flue-gas desulfurization, **2**-7, **2**-32
 costs of, **2**-35
 disposal of wastes from, **2**-35
 effectiveness of, **2**-33
 EPA requirements in, **2**-35
 maintenance problems of, **2**-35
 (*See also* Desulfurization systems, flue-gas)
Fluid inventory, **2**-29
Fluidized-bed combustor (FBC):
 control of, **2**-36
 corrosion in, **2**-38
 costs of, **2**-36
 effects of superficial velocity, **2**-36
 fuel feed to, **2**-40
 operating characteristics of, **2**-39
 usage of: with sludge disposal, **2**-39
  with steam generators, **2**-36
Fluorocarbons, **4**-20
Fly ash, **2**-19
Flywheels, energy storage by, **8**-15 to **8**-19
 mass-efficiency factor of, **8**-16
 NASA concept for, **8**-18
 volume-efficiency ratio of, **8**-17
Forced-outage rate, **2**-6
Forests, **6**-9
FPC (Federal Power (Commission), **3**-5
Fractional collection efficiency, **9**-18, **9**-28
Francis turbines, **5**-10
Fuel elements, fission reactor, **2**-24, **2**-29
Fuel oil, price of, **3**-21
Fuel-to-electricity conversion ratio, **4**-75
Fuels:
 alternate, **4**-32
 biomass, **6**-9, **6**-11,
 costs of, **3**-3, **3**-17, **4**-82
 nuclear (*see* Nuclear fuel)

Fuels *(cont.)*
 solid wastes, **4**-8, **7**-1

Gallium arsenide, **6**-23
Gamma (water value), **10**-13
Gas-cooled reactors (GCR), **2**-23
 high-temperature (HTGR), **2**-23
 steam generators for, **2**-25
Gas engine, **4**-73
Gas turbine generator, **4**-36, **4**-49
Gene, **9**-50
Generating units, economic loading of, **10**-8 to **10**-11
Generation equipment, restarting, **11**-12
Generators, electric, **2**-17, **2**-30
Geothermal power, **4**-20, **6**-3
 costs of, **6**-7
 environmental effects of, **6**-8
 resources of, **6**-3
 working fluids for, **4**-20
Graphite fuel elements, **2**-25
Graphite reactors, **2**-25
Gray, **9**-40
Greenhouse effect, **9**-9

Half-life of activation, **9**-45
 of fission products, **9**-48
 in liquid-metal reactors, **9**-47
 in water reactors, **9**-45
Head, **6**-3
 in hydraulic turbines and pumps, **3**-8, **5**-5, **5**-17
Header drums, **2**-4
Headers, **3**-17
Heat of vaporization, **4**-21, **4**-78
Heat balance, **4**-56
Heat exchangers:
 costs of, **3**-16
 stresses in, **2**-4, **3**-16
Heat flux:
 in boilers, **2**-5
 in condensers, **4**-20, **6**-34
 in reactor-fuel elements, **2**-25
Heat recovery, **4**-33, **4**-46, **4**-49, **4**-62, **4**-76
Heat storage, **4**-17
Heat transfer, **4**-20
 condensing, **4**-21
 in steam generators, **4**-21
Heat transport, **4**-14

Heat utilization standards, **4**-34, **4**-64
Heating, ventilating, and air-conditioning (HVAC) systems, **4**-63
Heating systems, **4**-69
  hot-water, **4**-69, **4**-76
  steam, **4**-69, **4**-76, **4**-78
Heliostats, **6**-18
Hot water for heating and air-conditioning loads, **4**-66, **4**-76
Hot-water heating system, **4**-69, **4**-76
Hot wells, **2**-3
Hydraulic turbines, **5**-8
  cavitation in, **5**-12
  costs of, **3**-8
  Deriaz, **5**-10
  Francis, **5**-12
  head in, **3**-8, **5**-5, **5**-17
  installation of, **5**-12
  Kaplan, **5**-9
  Pelton, **5**-10
Hydro and thermal generation coordination, **10**-12
Hydroelectric power plants, **5**-2 to **5**-20
  capacity of, **5**-9
  costs of, **5**-12
  dams for, **5**-2 to **5**-9
  maintenance of, **5**-14, **5**-17
  possibilities for small plants, **5**-16
  storage capacity, **5**-8
  tidal power, **5**-18
Hydroelectric resources, **5**-2

IDC (interest during construction), **3**-11
IHX (intermediate heat exchanger), **2**-26
Incinerators, **4**-8, **7**-1
Incremental costs, **10**-6, **10**-8
Incremental rates, **10**-6, to **10**-8
Industrial power generation, **1**-3 to **1**-27, **4**-3
Input-output curves, **10**-5, **10**-8, **10**-13
Insurance, **4**-82, **4**-86
Interest, **4**-82, **4**-84, **4**-86
Interest during construction (IDC), **3**-11

Intermediate heat exchanger (IHX), **2**-26
Inventory, fluid system, **2**-29

Kilowatthour (kWh), **4**-36

Lambda (system fuel cost), **10**-13
Latent heat energy storage, **8**-32
Lead-acid battery, **8**-19
Legislation on air pollution, **2**-6, **2**-34
Licensing, **3**-11 to **3**-14
Lines, transmission, **11**-7, **11**-8, **11**-14
Liquid-metal-cooled fast-breeder reactor (LMFBR), **2**-26
Lithium batteries, **8**-21
Load:
  allocation of, **10**-13
  division of, for minimum fuel cost, **10**-8 to **10**-11
Load factor, **3**-20
Load profile demands, **4**-79
Log-normal distribution, **9**-16
Losses due to transmission of power, **10**-13
Low-temerature cycles, **4**-18
Lubrication, **2**-2

Maintenance, **3**-18
Maintenance costs, **4**-82, **4**-85
Mariculture, **6**-36
Marine propulsion power system, **4**-40
Mass-efficiency factor of flywheels, **8**-16
Materials, energy requirements for, **6**-28
Methanation, **8**-36
Mined cavities, **4**-13
Municipal solid wastes, **4**-8, **7**-1

NASA flywheel energy storage system, **8**-18
Newcomen, Thomas, **2**-2
Nickel-cadmium battery, **8**-20
Niobium, **3**-16
Nitrogen, oxides of, **9**-6
Nuclear fuel, **3**-17
  disposal of, **9**-52
  reprocessing of, **9**-51
  spent, **9**-43
  wastes from, **9**-48

Nuclear generation, **10**-12
Nuclear reactors (*see* Reactors, nuclear)
Nuclear Regulatory Commission (NRC), **3**-21
Nuclear steam plants, **2**-22
Nuclear wastes:
  disposal of, **9**-52 to **9**-54
  high-level, **9**-43, **9**-48
  low-level, **9**-43
  pollution from, **9**-48 to **9**-50
Nucleate boiling, **2**-39

Ocean thermal energy conversion (OTEC), **4**-20, **6**-34
Oceans, **6**-29
Offshore platform combustion turbine generator application, **4**-42
Oil recovery, **4**-51
Oil wells, **6**-5
Operating expenses, **4**-82
  fuel costs, **33**-3, **33**-17, **4**-82
  interest, **4**-82, **4**-84, **4**-86
  maintenance costs, **4**-82, **4**-85
  property taxes, **4**-82, **4**-85
Optimization of the power cycle, **4**-27
Optimization considerations in turbine thermal cycle, **4**-52, **4**-59
Organic Rankine cycle, **4**-20
  turbines for, **4**-20
OTEC (ocean thermal energy conversion), **4**-20, **6**-34
Otto cycle, **4**-73, **4**-75
Overhaul, costs of, **3**-2
Oxides of carbon in atmosphere, **9**-8
  removal of, **9**-24
Oxides of nitrogen, **4**-75, **9**-6
Oxides of sulfur in atmosphere, **9**-5
  removal of, **9**-19

Part-load efficiency, **5**-10
Part throttle operation, **5**-10
Particle separators, **2**-6
Particulate matter in the atmosphere, **9**-14
  collection efficiency for, **9**-18
  distribution of, **9**-16
  removal of, **9**-27
Peak shaving, **8**-5

Peaking plants, **8**-4
Pelton wheel, **5**-10
Penalty factor for transmission losses, **10**-14
Penalty-factor computers, **10**-15
Photosynthesis, **6**-9, **6**-12
Photovoltaics, **6**-21
Pinch point, **4**-12
Planning, **4**-27
Plant availability, **4**-75
Plant design, **4**-70
Plant siting, **4**-3, **4**-12
Pollution, **9**-5 to **9**-50
   from fossil power plants, **9**-5 to **9**-19
   from nuclear power plants, **9**-44 to **9**-47
   from nuclear waste, **9**-48 to **9**-50
   thermal, **9**-34
   (*See also* Air pollution)
Polymerization, **4**-20
Polyvinylchlorides (PVCs), **4**-9
Population, urban, **4**-3
Power:
   fuel cells, **10**-4
   geothermal (*see* Geothermal power)
   hydro, **10**-12
   nuclear, **10**-12
   purchase and sale of, **10**-16 to **10**-19
   solar, **6**-8, **10**-4, **10**-12
   thermal, **10**-4, **10**-5, **10**-12
   water, value of, **10**-13
   wind, **10**-4, **10**-12
Power cycles, **4**-33
   bottoming, **4**-18, **4**-33 to **4**-35
   combined, **4**-33, **4**-47
   optimization of, **4**-27
   regenerative, **2**-2, **4**-61
   simple, **4**-34, **4**-36, **4**-45
   topping, **4**-33, **4**-35, **4**-60, **4**-81
Power exchanges, **10**-4, **10**-16
Power factor, **11**-8
Power parks, **4**-13
Power plant siting, **4**-12 to **4**-13
Power system reliability, factors affecting, **11**-4 to **11**-6
Power towers, **6**-18
Precipitators, electrostatic, **2**-6, **2**-33, **9**-24 to **9**-30
Preliminary study in plant design, **1**-3, **4**-70

Pressure drop:
   in fluidized beds, **2**-37
   in furnaces, **2**-19
   in steam piping, **2**-3
Pressure ratio, **2**-2
Pressurized fluidized-bed combustor, **2**-39
Pressurized-water reactors (PWR), **2**-22
   costs of, **3**-3, **3**-15, **3**-17
   crew size for, **3**-19
   safety of, **2**-22
Primary coolant circuit, **2**-24
Process heat and power, **1**-3, **4**-3
Process plant design, **1**-3 to **1**-47
Products of fission, **9**-48
Project financing, **4**-82
Propeller turbines, **3**-10
Property taxes, **4**-82, **4**-85
Pulverized-coal-fired furnaces (PCF), **2**-4
Pump, **4**-65
Pumping-power–heat-transport ratio, **4**-14
Purchase and sale of power, **10**-16 to **10**-19
PVCs (polyvinylchlorides), **4**-9
Pyrite ore roasting, **2**-39

Rad, **9**-40
Radiation:
   biological effects of, **9**-49
   in the environment, **9**-37
Radioactivity:
   in the environment, **9**-37
   units of, **9**-39
Rainfall, runoff from, **5**-18
Rankine cycle, **2**-2
   organic, turbines for, **4**-20
Rates, incremental, **10**-6 to **10**-8
Ratings, transmission-line, **11**-8
RBE (relative biological effectiveness), **9**-40
Reactors, nuclear, **2**-22 to **2**-30
   costs of, **3**-3, **3**-4, **3**-15, **3**-17
   crew size for, **3**-19
   radiations from effluents, **9**-44 to **9**-47
   safety of, **2**-22
   wastes from, **9**-43, **9**-48 to **9**-54
Reciprocating engines, **2**-2, **4**-73
Regenerative power cycle, **2**-2, **4**-61

Regenerator, **4**-17, **4**-25
Regulation error, **11**-6
Regulatory requirements, **4**-27
Reheat, **2**-4
Relative biological effectiveness (RBE), **9**-40
Relays, underfrequency, **11**-9
Reliability, power system, factors affecting, **11**-4 to **11**-16
Rem (roentgen equivalent man), **9**-40
Reprocessing of nuclear fuel, **9**-51
Reserve, spinning, **11**-4 to **11**-7
Restarting generation equipment, **11**-12
Restoration, load, **11**-11
Return on investment, **4**-72
Revenues, **4**-86
   depreciation, **4**-87, **4**-91
   electricity, **4**-86
   energy investment tax credit, **4**-92
   thermal, **4**-87, **4**-89, **4**-90

Safety, reactor, **2**-22
Satellite power plants, **6**-25
Scale effects:
   on costs: of biomass fuel, **6**-9, **6**-11
   of power plants, **3**-5
   of solar power, **6**-25
   of turbines, **3**-8
   on efficiency of energy transmission, **6**-30
   on OTEC plants, **6**-34
   on windmills, **6**-30
Scrubbing of flue gases:
   dry process, **9**-21
   wet process, **9**-19
Seawater:
   distillation of, **4**-7
   nutrients in **6**-36
Secondary coolant, **2**-24, **2**-29
Selective catalytic reduction (SCR), **4**-53
Service life, **2**-35
Sewage distillation, **4**-7, **7**-3
Sewage sludge, **4**-8, **7**-3
Shaft seals, **2**-16
Sievert, **9**-40
Silicon solar cells, **6**-21
Silver-zinc battery, **8**-20
Single alkali scrubbing, **9**-22

Sitting of power plants, **4**-12 to **4**-13
Sizing selection criteria, **4**-81
Slide rule, economic loading, **10**-10
Sludge, **2**-35, **4**-8, **7**-3
Sodium-cooled reactors, **2**-26
   radioactivity in, **9**-47
Sodium-sulfur battery, **8**-21
Soil erosion, **6**-12
Solar-energy, **6**-8, **10**-4, **10**-12
   for building heating and hot water, **6**-12
   maps for solar flux, **6**-20
Solar-energy collectors, **6**-13
Solar-energy conversion:
   via biomass, **6**-11
   costs of, **6**-11, **6**-19, **6**-21, **6**-25
   photovoltaic, **6**-21
   satellite systems, **6**-21
   thermal systems, **6**-12
Solvent refining, **2**-7
Somatic effects of radiation, **9**-49
Soyland compressed-air energy storage system, **8**-14
Spent nuclear fuel, **9**-43
   decay heat from, **9**-48
   disposal of, **9**-52
   reprocessing of, **9**-51
Spinning reserve, **11**-4 to **11**-7
   location and makeup, **11**-7
   response, **11**-6
Stack gas, **2**-32
Stacks, tall, **2**-6
Stages, number of:
   in feedwater heaters, **2**-3
   in turbines and compressors, **4**-26
State depreciation allowance, **4**-91
Steam as a working fluid, **4**-20, **4**-23
Steam engines, **2**-2
Steam generators, **2**-4, **2**-19, **2**-25, **2**-28
   for fluidized-bed combustors, **2**-36
   for nuclear plants, **2**-23
   reliability of, **2**-6
Steam heating system, **4**-69, **4**-76, **4**-78
Steam plants, **1**-3 to **1**-14, **2**-2 to **2**-45
   costs of, **3**-3, **3**-12
   efficiency of, **2**-13, **2**-25, **2**-30
   historical development of, **2**-2
   size of, effects of, **2**-30

Steam-system design, **1**-3 to **1**-14
Steam turbines, **2**-2, **4**-47, **4**-56, **4**-68
   cross-compound, **2**-4
   double-flow, **2**-3
   effects of size on, **2**-6, **2**-30
   high-, intermediate-, and low-pressure, **2**-3
   tandem, **2**-3
Stokers, **2**-4
Straight-line depreciation method, **4**-91
Stress, creep-buckling, **6**-7
Studies, **4**-26
   cost-benefit, **4**-26
   payback analysis, **4**-26
   preliminary, for plant design, **1**-3, **4**-70
   technical feasibility, **4**-26
Sulfur dioxide, **2**-39, **9**-19
Superconductive magnetic energy storage, **8**-21
Supercritical steam plants, **2**-3
Superheater, **2**-4, **2**-18
Supplemental firing, **4**-52, **4**-61
System fuel cost (lambda), **10**-13
System reliability, factors affecting, **11**-4 to **11**-16

Tall sacks, **2**-6
Tandem turbines, **2**-3
Temperature:
   distribution of, in reactor fuel elements, **2**-24
   effects of, on equipment, **11**-7
Tennessee Valley Authority, (TVA), **2**-12, **2**-22
   Bull Run plant, **2**-3, **2**-8
   hydroelectric power system, **5**-6
   Sequoyah nuclear plant, **2**-22
Thermal capacity:
   of nuclear reactors, **2**-22
   of systems, **4**-12, **4**-14
Thermal efficiency, **4**-33, **4**-75
Thermal energy, **4**-33
Thermal energy storage, **8**-7
   chemical, **8**-35
   latent, **8**-32
   sensible, **8**-26
Thermal matching of project to system demand, **4**-81
Thermal pollution, **9**-34
Thermal power plants:
   coordination of hydro and, **10**-12

Thermal power plants *(cont.)*
   efficiencies, **10**-5
   fuel supplies for, **10**-4
Thermal revenue, **4**-87, **4**-89
   for cooling, **4**-90
   for heating, **4**-89
Thermodynamic cycle efficiency, **2**-5, **2**-24
Thrust loads, **2**-3
Tidal power, **5**-18
Toluene, **4**-21, **4**-26
Topping power cycle, **4**-33, **4**-35, **4**-60, **4**-81
Topsoil, loss of, **6**-12
Tramp uranium, **9**-44
Transactions:
   for economic interchange of power, **10**-4 to **10**-16
   for purchase and sale of power, **10**-16 to **10**-19
Transmission-line ratings, **11**-8
Transmission lines:
   alternate arrangements of, **11**-14
   and station capability, **11**-7
Transmission losses, **10**-13
   penalty factor for, **10**-14
Tritium in reactors, **9**-46
Turnaround efficiency for energy storage:
   chemical, **8**-38
   latent heat, **8**-34, **8**-35
   pumped-hydro, **8**-10
   thermal, **8**-29
TVA (*see* Tennessee Valley Authority)

Underfrequency relays, **11**-9
Underground pumped-hydro energy storage, **4**-13, **8**-9
Uranium, tramp, **9**-44
Utilities, electric, **2**-2, **3**-2, **4**-27

Value, decremental, **10**-16 to **10**-19
Valves, **2**-4, **4**-64
Variable costs, **10**-4

Waste heat, **4**-3, **4**-11
Waste heat recovery, **4**-33, **4**-46, **4**-49, **4**-62
Wastes:
   nuclear (*see* Nuclear wastes)
   solid, municipal, **4**-8, **7**-1
Water:
   as a heat-transfer fluid, **4**-20

Water *(cont.)*
   purification of, **4**-7
   as a working fluid, **4**-20
Water desalting, **4**-7
Water supplies for hydro generation, **10**-4, **10**-12, **10**-13
Water value (gamma), **10**-13
Waterwall, **2**-4
Waterwheels, **5**-2
Wave-energy conversion, **6**-31, **6**-34
Wear, **2**-2
Welds, **2**-4

Welds *(cont.)*
   costs of, **3**-17
   heliarc, **3**-17
   inspection of, **3**-17
Wellman-Lord SO_2 recovery process, **9**-23
Wind energy, **6**-28
   maps for availability of, **6**-32
Wind power, **10**-4, **10**-12
Windmills, **6**-29 to **6**-31
Wisconsin superconductive energy storage, **8**-25

Wood as fuel, **6**-9
Worker productivity, **3**-14
Working fluids:
   low-temperature cycles, **4**-18
   Rankine cycle, **4**-20
Workweek, extended, **3**-14

X-ray inspection, **2**-4
X-rays, **2**-4

Zinc-chlorine battery, **8**-21